#22509265 96.6
N1

```
D1737604
```

WATER RESOURCES
CENTER ARCHIVES

JAN

UNIVERSITY OF CALIFORNIA
BERKELEY

Sea Levels, Land Levels, and Tide Gauges

K.O. Emery   David G. Aubrey

Sea Levels, Land Levels, and Tide Gauges

With 113 Illustrations

Springer-Verlag
New York Berlin Heidelberg London
Paris Tokyo Hong Kong Barcelona

K.O. Emery
Woods Hole Oceanographic Institution
Coastal Research Center
Woods Hole, MA 02543
USA

David G. Aubrey
Woods Hole Oceanographic Institution
Coastal Research Center
Woods Hole, MA 02543
USA

Library of Congress Cataloging-in-Publication Data
Emery, K. O. (Kenneth Orris), 1914–
  Sea levels, land levels, and tide gauges / K.O. Emery, David G. Aubrey.
    p. cm.
  Includes bibliographical references and index.
  ISBN 0-387-97449-0
  1. Sea level. 2. Subsidences (Earth movements) 3. Tide-gauges.
I. Aubrey, David G. II. Title.
GC89.E54 1991
551.4'58–dc20   90-49743

Printed on acid-free paper.

© 1991 Springer-Verlag New York Inc.
All rights reserved. This work may not be translated or copied in whole or in part without the written permission of the publisher (Springer-Verlag New York, Inc., 175 Fifth Avenue, New York, NY 10010, USA), except for brief excerpts in connection with reviews or scholarly analysis. Use in connection with any form of information and retrieval, electronic adaptation, computer software, or by similar or dissimilar methodology now known or hereafter developed is forbidden.
The use of general descriptive names, trade names, trademarks, etc., in this publication, even if the former are not especially identified, is not to be taken as a sign that such names, as understood by the Trade Marks and Merchandise Marks Act, may accordingly be used freely by anyone.

Typeset by Publishers Service of Montana, Bozeman, Montana.
Printed and bound by Edwards Brothers, Inc., Ann Arbor, Michigan.
Printed in the United States of America.

9 8 7 6 5 4 3 2 1

ISBN 0-387-97449-0 Springer-Verlag New York Berlin Heidelberg
ISBN 3-540-97449-0 Springer-Verlag Berlin Heidelberg New York

With appreciation for the long forebearance

of our spouses, Phyllis and Sandy

Preface

This book reinterprets and discusses tide-gauge records to investigate the eustatic rise of sea level caused by Holocene climatic warming and melting of ice sheets and glaciers thought by some to have been augmented since the beginning of the Industrial Revolution by the greenhouse effect of gases from burning of fossil fuels, making of cement, and other sources. This new view is necessary, because about as many tide-gauge records of the world show a fall of relative sea level as show a rise. A tide-gauge record at any site could indicate rising or falling *land* level as well as changing sea level. A rise of land level by coastal rebound in areas of now-melted ice sheets has long been accepted by scientists and the public, but less widely recognized is the concurrent post-glacial sinking of a ridge peripheral to the ice sheets. Local sinking of large deltas also has been observed, because their sediments are deposited with a high percentage of interstitial water and are compacted easily. During recent decades, sinking also has been noted in areas where fluids have been pumped excessively from the ground. Volcanic regions are known to exhibit both upward and downward changes in level probably because of movements of magma in chambers beneath the volcanoes; some of these changes occur along seacoasts of the world. Less well noted, however, are vertical movements of land levels in modern times because of crustal cooling of rifted continental crust and because of underthrusting of oceanic crust beneath continental margins. Adding to this geological complexity are the many space and time scales of ocean movement owing to vastly differing dynamics (meanders in western boundary currents, steric changes, shelf waves, tides, and surface waves, for instance).

The book begins with an introduction that describes recorded ancient water movements, continues with a description of the more precisely measured early-to-modern variations of tides and currents, and concludes with a description of the stages in development of tide-prediction tables. The next chapter addresses the causes and time scales of different kinds of vertical shifts of relative sea levels. Next is a discussion of plate tectonics and how plate movements have affected past emergence and submergence of continents, and the nature of resulting control exerted by these movements on the chemistry of seawater and the distribution and evolution of plant and animal life both on the land and in the ocean. Background information about past shifts in levels of the ocean and land derived from ocean-floor cores, archaeology, radiocarbon dating, and historical records sets the scene for discussion of tide gauges, whose recording began in earnest less than a century ago.

Most tide gauges were installed within seaports to facilitate the entrance and exit of commercial shipping, chiefly sailing ships that were at the mercy of flooding and ebbing currents. During the early part of the present century a few scientists began to study long-term tide-gauge records to measure temporal changes of relative sea level, some inferring a rise of sea level and others a change of land level, as discussed in a chapter on early studies of the records. These studies were hampered (and still are) by the poor worldwide distribution of tide gauges: mainly, they are concentrated along the coasts of industrial nations of the northern hemisphere. By about 1970 many scientists and the mass media began to advocate the possibility that heating of the atmosphere by the greenhouse effect of industrial waste gases might have increased the rate of melt of ice sheets and caused a faster rise of sea level that would be destructive to human habitation, industry, and agriculture along coasts of the world. Accordingly, some climatologists studied tide-gauge records in a search for a possible increase in eustatic rise of sea levels. They mainly ignored the vertical movements of the land beneath the tide gauges and tried various strategems for averaging records of indicated relative sea-level change to avoid some effects of poor geographical distribution of useful tide-gauge records and to minimize differences within groups of records.

We chose an alternate method of study: the investigation of the same differences in direction and rates within regional groups of tide gauges. Immediately obvious was the presence within the regions of systematic changes, some of which are related to deglaciation, others to sinking of cooling rifted continental crusts, and still others to plate subduction. In a sort of tour of the world's coasts we classified the indicated land movements according to inferred origin, finding good correlation between direction and rates of land movements and the regional geology. In other words, the noise in the tide-gauge records caused by land movements that obscure the signal for change of sea level is a source of information about plate tectonics and the role of tectonism in causing modern vertical movements of the land.

In a chapter on significance of results, the direction and rates of relative sea-level change for different causes of land movements were compared and found to differ in a systematic way for different coasts of the world. This knowledge provides some insight into likely future changes of level and their impacts on society. For example, protection by sea walls along all coasts is unnecessary because some coasts are rising, not sinking. In addition, much shore destruction is caused not by natural sinking of lands or eustatic rise of oceans, but instead it is among the unexpected results of human intervention through engineering works. Some of these works cause coastal subsidence owing to extraction of fluids, others reduce the natural sediment supply to beaches and thus accelerate coastal erosion, and others (dams and barrages) alter the sediment supply to deltas, thereby accelerating relative sea-level rise. Construction of groins and breakwaters often traps beach sand in its movement along coasts, thereby causing erosion of beaches that are located down-current from the structures.

Understanding of the cause of changes in level of the land and the ocean is paramount in importance to the proper development of coastal regions. Results of this study provide guiding data for scientific, engineering, or policy solutions to coastal flooding, by helping to assure that the true causes of relative subsidence are understood along each coastal compartment. Although we still cannot quantify unambiguously the contributions to relative sea levels from tectonism and oceanographic variability, we can describe the apparent dominant factors for most of the globe. Hopefully, our study can provide some impetus to gaining better understanding and use of geological and oceanological controls rather than the usual reliance on legal "remedies" or of trying to force nature by engineering means. An old Newfoundland skipper said, "We don't be *takin'* nothin' from the sea. We has to sneak up on what we wants, and wiggle it away" (Mowatt 1958).

K. O. Emery
D. G. Aubrey

Acknowledgments

This book has been in progress for about a decade, and there have been numerous occasions to discuss its objectives with many individuals and organizations. These discussions were supplemented by our publication in scientific journals of articles that were written for regions having densely spaced tide gauges (for example, Fennoscandia, Japan, and the United States) or having special geological situations (Australia, India, South America). This early work aroused some discussion that led to strengthening of later analyses. In essence, we have concluded that 'noise' in the records produced by tectonic movements and both meteorological and oceanographical factors so obscures any signal of eustatic rise of sea level that the tide-gauge records are more useful for learning about plate tectonics than about effects of the greenhouse heating of the atmosphere, glaciers, and ocean water. To these early discussants and writers on related aspects of the subject, we perhaps owe more thanks and appreciation than to those who fully agreed with our views. Among them were T. P. Barnett and Roger Revelle of the Scripps Institution of Oceanography; R. J. N. Devoy of University College, Cork, Ireland; V. Gornitz of Goddard Space Flight Center, Institute for Space Studies, New York; A. L. Bloom of Cornell University; N. C. Flemming of the Institute of Oceanographic Sciences, Godalming, England; N.-A. Mörner of the University of Stockholm, Sweden; W. A. Peltier of the University of Toronto, Canada; P. A. Pirazzoli of the Laboratoire de Géographie Physique, Montrouge, France; and P. S. Roy of the University of Sydney, Australia, some of whom favored only limited tectonism. To these people and the anonymous reviewers we express sincere appreciation for their interest, thoughts, and suggestions.

Special thanks are due D. T. Pugh formerly of the Permanent Service for Mean Sea Level at Merseyside, England, D. B. Enfield of University of Oregon, You Fanghu of the Institute of Oceanology at Qingdao, China, and V. Goldsmith of Hunter College, New York, for providing magnetic tapes of tide-gauge records of the world and tabulations of some records from South America, China, and Israel, respectively. We also thank our many colleagues and students at Woods Hole Oceanographic Institution for discussions on these topics, including B. V. Braatz, J. D. Milliman, A. Solow, Elazar Uchupi, and A. J. Withnell. We gladly acknowledge the aid of others in translating the summary chapter into various languages (French—Lucienne Taillebot, J. P. Eliet, Maureen Eliet; German—W. G. Deuser; Hebrew—David Neev; Japanese—Susumu Honjo; Russian—A. P. Lisitzin; Spanish—Carlos Palomo, Juan Acosta, Yuri Budenko).

Finally, we happily received financial aid from the Woods Hole Oceanographic Institution's Ocean Industry Program and Coastal Research Center, the A. W. Mellon Foundation, National Science Foundation under Grant Number OCE- 8501174, NOAA National Office of Sea Grant under Grant Number NA83-AA-D-0049, and Aubrey Consulting, Incorporated, the latter for covering costs of photography and xerographic duplicating of drawings and texts. The Woods Hole Oceanographic Institution also aided by providing pension funds for Emery. Pamela Barrows edited the final computer disk to fit Springer-Verlag's format. The Editorial Department of Springer-Verlag provided encouragement during the late phases of completion of the manuscript.

Contents

Preface vii
Acknowledgments ix

1. Introduction 1
   Objective   1
   Early Knowledge of Relative Sea-Level Changes   2
     Prehistoric   2
     Epic Flood   2
     Charybdis and Maelström   3
     Periodic Movement—Tides   4
     Early Substitutes for Laws of Nature   4
     Identification of the Moon as the Cause of Tides   5
   Tide Tables   6
     Developments in Europe   6
     Developments in China   7
     The Chinese Tables of A.D. 1056   8
     Modern Prediction Tables   10
   Devices for Measurement of Water Level   12
     Fixed Scales   12
     Mechanical Recorders   13
     Electronic–Digital Recorders   13
     Satellite Altimetry   14
   Frequency Realms of Tide-Gauge Records   14
   Semidiurnal and Diurnal Tidal Frequencies   17

2. Causes of Relative Sea-Level Changes 23
   Sea Levels Versus Land Levels   23
   Plate Tectonics/Sea-Floor Spreading   23
     General   23
     Changes in Spreading Rates   23
     Changes in Areas of Oceans and Land   23
     Changes in Direction of Plate Motions   24
     Other Thermal Effects   24
   Deep-Ocean Sedimentation   24
   Isostasy   24
     General   24
     Glacio-Isostasy   24
     Hydro-Isostasy   25
     Sediment-Isostasy   25
   Glacial Surge/Melting   25
   Ocean Surface Topography   26
   Geoid   26
   Geological Faulting   27

Sediment Compaction and Subsidence   31
   Processes   31
   Examples of Subsidence   33
Sinkholes   40
Climatic Effects   40
   Overview   40
   El Niño/Southern Oscillation   41
   River Runoff/Floods   45
Steric Ocean Response   46
   General   46
   Temperature Effects   46
   Salinity Effects   46
Long-Period Tides   47
   General   47
   Long Time Scales   47
   Amplitude Variation Through Time   48
   Changes in Hydrodynamical Conditions   48
Shelf Waves and Seiches   50
Tsunamis   51
Surface Gravity Waves   51

3. Ancient to Modern Changes in Relative Sea Levels 53

Source Materials   53
   Background   53
   Rocks on Continents   53
   Sediments and Rocks on the Ocean Floor   54
   Pleistocene Glacial Deposits and Ice   54
   Shoreline Classification   55
   Dating of Shore Fossils   56
   Archaeology   56
   Tide Gauges   57
Phanerozoic   57
   General   57
   Seismic Stratigraphy   57
   Volumes of Igneous Rocks   59
   Sequence of Plate Movements   59
Late Archaean to Present   61
   Orogenies   61
   Relative Sea Levels   62
   Tides   62
   Continental Glaciation   62
   Isotopes in Marine Sediments   63
   Evolution of Life   63
Late Pleistocene   64
   Oxygen Isotopes   64
   Solar Insolation   64
Latest Pleistocene and Holocene   65

4. Previous Studies of Relative Sea Levels from Tide Gauges 69

Data Base   69
Previous Methods and Interpretations   69
   United States   69
   Europe   72
   World (Through 1980)   72

        World (After 1980)   76
        Summary   78

5. **Detailed Mapping of Tide-Gauge Records in Specific Regions** 81
     General   81
     Northern Europe   82
     Southwestern Europe   88
     Mediterranean Sea   89
     Africa   93
     Antarctica   95
     India   96
     Southeastern Asia   100
     Australia   101
     New Zealand   104
     Eastern Asia   105
     Japan   107
     Northern Eurasia   114
     Western North America   114
     Central America   118
     South America   123
     Caribbean Sea   125
     Gulf of Mexico   125
     Eastern North America   127
     Islands of the Atlantic Ocean   134
     Islands of the Indian Ocean   137
     Islands of the Pacific Ocean   139
       Island Arcs   139
       Hot-Spot Chains   141
     Long-Term Tide-Gauge Records of the World   143

6. **Significance of Tide-Gauge Records** 151
     General   151
     Glacial Loading and Unloading   151
     Subduction and Rifting   152
     Volcanism   153
     Faults and Folds   154
     Deltaic Loading   154
     Extraction of Fluids   155
     "Stable" Coasts   155
       Observation   155
       Interpretation   156
       Eustatic Sea-Level Change   158
     Statistical Summary   160

7. **Future Eustatic Sea-Level Change** 163
     Climate Change Scenarios   163
     Sea-Level Change Scenarios   165

8. **Impact of Sea-Level/Land-Level Change on Society** 167
     General   167
     Physical, Chemical, and Biological Impacts of Rising Relative Sea Level   169
     Socio-Economic Impacts   174

9. Summary 175

Appendix I 193
Appendix II 205
Bibliography 207
Author Index 227
Subject Index 231

1. Introduction

Objective

Relative sea levels have strongly influenced the history of mankind and the evolution of this planet. On a geological time scale, the waxing and waning of ocean levels in response to early Earth processes and continuing movements of continents have left a rich record in the form of sedimentary deposits. These deposits provide insight into the effects of future ocean level changes on the continued evolution of the Earth.

On a historical time scale, fluctuations in water levels have been important in the evolution of cultures. Coastal cities have foundered or expanded as relative sea levels have shifted. That the future of our coastal areas may be impacted by accelerated rises in ocean levels is but an extension of past conditions that have exerted increasing control over cultural evolution. Recent United States census figures (Culliton et al. 1990) show that nearly 50 percent of the residents of the country reside in coastal counties, demonstrating our reliance on the ocean.

The lowering of ocean levels during the early Wisconsinan (Würm) glacial stage prior to about 100,000 years ago may have permitted migration of Asians to North America via the land bridge that existed across the Bering Strait. Subsequently, between 20,000 and 10,000 years ago, renewed exposure of this land bridge allowed a later migration into North America. Drowning of the earlier land bridge by higher sea levels during the intervening interglacial stage separated the migrants into two distinct groups that developed along different paths. Thus, human development on North America, for instance, appears to have been linked tightly with relative sea-level changes.

Entire cultures have depended on maritime activities, exposing themselves to the whims of the oceans. A major civilization in the Mediterranean Sea existed between about 3000 and 1400 B.C.: the great Minoan sea power. The demise of the Minoans has been ascribed to a series of eruptions on the island of Thera (Santorini), accompanied by tsunamis and followed by Mycenian invasion (Mavor 1969; Toynbee 1976; Cottrell 1979; McCoy 1980). Ancient voyages of conquest or expansion by many civilizations were at the mercy of the oceans. "Discovery" of the New World by western civilizations (Celts, Phoenicians-Iberians, Egyptians, and other Mediterranean countries) occurred after long, hard ocean crossings, and landfall locations were subject to the vagaries of the winds and currents (Emery and Uchupi 1984, p. 2–19; Richardson and Goldsmith 1987).

Coastal inhabitants have had to adapt to various types of relative sea-level impacts. While tsunamis and destructive storm surges have impacted many cultures (Minoan, Dutch, and Bangladeshian, to name a few), other societies have been affected by slow but inexorable rise or fall in ocean levels. Ancient Roman structures built on the shores of what is now Israel are submerged at Caesarea (Oleson 1989), the submergence commonly viewed as a result of a rise in level of the Mediterranean Sea relative to land; however, adjacent sites appear to indicate lowering of relative sea levels, suggesting considerable along-coast vertical movements caused by neotectonism (Emery et al. 1988a; Neev et al. 1987). At the Italian coast near Naples, the Temple of Serapis once was submerged for a long time, as indicated by mollusk borings on temple pillars. However, the temple now stands above sea level, as it did when constructed, reflecting a cyclical rise and fall in level of the land relative to the level of the Mediterranean Sea. On the other hand, the Scandinavians have experienced continued emergence of much of their coastline owing to unloading by melt of glacial masses emplaced during the last glaciation. This rebound has stranded fishing villages and seaports inland far from the sea, necessitating continued adaptation to the receding ocean. Other examples exist of adaptation to rising or falling ocean levels. In Guangzhou, China, an ancient lighthouse now stands more than 2 km inland from the riverfront, due to shifting sedimentation patterns and falling relative ocean levels (Inman 1980; Wang and Aubrey 1987). Well known is the continual struggle of the Dutch to preserve their lowlands, ending with massive dike and water-control projects in the Scheldt Estuary.

The future portends a continuation of the need for man to adapt to changing water levels. Bangladesh is an extreme example of this need, as this low-lying southern Asian country lies at the mercy of typhoon surges and upland monsoonal floods that periodically ravish the land. A storm surge and typhoon in 1971 caused nearly 270,000 deaths in the delta region of the country (Murty 1984). The years 1986 to 1989 brought continued monsoonal rains that caused widespread flooding when rivers overflowed their banks. The contemplated building of river flood-control structures on the major Bangladesh rivers (Ganges, Brahmaputra, and Megna rivers) probably would restrict sediment flow to the delta; however, this sediment is an essential part of the delta balance, compensating for subsidence that occurs as deltaic sediments settle and consolidate. A related situation occurs in Bangkok, Thailand. Here, groundwater withdrawal has submerged parts of the city at rates of nearly 10 cm/y. (Asian Institute of Technology 1981), about 100-fold faster than rates commonly ascribed to global sea-level rise. Since Bangkok lies only about 1 m above river levels, this drastic subsidence may lead the city back to the times when canals, not roads, were the primary avenues for transportation.

Recent concerns over the impact of man's activities on global climate have increased awareness of the effects of sea-level rise on coastal communities. Projections for sea-level rise during the next century range from 15 cm to about 380 cm, a range that reflects uncertainty in the primary causes (atmospheric warming) and secondary effects (melting of ice sheets and smaller glaciers and warming of the ocean, which causes thermal expansion). As this concern over climate change is increasing, it is appropriate to evaluate what we know about global sea levels throughout past geological times and into historical times. Our improved understanding of the other contributions to relative sea levels (tectonic movements of the coastal land, isostatic loading and unloading, pore-water withdrawal, oceanic variability, etc.) may permit improved monitoring of ocean level changes in the future and better determination of dates when accelerations in global sea levels might have occurred. This review of knowledge of relative sea levels is the primary focus of this book.

Early Knowledge of Relative Sea-Level Changes

Prehistoric

The peoples who have been most concerned with relative levels of the ocean surface and the land surface are those who live along the shore where a rise or fall of water level with tides, seasonal storms, or tsunamis causes the ocean to cover or expose a wide belt along the shore. Some ancient peoples who subsisted on gathering of mollusks and other bottom-living animals and on catches of fish along the shore were the most dependent on the changing position of these interfaces between earth, water, and air. More advanced peoples who constructed piscinas, salt-making ponds, water wells, piers, docks, warehouses, and homes along the shore used or knew well the limits of the tides. Their structures were more subject to damage by episodic storm waves and tsunamis than by the more frequent but more predictable and usually lower amplitude changes in level caused by the tide.

The earliest prehistoric contacts between humans and the shore are unrecorded and lost to the ocean and conjecture. Later, agriculture and trade permitted or required the establishment of permanent villages. In these villages an elite class kept verbal accounts of important events before written accounts were developed. Later, some of the verbal accounts were recorded on whatever medium was available at the time and place. Still later, these early records were edited and modified according to thoughts, embellishments, and rationalizations that developed long after the event. A few of these ancient examples are included here.

Epic Flood

An early example of this complex accounting of events is the Flood in the Sumero–Akkadian epic of Gilgamesh, legendary king of Uruk about 2750 B.C. The earliest written record of the *Gilgamesh* was found on 12 cuneiform tablets written about 1200 B.C. and found at Nineveh about A.D. 1850 (Gardner and Maier 1984). They describe a flood experienced about 4000 B.C. by Gilgamesh's ancestor Utnapishtim, who had been instructed by a god (Ea) to build a ship and load it with the seed of all living things, cattle, beasts, and his family. The storm lasted six days and nights, and afterward the ship supposedly was stranded on Mt. Nisir, about 350 km south of Mt. Ararat. The biblical story (Genesis 5: 3–29; 7; 8) is similar but that Flood is dated to 2348 B.C. (1656 years after Creation at 9 am on 23 October 4004 B.C., according to the Bishop of Ussher!; Adams 1982). This story has much greater detail. Noah, his wife, his three sons, their wives, and two of "every beast after his kind, and all the cattle after their kind and every creeping thing that creepeth upon the earth after his kind, and every fowl after his kind, every bird of every sort" came aboard the Ark (*Genesis* 7: 14, Authorized King James version). After seven days, rain began and lasted 40 days and nights (40 is a common large number in the Old Testament). After 150 days afloat the Ark came to rest on Mt. Ararat, and 197 days later (a total of 428 days aboard ship) the cargo was discharged, and Earth was repopulated. A briefer description of the flood given in the Koran (Sura XI) includes as passengers the family of Noah and "a pair of every kind."

A detailed comparison of the *Gilgamesh* and the biblical stories suggests a common source (Keller 1956, p. 35–39). Note that Abraham (the patriarch of Israel) emigrated from Ur on the lower Euphrates River near Uruk where Gilgamesh had been king. Abraham left Ur about 1950 B.C. and he or his entourage could have brought the account of the much earlier Gilgamesh Flood, later to be included in the Old Testament. In both accounts divine guidance was present, with the main object being the elimination of sinners (those who did not follow the injunctions of the priesthood). In fact, lack of knowledge about natural processes led to a general belief in divine causes for a wide variety of events such as earthquakes, storms, droughts, and plagues—a belief that continued through Greek times and even to the present, especially now in areas of fundamental Creationist religious influence.

The biblical account has many problems. If the Ark contained all existing species of animals, how did those from Australia, North America, and South America arrive; how did frigid zone animals survive; how could foods having a weight many times that of the animals and from very special environments have been collected and preserved; why are not the hundreds of thousands of extant species of insects and beetles mentioned; what about all land vegetation? If the Flood were worldwide, the ocean volume would have trebled to reach the top of Mt. Ararat. Dilution of seawater to this extent by rain would have killed essentially all marine life (both plant and animal) and left widespread brackish-water sediment; yet the sediments of lakes and ocean continue from more ancient times without a break in the record of fossils or of stable oxygen and carbon isotopes ascribable to a Flood. Why is there no record of such a Flood in Egypt and China with their long recorded histories? Divine intervention continues with the covenant of the rainbow, as though light refraction did not exist prior to the Flood.

Details in the biblical record of the Flood preclude its being a late stage in the main return of meltwater from Pleistocene glaciers. Sea-level data indicate that this return was rapid from about 15,000 years ago (when the level was about 130 m below the present) to about 5000 years ago (when the level reached about 5 m below the present one). The stated date of the Flood is after the end of the most rapid rise of post-glacial sea level, but the more rapid advance of the seashore during previous thousands of years would have affected the mostly nomadic peoples who then lived on what is now the continental shelf. Whether the Flood tradition reported in legends of primitive peoples actually existed is unknown to us, but probably it was held only by peoples who then lived along the ocean shore.

Rather than a divine source for the Flood, a more prosaic explanation is the presence of severe rains and floods accompanying a typhoon, or Indian Ocean cyclone, such as those that episodically drown hundreds of thousands of people on the lowlands of Bangladesh and eastern India. The floodplain of the Tigris and Euphrates Valley is similarly low, so that a flood of 15 cubits (22.5 feet; 7.6 m—Genesis 7: 20) depth above even the low hills of the plain would have been a very serious event.

Charybdis and Maelström

The blind Greek poet, Homer, from the Ionian island of Chios, lived in the 9th or 10th century B.C. and assembled the story of Troy VIIA that fell to the Greeks about 1260 BC. His *Odyssey* (Butcher and Lang 1930, p. 140–146) relates the adventures of Odysseus (Latin: Ulysses) whose 10 years of travels and shipwrecks in the Mediterranean Sea were caused by storms produced by Poseidon to avenge the blinding of his son Polyphemus. Polyphemus had imprisoned Odysseus and 12 of his men in a cave on an isle of the Cyclops and had eaten 6 of them. The survivors from the cave and those who had remained aboard the ships escaped and soon came within sight of their home island, Ithaca, when Odysseus' men opened a leather bag in which the evil winds of the world had been imprisoned. The resulting storm sank 11 of his 12 ships with all of their men.

The last ship, with Odysseus, landed on the isle of Aea where the goddess Circe turned his men into swine. After a year, Circe released the crew and ship, giving Odysseus instructions on how to avoid the evil Sirens and later to sail close to the cave of Scylla in Messina Strait between Italy and Sicily in order to avoid the whirlpool Charybdis that was shadowed by a great fig tree on the opposite side of the strait. She reported that the

> mighty Charybdis sucks down black water, for thrice a day she spouts it forth, and thrice a day she sucks it down in terrible wise. Never mayest thou be there when she sucks the water, for none might save thee from thy bane, not even the Earth-shaker! Odysseus' ship began to sail up the narrow strait lamenting. For on the one hand lay Scylla, and on the other mighty Charybdis in terrible wise sucked down the salt sea water. As often as she belched it forth, like a cauldron on a great fire she would seeth up through all her troubled deeps, and overhead the spray fell on the tops of either cliff. But oft as she gulped down the salt sea water, within she was all plain to see through her troubled deeps, and the rock around roared horribly and beneath the earth was manifest swart with sand, and pale fear gat hold on my men. Toward her, then, we looked fearing destruction; but Scylla meanwhile caught out from my hollow ship six of my company, the hardiest of their hands and the chief in might. . . . And there she devoured them shrieking in her gates, they stretching forth their hands to me in the dread death-struggle.

Thus, the ship escaped Charybdis, the first to do so since Jason and his *Argo*. The ship then landed on the isle of Thrinacia where, unfortunately, the men killed and ate several of Helios' (god of the sun) sacred cattle. When the ship left the island Helios took revenge by calling up a storm that broke up the ship and drowned all the crew. Only Odysseus survived, unable to prevent the wrecked hull from drifting back to Charybdis at sunrise.

> Now she had sucked down her salt sea water, when I was swung up on high to the tall fig-tree whereto I clung like a bat, and could find no sure rest for my feet nor place to stand, for the roots spread far below and the branches hung aloft out of reach, long and large, and overshadowed Charybdis. Steadfast I clung till she should spew forth mast and keel again; and late they came to my desire. At the hours when a man rises up from the assembly and goes to supper, one who judges the many quarrels of the young men that seek to him for law, at that same hour those timbers came forth to view from out Charybdis. And I let myself drop down hands and feet, and plunged heavily in the midst of the waters beyond the long timbers, and sitting on these I rowed hard with my hands. But the father of gods and of men suffered me no more to behold Scylla, else I should never have escaped from utter doom. Thence for nine days was I borne, and on the tenth night the gods brought me nigh to the isle of Ogygia.

The periodic alternation of Charybdis sucking water and then expelling it is a reasonable description of a tidal whirlpool developed at the maximum ebb and maximum flood of the tide through the Strait of Messina. Even though the tidal range in the Mediterranean Sea is small, the large area of sea partly dammed off by the large island of Sicily causes a large volume of water to flow through the strait between tides, thus producing the whirlpool. Danger from the whirlpool must have been important for small boats, but it did not affect an aircraft carrier aboard which Emery passed Charybdis in 1948.

The Maelström is another famous whirlpool, located between Lofoten Peninsula and Mosken Island off northwestern Norway. It was shown by Olaus Magnus in his Carta Marina of 1539, and its ship-catching reputation served as the basis for a short story by Edgar Allan Poe (1966, p. 113–120). In Poe's story the Maelström was well known to the local fishermen who knew its tidal relationship and avoided the area at times of maximum tidal current. Three brothers on 10 July of a year in the 19th century had loaded their smack with fish caught off the island when they were struck by a "hurricane" that carried away both masts, to one of which one brother had lashed himself. Because of the dismasting, the boat became trapped by the developing whirlpool. She was drawn across a peripheral belt of foam where the roaring of the water was drowned out by a kind of shrill shriek. Beyond the foam was a steep (more than 45°) curving wall of water, the side of the whirlpool that sheltered the dismasted smack from the windstorm.

As the smack sped erratically round and round the whirlpool, one brother noted that with every round the smack and the wrecks of other boats descended lower and lower toward the turbulent bottom of the funnel. In contrast, smaller bodies that were spherical or barrel-shaped descended more slowly, providing hope of survival by abandoning the smack and clinging to a barrel. After failing to persuade a second brother to do the same, the one brother leaped overside and clung to a barrel. An hour or so later, the boat and its passenger reached the bottom of the funnel and were destroyed, but the barrel descended only part way, and a little later when the whirlpool ceased turning, the barrel floated on a calm windless sea, and the surviving brother soon was picked up by a fishing boat.

Events at some other whirlpools are worth recounting. A fleet of 50 curraghs (skincovered wicker construction) was lost by Breccan, son of Niall king of Ulster about A.D. 400, in a great tidal whirlpool off Rathlin Island (Northern Ireland near the Giant's Causeway) that subsequently was named Breccan's Caldron. Another one off Wales at Bardsey Island damaged the ship *Gwennan Gorn* that was built by Madoc (son of Owain Gwynedd), who was a skilled sailor and a legendary discoverer of America about A.D. 1170. Afterward the strait at the island became known as Currents of Gwennan's Bane (Hakluyt 1600, v. 5, p. 79–80; v. 6, p. 58; Deacon 1966, p. 83–85). All of these whirlpools lie between the mainland and an island between which a large volume of water must flow during both ebb and flood of the tide. Thus, the situation and cause of all are similar (produced by tidal currents), and thus the whirlpools are predictable as to frequency and time of activity. Evidently, tidal whirlpools were considered a menace to navigation for the small ships of ancient and medieval times. Still, on 28 October 1989 a radio station in Boston (WGBH) reported that a whirlpool off Burma sank a double-decker ferry, with the loss of about 1000 people.

Periodic Movement—Tides

Earliest recorded European recognition of tides (periodic ebb and flow) are those by the Greek historian Herodotus during a visit to the Red Sea about 450 B.C. (Carter 1958, p. 96), by the Greek naturalist Aristotle at the Strait of Messina about 325 B.C. (Ross 1913, v. 4, p. 834), by the Macedonian conqueror Alexander III at the Indus Delta in 325 B.C. starting his return from India (Bunbury 1959, v. 1, p. 447; Fox 1974, p. 385), and by the Massalian Greek astronomer Pytheas (Lallemand 1956) during his voyage to Britain and beyond in 310 B.C. Similar recognition of the periodic flow of the tidal bore along the Qiantang River near Hangzhou was reported in China at least as early as 140 B.C. (Needham 1959, p. 485).

Early Substitutes for Laws of Nature

Earliest explanations by humans for the occurrence of natural disasters or even for regular climatic events included the whimsy, revenge, or punishment exerted by all-powerful gods. Effects on humans of these godly acts were well recorded by Homer in his *Iliad* and *Odyssey* in the framework of Greek civilization. At that time humans essentially were concerned about *who* did these deeds to them.

The next stage may have been assignment of free will to the various media themselves: holding earth, air, water, and fire (and other forms of matter) responsible for their own choices and actions. Several famous examples relative to waves, currents, and tides can be cited. First, after Darius was beaten by the Greeks at Marathon in 490 B.C., his son Xerxes came to the throne of Persia in 486 B.C. determined on revenge. Assembling a cosmopolitan army of 5 million men and 700 to 800 ships, he set forth in 480 B.C. According to Herodotus (Carter 1958, p. 422–423), Xerxes had his army and navy build a pontoon bridge across the Hellespont (narrowest part of the strait between the Black Sea and the Mediterranean Sea). When a storm destroyed the bridge, Xerxes ordered the sea to be flogged with 300 lashes, fettered, and branded with a hot iron. After beheading the officers who had built the bridge, he had a stronger double-pontoon bridge built. One side was along a cable atop 360 50-oared ships and triremes and the other along a cable atop 314 ships, with the ship bows facing down the stream of the Hellespont so as to offer the least resistance to flow of the current. Tree trunks, brush, and earth covered the gap between the two lines of ships. A successful crossing was followed by defeat of the Greek guards at Thermopylae. Then Xerxes' forces captured and plundered Athens, only to be defeated decisively in the naval battle of Salamis in late September 480 B.C. Retreat turned into rout in order to re-cross the pontoon bridges before their destruction by the Greeks. Afterward, Asians had little role in the European side of the Dardanelles until the fall of Constantinople in A.D. 1453.

Second, a well-known river bore (tidal wave that periodically moves upstream from the high tide in the ocean) has long occurred on the Qiantang River near Hangzhou, China. In this area is an important river crossing that sometimes is blocked by passage of the bore. Such a blockage impeded the crossing of Qan Miau, a local Chinese king of about A.D. 950, who thereupon ordered the Qiantang bore to stop. When it ignored him, he had several hundred heavy arrows shot into the front of the bore, whereupon it reportedly did stop (Shi E 1252, p. 16).

Third, Cnut (Canute) of Denmark accompanied his father, Sweyn, in the Danish invasion of England to avenge the murder by Ethelred the Unready of many Danes who had peacefully settled in England. Numerous battles with Edmund Ironsides (Ethelred's son) and others established Cnut's control by A.D. 1016 over much of England and part of Scotland, as well as Denmark and Norway, and Cnut was prominent enough to attend the coronation of Conrad II at Rome on Easter Day 1027 (Churchill 1968, p. 26–27). Cnut's reign was orderly and according to the laws; his was the power of a great and respected man. It is strange that he is best known for his command in 1030 to stay the tide and not touch his throne placed on the beach. Surely, he and others in England well knew by then that the tides were associated with the moon, and Cnut was too worldly and experienced to be so naive as to believe that the tide would obey him. Presumably, he was trying to convince his barons that his power was limited and that his army was too small to start an unnecessary war that they wanted. In one version of the story he had the barons aligned between him and the sea so that they would be wetted first.

Both the divine and the volution approach were found so unsatisfactory that more rational causes began to be sought, especially during the Greek age of enlightenment, the Roman age of law and order, and the Renaissance after the Dark Ages. This is a transition from *who* or what to *how* and why. An early approach to scientific explanation for the cause of rise and fall of tides in all regions was animistic or naturalistic: by breathing and exhalation, drinking and spouting, or inflow and outflow to and from mythical beasts or deep caverns, by expansion and contraction of ocean water, by evaporation and condensation, or by some kind of activities that occurred at times that were related to positions of the moon and/or sun. According to Needham (1959, p. 494), even Leonardo da Vinci tried to calculate the size of the world lung!

Identification of the Moon as the Cause of Tides

Correlation of the tides with the positions of the sun and moon could, and, in fact, did, occur without real knowledge of gravitation—because it was only 300 years ago, in his *Principia* of 1687, that Isaac Newton established the real cause as gravitational attraction between Earth, Moon, and Sun long after tide tables had been invented and used in practical ways.

Precise correlation of the tides with positions of the moon and sun required the interest of astronomers or philosophers, who apparently were widespread during ancient times but whose thoughts were recorded in only a few regions: inland parts of the Middle East (upper Euphrates Valley), Egypt and southern Europe (where tides are small), India, Central America, and China. Most ancient knowledge about the moon and sun was calendrical in nature, but it could be applied to tides under favorable conditions. Some idea of the antiquity of such calendrical knowledge is provided by Marshack's (1972a, p. 50–53; 1972b) interpretation of holes cut in a bone tool by a Cro-Magnon man in south-central France about 33,000 years ago; the holes correspond with transitions of moon phases during 11 months. Marshack also illustrated patterns of holes or slashes made by later stone-age humans on many bones from Europe, Asia, and Africa, most having markings grouped as though recording complete lunar cycles. Presumably, the markings identified dates of ceremonies, such as our ancient lunar-dated Easter holiday (holy day), or dates related to the hunting of certain game animals. American Indians also made similar records but mostly on perishable wood or leather. Many Indian tribes used the lunar cycle for timing of ceremonies, but some began the cycles at new moon and others at full moon. There was no precision in identifying the number of days in a lunar cycle or the number of these cycles in a solar year, nor is there evidence of recognition of a relationship between lunar cycles and ocean tides.

Who can believe that the Phoenicians and Polynesians did not know much about tides and their relation to the moon and sun? Records of the former were destroyed by conquering armies that burned libraries at Tarshish in 530 B.C., Carthage in 146 B.C., and Alexandria in 47 B.C. Destruction of most ancient Chinese books and records also occurred about 215 B.C. when ordered by Emperor Qin Shi Huang; other Chinese books were burned by Kublai Khan and his illiterate hordes about A.D. 1300. In the New World during the 16th and 17th centuries all but a few of the beautiful Maya and Aztec books (codices) were burned, their contents (including astronomical and historical records—Aveni 1981) being considered works of the devil by overzealous Spanish priests who could not read them. Knowledge accumulated by the Polynesians and other non-literate peoples was passed on by word of mouth until irretrievably forgotten, as illustrated by stick charts of ocean waves developed by the Marshall Islanders (Emery 1987).

On the Indian subcontinent, some knowledge of the tides dates back to at least 2400 B.C., according to recent excavations in the district of Ahmedabad that have revealed a marine dockyard. The earliest Indian reference that associates the tide with the sun and moon occur in the Samaved of the Indian Vedic period, 2000 to 1400 B.C. (Pannikar and Pannikar 1971; Pugh 1987).

Our awareness of ancient knowledge about tides required four coincident factors within the same region: 1) presence of conspicuous tides, 2) observation of tides by astronomers (philosophers), 3) recording of observations, and 4) preservation of observations and thoughts. All of these factors appear to have been available in only two main regions: Mediterranean lands through visits by learned men to shores of the world ocean, and China where multitudes gathered to view several spectacular tidal bores (waves of water driven upstream during rising tides). Transfer of tidal knowledge from the Middle East must have occurred through long voyages of

Arab traders even before Mohammed's hejira (A.D. 622), as illustrated by sea stories in *Tales of Sinbad*. Some factual evidence remains in journals, tables, and treatises dating back at least to 215 B.C., but only a few of these have been translated from Arabic (Aleem 1980). A Chinese expedition in A.D. 1421 reported that astrologers at Aden made accurate predictions of tides and eclipses (Mills 1970, p. 50). Extension of tidal knowledge to western Europe and Japan accompanied the rise of commercial shipping.

Accurate tide-prediction tables could be prepared on the basis of observations only if they included four main stages: 1) general relation of tide to moon, 2) quarter rule (function of sun relative to new, first-quarter, full, and third-quarter moon phases), 3) meridian rule (delay of high tide after moon's passage through local meridian), and 4) organization of data into a simple form. The first stage was reached early in China where tidal relation to the moon was considered such a commonplace thing that it was mentioned only in passing within some remaining literature that is summarized by Needham (1959, p. 483–494). For example, early fortune telling, according to a Zhou Dynasty oracle (prior to 221 B.C.) included a sign "kan" that represented both water and the moon. A poem of about 150 B.C. by Mei Chen described viewing of the bore of the Changjiang (Yangtze River) during full moon; another about A.D. 50 by Wang Chong mentioned that the rise of the bore in the Qiantang River follows the waxing and waning moon. A writing (4 B.C.) on bamboo described the tides of the Changjiang as coming daily from the sea. Finally, Yang Quan (A.D. 222–280) wrote that the moon is the spirit of water; tides are large and small because the moon is sometimes full and sometimes incomplete. Tides were discovered later but recorded sooner thereafter in the Middle East–Europe region possibly because they were newly observed and thus remarkable and worthy of being recorded.

The second stage (quarter rule) was reached in Middle East–Europe about 140 B.C. by Seleucas the Chaldean astronomer in the Persian Gulf (Sarton 1927, v. 1, p. 183–184), and then in 90 B.C. by the Greek naturalist Poseidonius during a visit to Gades in the Atlantic Ocean just outside the Strait of Gibraltar, according to Strabo (Hamilton and Falconer 1854, v. 1, p. 258–261). It also was summarized well by the Roman naturalist Pliny in A.D. 77 from works of others (Bostock and Riley 1855, v. 1, p. 125). The same stage may have been reached in China as early as 150 B.C. by Mei Chen and by A.D. 50 by Wang Chong, but it was clearly and directly stated only in A.D. 770 of the Tang Dynasty when the great scientist Dou Sumong of the eastern Zhejiang Province wrote his now-lost *On Tides* (Needham 1959, p. 489; Xui 1978). Dou's work soon was followed by others also of the Tang Dynasty. Fong Yan (A.D. 800) in his *Records of Things Seen and Heard* noted the daily lag of high tide in Jiangsu Province north of Shanghai: "Generally speaking, there were two high tides every day, one at daytime and another at night. If high tide hour was in the morning on the first of a moon month, it became night tide in the middle of the month. But the night tide changed to morning tide at the same time also. In this way the two tides exchange their hours alternately at the end of the month." Li Jipu in his *General Geography of the Yuanhe Reign-Period* of A.D. 814 (Needham 1959, p. 490) stated in reference to the Qiantang River bore that the tide is highest on the 3rd and 18th days of the lunar month and lowest on the 10th and 25th days.

The third stage (meridian rule) that noted the delay of high tide after meridional transit of the moon again was stated earlier in Europe (Needham 1959, p. 492–493). Tidal delay also is known as the establishment of the port—a recognition of its value for port entries and departures of European commercial sailing ships. This meridian rule first was recorded by Poseidonius in 90 B.C., repeated by Pliny in A.D. 77 (Bostock and Riley 1855, v. 1, p. 124–128), apparently forgotten until rediscovered by Saint Bede the Venerable (Sarton 1927, v. 1, p. 503–511) in northern Britain about A.D. 625 partly by his reading of Pliny, and clearly expressed by Gerald the Welshman in his travel book of A.D. 1188 (Sarton 1931, v. 2, p. 417–418). From St. Bede's book was largely derived *The King's Mirror* written by an unknown Norse author about A.D. 1250 that set the daily increase in retardation of the meridian transit of the moon at 48 minutes after that of the sun (Taylor 1971, p. 82). In China the real statement about establishment of the port first came in a book written about A.D. 1086 by Shen Gua (Needham 1959, p. 492).

Tide Tables

Developments in Europe

Near the beginning of the 12th century A.D. increased population and trade in western Europe demanded the use of ships larger than the open-decked ones used by the Vikings for European raids and Icelandic trade, and by the Normans in their transports to the Battle of Hastings (Landström 1969; Scammell 1981). These larger ships, especially those of the Hanseatic League, had several decks and considerable tonnage, so that they were far too large to be rowed and beached at night. Their deep drafts and reduced maneuverability required a good knowledge of the ocean on the part of the ships' masters. Much tidal information was committed to memory, including tide heights, tide times relative to moon (port establishments), and times and places of tidal currents. Note, for example, Geoffrey Chaucer's description of the ship-captain member of the group of pilgrims in his A.D. 1390 *Canterbury Tales* (Coghill 1978, p. 30—Prologue) as being skillful in reckoning his tides, currents, moons, and harbors. Arrival of high tide at a port was mentally computed in terms of memorized time of high tide at the port after meridional passage of the new moon corrected for the number of days after the last new moon, and these times were given in terms of direction on a compass card where any one of 32 compass points equals 45 minutes (24 hours divided by 32), the then assumed retardation per day of the moon behind the sun (Taylor 1971, p. 133–134).

Some of this tidal knowledge is assembled in the earliest known English pilot-book copies about 1470 but written perhaps a century earlier (Taylor 1971, p. 131–138). A tide table, probably by Matthew Paris about A.D. 1250, showing the time of *Fflod at london brigge* is more an astronomer's computation than a sailor's guide. It gives the time of high tide not for calendar days of month and year but with reference to the moon's age. A more practical one was made in 1375 by Abraham Cresques as part of the Spanish Catalan Atlas—a circular diagram showing the moon times of high and low tides at 14 seaports of western Europe. Even better tide tables of similar form and purpose but combined with small charts were made for the Brittany coast of France early in the 16th century (Taylor and Richey 1969, p. 139–142; Taylor 1971, p. 170, 180). These tide tables were more concerned with the time of high tide than with the actual height of high or low tide.

Developments in China

The development of tidal knowledge and tide prediction in China is poorly known in the West. Most of the original articles published centuries ago in China are available only with greatest difficulty, so, although reference is made to original authors and dates, reliance is placed on secondary syntheses by Needham (1959, p. 483–494), Xui (1978), Xui and Li (1979), Xui et al. (1980), and Yang et al. (1989).

Chinese knowledge of the tides was influenced greatly by studies of the bore on the Qiantang River below the city of Hangzhou. Best viewing of the bore is at Yanguan about 55 km downstream from Hangzhou at Lat. 30°25′N., Long. 120°33′E. Modern tide tables and ephemeri (Nautical Almanac Office 1979; National Ocean Survey 1980, p. 318) show that here the bore lags 2 hr 12 min behind the local meridional transit of the moon. This lag corresponds with an 18 h 43 min lag after the moon's meridional transit over Greenwich, England—the standard international reference point. The modern tide table indicates that the bore originates 20 km downstream from Yanguan near the time of low tide there or a little before high tide at Changjiang Approach, and it advances up the river at about 28 km/hr, progressing as far as Hangzhou only at spring tides. According to Needham (1959, p. 484), the total height of the bore above low tide at Yanguan usually reached 6.1 m, but according to the National Ocean Survey (1980, p. 79, 318) the spring range maximum is 5.8 m.

Dou Sumong (A.D. 770–Tang Dynasty) in his lost *On Tides* developed a graphical method (Xui 1978) of showing the daily lag of tides at a given coastal point by constructing an *X-Y* plot in which the *X* coordinate was divided into nine intervals, each representing one-ninth the progression of the illuminated part of the moon through the lunar month of 29 to 30 days. On the *Y* axis he plotted the time at which high tide arrived for each of the moon intervals, connecting the high-tide points by diagonal lines; the same was done for low tides. This plotting method served as a kind of tide-prediction table.

Xou Ouyang (about A.D. 1010) of the Northern Song Dynasty noted the table in his *Summary of Ancients*, as did a former mayor of Qiantang, Zhang Junfang (about A.D. 1020), in his now-lost book *Chausho (On the Tides)*. Wang Xiongfu (A.D. 1150) of the Southern Song Dynasty recorded in his Catalogue of Geographical Monument Inscriptions the discovery of a stone monument carved with Dou's *On Tides* and *Everyman's Tidal Prediction Table* and confirmed its date as between A.D. 765 and 779. After Wang's listing of it, the monument again was lost, so that we have no example of this early tide table, although we do know the general principles of its construction.

Further evidence of the early use of tide-prediction tables in China is provided by the later Tang Dynasty date of Lu Shau's (A.D. 850) *Essay on the Tide*, from which we know that early tide tables related spring and neap tides to phases of the moon. Unfortunately, these tables did not survive either. Dou Sumong's and Lu Shau's tables preceded the earliest known tables of Europe by about 600 and 500 years, respectively. The earliest extant Chinese tables lag earliest Chinese tables by 200 years, but they still are 300 years prior to the European ones.

During the early Song Dynasty there began a renewed interest in Chinese tide-prediction tables. A book by Zhang Junfang about A.D. 1020 described the bore of the Qiantang River using a table similar to the diagram described by Dou Sumong, but, instead of nine moon-position divisions of the lunar month, Zhang used 12 for the X coordinate and 100 for tidal time on the Y coordinate. The 100 units were named ke, each of 14.4-min duration. Also in A.D. 1025 Yan Su's lost *Illustrated Discourses on the Tides* included a detailed tide table for Ningbo at the ocean end of Hangzhou Bay, evidently for the use of ships rather than of sightseers (Needham 1959, p. 492). This tide table also was recorded in A.D. 1026 on now-lost monuments. Still other tide tables reportedly were published about A.D. 1050 for the Changjiang (Fig. 1) and Zhu Jiang (Pearl) estuaries (at Wushau near Guangzhou [Canton]) by Yui Qing in his now-lost *Preface to Tidal Maps* of A.D. 1050 (Xui et al. 1980). Those for the Changjiang were for a narrow stretch of river between Yangzhou and Zhenjiang named Haimen (Sea Gate). This is the same region where the poet Mei Chen mentioned about A.D. 150 a tidal bore then famous in poems (named Guanglin Chau), but that had disappeared by about A.D. 900 probably owing to changes in the Changjiang estuary (Chen et al. 1979; Ren and Zeng 1980).

Interest in the Qiantang bore continued to be so great that tide tables were inscribed at an unknown date on the Zhejiang Ting pavilion on the bank of Qiantang River (presumably at Yanguan), according to later records. This inscription probably is the source of the data edited in A.D. 1056 by Major General Lui Chang Ming, general monitor of the tax bureau of the Zhejiang province and preserved in Shi E's A.D. 1252 *Shunyou Reign-Period Records of the Hangzhou District* (Needham 1959, p. 492). The brevity and simplicity of these Song Dynasty tables are such as might be expected of a wall inscription for the benefit of tourists and travelers waiting to

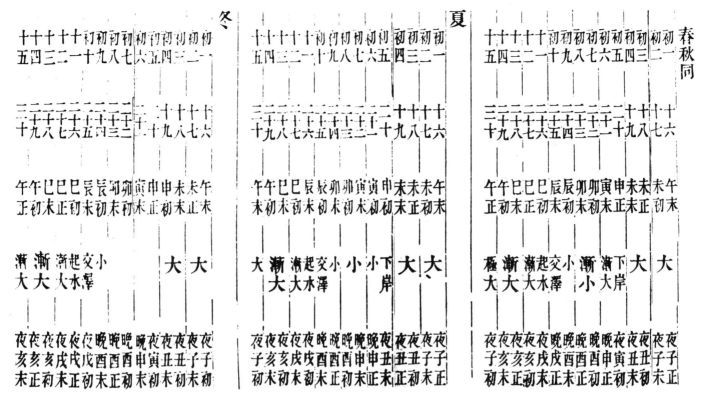

FIGURE 1. Reproduction of *Sishi Chaohou Tu* (Tide Tables for Four Seasons) for tidal bore at Yanguan, below Hangzhou on the Qiantang River. From Shi E (about A.D. 1252), but dating from before A.D. 1056, as discussed in text and by Yang et al. (1989). The table was printed in movable type two centuries before Gutenberg's Bible; very coarse (straw) paper was used. Reproduced with permission from the American Society of Limnology and Oceanography.

see the bore or to cross the river. Both Yanguan and Haimen were important river crossing points for army transport. Peraps an incentive for the A.D. 1056 tables was a rise in popular interest about astronomy following the very bright supernova in the Crab Nebula observed and recorded in China (Needham 1959, p. 426–428) and also possibly in southwestern United States (Brandt 1981).

As in Europe, more detail and precision in the tables were added during later centuries, perhaps especially during the early Ming period of ocean-going ships. Between A.D. 1405 and 1433 Admiral Zhan Ho's seven expeditions voyaged as far as Indonesia, Persian Gulf, and eastern Africa under the aegis of the early Ming Dynasty Emperor Ming Chen Zhu, as recorded in 1433 by the Moslem Ma Huan (Mills 1970), who participated in two of them. Ships were large enough to carry 800 men plus cargo, according to some reports. After 1433 China ceased to be a sea power, and tide tables were developed little further.

Interest in tide prediction for Japan followed and probably was derived from that of China. There it primarily was for commercial use and was associated with the need for knowledge about swift tidal currents in several of the straits within the Seto (Inland Sea). An observatory on Shikoku Island was established during the 12th-century Kamakura Period to study and make predictions for tidal currents among other oceanic phenomena. The observatory is now the Kampira Shrine.

The Chinese Tables of A.D. 1056

The Song Dynasty tide-prediction tables of Table 1 probably are the oldest surviving ones in the world. They pre-date A.D. 1056 by an unknown number of years, 1056 being only the date when the tables were edited by Lui Chang Ming (Needham 1959, p. 492). Ideas developed during previous centuries are incorporated in the tables. They are contrived so well that we believe it interesting and informative to apply them to the 20th century and to compare their predictions with those from modern tide tables. The tables (Fig. 1) are briefer than appears in translation, because some of the Chinese characters require several English words to denote moon time and general height of the Qingtang bore. A few gaps in the tide-height notations were filled by a later recording of the 1056 tables by Qian (1268).

The tide tables of Table 1 are based on the lunar calendar that contains 29.53 days for a complete lunar cycle from new moon to new moon. Months alternate with 29 and 30 days each. It was recognized early that 12 lunations totalled fewer days than a solar year (354.36 versus 365.25), and, accordingly, in 1260 B.C. King Wu Ding of the Shang Dynasty added a 13th month when necessary to keep the lunar calendar in accord with solar years. Later, one month was added to years 3, 5, 8, 10, 13, 15, and 18 of each 19-year cycle, with rare further additions more or less equivalent to the finer adjustments incorporated in the change from Julian to Gregorian calendars in Europe during A.D. 1582 (Moyer 1982).

Lunar months sometimes begin with a solar eclipse, and at the middle of the month there may be a lunar eclipse. Recording of eclipses on oracle bones began at least as early as 1361 B.C. in the Shang Dynasty, and more systematically with predictions in 720 B.C., long before Ptolemy wrote his *Almagest* about A.D. 140 at Alexandria (Needham 1959, p. 411–422). Despite the complications regarding nonconformity of lunar months to solar years, by definition the first day of each month in Table 1 corresponds with a new moon and sometimes with a solar eclipse. By this means the daily times of high tide expressed by the lunar calendar can be compared with times provided by modern tide tables that are based on the solar calendar. In the Chinese system, the changes of seasons occur at the middles of the 2nd, 5th, 8th, and 11th lunar months.

Spot-check comparisons of the A.D. 1056 tide (bore) predictions were made with the 1981 predictions from modern tide tables that are derived from several months of observations coordinated with meridional passages and other aspects of moon and sun. More thoroughly, predictions from both tide tables were plotted for the month of December, because highest tides occur at the latitude of Yanguan during December and June solstices, and they occur during the daytime for the December solstice. High tides and low tides at Changjiang Approach were plotted from National Ocean Survey (1980, p. 79) using local standard time (120°E meridian) for a period long enough to include a complete lunar cycle, 26 November to 26 December (solar dates), and a few days beyond (Fig. 2). The points (date, time, and height) for high and low tides are connected by short straight lines—yielding a jagged tide line rather than the normal sine curve. Next (Fig. 2C), the high-tide times and heights for Yanguan were transferred from Changjiang Approach using the +4 h 33 min time and +1.1 m height corrections given by National Ocean Survey (1980, p. 318). Levels of both semidiurnal high tides (bores) are indicated; the tide tables provide no data for low tides at Yanguan, presumably because low tide there is controlled only by river discharge and hence is not subject to astronomical prediction. Figure 2D shows the local dates and times of the phases and other events of the moon and of sunrise and sunset during the chosen span of tide curves.

The first day of a lunar month begins with the first midnight after the new moon; thus the first day of the 12th lunar month is 27 November. This lunar month and a few days of the preceding and following months are indicated on Figure 2E. They are parts of the winter season, thus identifying the

TABLE 1. English translation of tables in Figure 1 (turned 90° to left).

Date in lunar month		Time of day	Tide	Time of Night
\multicolumn{5}{c}{For spring and fall (they are the same)}				
1	16	End 7th TEB	Very high	Middle 1st TEB
2	17	Beginning 8th TEB	Very high	End 1st TEB
3	18	Middle 8th TEB	Very high	Beginning 2nd TEB
4	19	End 8th TEB	Very high	End 2nd TEB
5	20	Middle 9th TEB	High	Beginning 3rd TEB
6	21	End 9th TEB	Fairly high	Middle 9th TEB
7	22	Beginning 4th TEB	Low	Beginning 10th TEB
8	23	End 4th TEB	Low	Middle 10th TEB
9	24	Beginning 5th TEB	Very low	End 10th TEB
10	25	End 5th TEB	Lowest	Middle 11th TEB
11	26	Beginning 6th TEB	Low	End 11th TEB
12	27	Middle 6th TEB	High	Beginning last TEB
13	28	End 6th TEB	High	Middle last TEB
14	29	Beginning 7th TEB	High	End last TEB
15	30	Middle 7th TEB	Highest	Beginning 1st TEB
		For summer		
1	16	End 7th TEB	Very high	Middle 1st TEB
2	17	Beginning 8th TEB	Very high	End 1st TEB
3	18	Middle 8th TEB	Very high	Beginning 2nd TEB
4	19	End 8th TEB	Very high	Middle 2nd TEB
5	20	Beginning 9th TEB	High	End 2nd TEB
6	21	Beginning 3rd TEB	Low	Middle 9th TEB
7	22	End 3rd TEB	Low	End 9th TEB
8	23	Beginning 4th TEB	Low	Beginning 10th TEB
9	24	End 4th TEB	Low	Middle 10th TEB
10	25	Beginning 5th TEB	Lowest	End 10th TEB
11	26	End 5th TEB	Low	Beginning 11th TEB
12	27	Beginning 6th TEB	High	End 11th TEB
13	28	End 6th TEB	High	Beginning last TEB
14	29	Beginning 7th TEB	High	End last TEB
15	30	End 7th TEB	Very high	Beginning 1st TEB
		For winter		
1	16	End 7th TEB	Very high	Beginning 1st TEB
2	17	Middle 8th TEB	Very high	End 1st TEB
3	18	End 8th TEB	Very high	Beginning 2nd TEB
4	19	Beginning 9th TEB	Very high	End 2nd TEB
5	20	Middle 9th TEB	High	Beginning 3rd TEB
6	21	End 3rd TEB	Fairly high	End 3rd TEB
7	22	Beginning 4th TEB	Lower	Beginning 10th TEB
8	23	End 4th TEB	Low	Middle 10th TEB
9	24	Beginning 5th TEB	Low	End 10th TEB
10	25	End 5th TEB	Lowest	Beginning 11th TEB
11	26	Beginning 6th TEB	Low	Middle 11th TEB
12	27	Middle 6th TEB	High	End 11th TEB
13	28	End 6th TEB	High	Beginning last TEB
14	29	Beginning 7th TEB	High	Middle last TEB
15	30	Middle 7th TEB	High	End last TEB

Note: Tide cycle for days 16 to 30 repeats that for days 1 to 15.
Twelve Earthly Branches. Each 24-hour solar day was divided into 12 TEB; each TEB = 2 modern hours; 1st TEB began at 0000 hours (midnight). 12 TEB = 100 *ke*; 1 *ke* = 14.4 modern minutes. Points where TEBs are transferred between day and night columns are closest TEB to sunrise and sunset.

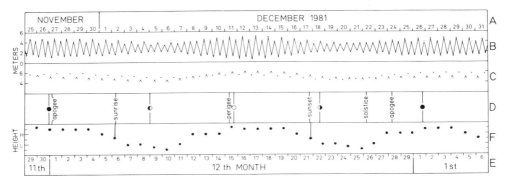

FIGURE 2. Comparison of A.D. 1056 bore predictions at Yanguan with those from modern tide table. Reproduced with permission from the American Society of Limnology and Oceanography.
A. Solar calendar for tide prediction from 1981 tide tables (National Ocean Survey 1980, p. 79, 318).
B. Times and heights of 1981 high and low tide predictions at Changjiang River Approach.
C. Dates and times of high tide (bore) at Yanguan derived from B. Small wings indicate those particular high tides that are predicted by F.
D. Dates and times of moon phases and solar events: new, first quarter, full, third quarter, perigee (moon closest to Earth in orbit), apogee (moon farthest from Earth), and December solstice (sun relatively farthest south), plus sunrise and sunset (from National Ocean Survey 1980, p. 355, inside back cover) for days so indicated in A.D. 1056 tables.
E. Dates of lunar calendar month (beginning on the midnight after appearance of new moon) and lunar seasons (from Don 1956). The latter control which tide-prediction tables of Table 1 are to be used.
F. Times and general heights of bore at Yanguan from A.D. 1056 data in Table 1 for comparison with those from 1981 tables in C.

section of the A.D. 1056 tide tables (Table 1) to be used. From this table the times and general heights of the bore at Yanguan were plotted (on Fig. 2F) with respect to date and twelve earthly branches (TEB) of Figure 2E.

Comparison of the predictions in the A.D. 1056 tables with those of the 1981 ones shows that the fit of tide (bore) heights is excellent for the middle of December—during the span of the highest perigean tides, but that the A.D. 1056 tables fail to shift to the higher of the daily apogean tides. Nevertheless, the general trends of the bore heights are similar in Figures 2C and 2E. The times of the bore at Yanguan during the span of Figure 2 according to the A.D. 1056 tables range from 2 h 51 min before to 1 h 15 min after the times given by the modern tables. An opportunity for Emery to check the A.D. 1056 tide table in another way was provided on 14 April (2 March on lunar calendar) 1983 during the *International Symposium on Sedimentation on the Continental Shelf* at Hangzhou. A field trip to Yanguan brought several hundred members of the symposium, who then awaited the tidal bore. The bore arrived at 1319, in contrast with a prediction of 1508 from the modern tide tables and a prediction of 1620 standard time from the ancient tables. All are expressed in Chinese Standard Time (at the 120° E meridian). Clearly, both ancient and modern predictions have room for improvement. The difference between actual time and time predicted from the ancient tables may be due to extensive reclamation of farm land and to other engineering modifications of the Qiantang Estuary since about 1970; further modification including a tidal barrier is planned to eliminate the bore entirely. The reason for the difference between actual time and time predicted from the modern tables is even less certain and solution of the question would require observations recorded during a year or more.

Modern Prediction Tables

Tide-prediction tables that combine the times, dates, and heights of high tides and low tides began to be published for several European seaports early in the 19th century by local private enterprises. In 1833 the British Admiralty began to publish them, followed by the French Hydrographic Service in 1839, the United States Coast and Geodetic Survey in 1853, and by bureaus of other countries (Marmer 1926, p. 168). A page from the Coast and Geodetic Survey (1970a) tide-prediction tables at Boston for the last three months of 1970 is illustrated in Table 2. Supplementing these tide tables are others printed and sold by private sources, an example of which is the *Eldridge Tide and Pilot Book* for waters off northeastern United States; it has been published continuously since 1875 with information on tides, currents, and much other nautical material of interest to fishermen and ship captains.

Easier interpolation of tidal heights and times intermediate between those of high and low level as given by Table 2 is provided by curves. Hand-drawn curves based on the tabulated predictions by the National Oceanic and Atmospheric Administration (successor to the Coast and Geodetic Survey) have been prepared and distributed by various companies, including the Rapid Blueprint Company of Los Angeles beginning before the 1950s. Recently, computer-drawn tidal curves for Massachusetts, northern California, southern

TABLE 2. Boston, Mass., 1970: Times and heights of high and low waters.

	October						November						December				
Day	Time H.M.	Ht. Ft.	Day	Time H.M.	Ht. Ft.	Day	Time H.M.	Ht. Ft.	Day	Time H.M.	Ht. Ft.	Day	Time H.M.	Ht. Ft.	Day	Time H.M.	Ht. Ft.
1 Th	0518 1124 1736 2348	0.2 9.7 −0.1 9.4	16 F	0536 1148 1806	−0.9 11.6 −1.8	1 Su	0000 0600 1212 1836	9.0 0.6 10.2 −0.6	16 M	0048 0642 1300 1924	9.2 0.3 10.5 −0.7	1 Tu	0024 0624 1236 1900	8.9 0.4 10.5 −1.0	16 W	0118 0706 1324 1942	8.7 0.7 9.8 −0.3
2 F	0554 1200 1818	0.3 9.9 −0.2	17 Sa	0024 0618 1236 1854	10.2 −0.6 11.3 −1.5	2 M	0042 0642 1254 1918	8.9 0.7 10.2 −0.5	17 Tu	0136 0730 1348 2012	8.8 0.7 10.0 −0.2	2 W	0112 0712 1324 1954	8.9 0.4 10.5 −0.9	17 Th	0200 0754 1406 2024	8.5 0.9 9.5 0.0
3 Sa	0030 0630 1236 1854	9.2 0.5 9.9 −0.2	18 Su	0112 0706 1324 1948	9.8 −0.1 10.8 −0.9	3 Tu	0130 0730 1342 2006	8.7 0.8 10.1 −0.3	18 W	0230 0824 1436 2100	8.5 1.1 9.5 0.2	3 Th	0206 0806 1418 2042	9.0 0.5 10.3 −0.7	18 F	0242 0842 1454 2112	8.4 1.1 9.1 0.4
4 Su	0106 0706 1318 1936	9.0 0.7 9.8 −0.1	19 M	0200 0800 1418 2036	9.2 0.5 10.2 −0.3	4 W	0218 0818 1430 2100	8.6 1.0 9.9 −0.1	19 Th	0318 0912 1530 2154	8.2 1.5 9.0 0.6	4 F	0300 0900 1512 2136	9.0 0.5 10.0 −0.5	19 Sa	0330 0930 1542 2200	8.3 1.3 8.6 0.7
5 M	0148 0748 1400 2024	8.7 1.0 9.7 0.1	20 Tu	0254 0848 1506 2130	8.6 1.1 9.6 0.3	5 Th	0318 0912 1530 2200	8.5 1.1 9.7 0.0	20 F	0412 1006 1624 2242	8.0 1.7 8.6 0.9	5 Sa	0354 1000 1612 2236	9.2 0.5 9.7 −0.3	20 Su	0418 1024 1630 2248	8.3 1.4 8.3 0.9
6 Tu	0236 0836 1448 2118	8.5 1.2 9.6 0.3	21 W	0348 0948 1606 2230	8.2 1.6 9.1 0.8	6 F	0412 1018 1630 2300	8.5 1.1 9.6 0.1	21 Sa	0506 1106 1718 2336	8.0 1.7 8.4 1.1	6 Su	0454 1106 1718 2336	9.4 0.4 9.4 −0.2	21 M	0506 1118 1724 2336	8.4 1.4 8.0 1.1
7 W	0330 0930 1548 2218	8.2 1.4 9.4 0.5	22 Th	0448 1042 1700 2330	7.9 1.9 8.7 1.1	7 Sa	0518 1124 1736	8.8 0.9 9.5	22 Su	0554 1206 1812	8.2 1.6 8.2	7 M	0554 1212 1818	9.7 0.1 9.2	22 Tu	0554 1212 1818	8.5 1.2 7.8
8 Th	0430 1030 1648 2318	8.1 1.5 9.4 0.5	23 F	0548 1148 1800	7.8 1.9 8.6	8 Su	0000 0618 1230 1842	0.0 9.2 0.5 9.6	23 M	0030 0642 1300 1906	1.1 8.4 1.3 8.2	8 Tu	0030 0648 1312 1924	0.0 10.0 −0.2 9.1	23 W	0024 0642 1306 1912	1.3 8.7 1.0 7.8
9 F	0536 1136 1754	8.3 1.3 9.6	24 Sa	0024 0642 1248 1900	1.2 8.0 1.7 8.5	9 M	0100 0712 1330 1942	−0.2 9.8 −0.1 9.7	24 Tu	0118 0730 1348 1954	1.1 8.8 0.9 8.3	9 W	0130 0748 1412 2024	0.0 10.4 −0.6 9.1	24 Th	0118 0730 1354 2000	1.3 9.0 0.6 7.9
10 Sa	0024 0636 1242 1900	0.2 8.7 0.9 9.8	25 Su	0118 0736 1342 1954	1.1 8.3 1.4 8.6	10 Tu	0154 0812 1430 2036	−0.4 10.4 −0.7 9.9	25 W	0200 0818 1436 2042	1.0 9.1 0.5 8.4	10 Th	0224 0842 1506 2118	0.0 10.6 −0.9 9.1	25 F	0206 0818 1448 2054	1.1 9.4 0.2 8.1
11 Su	0124 0736 1348 2000	−0.1 9.3 0.2 10.2	26 M	0206 0818 1430 2036	0.9 8.7 1.0 8.8	11 W	0248 0900 1524 2136	−0.5 10.9 −1.2 9.9	26 Th	0248 0900 1518 2130	0.9 9.5 0.1 8.5	11 F	0318 0930 1600 2212	0.0 10.8 −1.1 9.1	26 Sa	0254 0906 1536 2142	0.9 9.8 −0.3 8.3
12 M	0218 0830 1442 2054	−0.6 10.0 −0.5 10.5	27 Tu	0248 0900 1512 2124	0.7 9.1 0.5 8.9	12 Th	0336 0954 1612 2224	−0.6 11.2 −1.5 9.9	27 F	0330 0936 1600 2212	0.7 9.8 −0.3 8.7	12 Sa	0406 1024 1648 2300	0.1 10.8 −1.2 9.1	27 Su	0342 0954 1624 2230	0.7 10.2 −0.7 8.6
13 Tu	0312 0924 1536 2148	−0.9 10.7 −1.2 10.7	28 W	0324 0936 1554 2200	0.6 9.5 0.1 9.0	13 F	0424 1042 1700 2312	−0.5 11.3 −1.6 9.8	28 Sa	0412 1024 1648 2254	0.6 10.1 −0.6 8.8	13 Su	0454 1106 1730 2348	0.2 10.7 −1.1 9.0	28 M	0430 1042 1712 2318	0.3 10.5 −1.1 8.9
14 W	0400 1012 1630 2242	−1.1 11.3 −1.7 10.8	29 Th	0406 1012 1630 2242	0.5 9.8 −0.2 9.1	14 Sa	0512 1124 1748	−0.3 11.2 −1.5	29 Su	0454 1106 1730 2342	0.5 10.3 −0.8 8.9	14 M	0536 1154 1818	0.3 10.5 −0.9	29 Tu	0518 1130 1754	0.1 10.8 −1.3
15 Th	0448 1100 1718 2330	−1.1 11.6 −1.9 10.6	30 F	0442 1054 1712 2318	0.5 10.0 −0.4 9.1	15 Su	0006 0600 1212 1836	9.5 −0.1 10.9 −1.2	30 M	0536 1148 1812	0.4 10.5 −0.9	15 Tu	0030 0624 1242 1900	8.8 0.5 10.2 −0.6	30 W	0006 0606 1218 1842	9.2 −0.2 10.9 −1.5
			31 Sa	0524 1130 1754	0.5 10.1 −0.5										31 Th	0054 0654 1312 1936	9.5 −0.3 10.9 −1.4

Time meridian 75°W. 0000 is midnight. 1200 is noon. Heights are reckoned from the datum of soundings on charts of the locality that is mean low water.

FIGURE 3. Mechanical tide recorder of the kind used by the U. S. Coast and Geodetic Survey in 1922. Photograph provided by S. D. Hicks of National Oceanographic and Atmospheric Administration.

California, and Puget Sound became available as booklets (Born 1989). These *Tidelogs* are especially useful, because they include useful correlative astronomical data for moon, sun, planets, meteor showers, and eclipses. Each booklet is for one station with height and time corrections for application to other nearby parts of the region.

Devices for Measurement of Water Level

Fixed Scales

The early interest in tide prediction was far more concerned with the timing of the tide than with its height. Identification of responsible astronomical factors did not require precise knowledge of the variations of tide level throughout a day, a month, or a year. Nevertheless, one can be reasonably certain that tide levels were measured at whatever intervals might have been required at the time by reference to tide staffs attached to pilings, wharves, or other structures. Simplest would be sticks having marks painted or cut at convenient intervals in cubits, feet, inches, centimeters, or other height units. Such are common now where only a rough idea of the water level is needed, such as on lakes after seasonal rains or at the ocean shore during past hurricanes or other severe storms. Probably such installations lasted only a few decades, and only the records that may have been transferred to paper may still remain, although no detailed ones older than the 19th century are known to us. Tide staffs still are used, however, as daily checks on tides measured with even the most sophisticated measuring devices in advanced countries. They also are used by hydrographic surveyors to reference depth measurements to a known datum; the staff is read visually at specified time intervals, such as daily.

At many coastal villages or even cities the waterfront also has prominent markings inscribed in stone or even cast in bronze to show water levels reached by historically important storms. Still other cities have ancient benchmarks cut into rock cliffs or walls whose changing elevations above sea level have been recorded periodically as early as 1682 in the Netherlands (van Veen 1945, 1954), 1704 in Scandinavia (Mörner 1979), and soon followed by others elsewhere in northern Europe (Witting 1918, 1922; Bergsten 1954; Rohde 1975, 1977; Sjöberg 1984). Similar early tide staff or benchmark readings were made in England, France, Spain, Italy, and at the Dead Sea. Archaeological interest is attached to functional indicators of water level on ancient stone structures along waterfronts of Greece, Rome, Israel, Egypt, India, and elsewhere. Most are associated with spillways for flushing of harbors, with piscinas, baths, and salt pans, with levels for the top surfaces of wharves and roadways, and for the entrances of ship construction and repair facilities, and with bases and other features of walls and mooring facilities for ships. Some examples are discussed by Flemming (1968), Raban (1983), and Neev et al. (1987).

The reading of tide staffs cannot be precise when waves are present, unless a stilling well is used—an enclosure around the tide staff that has a small hole to damp out the disturbing effects of passing waves. Moreover, a timing device more accurate than an hourglass is needed to benefit from precision in reading water levels. For these reasons it is doubtful that pre–18th-century tide measurements could enable the drawing of tide curves of the sort that we are accustomed to using.

Mechanical Recorders

About the beginning of the 20th century, mechanical tide recorders began to be developed and used at many tide-gauge stations to record information that was not practicable or economical to obtain by human observation alone. One type of recorder was and is a spring-wound drum that is covered by a sheet or roll of paper that slowly turns past a scribing bar, along which moves a pencil or a pen that continuously marks the moving paper at a position controlled by the water level (Fig. 3). A weighted float in a stilling well is attached to a wire cable that is counterweighted and passes over a drum and axle to control the position of the pencil/pen along the scribing bar. During its use, the recorder was improved for increased precision and reliability. Other variations use electric motors in place of wound clockwork, and some of them replaced mechanical measures of the water level with pneumatic, electric, or acoustic sensors in order to avoid limitations from inertia and wear. A common sensor used especially in Great Britain is a pressure-measuring device that uses hydrostatic pressure to estimate water depth. These devices originally developed during the 17th century by Robert Hooke and Edmund Halley have been improved recently. A basic pneumatic system (Fig. 4) commonly known as a bubbler gauge improves long-term recording of the tide. Various analog recording devices have been used with these gauges, including a strip chart or a mercury manometer having a float (similar to that of a stilling well). Regardless of the sensor, the recording was analog.

Electronic–Digital Recorders

With time, the drawbacks of analog recording caused it gradually to become replaced by digital recording after about 1980. Analog records have to be digitized, checked, and then processed, and they are subject to the vagaries of recording

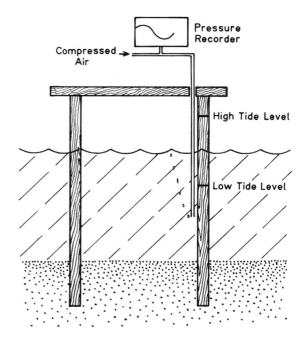

FIGURE 4. Schematic drawing of a bubbler tide gauge.

systems (dry inkwells, leaky pens, etc.). Accompanying digital recording were improvements in measurement techniques.

Among the first automatic recording systems were punched paper-tape devices that can be read (scanned) automatically to provide a digital output. An example from the United States is the National Oceanographic and Atmospheric Administration's ADR system (for Automatic Data Recording). This system provides a dramatic reduction of the time lag between acquisition and publication of tide-gauge data. Soon thereafter, some tide gauges were fitted with magnetic cassette tape drives to record the data. The magnetic tapes store a considerable volume of data (to 14 megabits), so the recording medium no longer was the factor that controlled the frequency of measurement or servicing many tide gauges. These magnetic tapes are read in a central processing laboratory. Great Britain adopted this technology relatively early. More recent improvements include a combination of a local recording medium (such as a paper tape or magnetic tape) combined with satellite telemetry to a central recording area. The National Oceanographic and Atmospheric Administration currently operates with the Geostationary Orbiting Earth Satellite (GOES) satellite, sending tide-gauge data several times a day and retrieving it at a central receiver. Real-time (or near-real time) data acquisition is now available for dozens of stations in the United States and elsewhere. The automatic telemetry of data (via satellite or ground link) permits instantaneous checking of data quality, so that gauges can be fixed soon after they are noted to be offline. This method contrasts with the loss of data under previous conditions when errors or problems were encountered much later during specified servicing intervals.

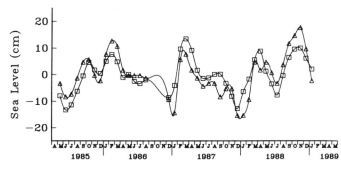

FIGURE 5. Sea level at Majuro Atoll in the western tropical Pacific Ocean measured by GEOSAT altimeter from crossover differences (smooth line) compared with monthly mean values from Majuro tide gauge (diagram by courtesy of R. E. Cheney, National Oceanic and Atmospheric Administration).

The new data-recording and transmission capabilities introduced new technologies for measurement. The United States adopted an acoustic travel-time sensor for the Next Generation Water Level Measuring System (NGWLMS). This system measures the time for an acoustic pulse to travel to and from a reflector (the water surface) in a stilling well. Although a stilling well is not required, it makes the pulse easier to detect. Other new techniques for measuring and recording tides include the use of airborne or satellite-borne sensors.

Satellite Altimetry

Satellites offer an opportunity to obtain global scale coverage of the ocean surface at regular time intervals, a capability that is lacking in conventional water level measuring techniques. Satellite altimetry relies on transmission of an electromagnetic pulse from a satellite to the water surface that reflects it back to the satellite. The two-way travel time of the pulse provides an estimate of the sea surface elevation at that time. Accuracy of the record is a function of the satellite ephemeris, resolution of the electromagnetic pulse, clock resolution, and knowledge of the orbit of the satellite. The method is accurate for measuring relative differences in ocean level, but less accurate for measuring absolute sea levels.

For a satellite to be able to resolve a 1-cm signal in sea level, it must be able to be timed to an accuracy of 10^{-10} or better for typical orbit heights of the altimetry satellites (about 1000 km above the Earth's surface). For comparison, one can examine how accurate past or existing satellite altimeters have been (Pugh 1987). Skylab S-193, the first satellite to fly an effective altimeter, had a range resolution of approximately 1 m. The GEOS 3 satellite, launched in 1975, had a range resolution of about 50 cm, whereas the SEASAT altimeter in 1978 had a range resolution of less than 10 cm. Improvements also were made in the ability to target smaller sea surface areas (footprints) for the return of the electromagnetic signal, with the footprint being reduced from 3.6 km to 2.5 km (over calm seas) from the GEOS 3 altimeter to the SEASAT altimeter. Future altimeter missions are being planned as part of the TOPEX/Poseiden experiment, to be launched during the early 1990s. Future improvements are expected to include resolutions of better than 10 cm with a repeat cycle of three days for ground coverage. Accuracy also is limited by uncertainties in the orbits of the satellites. To overcome these problems, sophisticated error computations are calculated and corrections are implemented. Some corrections can be made to the satellite orbit based on physics, and tidal corrections can be made using deep-ocean tidal models, but uncertainties in gravitational anomalies and atmospheric drag make precise determination impossible. To help resolve these issues, satellite positions and velocities are monitored with respect to ground stations, using laser ranging and radio ranging. Relative accuracies in these tracking techniques are high, whereas absolute certainties are low, particularly over large spatial scales.

Uses for satellite altimetry vary considerably. Their value as a platform for monitoring long-term sea-level changes is limited because of limited resolution and lack of precise orbit calculations, among other reasons. However, they have proven useful for monitoring large-scale currents such as the Antarctic Circumpolar Current (Fu and Chelton 1985), oceanic eddies (Cheney and Marsh 1981), and the El Niño (Cheney and Miller 1988), by monitoring dynamic height changes (Cartwright and Alcock 1981). Figure 5 compares averaged GEOSAT-altimeter data with averaged tide-gauge data, indicating a rms difference of only 4 cm during a 4-year period (Cheney et al. 1989). Altimeters also have proven useful for measuring surface tides (Woodworth and Cartwright 1986). After nearly five years of successful operation, the GEOSAT mission ended on 5 January 1990 and is to be replaced by a more sophisticated satellite, SALT, during 1990. Future improvements may permit satellite altimetry to record tides in the open ocean and in coastal regions where conventional fixed tide-gauge information is not available. Such recordings could lead to far better than present knowledge about the tides and changes in relative sea levels.

Frequency Realms of Tide-Gauge Records

Relative sea-level changes span a very wide range of frequencies, from longer than 100 m.y. to shorter than 1 sec (Fig. 6), in which different parts of the total range are widely recognized as belonging to specific physical phenomena. Lowest frequency is geological–related to plate movements, wherein the half-life of an ocean such as the Atlantic appears to be about 200 m.y., so that the complete lifetime of an ocean and the margin of a companion continent may average about 400 m.y. During this time span a continent tends to broaden, undergo wide epicontinental uplift, develop belts of folding, and exhibit peripheral sinking by cooling of rifted crustal rocks and by compaction of post-rift sediments. Some of the folding belts are likely to have a much shorter lifetime–even

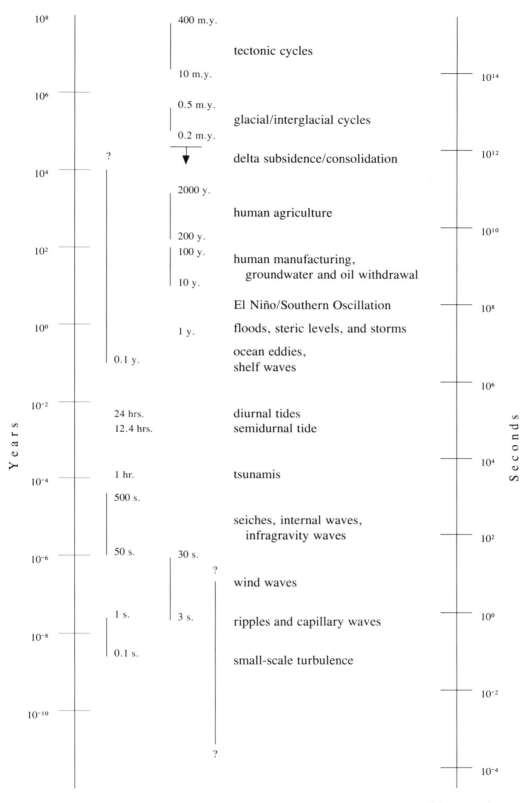

FIGURE 6. Frequency realms of relative sea-level change showing the estimated parts of the range that are covered by the most important agents and processes of change.

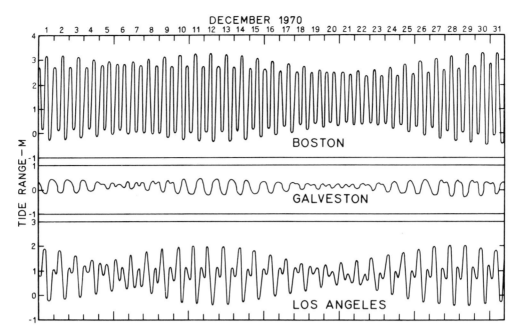

FIGURE 7. Curves of predicted tide for December 1970 that are characteristic of Atlantic, Gulf of Mexico, and Pacific coasts of United States. Curve for Boston prepared from data given in Table 2. From Emery and Uchupi (1972, fig. 191) with permission from the American Association of Petroleum Geologists.

as short as 10 m.y., especially where extrusive igneous emplacement is the main product of tectonism.

Relatively short-period (0.5–0.1 m.y.) geological causes of changing sea level are typical of glaciation, which can cause both lowering of sea level through withdrawal of water to precipitate on land as snow to make ice that weights the Earth's crust and causes lowering and rising of land levels. The reverse movements occur during interglacial stages. Glacial and interglacial stages alternate in groups, as illustrated especially between late Miocene and Pleistocene (between 10 and 0.01 m.y.). Periodicity also occurred but is poorly known for earlier glaciations that centered around Eocene–Oligocene (40 m.y. B.P.), Permian (280 m.y. B.P.), Ordovician (450 m.y. B.P.), and Proterozoic (800, 2000, and 2800 m.y. B.P.) times. Inferences about these changes in relative sea level are based largely on seismic profiles corroborated by drill holes, becoming a standard tool in the search for oil and gas reservoirs and increasingly for studies of erosion and deposition on the deep-ocean floor and effects on environments for deep-ocean benthic communities.

Periods or episodes of a few thousands to a few hundreds of years that affect local to regional land levels or water levels come within the spans of human activities in agriculture. These activities produce accelerated soil erosion and runoff that have their counterparts in deltaic growth and in increased water levels in lakes—for example, at the Dead Sea after the Arab conquest, followed by the present fall of the Dead Sea level caused by withdrawal of water from the inflowing Jordan River for agricultural purposes in both Jordan and Israel.

More intense and local human activities in mining have a shorter periodicity than agriculture, thus affecting land levels and shoreline positions. Examples are provided by hydraulic mining for alluvial gold in central California, mining of taconite in the Lake Superior area, mining of coal in many coastal and offshore sites (Nova Scotia, Japan), and mining of diamonds in alluvium of southern Africa. Extraction of groundwater for irrigation and other uses (Venice, southern California, Texas, Florida, Bangkok) causes subsidence of the land surface, as also does extraction of oil and gas (southern California, Texas, Venezuela, and many other coastal regions). These changes, owing to fluid or gas withdrawal, may persist for between about 200 and 20 years. The spans for other human activities such as manufacturing (with waste dumping), harbor building, and repeated dredging, river leveeing, and extension of levees into the ocean as jetties are on the order of decades to centuries.

An especially common periodicity for water-level variations is one year, a natural period because of control by geometry of the solar system as expressed by seasonal changes of temperatures, salinities, ice formation, and distributions of floods and high waves.

A primary diurnal tide frequency of 24.8 hours is a function of the lag of the moon behind the sun in reaching successive meridian transits during rotation of the Earth. The same mechanism is responsible for a principal semidiurnal (12.4-h) tidal period.

Still shorter periods are typical of surface wind waves, ranging from about 30 sec to 3 sec (longer for longer fetches,

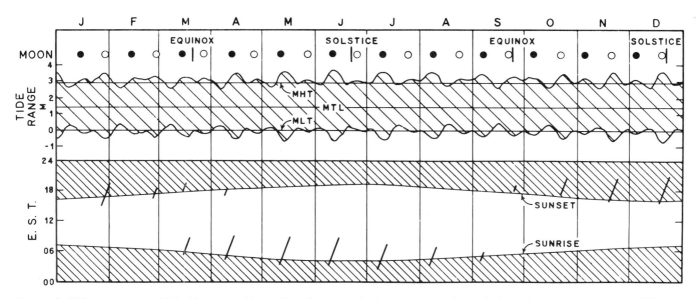

FIGURE 8. Tide range (mean high tide, mean tide level, and mean low tide) during 1964 for Boston relative to phases of moon, sunrise, and sunset. Lowest (lower than −30 cm) tides occur in the night during summer and day during winter. From Emery and Uchupi (1972, fig. 192) with permission from the American Association of Petroleum Geologists.

longer durations of wind, and higher wind speeds). Even shorter periods are typical of ripples controlled by capillarity; however, ripples and most waves are shorter than the periodic phenomena usually measured by tide gauges.

Examination of Figure 6 shows that the different frequency ranges of the various agents and processes that affect sea level are discontinuous, with breaks between them. These discontinuities are unlike the continuous spectra believed to be followed by electromagnetic phenomena, perhaps because the ocean comprises only a restricted part of the physical universe. This many-order range of periodicity in movements of the land surface and the ocean surface is incorporated within continuous measures of relative sea level. Each can be filtered or concentrated for close examination by choice of methods, durations of measurement, and methods of processing. When frequency ranges are compared with natural processes and geographical distribution, the causes for specific frequencies commonly can be determined. Unfortunately, in the time spans covering years to many decades, the causes for changes in ocean level are ambiguous because of overlap in time scales of different processes. Thus, it is difficult to identify a unique cause of many changes in tide records on the time scale of decades, whereas at higher and lower frequencies such separation is more easily possible.

Semidiurnal and Diurnal Tidal Frequencies

Among the earliest measurements of relative change in level of the ocean surface were those associated with the tides. Information about tides was needed to determine whether and when a ship might cross bars at river and harbor entrances, pass above submerged obstacles, be able to approach and stay at dockside, or be hauled safely onto the beach. Additionally, the current produced by flooding (rising) and ebbing (falling) tide frequently was sufficient to complicate a ship's steerageway that depended on winds. Essentially, the question was that of the time of high and low tide at any particular harbor or other place important to shipping. Knowledge of the time of local high tide (with respect to meridian transit of the moon) coupled with familiarity with bottom depths and current patterns was required for ensuring navigation and safety of a vessel controlled by a skilled local pilot.

Continued and more sophisticated need for tidal knowledge must soon have led to need for information about tidal range, especially for the twice-monthly spring tides. Thus was born the need for the quarter rule (the changing relative roles of the sun and moon during a lunar cycle). The daily and monthly progression of the phase angle between the sun and moon as measured from the Earth is responsible for the two spring and two neap tides each month with their gradual growth and decay. Longer, seasonal changes are related with changing tilt of the earth's axis with the ecliptic (plane of earth's orbit), with changing moon–earth distance, and other predictable changes in geometry of the Earth–Moon–Sun relationship. Some of these changes and differences in tide range are illustrated in Figures 7 and 8 that were drawn from predictions of semidiurnal high and low tide times and levels published by the Coast and Geodetic Survey (1964, 1970a,b). High semidiurnal symmetrical tides dominate on the Atlantic coast, low diurnal ones on the Gulf Coast, and markedly asymmetrical semidiurnal ones (mixed tides) on the Pacific coast of the United States (Fig. 7) and in other separated regions of the world ocean. On an annual basis Figure 8

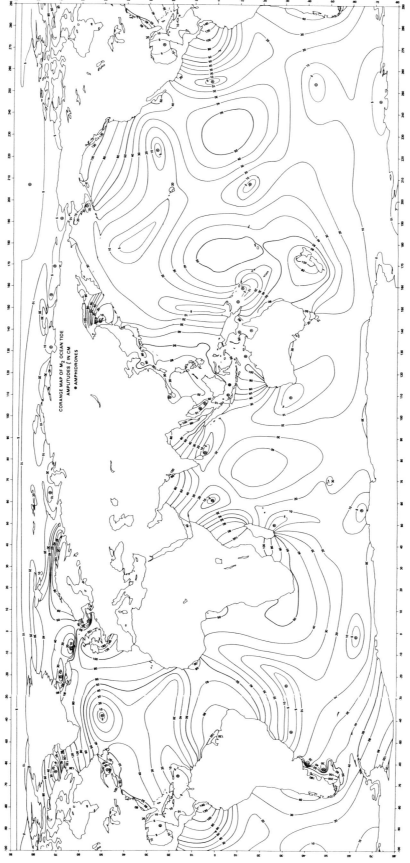

FIGURE 9. Chart of predicted semidiurnal (M_2) tidal range of the world ocean (from Schwiderski 1979) with permission from the Naval Weapons Center.

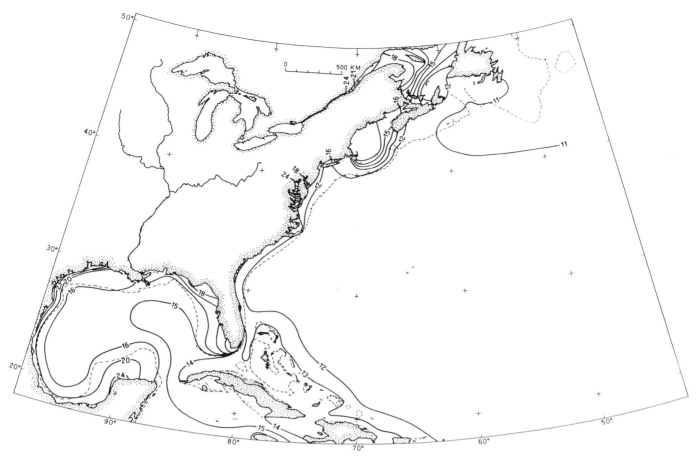

FIGURE 10. Map of high tide along eastern North America at 1-hour intervals after moon's transit of meridian at Greenwich, England. From Emery and Uchupi (1972, fig. 194) with permission from the American Association of Petroleum Geologists.

reveals the changing envelope of tidal range as the Earth–Moon–Sun geometry changes, with maximum tidal range at new moons during summer and full moons during winter, and lowest tides occurring at night during summer and day during winter. Many other geometrical relationships are evident on close examination of tide-prediction tables, as illustrated and discussed by Redfield (1980).

On a broader scale the times and heights of tides can be plotted at positions of tide-gauge stations to produce world or regional cotidal and coamplitude charts of the advancing tidal wave. These charts generally are based on a single tidal constituent, such as the principal lunar semidiurnal tide alone, because complications introduced by other tidal constituents obscure the picture of co-oscillation that is dependent on the dimensions of the oceans and the period of the tidal constituent. Many different charts for entire oceans have been prepared during the past, because different interpretations are permitted by the general restriction of tide-gauge stations only along the mainland coasts and by the different solutions to the theories from which these charts can be derived. A recent world ocean chart by Schwiderski (1979) is illustrated in Figure 9, but others are likely to appear when tidal information from the open ocean becomes available from radar altimetry from satellites. Typical are patterns of amphidromic points around which cotidal lines swing counterclockwise in the northern hemisphere and clockwise in the southern hemisphere. Tidal amplitudes are lowest near these amphidromic points.

Larger scale cotidal charts reveal a considerable slowing of the advancing high tide as it crosses progressively shallower waters from the deep-ocean floor, over the continental shelf, and into estuaries and lagoons. This slowing, or retardation, shown by Figure 10 for the western North Atlantic Ocean clearly shows the importance of knowledge of the establishment of the port (lag of high tide behind meridian transit of the moon) whether calculated from the time of the moon's transit actually observed aboard ship, or, as now is usual, taken from pre-calculated printed tide tables. Moreover, the compaction of cotime lines along a meridian as in Chesapeake Bay contradicts the frequently heard statement that tide is highest when the moon is overhead—it can be high, low, or at any intermediate level. The reason for the slowing of the tide as it enters shallower water is that the tide is a wave that is long compared with the water depth. Therefore, to a first-

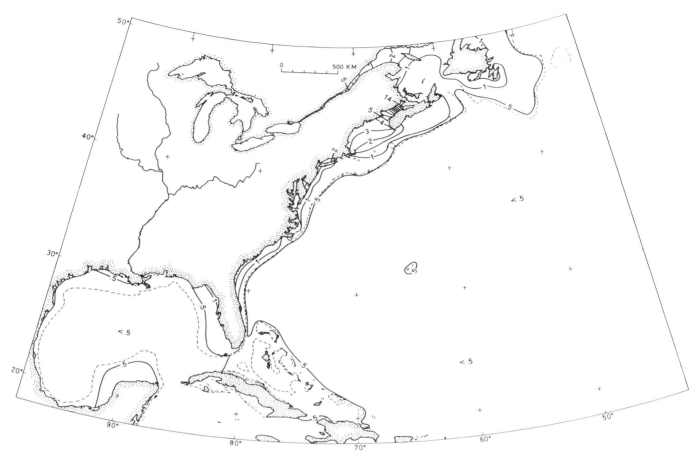

FIGURE 11. Map of mean spring tidal range along eastern North America. Contours are in meters, with extra contours for 0.5 and 1.5 m. From Emery and Uchupi (1972, fig. 193) with permission from the American Association of Petroleum Geologists.

order approximation its speed equals the square root of gravity times the depth. Speeds over water depths of 1000, 100, 10, and 1 m are 356, 113, 36, and 11 km/h, respectively. In extremely shallow water water (as defined by the Ursell number; Ursell 1953), the speed is affected by finite amplitude effects and no longer obeys the first-order model.

The coamplitude chart (Fig. 11) for the same area as Figure 10 shows that the range between low and high tide is about 0.5 m in the open ocean of the western North Atlantic. As the wave enters shallower water above the continental shelf, it slows and the water becomes compacted into a narrower belt so that the wave height increases commonly to 1 or 2 m by the time the wave reaches the shore. Where the shelf is very wide and shallow, so much energy is lost by interaction with the bottom that the height range again diminishes. In fact, the very wide and shallow shelves of the Arctic Sea are nearly tideless for that reason. In contrast, where the length of embayments happen to be such that the natural period of water moving from end to end approximates the period of the tide, resonance causes amplification of the tidal range to a maximum of 15.4 m at spring tides. Mean high tides range to 13.4 m at the Bay of Fundy; 11.0 m—southern Argentina; 10.7 m—Davis Strait; 10.0 m—western England; 9.1 m—Hudson Strait; Cook Inlet, and Magellan Strait; and 8.5 m—northern France (Marmer 1926, p. 216). These and other regions of especially high tides tend to be concentrated at high latitudes.

The astronomical tidal spectrum consists of a large number of constituents, organized into species, whose mutual interactions represent a complex physical problem (Munk and Cartwright 1966; Gallagher and Munk 1971). The primary frequencies of interest are integral linear combinations of six basic components related to celestial mechanics of the Earth–Moon–Sun system:

f_1^{-1} = 1 mean solar day = period of Earth's rotation relative to the Sun
f_2^{-1} = 1 lunar month = 27.3217 mean solar days period of Moon's orbital motion
f_3^{-1} = 1 year = 365.2422 mean solar days = period of Sun's orbital motion
f_4^{-1} = 8.85 Julian years = period of lunar perigee
f_5^{-1} = 18.61 Julian years = period of regression of lunar nodes
f_6^{-1} = 20,942 Julian years = period of solar perigee.

Some authors (Pugh 1987) provide a seventh component (mean lunar day = 1.0351 mean solar days); however, this frequency when combined with the sidereal month is a linear combination of the mean solar day and the tropical year: $f_{lunar} = f_1 - f_2 + f_3 = 0.966 f_1$. Interactions between these basic frequencies result in other energetic tidal frequencies, as expressed by other linear combinations. Tidal frequencies are divided into species that are separated by one cycle/lunar day, into groups that are separated by one cycle/month, and into constituents that are separated by one cycle/year. Table 3 shows some of the more common species and constituents.

Compound constituents and harmonic constituents, which are linear combinations of basic frequencies, can be generated through non-linear celestial and fluid mechanics. For example, MS_f consists of both a weak astronomical term and a potentially larger hydrodynamic term arising from M_2–S_2 interactions. Some terms are purely fluid mechanical, such as MS_4, which arises solely from non-linear hydrodynamic interactions of M_2 and S_2. Similarly, M_4, MN_4, and M_6 have no equilibrium (astronomical) tidal argument, but reflect non-linear generation in ocean basins of various scales. Magnitude and phase of each observed constituent provide insight into hydrodynamic processes. Further detail of the theory of tidal motions is provided by Pugh (1987).

Even a few months of recorded water levels provide general evaluation of the main harmonic components and their relative effects at a given locality. Improvements in prediction accuracy occur 1) with analysis of longer recordings to eliminate effects of "noise" caused by local weather, 2) with addition of further harmonic components to account for

TABLE 3. Tidal species and constituents of common interest.

Species	Constituent	Period
Fortnightly	MS_F	1.4 solar days
Diurnal	O_1	25.8 solar hours
	K_1	23.93 solar hours
Semidiurnal	N_2	12.66 solar hours
	M_2	12.42 solar hours
	S_2	12.00 solar hours
Ter-diurnal	MK_3	8.18 solar hours
Quarter-diurnal	MN_4	6.27 solar hours
	M_4	6.21 solar hours
	MS_4	6.10 solar hours
Sixth-diurnal	M_6	4.14 solar hours
	S_6	4.00 solar hours

FIGURE 12. Harris-Fischer tide-predicting machine operated by the U. S. Coast and Geodetic Survey from 1912 to 1966. Photograph provided by S. D. Hicks of National Oceanographic and Atmospheric Administration.

effects of local seiches, or reflections of long-period waves caused by local topography, and 3) with additional harmonic constituents for longer term changes of Earth–Moon–Sun geometry or more complex tidal interactions. Harmonic analysis by hand is time-consuming, and after the method was devised by Lord Kelvin (William Thompson of the University of Glasgow) in 1867, he invented a machine in 1872 to combine the various component tides. An improved and more complex machine was made by William Ferrell (U.S. Coast and Geodetic Survey) in 1882, principally for prediction of semidiurnal tides. Desirable features of both machines were combined and automated by R. A. Harris and E. G. Fischer (U. S. Coast and Geodetic Survey), and this machine (Fig. 12) was used from 1912 until replaced by computers in 1966 (Hicks 1967). It included harmonic analysis for 37 constituents, but for most localities 20 to 30 are sufficient for desired accuracy of prediction. Because the principles of tidal current prediction are the same as for tidal height prediction, the same machine was used for both purposes. Presently, harmonic analysis is carried out by computers of all sizes (from mainframes to microcomputers), and commonly they calculate more than 200 different constituent amplitudes and phases (Dennis and Long 1971; Schureman 1976; Foreman 1978). There now exist several methods for performing harmonic analysis, including least-squares tidal harmonic analysis, the admittance method of Munk and Cartwright (1966), and Fourier harmonics.

Regular measurements of tide time and height were beyond the capability of hydrographic engineers little more than a century ago, and even greater were the difficulties of calculating future tides from observed ones. They still are beyond the interests and abilities of bureaus in most coastal nations at present, as witnessed by the poor geographic distribution of tide-gauge stations. Of about 250 stations having acceptable records, 65 percent are along the coasts of United States, Japan, and Scandinavia. Few are available for South America, Africa, southeastern Asia, and communist nations, but the lack of data from the last appears to be caused by unwarranted desire for national security.

Although tide-gauge data are far less important to ship navigation now than in the time of sailing ships, the same information is being used for purposes not conceived only decades ago. Among these purposes are estimates of tidal currents—more important than water depth for supertankers, measures of activity of crustal plates of the Earth, geoid changes, climate, rate of post-glacial rise of sea level, engineering of harbors, groundwater losses, and municipal planning. These multiple uses probably far exceed the value of tide prediction for navigation in harbors.

2. Causes of Relative Sea-Level Change

Sea Levels Versus Land Levels

Relative sea levels (RSL) have changed constantly throughout geologic time as the volume of ocean waters has fluctuated, the shapes of the ocean basins have changed, and the land masses have been broken apart, been welded together, and have emerged or submerged. The challenge is to separate changes in level of the land from changes in level of the ocean surface in an unambiguous fashion, a task that only partly can be successful given the broad spectrum of processes affecting relative sea levels, especially those processes that contribute to the low-frequency portion of the spectrum. To separate land-level changes from sea-level changes effectively, a diverse mixture of geological and geophysical approaches must be integrated. Some contributors to relative sea-level change and their approximate magnitudes are summarized here in a general way, with documentation for some of the more important processes concentrated in later and more pertinent sections of the book. From this summary one can differentiate between processes presently affecting relative sea level and those that were important only during the past.

Plate Tectonics/Sea-Floor Spreading

General

Plate tectonics and sea-floor spreading exert significant control on relative sea-level changes, on both geological and historical time scales. The volume of ocean basins changes with time because the crust of the earth separates and collides when plates move with respect to one another. These plates can become larger through formation of new crust along ocean-spreading belts, or they can become smaller through ocean/continent collision (at trenches/volcanic island arcs) or through continent/continent collision (forming folded mountains). When newly formed, oceanic crust cools, thins, and subsides, lowering sea levels worldwide, although perhaps causing local rise in relative land level near the belt of subsidence. Changes in ocean volume can occur either from a change in average depth of the oceans or from a change in area of the oceans. Several subprocesses can be defined within this category.

Changes in Spreading Rates

At any given geological time there is a unique areal distribution of oceanic crust and continental crust. When the rate of spreading changes, or the rate of consumption of crust varies because of changes in relative plate motions, the relative proportion of young to old oceanic crust changes. As this proportion changes, so does the hypsometric curve of the oceans, thus altering the average depth of the oceans. When increases in spreading rates occur, the average height of the ocean floor increases and average water depth decreases, with the converse being true for decreases in spreading rates. When the area of the ocean floor remains constant and the average depth of ocean floor decreases, relative sea level increases; when the average depth increases, relative sea level falls. This mechanism requires a change in rate of spreading; it is not valid for the steady-state situation. Pitman (1978, 1979) developed several models that clarify the magnitude of this effect. He calculated that a maximum rate of sea-level change resulting from changes in geometry of the oceanic ridge system would be about 1 cm/1000 years, or 0.01 mm/y.

During part of the Cretaceous Period, sea level was approximately 350 m above present ocean levels, corresponding to a time of rapid sea-floor spreading (110 to 85 m.y. ago). This sea-level rise would be sufficient to cover 35 percent of the world's present subaerial continental surface area. The Cretaceous was perhaps the time of fastest sea-floor spreading; since then, the spreading rate has been decreasing. Relative sea level is inferred to have responded by falling rapidly during the Oligocene and falling more slowly during early Miocene time.

Changes in Areas of Ocean and Land

As the Earth evolves, the relative areas covered by oceanic and continental crusts vary; hence, the total area of ocean basins changes. One example is the episodic welding of microcontinents onto the major continents along ocean margins, creating the numerous suspect, or exotic, terranes along these continental margins. In addition, the landmass of Pangaea and probably those of earlier megacontinents were concentrated in a different part of the globe than occupied by our present more widely distributed continental fragments.

As the ratio of continental to oceanic crustal area changes, so does the level of relative sea level. By determining through time the product of the average depth of the oceans and the average area of the oceans, one could derive the geological relative sea-level fluctuations. Not all the necessary data exist, however, so quantification of this effect is difficult.

Changes in Direction of Plate Motions

Analogous to plate-tectonic factors, changes in direction of relative plate motions must influence relative sea level. Changes in direction of plate motions modify the average age of consumed crust and thus affect the hypsometric curve. Also modified are rates of convergence (subduction and mountain building) and of divergence (spreading and rifting) of the plates. All of these factors contribute to changes in volume of ocean basins, and hence to changes in relative sea levels. No quantification of this effect is available, in part because this process is difficult to separate from other sources of relative sea-level change.

Other Thermal Effects

Other thermal effects besides the cooling of new oceanic lithosphere can change the volume of ocean basins. For instance, volcanism especially in the Pacific Ocean has altered the average depth of this basin. Hot spots and aesthenospheric "bumps" may partly be responsible for these effects; certainly they have had a major role in the Pacific (Line Islands, Hawaiian Chain) as well as the Atlantic (Bermuda Swell, Canary Islands) oceans. Seamount frequency and volume calculations can be used to estimate the magnitude of this effect.

Deep-Ocean Sedimentation

Deep-ocean sedimentation impacts the height of relative sea level because sediments deposited on the ocean bottom raise sea level. Similarly, dissolution of carbonate or siliceous sediments can lower global relative sea level. Although sedimentation rate expressed in mm/year or mm/millennium is misleading because of variations in water content of sediments and depth of burial through time, it can be used as a first-order indication of impact on relative sea level. A generous estimate of average global ocean sedimentation rate is 1 cm/1000 y. Most rates of deposition are measured near the sediment surface where the water content averages about 50 percent by wet weight. At a grain density of 2.76 and a water density of 1.05, a 1-cm thickness of wet sediment contains 0.276 cm^3 of granular material (Emery and Uchupi, 1984, p. 682). The rate of deposition of 0.276 cm^3/1000 y. means that the deposited solid grains would displace an equal volume of water and cause a rise of sea level of 0.003 mm/1000 y. and even less if the weight of the deposit caused isostatic sinking of the ocean floor. This is only a small fraction of the relative sea level change at present.

Isostasy

General

Isostatic adjustment is the process by which the Earth attains gravitational equilibrium with respect to superimposed forces. The crust continuously seeks to remain in isostatic equilibrium, such that at some depth the total integrated vertical load at any point is the same as that at adjacent points. If a gravitational imbalance occurs, the crust rises or sinks to correct that imbalance. Details of the dynamics governing such adjustment are complex, and generally they are modeled either analytically or numerically. The latter calculations are required to take into account the full complexities of time history of loading, as well as the tangential and radial variability in this loading on the crust. The age/depth relationship for oceanic crust (Sclater et al. 1976) reflects isostasy because the density of crust varies with temperature.

Changes in crustal loading leading to isostatic adjustment have several different causes. Glacial cycles alternately impose and remove large masses of ice on continental areas. The latest Wisconsinan glaciation that created the former Fennoscandian and Laurentide ice sheets is an example of this process. Water loading on continental shelves is an additional factor. As sea level rises, the shelf becomes loaded with water, causing the shelf to become depressed toward the ocean. This response is the origin of the commonly cited "hinge zone," where the outer shelf is depressed more than the inner shelf. A related process is loading by sediments. As ocean levels rise, more sediment may be deposited on continental shelves, causing additional depression of the shelf surface.

Glacio-Isostasy

The largest source of geologically rapid isostatic adjustment is from waxing and waning of glaciers. These glacial cycles have occurred on time scales of 100,000 years with superimposed shorter period fluctuations, beginning during the Oligocene (about 40 m.y. B.P.). The latest major interglacial stage occurred about 120,000 years ago, followed by the Wisconsinan or Würm glaciation. Since then stadials (low stands of sea levels associated with minor advances of ice) and interstadials (high stands of sea levels associated with minor retreat of ice) have left their signature on the relative sea-level record. Interstadials occurred approximately 122,000, 103,000, and 82,000 y. B.P., with another interstadial possibly at 30,000 y. B.P. Evidence for these minor glacial cycles comes from a variety of sources, including raised or submerged terraces, deltas, coral reefs, and submarine canyons (Chappell, 1974).

As glaciers grow, they impose an overburden on the Earth's surface. This growth results in flexure of the lithosphere and flow of the aesthenosphere. The region below the central mass of the ice sheet warps downward because of glacial loading, while the surrounding area rises due to aesthenospheric flow and flexural warping. This latter process forms a peripheral bulge associated with glaciers. Around Scandinavia, this peripheral bulge reached France, Germany, and Poland, while in the United States it extended along much of the Atlantic coast, with a high centered near Cape Hatteras.

When a glacier melts, the depressed area under the dominant glacial mass rebounds and the peripheral bulge sinks. This growth and retreat cycle creates a complex history of relative sea levels, the details of which depend on the time history of loading and unloading of the ice sheet and on the distance from the ice sheet to the shoreline. Numerical modeling has provided interesting results of such processes (Clark et al. 1978; Chappell 1983; Peltier 1985), as has analytical modeling (Walcott 1972a, b). What is needed is a self-consistent model that takes into account the loading effects of water added to shelves and the direct gravitational attraction due to the presence of large ice sheets themselves. Although some isostatic models are self-consistent (Peltier 1980, 1987), they are limited in their use of radial symmetry (they cannot differentiate between oceanic and continental crust, a potentially serious limitation considering that most tide gauges are near the boundary between these two crusts), they have only coarse resolution, and they rely for calibration on tide-gauge and radiocarbon data that are impacted by other tectonism in addition to isostatic processes.

The magnitude of this glacio-isostasy can be illustrated by recent examples. In Fennoscandia, a rebound of 800 m may have occurred during the past 14,000 to 10,000 years (Mörner 1978), for an average rate of 50 to 70 mm/y. crustal rise. In the northern United States and Canada, similar rates may be present. Oldale and O'Hara (1980) estimated that rebound in Baffin Island reaches about 4 m/century, whereas larger areas reach 1 m/century. These rates are equivalent to 40 mm/y. and 10 mm/y., respectively — the same order of magnitude as in Fennoscandia.

Rebound after ice melt is comparable in rate with sea-level rise owing to input of meltwater from glaciers. During the past 17,000 to 10,000 years, ocean levels rose between 60 and 120 m. This rise corresponded with an average range in rate of relative sea level change during the past interglacial epoch of between about 3.5 and 12 mm/y. It is thought by most workers that the maximum deglaciation occurred during approximately 10,000 years, at an average rate of sea-level rise approximating 8 mm/y. (Kennett 1984). About 7000 years ago sea level was 7 to 10 m below its present position, yielding an average rate of sea-level rise since then of about 1.0 to 1.4 mm/y.

Hydro-Isostasy

Loading of continental shelves and the ocean crust by water is a source of relative sea-level change. As glaciers wane and meltwater flows to the oceans, the increasingly submerged crust has a greater load on it. Numerical models, although generally assuming a radially symmetric earth, take this water loading into account (Peltier 1984). What they are unable to represent well are the effects of lateral (tangential) variations in crustal properties (such as along the continental/oceanic transition).

Chappell (1976) addressed the issue of hydro-isostasy on the northeastern shelf of Australia and indicated low rates of relative sea-level change. Clark and Lingle (1979) also presented theory and examples of hydro-isostasy.

Sediment-Isostasy

When sea level rises, sediment deposition on the shelf increases, but when sea level falls, erosion occurs there. This cycle of deposition and erosion, which takes place on various time scales, impacts the shelf elevation by varying the loading on the margin of the continental crust. As the load varies, the crust rises and falls, resulting in additional changes in relative sea level. Vail et al. (1977a, b) illustrated these depositional/erosional cycles. The cyclicity of these processes is subject to some debate, but their occurrence is widely acknowledged.

The magnitude of sediment loading has not been well established. Walcott (1972a, b) performed some analytical tests of this loading and estimated the flexure to be as much as 4 km for his hypothetical but realistic examples. If the sediment that caused the flexure were deposited during an interval of one million years (for example), it would produce a fall in relative sea level of 4 mm/y., the same order as changes in relative ocean levels today. Thus, in restricted areas such as deltas or marginal seas where very rapid accumulation might occur (in contrast with most other shelves where the sedimentation rate is much less), subsidence caused by sediment loading may be an important factor in relative sea-level rise. Deltas such as those of the Mississippi, Nile, Huanghe, and the Ganges-Brahmaputra rivers are examples of such situations where delta subsidence caused by sediment loading is significant.

Glacial Surge/Melting

As water is exchanged between the ocean basins and the major ice sheets, the level of the ocean rises and falls. The present major glaciers and ice sheets account for approximately 77 m of water equivalent spread over the global oceans (Table 4). During the past, the volume in glaciers has been estimated to have been at least as much as 150 m of water equivalent, and most likely nearer 200 m of water equivalent (the ocean basin area is 361×10^3 km^2; all sea-level rise equivalents are based on this area, ignoring the

TABLE 4. Sea-level equivalents of ice sheets and glaciers.

	Area (10^6 km^2)	Ice volume (10^6 km^3)	Sea-level equivalent (m)	Mass turnover time (y.)	Response time (y.)	Loss mechanism
East Antarctica	9.86	25.92	64.8	15,000	10^4–10^5	Calving
West Antarctica	2.34	3.40	8.5	–	10^2–10^3	Ice-stream
Greenland	1.7	2.4–3.0	6.–7.6	5,000	10^3–10^4	Melting/calving
Small ice sheets and mountain glaciers	0.54	0.12	0.3	50–1000	10–100	Melting

400,000 km^3 of ice is equivalent to 1 m global sea-level rise.
From Holtzscherer and Bauer (1954), Flint (1971), Ambach (1979), Radok et al. (1982), Giovinetto and Bentley (1985), Barry (1985), Ohmura and Reeh (in press).

small changes in basin area as continents became flooded or dried).

Glacial surge and melting may account for the interstadials and stadials during the past 120,000 years. Relative sea-level changes of tens of meters can arise from these processes. It has been hypothesized that during the previous interglacial (120,000 y. B.P.) the West Antarctic Ice Sheet collapsed, leading to a sea level 6 m higher than at present. This possibility, although controversial, was a rapid event, in contrast with the usually much slower, more continuous waxing and waning of glaciers. Recent glacier surges and retreats have taken place. Glacier Bay, Alaska, recently underwent rapid retreat. In about 1930, the Nordaustlandet ice sheet of Spitsbergen (Svalbard Archipelago) surged approximately 10 km; subsequently, the ice front retreated to its former position. Although these local events do not produce a measurable change in absolute global sea level, they are indicative of the types of scales and events perhaps responsible for stadials and interstadials.

Ocean Surface Topography

It has been well-documented in recent years that the ocean is not completely level. The causes for this unevenness are many, but one aspect that has been addressed is dynamic topography, which includes the effects of ocean currents on sea-level height. Ocean surface topography has become increasingly studied during the past decade as our ability to measure it directly using satellite altimetry has improved. In the Introduction, a comparison of tide-gauge data with GEOSAT data for sea levels at Majuro was presented (see Fig. 5). This comparison shows great similarity in the monthly average sea levels from satellite and tide gauge, a measure of the strength of dynamic topography and its fluctuations, which may impact tide-gauge records.

The Pacific Ocean stands higher than the Atlantic Ocean (on average), a fact attributed to differences in water density of the two water bodies. However, other forces impact the height of the water at any point, including the presence of major ocean currents. The topography across the Kuroshio Current and the Gulf Stream, for instance, exceeds 1 m. Fluctuations in ocean currents can cause the dynamic topography to fluctuate as well, a factor that can influence some tide-gauge recordings, particularly for island tide gauges. White and Hasunuma (1980) and Blaha and Reed (1982), for instance, documented changes in major North Pacific circulation patterns, which Aubrey and Emery (1986a) successfully related to tide-gauge records of Japan. For instance, White and Hasunuma found dynamic height fluctuations on the order of 3 to 4 cm, and from 1954 to 1974 they found a steady decrease in dynamic height of 2 to 4 dynamic cm (for an average rate of 1–2 mm/y.). Fluctuations in Kuroshio strength may contribute to these fluctuations in dynamic height, which then are recorded by the adjacent tide-gauge records.

Geoid

The geoid is the level surface of equal gravitational potential that the sea surface would assume in the absence of external disturbing forces. Because of uneven distribution of mass in the Earth, however, the geoid has undulations of more than 100 m about a reference ellipsoid of smoother character. In a cross-section through the Earth, the geoid is seen to be about 10 m higher than the reference ellipsoid at the North Pole, and 30 m lower at the South Pole. The greatest geoidal depression of -105 m corresponds to mass deficiency south of India, and geoid elevations of 73 m north of Australia (Pugh 1987) arise from local mass excess within the mantle (Fig. 13). The sea surface approximates the geoid in most areas, but departures of a meter or more from this equipotential surface arise from differences in water density around the globe, ocean currents that deflect the sea surface from the gravitational equipotential, and atmospheric forcing. The geoidal shape may change through time owing to changes in mass distribution within the Earth or to changes in the rate of rotation of the Earth. These geoidal changes will be mirrored by changes in sea levels.

Because these geoidal departures are large, and since the sea surface attempts to follow the geoid absent of any other external forces, migration of the geoid can cause significant changes in rate of rise of the ocean. The geoid may change because of isostatic loading and unloading, as when glaciers wax and wane and contribute to mass redistribution within the aesthenosphere. Mörner (1978) postulated geoidal changes on time scales of 100 to 1000 y. in the hydrosphere, aesthenosphere, core–mantle interface, and outer core. In

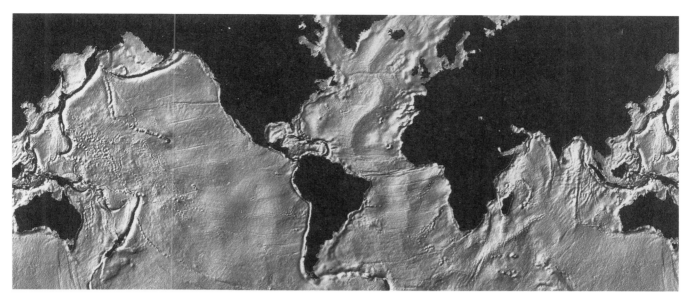

FIGURE 13. Surface topography of the world ocean derived from GEOS-3 and SEASAT satellite altimetry, mainly a response to the shape of the geoid and ocean dynamics. Provided by C. Koblinsky, NASA Goddard Space Flight Center.

particular, Mörner emphasized the role of rate of rotation, tilt of the spin axis, precession of the equinox, external gravity, mass redistribution above the core–mantle boundary, convection in the outer core, centration of the inner core, and core–mantle topography changes, all of which can contribute to coupling of the core and mantle. As the geoid correspondingly migrates, the level of the sea level must migrate similarly. The potential for geoidal changes to generate significant rates of sea-level change and causation by forces other than glaciation/deglaciation have yet to be proven in the refereed literature. However, it is clear that even slow migrations of the geoid's undulations of 100 m and more could contribute to local sea-level rise. For instance, if a geoid high of 100 m distributed across 1000 km migrated at a rate of $0.01°$ arc/century, the equivalent rate of relative sea-level rise would be on the order of 1 mm/y. What is needed is a proven mechanism to demonstrate why such features should migrate rapidly enough to contribute to local relative sea levels.

Geological Faulting

When adjacent portions of the Earth move relative to each other, rupture commonly occurs—a fault. Faults can be spectacular along plate boundaries or major lithologic breaks within the stratigraphic record, or less spectacular within relatively uniform strata. They occur where stresses build within an area, causing either slow relief of stress via local strain, or more sudden release by rupturing. Contrary to expectation by many scientists, faults are extremely common, especially in mountainous and coastal regions that are near the contact between dense oceanic crust and less dense continental crust. The latter regions contain many tide gauges whose records respond to the vertical components of the crustal movements.

It is useful to examine the frequency of faults on a continental scale. One way to do so is to examine the state of lithospheric stress in a large area. Maps of stress distribution can be derived from geologic data, focal mechanisms for earthquakes, and *in situ* stress measurements. Zoback and Zoback (1980) prepared such a map for the United States, showing stress provinces that have linear dimensions ranging from 100 to 2000 km. Some of the provinces have high levels of tectonism, earthquake activity, and heatflow; others are tectonically relatively quiescent. For instance, eastern and central United States are relatively quiescent, yet there is a major variation in principal stress orientation between the Atlantic coast and midcontinent. Coastal areas are not immune from high stress, and the southwestern United States is a good example of high tectonic activity with significant earthquake occurrence because of translation and convergent plate movements.

A global analysis of tectonic stress was completed recently by Zoback et al. (1989). They identified major patterns of stress concentrated along major plate boundaries with compressional stress regimes in interplate regions, and extension limited almost entirely to thermally uplifted areas (Fig. 14). High stress areas are concentrated along certain coasts: eastern North America extending to the Aleutian Islands, southeastern United States, western Europe, southern Asia, eastern Mediterranean, Australia, and the western Pacific margin. These regions also correspond with areas of densest tide-gauge coverage, so one would expect an impact of tectonic stress on tide-gauge records.

28   2. Causes of Relative Sea-Level Change

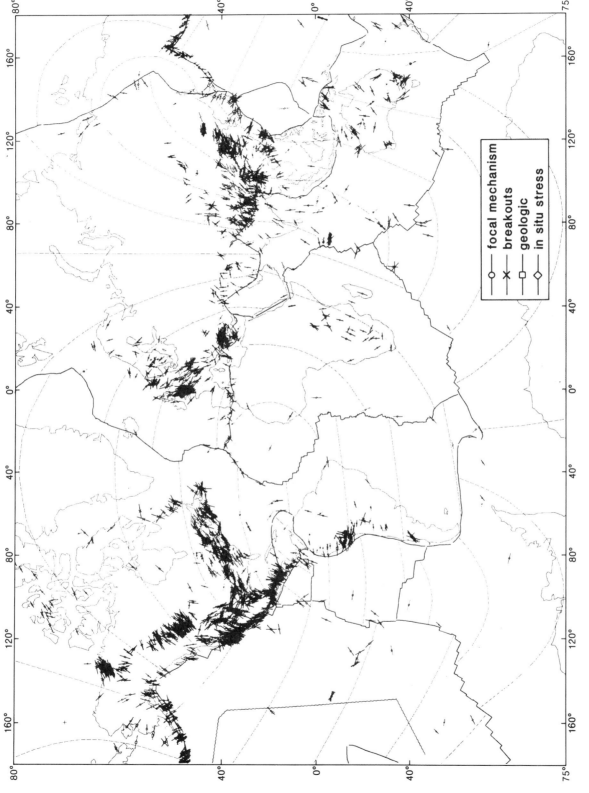

FIGURE 14. World crustal stress map. From Zoback et al. (1989). Reprinted by permission from Nature, v. 341, p. 291–298, copyright 1989 Macmillan Magazines Ltd.

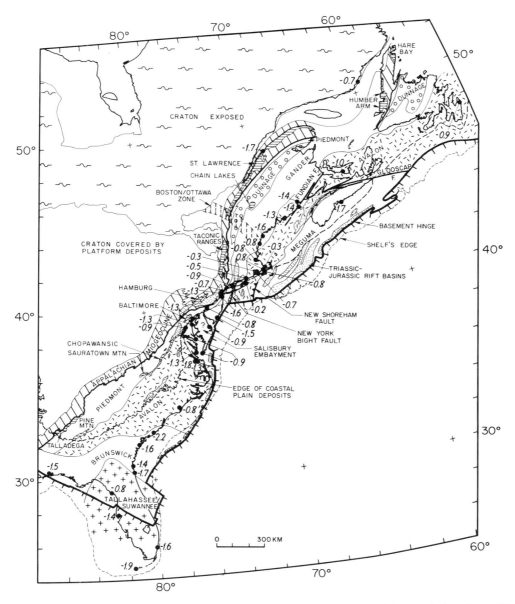

FIGURE 15. Tectonic map of the eastern North American margin, showing rates of relative sea-level change in mm/y. Negative numbers indicate subsiding land (or rising sea level). From Uchupi and Aubrey (1988), with permission from The University of Chicago Press.

Most continental margins consist of suspect, or exotic, terranes. The margins are melanges of bits and pieces of various continental fragments added and annealed throughout geological time. For instance, the margins along the North American Atlantic coast (Fig. 15) consist of complex melanges of primarily Paleozoic segments and those along the North American, Central American, and South American Pacific coasts contain Mesozoic and Cenozoic fragments (Fig. 16). Differential vertical land motion along these opposite coasts may result from non-uniform stresses caused by relative plate movements as well as unequal response by dissimilar and partly decoupled geologic terranes that form the margins. Different responses of unlike adjacent terranes to glacial loading/unloading undoubtedly contribute to the variation in signals observed in tide-gauge records.

Published examples of effects of earthquakes on tide-gauge records are somewhat rare. A Japanese example is perhaps best known, because Japan has a dense concentration of earthquakes because of its tectonic setting. Examination of Japanese tide-gauge records shows several episodes that might indicate vertical motion, but not unambiguously because of other tectonic and oceanographic influences on annual averaged data. Hourly or daily data would show much more clearly these shifts caused by earthquakes. The data should be examined preferably before corrections have been made to the local benchmark network; otherwise, shifts of the benchmarks may

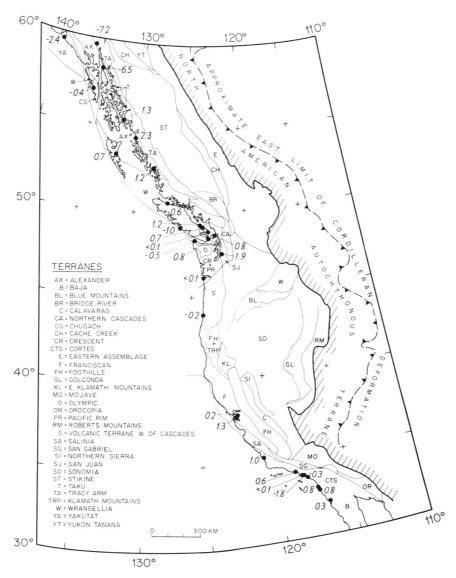

FIGURE 16. Tectonic map of the western North American margin, showing rates of relative sea-level rise in mm/y. Negative numbers indicate subsiding land (or rising sea level). From Uchupi and Aubrey (1988), with permission from The University of Chicago Press.

eliminate the records of relative sea-level changes on tide gauges. Perhaps the best example of earthquake motion on tide gauges is at Nezugaseki (west coast of Honshu Island, Fig. 17), showing considerable land submergence along a coast generally dominated by emergence. This area was impacted by the Niigata earthquake of 16 June 1964, whose epicenter was 40 km distant. Prior to the earthquake, mean sea level at Nezugaseki was 113.3 cm, whereas immediately afterward, mean sea level there was 13.4 cm higher (Yamaguti 1965). This difference reflects submergence of the sea bottom where the tide gauge was located. Measurements during the following weeks suggest that the impact of the Niigata earthquake on vertical motion at Nezugaseki was as much as 15 cm. Yamaguti also noted similar motions for the Kwanto and the Nankaido earthquakes on the tide-gauge record of yearly means at Aburatsubo. Satake and Shimazaki (1988) documented the detection by tide gauges of very slow earthquakes in Japan. Also, using the Niigata earthquake of 1964 as their example, they recorded the 7-magnitude earthquake from seismic oscillations generated within the Japan Sea, as well as from changes of relative mean sea level.

Western North America is another tectonically active area where local faulting can be expected to impact tide-gauge

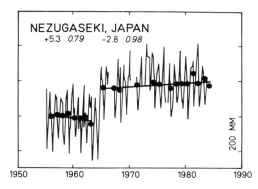

FIGURE 17. Tide-gauge record from Nezugaseki, Japan (Lat. 38°34′N; Long. 139°33′E), showing vertical motions indicative of faulting/earthquake influence.

recordings. For example, the 14-m uplift along a fault at Yakutat, Alaska, in 1899 (Tarr and Martin 1912) would have impacted tide-gauge recordings; however, the motion was not recorded there because the tide gauge was installed only after the earthquake. The 8.5-magnitude Good Friday earthquake of 1964 at Anchorage, Alaska, similarly would have impacted local tide gauges; the closest permanent one appears to have been in Cordova, 65 km southeast of Valdez in Prince William Sound and peripheral to the area of vertical land motion (Hicks 1972). During the earthquake of 28 December 1908 in Messina, Sicily, the tide gauge there showed a rise of 57 cm in relative sea level; the level continued to change in the same direction during ensuing years, although at a lower rate (Lisitzin 1974).

Many earthquakes or faulting motions cannot be recorded by tide gauges because of the methods used for linking tide gauges to local geodetic datums. If the sense and magnitude of vertical motion at a tide gauge are the same as for the nearby leveling net, the faulting may not be recognized or reported by the overseeing agency. Prior to reporting tide-gauge data to the Permanent Service for Mean Sea Level (PSMSL), the data are "corrected" to the local datum. If the local datum has shifted, the tide-gauge record is modified as well. Faulting may be reported only where there is strong differential movement between tide gauge and leveling net. At such a site, the leveling net indicates a different vertical displacement than that recorded by the tide gauge.

Sediment Compaction and Subsidence

Processes

Sediment compaction and land subsidence contribute to relative sea-level rise at many scattered locations around the globe and can dominate local tide-gauge recordings. Just as for tectonism, however, local compaction often is poorly understood and even more poorly mapped. When tide-gauge records are interpreted, the potential contribution from compaction should be acknowledged, even though measurements may be lacking. Subsidence should be suspected near large cities on river plains or deltas and at sites where fluid withdrawal for domestic and agricultural use or for hydrocarbon extraction is significant.

Sediment compaction may result from vastly different causes. In general, compaction can be thought of as a reduction in volume of relatively poorly packed sediments into a denser matrix. Compaction may be caused by vertical loading by other sediments, by draining of fluid from interstices (groundwater or hydrocarbon mining), by desiccation, by vibration, and even by adding water to certain types of soils (hydrocompaction). Here the term compaction is used in its general geological sense, and it is not distinguished from the individual processes of soil consolidation or compaction (in the civil engineering terminology). The specific processes associated with compaction are described in some detail in many publications, including readable summaries by Allen (1984) and Poland (1984). Following is a brief introduction to some of these processes.

Deep-rooted compaction caused by loading through accumulation of great thicknesses of fine-grained sediments in a short period of time can produce significant subsidence. For instance, between 300 and 500 million tons of sediment are deposited each year in the Mississippi River delta, leading to subsidence of delta deposits by as much as 6 to 8.5 m (Fisk et al. 1954). Compaction also can be caused by artificial filling. Artificial loading of New York's LaGuardia airport increased compaction of an 18-m thick layer of organic silt and clay. After 25 years of operation, the airport had subsided about 2.5 m. This subsidence created a need for expensive diking around the perimeter of the airport.

Draining of low-lying areas can accelerate compaction and enhance subsidence. An example of such compaction is in the polders of The Netherlands, where vast areas of delta have been reclaimed by water pumping and diking. Bennema et al. (1954) found that the clay deposits in The Netherlands compressed to about half their original thickness during a 100-year interval following land reclamation. Peat beds similarly are susceptible to subsidence caused by drainage, because peat has a water-retention capacity of between 300 and 3000 percent by dry weight. Besides physical compaction, peat undergoes irreversible biochemical changes on drying that reduce its volume. The Florida Everglades contains a large body of peat that has been subsiding following reclamation for agricultural purposes.

Sediments may compact under vibration, such as occurs during an earthquake. On a smaller scale, construction practices can cause compaction due to vibration. In Tianjin, China, a large earthquake in 1976 was implicated in accelerated local subsidence.

In contrast with subsidence caused by loss of interstitial fluids, some arid areas having sediments of extremely low density have subsided markedly when these sediments became wetted. This hydrocompaction produces rapid and irregular subsidence of the ground, locally ranging up to 5 m.

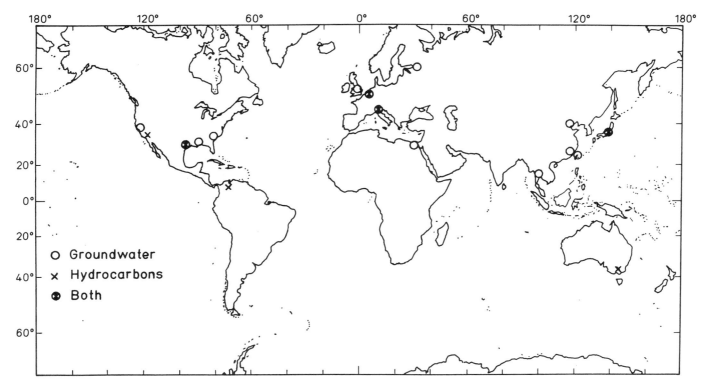

FIGURE 18. World map depicting sites of land subsidence due to sediment compaction resulting from pumping of water and/or hydrocarbons. Sites are differentiated by primary cause of subsidence.

Such hydrocompaction may be important in arid areas newly exposed to irrigation, for instance. One theory of hydrocompaction states that in dry soils, clay particles are loosely packed because of interparticle attractions; wetting of these particles changes the attraction and causes collapse of the loose packing. Lofgren (1969) reviewed the processes and cited examples of subsidence in the United States, Europe, and Asia. All areas where hydrocompaction is known are in continental interiors; thus, tide gauges are not impacted by this type of subsidence.

A primary cause of sediment compaction on a regional scale is withdrawal of groundwater for agriculture, industry, and households or for hydrocarbon exploitation. These occurrences are widespread and can cause subsidence in excess of 8 m, as at Long Beach, California. Generally, compaction is greatest in alluvial soils that have thick lenses of clay or peat (aquacludes). As the fluid pressure decreases, the effective stress on soil particles increases, causing the material to consolidate (compact). For granular soils, the consolidation is small, although some recent studies have shown that for great overburden pressures, an increase in effective stress can cause even quartz grains to break, thus enhancing compaction at depths of 1000 m and more (Roberts 1969). Allen and Mayuga (1969) indicated that in the Wilmington Oil Field at Long Beach, California, the compaction took place primarily in the sands, and that the sands are as compactable or more so than the shales. Generally, increases in effective stress have larger impacts on fine material, although in fine-grained material having low permeability some time is required for fluids to drain from sediment. Until the sediment drains and the fluid pressure comes to equilibrium, the sediment may continue to consolidate at a low rate, long after the original fluid withdrawal that caused the subsidence has ceased. This phenomenon produces a time lag between withdrawal of fluid and subsidence of the land, locally resulting in continued subsidence for years after the original land use.

Sinking of land because of withdrawal of pore fluids is widespread, and very likely many locations are undergoing subsidence without full recognition of its extent. Even the distribution of those areas where subsidence is well documented is dense (Fig. 18). The subsidence confuses the interpretation of tide-gauge records in several ways, as does tectonism. Tide-gauge installations are referenced to local geodetic benchmarks, and commonly they are re-surveyed once or twice each year. If the land under both tide gauge and benchmark is moving vertically at equal rates because of tectonism or land subsidence, the tide gauge records the subsidence as a rise in sea level. However, if the tide gauge and benchmark are moving differentially, as may occur in regions of tectonic subsidence or fluid withdrawal, the benchmark correction may increase or decrease the measured rate of relative sea-level rise at the tide gauge, depending on the direction of the gradient in rate of ground movement. Since the spatial scales of subsidence owing to groundwater or hydrocarbon withdrawal are small, the likelihood of having large spatial gradients increases compared to other tectonic processes. Thus, the interpretation of relative sea-level rise in the vicinity of large hydraulic subsidence becomes more difficult and requires close analysis. Since different countries relate tide gauges to different arrays of geodetic control, some tide gauges may be more affected than others by this uncertainty.

Examples of Subsidence

Since this book concentrates on coastal records of relative sea-level rise (such as tide gauges), a few examples of subsidence are presented, indicating rates of subsidence and where available comparison with nearby tide-gauge records that span the same time interval. A summary of known subsidence rates is presented in Table 5.

The Netherlands

The Netherlands, situated almost entirely on deltaic sediments of the Rhine River, has been subjected to continued subsidence, accelerated during the past half century by extensive man-made water works. Nearly 20 percent of the country lies below mean sea level, and 38 percent is below mean high water. To study subsidence in The Netherlands, a network of nearly 45,000 benchmarks has been established. Because there is no stable rock in the Netherlands, the benchmarks are subject to local variations in land level; underground benchmarks recently were anchored in Pleistocene deposits to minimize this local variation.

First-order levelings of the benchmark networks have taken place during the periods 1875 to 1885, 1926 to 1939, 1950 to 1959, and 1965 to 1980 (Rietveld 1984). In a study of the earliest leveling data, Edelman (1954) found irregular subsidence of as much as 100 mm in the northern part of the country, and emergence of about 40 mm in the south, indicating a tilting of The Netherlands along an axis through Amsterdam in a WSW–ENE orientation. Uncertainties in the quality of the benchmarks has caused some doubt about Edelman's interpretation. Later levelings show consistent subsidence, averaging roughly 1 to 2 mm/y., although some small scattered areas exhibit localized emergence. On a very local scale, in recently reclaimed areas, a net subsidence of 1.5 m can be expected due to dewatering and shrinkage of peat. The Netherlands also contains areas where coal, gas, salt, oil, or water are extracted, leading to increased subsidence. Examples include the large Groningen gas field, where subsidence of 250 mm is expected (Rietveld 1984).

Observations of water levels in The Netherlands have a long history. In 1682, observations either once or twice per hour were begun in Amsterdam, continuing uninterrupted until 1930, when the Zuider Zee was made into a lake insensitive to tidal motion. From 1870 to present, a network of nine tide gauges has existed in The Netherlands. However, because of the datum prescribed, the tide-gauge data yield only the subsidence of the primary Netherlands datum in Amsterdam relative to sea level (Rietveld 1984). The data are not processed to record relative sea level (tectonic shifts,

TABLE 5. Partial listing of regions of coastal subsidence due to human intervention, subsidence is caused by extraction of hydrocarbons (H) or withdrawal of water (W).

Location	Dates	Total Subsidence (m)	(mm/y.)	Cause
The Netherlands	–	–	2–10	W,H
London, England	1865–1931	0.18	2.7	W
Ravenna, Italy	1949–1977	1.2	41	W
Porto Corsini	1937–1972		8.2	W,H
Po Delta, Italy	1951–1956	–	300	H
Venice, Italy	1952–1969	0.14	8–17	W
Nile River delta	1923–1946	–	5	W
Bangkok, Thailand	1933–1981	0.6	10	W
Shanghai, China	1921–1965	2.3	98	W
Tianjin, China	1959–1982	2.15	261	W
Taipei Basin[a]	1895–1969	1.3	25	W
Tokyo, Japan	1900–1976	4.6	27	W
Nagoya	1961–1975	1.47	98	W
Osaka, Japan	1934–1968	2.8	82	W
Niigata Plain, Japan	1959–1974	2.0	133	H
Latrobe Valley	1920–1977	1.7	34	H,W
Santa Clara Valley	1920–1967	4	85	W
Wilmington, CA	1937–1968	8.8	285	H
Lake Maracaibo	1926–1954	3.4	117	H
Houston–Galveston	1943–1973	2.3	74	W,H
Goose Creek, TX	1917–1925	1	125	H
Baton Rouge, LA	1934–1965	0.3	10	W
New Orleans, LA	1924–1978	2.03	37	W,H
Savannah, GA	1933–1955	0.20	9	W

[a]Hwang and Wu (1969).

subsidence). Thus, The Netherlands tide-gauge data do not help interpret water-level history.

London, England

Subsidence in London has been recognized for more than 100 years (Poland and Davis 1969) to be superimposed on a regional tilting to the south. Wilson and Grace (1942) presented estimates of subsidence based on measurements and on soil-mechanics theory. They estimated subsidence of more than 180 mm during the period from 1865 to 1931. Extrapolating back to the date of initiation of groundwater pumping, a total subsidence caused by groundwater withdrawal was estimated to be 210 mm. This rate of subsidence is approximately equal to what is thought to be the global eustatic rate of rise during the past century, so in London the impact of rise in relative sea level may be about twice that of eustatic rise. Southend and Sheerness, the two tide gauges nearest London, are at the mouth of the Thames River and too far away to be affected by the local subsidence at London, and both have relative rates of sea-level rise of 1.5 mm/y.

Italy

Subsidence in Italy is attributed both to groundwater mining and to hydrocarbon extraction. The subsidence appears to be confined to river deltas, primarily in the greater Po River delta in the northwestern Adriatic Sea (Gulf of Venice).

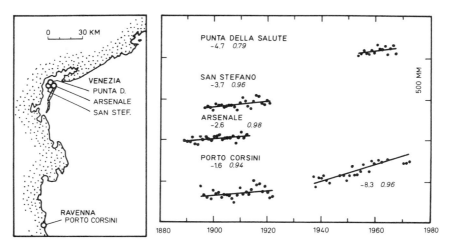

FIGURE 19. Tide gauge stations at Venice and Ravenna, Italy, showing land subsidence owing to natural processes locally accelerated by groundwater and methane withdrawal.

Po Delta. The Po River delta (Fig. 19) has experienced considerable subsidence during the 20th century as a result of extraction of water and methane gas from the deltaic sediments. Centered in part around the industrial area in Ravenna, the subsidence has influenced an area of 700 km^2. The natural rate of subsidence of the delta quoted by Carbognin et al. (1984a) is 5.1 mm/y. (a high rate, but not excessive for a delta). Recent rates of subsidence locally exceed 50 mm/y., more than 10 times that of natural subsidence, and they continue without abatement.

The past subsidence at Ravenna generally has been attributed to extraction of methane gas. However, Carbognin et al. (1984a) argued that because the gas fields are nearly depleted, the effects of extraction are limited now and probably were also in the past. Also discussed is the likelihood of subsidence owing to reclamation of marshlands, a contributing factor. The major subsidence currently appears to be related to withdrawal of groundwater for industrial and agricultural use. The local subsidence is of particular concern since 90 percent of Ravenna is at one meter or less above sea level, with 20 percent of this area being below sea level. Accompanying this subsidence has been shoreline retreat (at rates exceeding 5 m/y.; Carbognin et al. 1986) and saltwater intrusion into the aquifers. These problems are accelerating as the causes continue unchecked and as development extends even farther outward from Ravenna.

The nearest tide gauge to Ravenna is at Porto Corsini, also on the Po River delta (Emery et al. 1988a). Although the record quality is not very good, it is included for descriptive purposes. At Porto Corsini, subsidence was at a rate of 1.6 mm/y. from 1896 to 1922, accelerating to 8.2 mm/y. from 1937 to 1972. Although difficult to interpret given the short record lengths, the two records suggest substantially increased subsidence rates during the period when groundwater mining became more extensive.

Poland and Davis (1969) cited subsidence rates on the Po Delta north of Ravenna as high as 300 mm/y. from 1950 to 1956 compared with the natural rate of subsidence here of 1 to 3 mm/y. Possible causes for this subsidence have included dewatering of peat, rapid sediment loading of the area, land reclamation, and extraction of methane gas. Following a temporary shutdown of the gas industry in one area, subsidence ceased, suggesting that the pumping of methane was the primary cause of the subsidence. The region has considered stopping all pumping of methane to combat this expensive subsidence (Gortani, 1961).

Venice. Subsidence at Venice is well known and well publicized because of the threat to historical property and the arts. Subsidence here has been attributed to groundwater extraction, and successful measures have been taken to reverse this subsidence. Prior to 1952, groundwater pumping was inconsequential, and subsidence was at the natural rate of 1 mm/y. according to Carbognin et al. (1984b), citing Leonardi (1960) and Fontes and Bortolami (1972). However, existing tide-gauge records from 1886 to the early 1900s (Fig. 19) show that subsidence in Venice ranged from 2.6 to 3.7 mm/y. near the turn of the century. The historical rate of relative sea-level rise appears somewhat uncertain. After 1930, groundwater pumping became more extensive following construction of factories in Marghera (Gatto and Carbognin 1981; Carbognin et al. 1984b). From 1952 to 1969, subsidence became much more rapid, reaching rates of 14 to 17 mm/y. in Venice and Marghera, respectively. Between 1952 and 1969 average subsidence was 110 mm in the industrial zone and 90 mm in the city.

After 1969, efforts began to combat the subsidence. With use of artesian water declining because of use of water mains and other surficial sources, the land surface has begun to rebound. In Venice, the rebound may be as high as 20 mm since 1969, which demonstrates that land subsidence has been arrested. With the recovery of the piezometric head in the aquifers, the rebound should continue in the future. Total rebound, however, is likely to be only about 20 percent of the

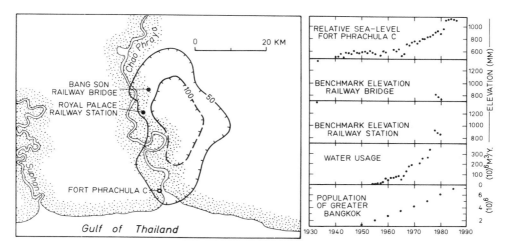

FIGURE 20. Map of Bangkok, Thailand, shows areas that have rapid land subsidence caused by withdrawal of groundwater. Contours indicate rate of subsidence in mm/y. Graphs at right indicate trends between 1930 and 1990 of rising relative sea level and sinking of land compared with the volume of groundwater withdrawal and increase of population. Data are from the Asian Institute of Technology (1981, 1982a, 1982b).

original subsidence, corresponding to about 30 mm. The latest tide-gauge recordings in Venice (Fig. 19) do not show the rebound unambiguously. Although the data are sparse (only 15 years of record), the record indicates continued subsidence from 1953 to 1967 of 4.7 mm/y. (slightly elevated compared to the records from the turn of the century); however, no data are available after 1967 in the PSMSL data file. The best record of rebound appears to be repeated geodetic leveling, which has occurred on nearly an annual basis since 1970.

Nile Delta, Egypt

Construction of the Aswan Low Dam in 1904 and the Aswan High Dam in 1964 have led to markedly decreased sediment loads to the Nile River delta. As the delta continues to subside because of consolidation of deeper strata, no longer will this subsidence be balanced by new sedimentation. A balance of compaction and sedimentation maintains the health of any equilibrium delta. As a consequence of the decreased sediment load, erosion of the coast has increased markedly (Inman et al. 1976; Milliman et al. 1989). As the sediment erodes, the delta continues to subside. Subsidence at present is documented by tide gauges at Port Said (Emery et al. 1988a) and by geological evidence from drill holes (Stanley 1988a). The evidence indicates that the Nile Delta in recent times has been subsiding at a rate of 5 mm/y., a rate that has been attained periodically during the past 7500 years when local sedimentation was not occurring to combat natural subsidence.

Bangkok, Thailand

Severe subsidence in Bangkok has been caused by factors common throughout the world: dramatically increasing population extracting increasing amounts of water from local aquifers. Subsidence in Bangkok first was observed in 1968, but little was done to quantify it until 1978. Subsequently, there has been a continuous series of studies addressing subsidence problems, providing not only observations but also specifying remedial actions. Some of the major studies include those by the Asian Institute of Technology (1981, 1982a,b). The area impacted by the subsidence (Fig. 20) encompasses most of downtown Bangkok and much of the surrounding area. More than 300 km² were subject to subsidence rates of 100 mm/y. or more, and 700 km² were impacted by subsidence rates of 50 to 100 mm/y. Once the subsidence was quantified and its alarming rate determined, remedial measures were initiated to reduce the rate of subsidence. Since much of Bangkok is only 1 m above sea level, the observed subsidence presented major problems, including:

Flooding, owing to overflowing water from the Chao Phraya River, heavy rain, and high tides
Damage to drainage systems, including backing up of water in sewers during high tides
Damage to structures because of uneven subsidence
Protrusion of wells from the ground and through houses
Difficulty of establishing reference elevations as benchmarks sank
Deteriorating groundwater quality because of intrusion of salt water
Effects on urban activities because of direct flooding impacts, disruption of the transportation system, and degradation of groundwater quality as septage and other runoff entered the groundwater aquifers.

Subsidence within Bangkok has been monitored with respect to benchmarks set in the 1930s. The first re-leveling useful for estimating subsidence came in 1978; since then, re-leveling occurred twice in 1978, twice in 1979, and once in 1981 (according to data available to us). Some leveling during 1954 provides useful additional information for local areas.

Maximum subsidence since the 1930s measured about 0.6 m, with most occurring during the past two decades. In the late 1970s, local subsidence reached rates of 100 mm/y. (Fig. 20). For instance, at the Royal Palace Railway Station and the Bang Son Railway Bridge, subsidence between 1933 and 1981 reached 0.6 m. However, measurements made between 1978 and 1980 show high subsidence rates at benchmark stations 2, 4, 10, and 18 – all exceeding 100 mm/y. Figure 20 also shows the time history of subsidence, sea-level rise, and groundwater pumpage for Bangkok, illustrating rapid acceleration during the post-1970 era. Bangkok currently is attempting to recover from this serious subsidence by drawing water from a broader area instead of concentrating waterworks within the city. Since recovery from subsidence is always incomplete, however, Bangkok had to institute many civil works to combat the high water levels and flooding.

Thailand has four tide gauges reported in the PSMSL data base. Only one of these, at Fort Phrachula C (see Fig. 15), is within the zone of influence of the subsidence, but it is near the margin. That tide gauge reveals a mean subsidence of 11.5 mm/y. during the period 1940 to 1984; however, a period of rapid subsidence began in 1965 and continues to the end of the record. This subsidence is at a rate of nearly 25 mm/y., double the previous rate.

China

Several areas of subsidence have occurred within China. Little documentation has been presented in the western literature, so available data are sparse. However, subsidence is a problem for cities built on alluvial material; because China has such a large river delivery system with abundant sediment input, there is abundant alluvium where cities have been built (Emery and Aubrey 1986a).

Shanghai. Pumping of groundwater from Shanghai has produced local subsidence. By 1965, an area of 121 km² had subsided more than 500 mm. Maximum subsidence in the area was 2.63 m during a period of about 44 years (Shi and Bao 1984; Su 1986; Chen in press). From 1956 to 1959, some areas were subsiding at a rate of 98 mm/y. By 1965, following implementation of measures to slow subsidence, the maximum rate declined to 23 mm/y. Much of Shanghai lies only 3 to 4 m above sea level, so even such a reduced subsidence can have dramatic impacts on frequency and magnitude of flooding. Data from the Shanghai tide gauge are available, although not included in the PSMSL data holdings (Emery and Aubrey 1986a). These data (spanning the interval 1954–1976) reveal local subsidence rates of 2.6 mm/y., an order of magnitude less than the rates cited by Shi and Bao (1984). The tide gauge may be distant from the center of subsidence (a circle roughly 6 km in radius), so that the tide-gauge rate may be affected only slightly if at all by the subsidence. Chen (in press) reported on two tide-gauge records in China from 1920 to present and documented the subsidence at Wushong and Huangpu Park in Shanghai.

Tianjin. Tianjin is one of the most important industrial cities in China, situated on the Bohai Gulf in northern China, about 120 km southeast of Beijing. In 1959, subsidence was first noticed in the area. Since 1959, it reached a maximum of 2.15 m by 1982, with a maximum rate of 261 mm/y., and expanded to an area of 7300 km² (Zhang and Niu 1986). Since Tianjin lies only a few meters (3–5) above sea level, such subsidence rates are of serious economic concern.

Groundwater mining began in the early 19th century in shallow aquifers and by 1923 expanded to include the deeper aquifers. Before 1949, there were only 51 deep wells, yielding only 40,000 m³ per day. Following 1958, extraction of groundwater accelerated dramatically, leading to increased rates of subsidence. Zhang and Niu (1986) also attributed the subsidence to the strong (magnitude 7.8 on the Richter Scale) earthquake of 1976, located 100 km away in Tangshan, Hebei Province. Following this earthquake, rates of subsidence appeared to accelerate.

The nearest tide gauge to Tianjin is on the coast, at Tanggu. For the period of record (1951–1978), the tide gauge indicated a rise in relative sea level of 1.5 mm/y. Since the Tanggu station is tens of kilometers from Tianjin, one would not expect the subsidence owing to water extraction at Tianjin to extend this far.

Japan

Japan has many documented instances of land subsidence caused by groundwater withdrawal (Fig. 21). Some of the major areas of concern are addressed separately here. Yamamoto and Kobayashi (1986) provided a map from the Environmental Agency that depicts the distribution of alluvial and diluvial materials in Japan, and those areas where subsidence has occurred. The latter map is similar to that of Figure 21.

Tokyo. Tokyo is located partly on the fans and deltas of small rivers that enter Tokyo Bay. It has experienced high groundwater use for most of this century, until 1975 when restrictions were put into place to control land subsidence. Because of these restrictions, the groundwater usage dropped by a factor of 10 between 1964 and 1975 (Yamamoto 1984a). Land subsidence was discovered after the earthquake of 1923, which struck the eastern portion of Tokyo. Geodetic leveling to examine the crustal disturbance was begun. The yearly leveling that followed identified general land subsidence, with an early finding that about 100 km² of land were impacted between the Sumida and Arakawa rivers. At present, maximum total subsidence in Tokyo is about 4.6 m, and the maximum rate is 27 mm/y. The total area of subsidence in the broader Kwanto region is now 2420 km² and the area where subsidence amounts to more than 10 mm/y. is about

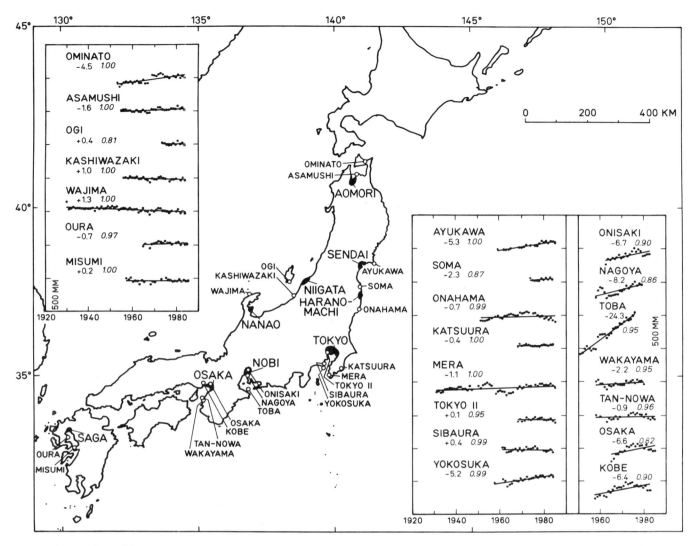

FIGURE 21. Regions of land subsidence in Japan owing to fluid withdrawal (water or hydrocarbons), and the land subsidence indicated by tide gauges nearest the chief areas of pumping.

100 km². Inaba et al. (1969) presented a useful summary chart of the history of subsidence-related observations and research.

Tide-gauge records for the Tokyo area (Fig. 21) do not document the local subsidence very well. East of Tokyo Bay, relative sea-level rise is slow, about 1 mm/y. The Tokyo tide gauge has only a short record, and it shows land rising slowly relative to sea level. Similarly, the Sibaura tide gauge shows land emergence. At Yokosuka, south of Tokyo Bay, the relative sea level is rising at a rate exceeding 5 mm/y. Clearly, local subsidence caused by groundwater withdrawal is not the dominant control on relative sea levels here; the effects of that withdrawal cannot be discerned in all tide-gauge records near Tokyo.

Nagoya. Nagoya is situated on the Nobi Plain in central Japan. Subsidence here was described by Yamamoto (1984c).

It was caused primarily by withdrawal of groundwater, which increased dramatically from 1925 to 1973; industry uses more than 60 percent of this groundwater. Early subsidence was measurable, but later subsidence swamped the earlier values. From 1961 to 1975, an area facing Ise Bay subsided 1.47 m. The total area subsiding exceeded 1140 km². About 363 km² had become lower than mean high sea level (about 1.1 m higher than mean sea level), 248 km² were lower than mean sea level, and 37 km² were lower than mean low sea level (1.4 m lower than mean sea level). This subsidence is being combatted by a variety of land-use restrictions, including restricted use of subsurface waters (more reliance on surface waters). These restrictions have led to general recovery of the piezometric head in the shallower aquifers since 1976.

Several tide gauges are near Nagoya (Fig. 21). The Onisaki and Nagoya gauges are within the area affected by groundwater pumping. Both show high rates of land subsidence—6.7

to 8.2 mm/y. There appears to be no slowing of this rate in later years of record. The tide gauges are outside the maximum zones of subsidence, so groundwater-caused subsidence may be compounded by other factors, such as tectonism. Subsidence at nearby Yokkaichi, 30 km southwest of Nagoya, is described by Komaki (1969).

Osaka. Industrial use of groundwater at Osaka beginning in the 1930s caused the water-table level to drop significantly, leading to local subsidence. Yamamoto (1984b) and Murayama (1969) stated that prior to 1928, local subsidence was relatively slight (6–13 mm/y., still high on a global basis!), a rate attributed to tectonic processes. According to Poland and Davis (1969), subsidence was noted first in 1885. Since that time, the rate of subsidence increased markedly. Until 1964, groundwater usage was high, although not uniform because of fluctuations in industrial demand. In 1964, restrictions were set in place to limit groundwater consumption and hence subsidence. Subsidence covers 540 km² at present, but the area is decreasing each year. Total subsidence from 1934 to 1968 exceeded 2.8 m, with highest subsidence observed near the Port of Osaka. Nearly 280 km² of the city itself were affected.

Several tide gauges surround Osaka (Fig. 21). Two at Wakayama and Tan-nowa show relatively low rates of relative sea-level rise (0.9 and 2.2 mm/y.). The Osaka gauge exhibits strongly variable sea-level rise. Prior to 1978, the rate of rise approached 28 mm/y. After 1978, the rate decreased considerably, and even showed a relative drop, perhaps documenting rebound from the recovery of the aquifer. At Kobe, west of Osaka, the rate of sea-level rise also is high, about 6.5 mm/y., suggesting either local tectonism or groundwater withdrawal. At Osaka and vicinity, land subsidence caused by groundwater withdrawal may be contaminating the tide-gauge signals, unlike the other areas examined above.

Niigata Plain. Withdrawal of methane gas from the Niigata Plain (the largest coastal plain in Japan) has led to increased subsidence in the region. Gas exploitation for industrial use peaked in 1958, accounting for nearly 60 percent of the total natural gas production in Japan. For a country poor in hydrocarbon resources, the Niigata Plain provides an important, although limited, source of natural gas. Subsidence of the coast was noticed as early as 1930, when geologists called it "pseudo-sinking" of the coast (Hirono 1969; Yamamoto 1984d). By 1955, subsidence was widely noticeable, especially in the harbor district. By then, subsidence rates had reached 240 mm/y. locally. The cause was identified as dewatering of the aquifers, as saline waters were extracted to recover the dissolved gases.

Examination of survey data show slow subsidence of about 5 mm/y. from 1898 to 1952 and rapid acceleration since 1955. The cause of the early subsidence is unknown, even though it was faster than at many other coastal areas. Natural gas production began about 1947 and increased rapidly during the 1950s. Accompanying this natural gas production was subsidence over about 430 km² of land. To combat this rapid subsidence, Niigata instituted a program of saline injection, whereby saline waters are injected into the strata after gas extraction. The success of this program has not been reported. Reduction of gas recovery is not a clearly viable alternative because of its scarcity as a resource in Japan.

One tide gauge is located in the general vicinity of Niigata, at Kashiwazaki (Fig. 21). Unfortunately, it is outside the zone of subsidence because it indicates that the land is rising at a rate of 1.0 mm/y. relative to sea level. A tide gauge just offshore of Niigata, at Ogi, also records a falling sea level, at a rate of about 0.5 mm/y. Neither site shows subsidence.

Latrobe Valley, Australia

The Latrobe Valley, in southeastern Australia (Victoria–Fig. 18), is unique in that its subsidence is caused primarily by mining of coal, although some contribution comes from extraction of oil and gas. The coal mining can take place only if the strata are dewatered to reduce the confining pressures within the coal region; this dewatering of the aquifer has caused widespread subsidence. Vertical movement here began in the 1920s with the first excavation of coal from the region and it reached a maximum of 1.68 m by 1977. Gloe (1984) predicted that subsidence may reach 3 m in the area as mining continues. This case is interesting because no mitigation is planned to reduce the magnitude of subsidence. No tide gauges exist in the area, although the region is only tens of kilometers from the shore. The subsidence is localized enough that it will not bias our estimates of relative sea-level rise for Australia.

Lake Maracaibo, Venezuela

Oil was discovered in 1917 at the northeastern shore of Lake Maracaibo. Extraction of hydrocarbons from the Lake Maracaibo area resulted in subsidence of 3.4 m from 1926 to 1956; since 1956 the rate of subsidence has decreased markedly. Kugler (1933) first reported subsidence in the oil fields along the eastern shore of the Lake. Kennedy (1961) described the subsidence from the perspective of well-casing failures, attributing the subsidence to decrease in pore fluid pressure and increase in effective stress.

United States of America

Numerous well documented cases of coastal subsidence are known in the United States. Some more notable examples are discussed below.

Santa Clara Valley, California. Of the many areas of subsidence in California, the Santa Clara valley is one of the few near the ocean (Fig. 18). Inland areas of subsidence are well documented but the Santa Clara valley debouches into San Francisco Bay and thus impacts relative sea levels in part of the bay. Poland and Ireland (1988) provided one of the latest

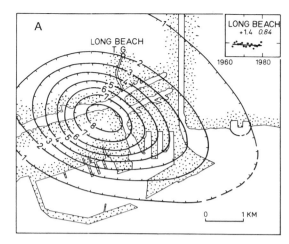
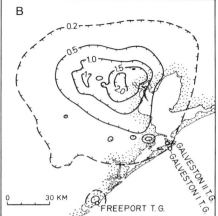

FIGURE 22. Subsidence contours in meters for two areas in the United States.
A. Wilmington Oil Field at Long Beach, California, for the period 1937 to 1968. Redrawn from Poland and Davis (1969). In: Review in Engineering Geology, vol. II, pp. 187–269, D. J. Varnes and G. Kiersch (eds.), Geological Society of America.
B. Houston-Galveston-Freeport area of Texas for the period 1943 to 1973. Redrawn from Gabrysch (1969).

(of many) updates of the Santa Clara subsidence history. From 1934 to 1967, portions of the valley had subsided by 2.7 m, at an average rate of 80 mm/y. Subsidence here stopped briefly around 1940 because of natural recharge (high precipitation rates) and again after 1971, largely because of man's activities (water import program). This was the first area in the United States where land subsidence caused by groundwater mining was recognized (Rappleye 1933). No tide-gauge records in the PSMSL archives are located near the Santa Clara valley, so this localized subsidence has not biased the estimates of sea-level rise from tide-gauge readings.

Wilmington Oil Field, Long Beach, California. One of the most commonly cited examples of land subsidence caused by extraction of hydrocarbons, Wilmington also is an example of a site where such subsidence has been arrested effectively. By 1958, subsidence had reached about 7.5 m, with a total volume of land subsidence of about 140×10^6 m^3, about the same as the 120×10^6 m^3 volume of oil that had been produced (Emery 1960, p. 318). By 1968, the Wilmington oil field had subsided nearly 9 m (equivalent to a mean rate of subsidence of 285 mm/y. in the zone of maximum compaction; Fig. 22A and Table 4). This subsidence caused major economic hardships in a highly industrialized area. To combat the subsidence, massive levees, retaining walls, fill, and elevation of structures were required. Repressurization of the subsidence zones began in 1958, resulting not only in increased recovery rates of hydrocarbons, but also arrest of subsidence (Poland and Davis 1969). By 1961 the land was essentially stable, and by 1964 the land had rebounded about 150 mm. Ultimate rebound expected because of repressuring is about 340 mm. The nearest tide gauge to Wilmington is in Long Beach, recording from 1963 to 1983; the next closest is Los Angeles, recording from 1923 to present. Although the Los Angeles tide gauge is near the oil fields of Wilmington, Baldwin Hills, and Inglewood, the subsidences are so localized that the tide-gauge recordings are not affected. At Long Beach, the land actually is rising relative to sea level; the tide gauge may be recording the rebound of strata following repressurization. Maximum elevation change since 1963 is approximately 37 mm, within the bounds expected from repressurization.

Houston-Galveston, Texas. Subsidence in Houston and Galveston has been known for many years (Fig. 22B). The U.S. Coast and Geodetic survey, for instance, noted a subsidence of 410 mm since 1943 (a rate of 50 mm/y.). A subsidence of 1.2 m was observed at Texas City by 1954. Gabrysch (1969) cited subsidence of 2.3 m in 30 years in Pasadena (Fig. 22). By 1973, the area impacted by at least 0.3 m of subsidence covered 6500 km^2. The subsidence appears to be related to extraction of both groundwater and hydrocarbons. Since 1976, groundwater withdrawal has stabilized and the rate of subsidence has decreased. With increased concern about subsidence, leading to the formation of the Harris–Galveston Coastal Subsidence District, the rate of subsidence should decline even more.

Tide gauges in the Houston–Galveston region document the local subsidence. All tide gauges of Freeport (14.0 mm/y.), Galveston I (7.5 mm/y.), Galveston II (6.4 mm/y.), and Sabine Pass (11.7 mm/y.) show the rapid subsidence characteristic of this section of the Gulf Coast. Sabine Pass is outside the zone of influence of the Houston–Galveston subsidence, but evidently it has an equally large subsidence rate. These tide gauges do not measure maximum rates of subsidence, which locally exceed 75 mm/y.

Goose Creek, Texas. Adjacent to the Houston–Galveston area, Goose Creek Oil Field is a coastal site where hydrocarbon extraction has contributed to local subsidence. Poland

and Davis (1969) cited this as the first instance of subsidence caused by fluid withdrawal identified in the literature. Development of the field began in 1917. Maximum subsidence was more than 1 m by 1925, and since the affected area surrounded the oil field, the cause of the subsidence was attributed to hydrocarbon extraction.

Mississippi River Delta, Louisiana. Subsidence in the Mississippi River delta has become a major issue lately because of tremendous loss of wetlands and increased incidence of upland flooding. It appears to have been caused by a wide variety of factors, including groundwater withdrawal, hydrocarbon extraction, river diversion, and sedimentation processes. The precise roles of these various elements have not been quantified, and at the moment they are the subject of intense scientific and legal debate. Subsidence in Baton Rouge has been attributed primarily to groundwater mining (Davis and Rollo 1969). Elsewhere, the situation is not so clear.

New Orleans has experienced tremendous subsidence through the years. At least two-thirds of New Orleans is built on reclaimed land. Peat there is 70 to 80 percent organic matter by dry weight, and submerged peat (below the water table) is approximately 85 percent water by weight (Snowden, 1986). Drainage and land filling cause three types of land subsidence: consolidation of drained peat and underlying clays, secondary compression of the peat and clay from loading of landfill and draining of peat, and oxidation of the drained peat. Saucier (1963) calculated the average rate of subsidence in New Orleans to have been 120 mm per century for the past 4400 years (a rate of 1.2 mm/y.), based on radiocarbon dating. Snowden provided subsidence curves for New Orleans, showing a net subsidence by 1978 of about 2.03 m since 1924 in one area of New Orleans. Such subsidence will continue in the future; its rate depends on remedial actions that can minimize continued subsidence.

Two of the longer term tide gauges in the Mississippi River delta show high rates of subsidence caused by fluid withdrawal and water management practices. The two stations, Eugene Island and Bayou Rigaud, document subsidence rates of 9.6 and 10.5 mm/y., respectively. Both stations have been reporting from the 1930s to present and they indicate an increase in rate of subsidence beginning about 1960.

Savannah, Georgia. Davis et al. (1963) reported subsidence in the Savannah region exceeding 200 mm caused by decline in artesian pressure in a local aquifer. Most subsidence occurred between 1933 and 1955, for a rate of about 9 mm/y. This subsidence area extended from Savannah for 8 km to the north and west. The nearest tide gauge to Savannah (Fort Pulaski) is east of Savannah, outside the zone of influence of the local subsidence. It shows a rate of relative sea-level rise of 2.9 mm/y., suggesting some local contribution to subsidence, perhaps by tectonism.

Sinkholes

Sinkhole development poses a problem in certain geological terranes, and occasionally it can occur at the coast, causing local subsidence that can impact either benchmarks or tide gauges (either impact could contaminate tide-gauge records). Sinkholes generally occur in carbonate rocks such as limestone, dolomite, and marble, although others are known in gypsum and halite. A perennial stream commonly exists and water moves through the carbonates via fractures, solution cavities, bedding planes, joints, and faults. In time, the roof of a cavity or cavern in the rock collapses because of progressive solution or from downward migration of sediments into solution cavities. The relative importance of these two modes of formation is not well quantified, but eventually a karst topography is produced.

Sinkholes can be either natural or induced. Natural sinkholes occur regardless of human intervention, requiring hundreds or even many thousands of years to form (Newton 1986). They occur because of progressive solution of bedrock and decline of the water table through natural climatic fluctuations (droughts). An induced sinkhole is caused at least in part by man's activities, and they can form within hours of man's influence. There are many causes: where groundwater level is lowered because of pumping, during construction because of increased weight of a structure above the soil and rock, and because of impoundment or diversion of surface water or other modifications of the land surface.

The distribution of sinkholes is rather localized, and there are no instances known to the authors where sinkholes have impacted tide-gauge records. Newton (1986) listed sinkholes in 19 of the 31 states east of the Mississippi River. The total number of sinkholes at about 850 documented sites in the eastern United States exceeds 6000. They are rare in glaciated areas, but Florida has a dense network of sinkholes along its northwestern margin. Thus, although not pervasive, they are not uncommon. Many other nations have sinkholes as well but the global extent has not been documented.

If a sinkhole were near a tide gauge it could affect the records in two ways. First, it could cause sinking of the tide gauge, resulting in an apparent rise of sea level. Second, if the sinking were only near survey benchmarks, the tide-gauge datum would be shifted and an apparent rise in sea level would be measured again.

Climatic Effects

Overview

Sea levels can be impacted by climatic effects that span a broad range of time scales (Fig. 6). Climatic changes can cause fluctuations in water volume because of steric effects or through variations in storage of freshwater in land ice, lakes, and groundwater. As the Earth's climate has changed continuously since the planet formed, contemporaneous fluctuations in sea levels have occurred. Climate influence on sea levels has not always been accepted. Davis (1899) and Johnson (1919) favored the notion that sea levels remained essentially constant for millions of years. Only more recently have the influences of climate and tectonism on local relative sea levels been investigated.

On the longest time scales, the climate has experienced gradual warming or cooling depending on the composition of the Earth's atmosphere and the heat flow from the Earth. These longer term signals are difficult to decipher in the sea-level record because of their long time scales and poor geological recording. For instance, the oldest ocean sediments are roughly Late Triassic in age; older fluctuations in climate are not recorded in even well preserved ocean records. For time scales of 1 to 100 million years, variations in the geometry and volume of ocean basins have caused drastic changes in ocean levels; this effect may overwhelm the record produced by more continuous but slower evolution in climate. For instance, since Late Cretaceous, relative sea levels have fallen approximately 350 m because of gradual cooling of the aging ocean basins and related effects (Pitman 1979). The best indicators of sea level on these time scales may be the sedimentary record; unfortunately such records do not permit simple deconvolution of the impacts of climate change versus change in ocean volume. The sedimentary record mirrors sea level, in part, but also local tectonic effects and fluctuations in source and supply of sediments to various shelves and deep-sea areas.

On time scales of hundreds of thousands of years, climatic variations are thought to be well known, appealing strongly to the Milankovitch cycle (Milankovitch 1938). The Milankovitch cycle addresses the orientation of the Earth's axis of rotation and the shape of its orbit and how these affect the insolation on the Earth's surface. The glaciation/deglaciation events during the past few million years may reflect variations in solar insolation. Hays et al. (1976), for instance, determined from stratigraphic records that cycles of 100,000 y., 40,000 y., and 20,000 y. dominate the power spectrum of climatic variation. These periods correspond to major solar forcing: a 41,000-y. period is exhibited by the variations in the obliquity of the Earth's axis (now at 23.5°, but slowly varying between 22.2° and 24.5°), whereas a 23,000-y. period is exhibited by the quasi-periodic precession of the earth's axis (controlling where summer and winter occur on the Earth's surface). Unexplained by the Milankovitch theory is the dominance of a 100,000-y. periodicity found by many investigators. Also unclear is the mechanistic link between the climate variables and the response of the ice sheets and the other non-linear climate elements (geosphere, cryosphere, hydrosphere, biosphere, and atmosphere). However, the Milankovitch cycle commonly is invoked as the most appropriate model to explain Cenozoic climatic cycles in the sedimentary record.

On a shorter time scale, fluctuations in climate are somewhat better understood, although the mechanistic linkages again appear to be uncertain. During the past several hundred years, the Little Ice Age created a dramatic fluctuation in global climate. This event, spanning about A.D. 1300 to 1800, was accompanied by lowered temperatures, with reporting of increased incidence of freezing temperatures throughout Europe and elsewhere. The globe still appears to be recovering from this Little Ice Age. The largest climatic fluctuation today is the El Niño/Southern Oscillation, an event that has global implications even though its effects are felt strongest in the Pacific Ocean region. The impacts of this major climate fluctuation are discussed below. In a later section the potential impacts of changing climate accelerated by man's activities are discussed.

El Niño/Southern Oscillation

The El Niño/Southern Oscillation (ENSO) is a quasi-periodic event that occurs every two to seven years. Although these perturbations have been described for centuries, the important forcing mechanisms that cause them only now are being clarified. El Niño ("The Child") gets its name from its occurrence in the southeastern Pacific Ocean around Christmas, when the waters become much warmer than normal, bringing with them exotic flora and fauna from equatorial regions. Aperiodically, however, the warmth can bring undesirable results of widespread mortality of fish and birds because deep water no longer is able to be upwelled to provide nutrients for fisheries and higher trophic levels of organisms. The Southern Oscillation (SO) relates the pressure fluctuations in the western Pacific Ocean with alternating high or low pressure over the southern Asia region. With a cycle of about 30 months, the SO reflects an alternation of regional anomalies between a low-pressure area in the Indian Ocean (Djakarta, Darwin, or Bombay are used as regional stations) and a high-pressure area in the southeastern Pacific (where Santiago and Easter Island are used as regional stations). Quinn et al. (1978, 1987) documented ENSO during the past four and a half centuries and Thompson et al. (1984) identified these events from ice accumulation records derived from the Quelccaya ice sheet in Peru during the past one and a half millennia. Michaelsen (1989) described ENSO occurrences in southwestern United States from tree-ring records of the past 400 years, documenting consistent fluctuations in amplitude and frequency having 80- to 100-year time scales. Casey et al. (1989) described the ENSOs and their relationships with radiolarians off the southern California borderland. Radiolarian assemblages respond to changes in oxygen content of the water that in turn are modulated by circulation associated with El Niños. Casey et al. also documented El Niño-like conditions in the sediments off northern California dating to at least 8 million years ago, and more modern El Niño conditions after at least 5.5 million years ago!

The ENSO has been implicated by some recent studies as impacting many aspects of human endeavors. On the Indian subcontinent, farming success is linked strongly with the Southern Oscillation index. In both the western and eastern Pacific Ocean, the fisheries harvest is tied to ENSOs, in the east because the stable cap of warm water prevents the normal upwelling of nutrient-rich deep waters off Peru. During the recent 1982–1983 El Niño, the many impacts on marine fisheries, birds, and other marine life were described by Barber and Chavez (1983). As with many other geophysical occurrences, the impacts of ENSO are being extended to many phenomena where direct physical linkage is obscured. For instance, Walker (1988) examined correlations between seismicity of the East Pacific Rise with the Southern Oscillation Index. The causes for the positive correlations are unclear, but several alternatives are presented. Shaw and Moore (1988) examined the relationship between magmatic heat and the El Niño cycle, postulating magmatic heat release as a possible heat source to drive the El Niños. Their calculations indicate that the volume rates of midocean magma production could generate repetitive thermal anomalies as large as 10 percent of the average El Niño sea surface anomaly, at intervals of about five years (comparable to the recurrence interval of El Niños). The Shaw and Moore model couples with the Walker model in that the former provides a mechanism for linking seismicity and ENSO. That mid-Earth processes and surficial climatic variability may be related on short time scales is a difficult concept to accept, but suggestions that such relations may exist have been proposed.

The impacts of El Niño on global weather patterns are significant. Large-scale global teleconnections within the atmosphere help couple Pacific climate with that of the remainder of the world. During the past, the tropics commonly have been considered as the generating source of weather and climate anomalies almost everywhere. However, the tight coupling of climate throughout the globe by means of these atmospheric teleconnections makes cause and effect difficult to separate. Certainly the feedback between oceans and atmospheres can have major regional as well as global impact. Kerr (1988a) discussed the El Niño of 1986–1987, focusing on weather impacts. That El Niño year was the warmest year of historical record, a fact attributed at least in part to the ENSO. Following that El Niño, Kerr (1988b) then reported on the mirror image to El Niño, La Niña, which is a pool of cooler than normal tropical Pacific waters (Philander 1985). La Niñas might portend temporary chilling of the globe and drastic differences in rainfall and storm patterns. For instance, El Niño is associated with drier than normal conditions in southeastern Africa, India, the western Pacific including Australia, and northern South America, and with wetter conditions in the central Pacific and southeastern United States, but La Niña might have just the opposite associations.

ENSOs have dramatic influence on oceanography of the Pacific Ocean. Cane (1983) described sea surface temperature exceeding 5°C above normal in the eastern equatorial Pacific, and sea level at the Galapagos Islands of 22 cm above normal during the 1982–1983 El Niño. In general, during the first year of an El Niño the sea-surface temperature (SST) anomaly extends across the south Pacific as the easterlies begin to diminish. The sea level rises along the eastern Pacific eventually reaching Canada, whereas it falls in the western Pacific. Westerly winds between Longitudes 100°W and 170°E increase, and the warm anomaly continues to move northward. During the second year the anomalies tend to be much weaker, and the spring (southern hemisphere) warming is followed by more rapid cooling toward a La Niña (which does not always follow an El Niño). Breaker (1989) described variability in sea-surface temperature along the central California coast and its relationship to El Niño. He found that as far north as Monterey, the sea-surface temperature field is dominated by El Niños, with the warming influence strongest in the fall and winter and weakest in the spring.

The impacts of El Niño/Southern Oscillation on Pacific sea levels have been well documented (Fairbridge and Krebs 1962; Wyrtki 1977; Enfield and Allen 1980; Chelton and Davis 1982; Huyer and Smith 1985; Chelton and Enfield 1986; Mitchum and Wyrtki 1988). Roden (1960, 1966) first examined interrelationships between sea levels, atmospheric pressure, and sea-surface temperature, finding a strong correlation between sea level and pressure but a weaker and variable one between sea level and sea-surface temperature. He identified significant energy in the period bands associated with El Niño, without referring specifically to El Niños. Fairbridge and Krebs (1962) determined strong relationships between sea levels throughout the Pacific and the Southern Oscillation index. They hypothesized a direct pressure relationship raising or lowering sea levels, but also other effects possibly caused by winds and temperature. These suggestions appear to be validated by later investigations of these phenomena. Wyrtki (1977) presented a detailed study of the 1972 El Niño, showing five phases of relationship with sea levels in the Pacific. Preceding an El Niño, stronger than normal equatorial winds caused sea level to rise in the western Pacific. After the winds peaked, sea level began to drop slowly in the western Pacific. When the easterlies collapsed, sea level was high along the eastern Pacific, initiating El Niño off Peru. This third phase consisted of an equatorial Kelvin wave that was associated with a strong drop in sea level in the western Pacific. The equatorial trough filled, the South Equatorial Current retreated to south of the equator, and the Pacific Countercurrent intensified. Later, a second peak in sea level occurred along the eastern Pacific including the coast of Central America, and there was a dramatically lower sea level in the western Pacific. Sea levels finally returned to normal, but at a more rapid pace than the onset of El Niño.

In a study of sea levels along the Pacific coasts of North and South America, Enfield and Allen (1980) found positive and negative sea-level anomalies related to El Niño and anti-El Niño cycles (the latter now termed La Niñas). Analysis of these anomalies shows a poleward propagation, with a phase speed estimate of 180 ± 100 km/day. Tropical temperature anomalies appear to be coherent with sea levels and El Niños. Along California, however, temperature anomalies lag several months behind both local sea level and the Peru tempera-

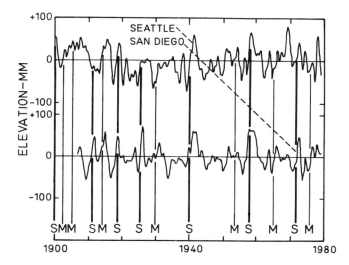

FIGURE 23. Relative monthly mean sea levels at Seattle and San Diego for the period 1900 to 1982. El Niño/Southern Oscillation occurrences are represented as strong (S) or moderate (M), according to the classification by Quinn et al. (1978). Both Seattle and San Diego sea levels respond in phase to this El Niño forcing, although ENSO response is but a small fraction of total sea-level variance. Tidal data for San Diego are not exactly as provided by PSMSL. A 100-cm jump in water level has been removed from data preceding 1926, providing a closer correspondence for records from San Diego and nearby La Jolla. From Emery and Aubrey (1986b), Journal of Geophysical Research, v. 92, p. 13941–13953, fig. 10, copyright by the American Geophysical Union.

ture anomalies. Chelton and Davis (1982) examined sea levels along the Pacific coast of North America and determined a poleward propagation of sea-level anomalies related to El Niños in the form of Kelvin waves at speeds of 34 km/day. Mitchum and Wyrtki (1988) provided a review of some of the developments leading to improved understanding of Pacific sea levels and their relationship to El Niños.

Emery and Aubrey (1986b) published monthly averaged sea levels at Seattle and San Diego, showing the correlation with strong and moderate El Niños (Fig. 23). In addition, they examined sea levels from La Jolla north to Yakutat, showing that El Niño influences sea level along the entire coast (Fig. 24). For the western Pacific Ocean, sea-level pressure at Darwin was related to sea level at Sydney–Fort Denison and the sea levels of all Australia (Fig. 25), exhibiting a strong correlation between sea level in Australia and atmospheric pressure at sea level. In general, sea level is high when the pressure at Darwin is low, and vice versa. Examination of the tide-gauge record shows that Sydney sea levels generally are low during El Niños (Fig. 26), in accord with the inverse correlation of sea levels between the western and eastern Pacific Ocean during El Niño.

Theories about the causes of El Niños are abundant, but recent work seems to be improving our understanding. Bjerknes (1961) proposed that weak trade winds in certain years could lead to cessation of upwelling off Peru, allowing warmer water to approach the coast and move southward along the coast. Wyrtki (1975) showed that El Niño is caused not by weakening of southeast trades but rather by higher speed trade winds that precede El Niño the two previous years. This increased wind stress increases the east-to-west gradient in sea-surface levels. As the trade winds relax, the warm water piled up along the western Pacific propagates eastward as an equatorially trapped Kelvin wave, eventually leading to accumulation of warm water off Peru and Ecuador. Cane (1984) modeled sea levels caused by El Niños, testing the hypothesis that the sea-level variations are caused by wind stress variations. He was able to model the timing and pattern of the observed sea-level anomalies but predicted systematically low amplitudes. In particular, Cane reproduced the twin peaks of El Niño, the first caused by weaker wind changes during the fall preceding El Niño and the second caused by massive collapse of the trade winds during the middle of El Niño year.

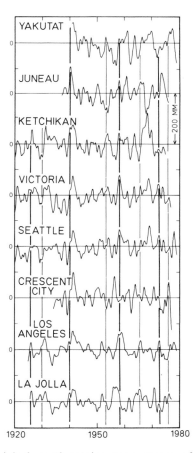

FIGURE 24. Triple 6-month running mean averages of monthly relative sea-level data along the Pacific coast of North America (mean monthly values have been removed to eliminate seasonal effects). This averaging is identical to that of Quinn et al. (1978) and was discussed in Aubrey and Emery (1986a). From Emery and Aubrey (1986b), Journal of Geophysical Research, v. 92, p. 13941–13953, fig. 5, copyright by the American Geophysical Union.

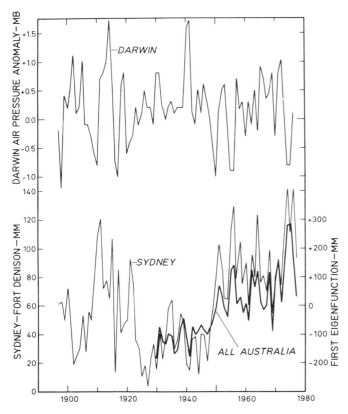

FIGURE 25. Correspondence of sea levels for Sydney–Fort Denison and for all Australia (by eigenanalysis–first temporal function), and atmospheric pressure at Darwin (the latter from Quinn et al. 1978). At Darwin high pressure often is accompanied by lowered sea levels at Sydney–Fort Denison and at Australia as a whole. Correspondence between pressure at Darwin and sea levels is more clearly evident on monthly time scales (not shown here). From Aubrey and Emery (1986b, fig. 9) by permission from Journal of Geology, the University of Chicago Press.

Cane and Zebiak (1985) provided simulations of El Niño and the Southern Oscillation using a coupled atmosphere–ocean model. This model reproduced some of the major features, including recurrence at irregular intervals. The major feedbacks modeled were between ocean dynamics, altering sea-surface temperature, and changing atmospheric heating and wind patterns; in turn, changes in surface wind altered the ocean dynamics. The model produced El Niño-like symptoms using purely deterministic interactions that are simpler than natural interactions. The determinism argues for improved predictive ability in the future, although the simplicity of this model suggests that more complex interactions may limit its predictability, however deterministic these complex interactions might be. Zebiak and Cane (1987) produced another coupled model of the ENSO, where they identified the equatorial heat content in the upper ocean as a critical element. The model correctly predicts a preference for a 3- to 4-year interval for these events. This model provided a potentially useful indicator for future El Niños: the above-normal equatorial heat content.

Chao (1988) related the occurrence of ENSO to the length-of-day (LOD) variation, concluding that most of the interannual LOD variation is caused by ENSO and that the transfer of ENSO's axial angular momentum to the solid Earth has a two-month lag. Graham and White (1988) modeled the ENSO as a natural oscillator of the Pacific Ocean–atmosphere system, including simple couplings between circulation of the tropical atmosphere, dynamics of the warm upper layer of the tropical ocean, and sea-surface temperatures in the eastern equatorial Pacific. Finally, Barnett et al. (1988) reported on the prediction of the El Niño of 1986–1987 using three different classes of computer models. All three successfully predicted the El Niño at lead times of 3 to 9 months. The three techniques included a statistical model, a dynamic model including wind stress and ocean transport on an equatorial β-plane, and a numerical model simulating the evolution of the coupled ocean and atmosphere of the tropical Pacific. That all three models were successful in predicting this relatively weak El Niño yields hope that such predictions will be more successful in the future.

The impacts of ENSO on tide-gauge records are poorly quantified. Available records indicate that El Niño produces large positive displacements of sea surface, but non-El Niño conditions produce smaller negative displacements. Other interannual fluctuations aside, this condition would lead to a positive bias of sea level. If a 10-cm increase in sea surface height were to occur during 18 months of every four years, the bias would be about 4 cm. If successive 4-year intervals (the approximate period of an El Niño) had signals of strongly varying strength (such as the difference between the 1982–1983 and the 1986–1987 El Niños), this bias could significantly affect estimates of long-term relative sea-level rise, particularly for short records. However, since there are many other causes of interannual variability, and since the time variability through decades of El Niños and other perturbations is unclear, the net bias on estimates of relative sea-level rise also is uncertain.

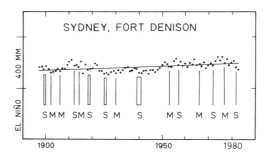

FIGURE 26. Yearly averaged relative sea levels measured at Sydney (as in fig. 25), Fort Denison during the period 1987 to 1982. Strong (S) and moderate (M) El Niños, as classified by Quinn et al. (1978), correspond with periods of lowered sea levels, although not all local minima in sea levels correspond with El Niño activity. Because tide-gauge data that are yearly averaged lose much of the definition of El Niños, this relationship is clearer using monthly averages. From Aubrey and Emery (1986b, fig. 7) by permission from the University of Chicago Press.

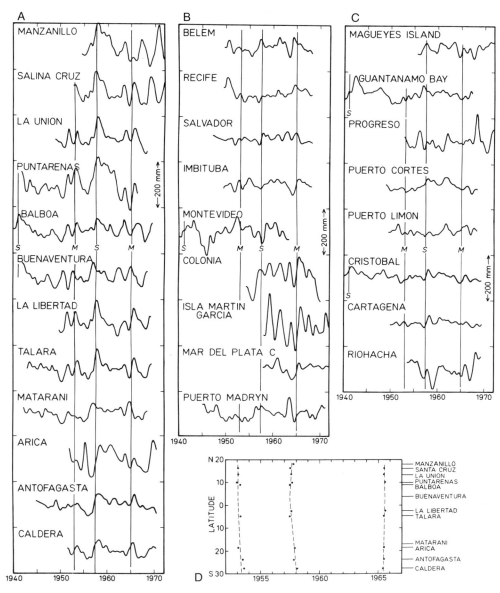

FIGURE 27. Monthly mean sea levels along the Central and South American coasts for years of record. Shown are dates of some stronger El Niños, as indicated by Quinn et al. (1978). Although El Niño high waters can be observed along the eastern Pacific coast, they are not evident along the western Atlantic coast. Note the highly variable tide records of Montevideo, Colonia, and Isla Martin Garcia, which are within the Paraná River mouth, reflecting considerable hydrological influence.
A. Pacific coast.
B. Atlantic coast.
C. Caribbean coast.
D. Variation in dates of arrival of El Niños at Pacific coast stations.

River Runoff/Floods

Tide gauges commonly are sited within harbors, which are particularly prone to impacts from river flooding. Because streamflow is highly variable, the tide-gauge records may experience considerable interannual variability. Cayan and Peterson (1989), for example, examined river flow in the western United States and showed how both seasonal and non-seasonal behavior of streamflow are correlated with atmospheric circulation. Since streamflow also is correlated in part with atmospheric pressure and large-scale events such as El Niños, separation of various contributions to the tide-gauge records from these different sources is difficult.

Meade and Emery (1971) investigated the relationship between mean annual variation in mean sea levels recorded by tide gauges and variation in river runoff within four coastal segments of the eastern United States between Canada and Mexico. Multiple regression analyses showed that variations in river flow accounted for 7 to 21 percent, and secular rise of sea level (or secular submergence of the coast) accounted for 29 to 68 percent of the variation in sea level during a 40-year period. This accounting left 10 to 50 percent of the sea-level variation that must be caused by other factors such as wind, water temperature, currents, atmospheric pressure, experimental error, and random statistical error. Roden (1960) also found that along the United States west coast, anomalies in sea level reflect anomalies in river discharge. This is particularly true at Astoria, Oregon, where the Columbia River debouches. Depending on the frequency band, between 15 and 50 percent of the sea-level variations could be related to variations in discharge.

In their study of South American sea levels, Aubrey et al. (1988) examined monthly tide-gauge records for the rivers entering the western Atlantic Ocean. Large monthly variations are shown clearly where tide gauges are within rivers (Fig. 27). Locations where interannual fluctuations are related to river flows include the entrance to the Paraná River along the Argentine and Uruguayan coasts, where tide gauges at Isla Martin Garcia, Colonia, and Montevideo show drastic swings reflecting changes in river flow. Tide gauges within river mouths thus are suspect for determination of relative sea-level rise, particularly for short records. Although these examples are limited to the Americas, similar examples exist throughout the world, including tide-gauge records at the mouth of the Ganges River in India, as described in a later section of this book.

Steric Ocean Response

General

Ocean levels respond to changes in density of the water column as the column attempts to come to isostatic and dynamic equilibrium. The sea surface is not perfectly level because it is influenced by isostatic processes, gravitational attraction, and dynamical processes. As an example, over ocean ridges or highs on the ocean bottom, sea level topography has as much as tens of meters relief (Emery and Uchupi 1984, fig. 110). Across the rapidly flowing Gulf Stream and Kuroshio, there also is a difference in mean sea-level elevation of about one meter. On an annual basis, the mean level of the water along our shorelines varies by several centimeters to more than 50 cm, as the winter cooling/freshening and summer warming/salinity rise influence the ocean surface. Along the Atlantic coast of the United States, for instance, temperature and salinity effects cause annual variations of 20 to 30 cm. Along the China coastline of the Yellow Sea, these variations can exceed 50 cm.

The steric response of the ocean is the reaction to changes in temperature and salinity of ocean waters. As an ocean approaches gravitational equilibrium, the height of a water column above a uniform pressure level at any location depends on the density of that water. Less dense water achieves a higher free surface, whereas more dense water is lower. Since warmer or fresher waters are less dense, these two variables can affect the level of the oceans. Increases of pressure with depth also control water density and must be included in appropriate calculations. On a seasonal basis sea-surface pressure impacts water levels as storms (low pressure centers) become more or less common and intense.

Temperature Effects

The density of pure water at 25°C and atmospheric pressure is equal to unity (1 g/cm^3) and is highest at 4°C, but density of sea water is not constant. Sea water density can be represented by $\rho(S,T,P)$, showing that its density is a function of salinity, temperature, and pressure. However, we often use another form of density, sigma, defined as: $\Sigma S,T,P = (\rho[S,T,P] - 1)\,1000$. Thus, a density of 1.02800 has a sigma value of 28.00, a more convenient notation. We normally use the reciprocal value of density in these calculations, which is the specific volume of water: $\delta(S,T,P)$. Tables of this value have been compiled to simplify its use (Knudsen 1901; Sverdrup et al. 1942, p. 1053–1059). These tables can be applied to different levels of the water column to estimate the impact of warming on ocean levels.

Pattullo et al. (1955) examined these steric effects, finding that the average range of thermal seasonal departures is 11 cm. Small thermal departures occur in some equatorial and most polar regions. The maximum ranges of about 25 cm were found in the Sea of Japan and in regions north of Bermuda. All of these values are derived from monthly means, not individual observations.

Salinity Effects

Haline departures have a similar impact. Pattullo et al. (1955) found that two thirds of the world ocean experiences seasonal haline effects on sea level of 5 cm or less. These were especially weak around Bermuda and in ocean areas east and southeast of Japan. Large salinity departures were observed in the Bay of Bengal (a range of 41 cm) and along the continental slope off eastern Asia in the region between Taiwan and Hokkaido.

In general, the steric departures in equatorial regions are mainly thermal and relatively small. Steric departures in the subtropics are mainly thermal and large. In the subpolar latitudes, thermal departures are small but sea-level variations are large. Whether or not these are haline effects is difficult to determine because observations of salinity commonly are lacking.

The history of salinity and temperature of the oceans is complex and has not been unraveled fully, although paleoceanographic studies have begun to clarify past oceanographic conditions. It is not yet possible, given the sparse data available, to quantify the steric response of the oceans on time scales of millions of years. It is problematical, even, to determine steric trends for time scales of many decades because the data base is sparse, biased, and non-uniformly sampled. For example, Barnett (1984; 1988) could identify no trends in ocean surface warming during the past half century, largely because of changes and biases in instrumentation. Roemmich and Wunsch (1984) found a warming of the North Atlantic Ocean through comparison of two repeated hydrographic transects, one made during the International Geophysical Year in 1957, the second in 1978. Their comparison showed warming of the North Atlantic Ocean to a depth of 3500 m, far deeper than most models of steric response have assumed. This warming depth must be caused by dynamic processes such as large-scale ocean circulation and not just to heating from above. Whether the warming of the North Atlantic represents a global effect is unknown because equivalent measurements are lacking in any other ocean. It may represent a redistribution of North Atlantic water rather than a response to a global temperature rise. To resolve the cause of the differences between these two repeated transects one must await similar comparisons between other transects in this and other ocean basins. More recently, Levitus (1990) reviewed the available data for the entire North Atlantic Ocean.

Long-Period Tides

General

Tides are complex features having a broad spectrum of space and time scales. Some of these scales were presented in an earlier section (Introduction: Frequency Realms of Tide-Gauge Records). Included there was a list of dominant tidal time scales (see Fig. 6), ranging from 1 mean solar day (period of Earth's rotation relative to the Sun), to 1 lunar month (period of Moon's orbital motion), to 1 year (period of Sun's orbital motion), to 8.85 Julian years (period of lunar perigee), to 18.61 Julian years (period of regression of lunar nodes), to 20,942 Julian years (period of solar perigee). Tidal motions are not only at these frequencies but also at integral sums and differences of all of these frequencies. Such compound tides can emanate not only from astronomical forcing but also from non-linear hydrodynamics within ocean basins. Thus, the list of possible tidal motions that can impact tide-gauge records is long. These relationships were encoded using the method of harmonic analysis developed by Sir William Thompson (Lord Kelvin) about 1867; in fact, the principle of superposition of harmonic constituents was introduced first by Eudoxas as early as 356 B.C. (Schureman 1976).

The complexities of tidal harmonics have been treated rigorously and thoroughly by Schureman, who also described the sources of the various six primary tidal frequencies described above. He also listed the frequencies and pertinent tidal data for nearly 200 different tidal constituents! Of most interest for understanding rise of relative sea-level rise is long-period tidal action, having periods much greater than the usual semi-diurnal and diurnal tidal rise and fall of the ocean (these latter signals are well filtered by the rapid sampling and considerable averaging of tide-gauge records). The longer term tides can introduce variability in three primary ways: because of their long time scales relative to the length

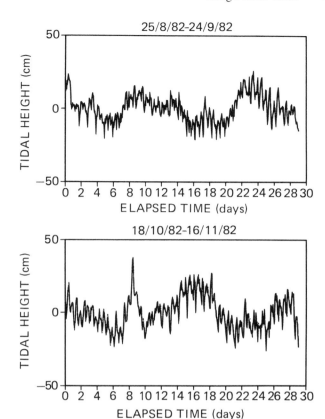

FIGURE 28. Fortnightly tidal variation caused by nonlinear tidal interactions between the M2 and S2 tides at Nauset Beach, MA. Top and bottom panels are for two different months in 1982. From Aubrey and Speer (1985), with permission of Estuarine, Coastal and Shelf Science.

of tide-gauge records, because their amplitude may vary over long time scales due to astronomical factors, and because local harbor conditions may change and alter the mean water elevation at the tide gauge.

Long Time Scales

The five most commonly discussed and most energetic long-period tides have little impact on the times of high and low waters (factors of primary interest to navigators), but they do affect the mean level of the water from day to day. These five are the lunar fortnightly (Mf), the lunisolar synodic fortnightly (MSf), the lunar monthly (Mm), the solar semiannual (SSa), and the solar annual (Sa) tides. The first three are normally small, but the last two may be large. However, compound tides caused by non-linear hydrodynamic interactions between M2 and S2 can create locally significant fortnightly tidal variations in extremely shallow waters (Fig. 28), larger than the longer term constituents mentioned here (Aubrey and Speer 1985). The intense averaging of high-frequency (hourly) observations effectively obscures much of the fortnightly tide, but not as effectively as the averaging filters the semi-diurnal and diurnal tides. By averaging over an entire year, the fortnightly constituents are reduced to less than 1 mm in amplitude; for monthly means the fortnightly tide is attenuated to several millimeters.

The even longer term constituents, such as the 18.61-year lunar nodal tide, contribute little to the vertical tide but they appear to impact the horizontal tide more energetically. However, at high latitudes the changes in mean sea level caused by the nodal tides can reach 40 to 50 mm. Maximov (1959) estimated a latitudinal dependence for the lunar nodal tide as follows:

Latitude (N or S)	0°	10°	20°	30°	40°
ΔH (mm)	22	20	14	5	−5

Latitude (N or S)	50°	60°	70°	80°	90°
ΔH (mm)	−17	−27	−36	−41	−44

Most important for discussions of relative sea-level changes is the solar annual tide (Sa). This tide is controlled by both astronomical and local factors, such as steric changes in sea level. These patterns are difficult to separate and are often discussed together. However, the astronomical contribution to the annual tide is much smaller than the meteorological and oceanographic contributions (Pattullo et al. 1955). The equilibrium atmospheric contribution to Sa has been quantified by Maximov (1965b) and Lisitzin (1974) as:

Latitude (N or S)	0°	10°	20°	30°	40°
ΔH (mm)	1.1	1.0	0.7	0.3	−0.3

Latitude (N or S)	50°	60°	70°	80°	90°
ΔH (mm)	−0.8	−1.3	−1.7	−2.0	−2.1

The magnitude of the Sa tide is important here because this tide is long compared with the averaging interval for the annual tide-gauge records generally used for examination of sea-level rise. Variations in the annual solar tide are not well filtered by this averaging, so that changes in Sa can cause spurious fluctuations in the annual averages (although generally not to the long-term trends). Pattullo et al. (1955) examined the seasonal oscillation in sea level throughout the globe, focusing on steric oscillations. For open ocean stations (islands), the steric signal is weak generally but not everywhere. Most continental stations have steric changes that range from 10 to 30 cm. The most extreme steric signals (Fig. 29) are from the Bay of Bengal (a 120-cm oscillation each year), the Bohai Gulf region (a nearly 60-cm oscillation), and the Persian Gulf (a 50-cm oscillation). Such large oscillations are sensitive to year-to-year climatic variations, which contribute significantly to mean annual sea-level changes and contaminate the records for interpreting global sea levels or local tectonics. Lisitzin (1974) similarly examined seasonal sea-level fluctuations and identified major steric contributions globally. The largest seasonal signature identified by Lisitzin has a range of 148 cm, at Moulmein (Lat. 16°29′N; Long. 97°37′E) near the head of the Bay of Bengal.

The latitudinal dependence of the equilibrium and observed SSa (solar semi-annual) tide is given by Maximov (1965a) and Lisitzin (1974) as:

Latitude	80°S	60°S	40°S	20°S	0°
ΔH (mm): theory	−12.4	−8.1	−1.6	−4.2	−6.5
ΔH (mm): observed	−71	−51	−34	2	56
ΔHobs/ΔHtheory	5.7	6.3	21.2	0.5	8.6

Latitude	20°N	40°N	60°N	80°N
ΔH (mm): theory	4.2	−1.6	−8.1	−12.4
ΔH (mm): observed	39	−11	−47	−61
ΔHobs/ΔHtheory	9.3	6.9	5.8	4.9

Clearly, the SSa tide has a larger equilibrium argument than the Sa equilibrium tide. However, the SSa tide is attenuated more strongly by yearly averaging than is the Sa tide, so its impact on tide-gauge records must be less than indicated above. Lisitzin (1974) also provided the equilibrium theory and observed amplitudes of other long-period tides.

Amplitude Variation Through Time

Tidal prediction must consider the changes in tidal amplitude and phase that occur because of variations in the forcing through time. These variations are taken into account by incorporation of astronomical factors that vary with time, and are used in major tide prediction programs. Nodal factors and other variables are specified to extract the dominant time variability. Practically, such time variability is not significant when using yearly mean sea-level estimates because these estimates are attenuated significantly by yearly averaging.

Changes in Hydrodynamical Conditions

Tide gauges commonly are located in ports and harbors, where information on tidal stage has particular concern for navigation. However, ports and harbors have complex tidal circulations and storm surge characteristics, which depend on their water depth, connection to the open ocean, orientation with respect to storm winds, planform geometry, and other factors. As ports and harbors develop, some of these characteristics change, and along with them the mean water level may change. Examples of harbor changes that may impact mean sea levels are many. In some harbors, basin areas have become larger with port expansion, smaller with land reclamation, or have become more or less enclosed. For example, harbor dredging is a common response to need for increased deep-water access or increased dockage or turnaround space. Kings Bay Naval Submarine Base in southern Georgia (United States) is an example of how the northern half of Cumberland Sound has been altered substantially by need for harbor expansion for military use. San Francisco Bay is an example where an entire estuary has shrunk because of land reclamation; fully 90 percent of the marshes have disappeared since 1950 because of aggressive land reclamation (Nichols et al. 1986). The major estuaries of The Netherlands provide examples of still greater land reclamation and alteration of basin geometry. Finally, Long Beach Harbor,

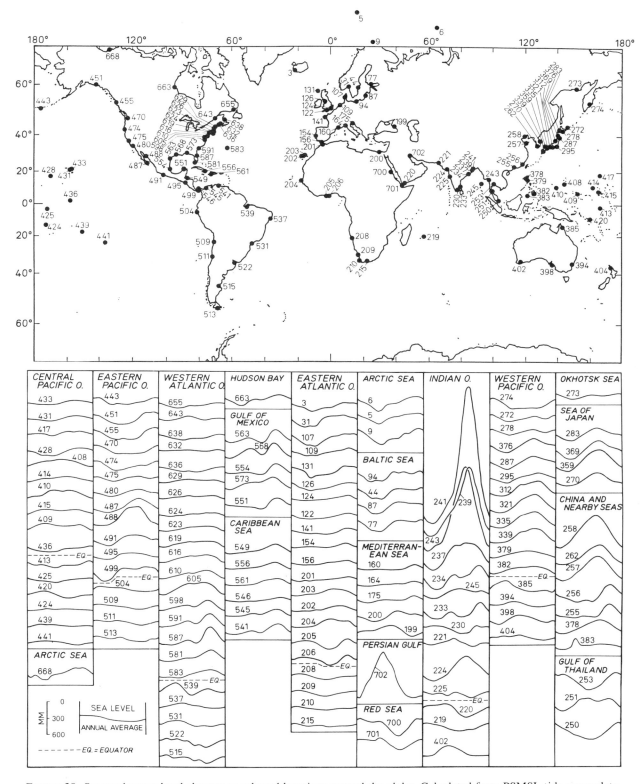

FIGURE 29. Seasonal water level changes at selected locations around the globe. Calculated from PSMSL tide-gauge data.

California, is an example of progressive enclosure as the breakwaters protecting the harbor have been extended. This enclosure changes the hydrodynamics of the harbor, altering the mean water elevation within it. Another common border modification involves altering the entrance channel (inlet) configuration, commonly by adding jetties and dredging. Such changes at the inlet impact tidal stages throughout the embayment served by the inlet.

The dominant physics responsible for causing changes in water level include changes in tidal mean set-up (water sloping up into the bay) within a harbor (in the frequency realm known as zero frequency tidal interactions) and changes in freshwater/saltwater exchange (and thus changes in the steric balance within a harbor). The impact of changes in tidal mean set-up can be significant, particularly compared with the eustatic sea-level signal, thought to be on the order of 1 mm/y. Tidal set-up within a harbor can reach tens of centimeters, swamping the relative sea-level contribution. Changes in steric balance are partially caused by changes in saltwater/freshwater mixing processes. An example of this change is Savannah Harbor (Georgia), where modifications have allowed the salt wedge to migrate several kilometers upstream, causing severe shoaling in a previously stable channel segment and also changing the mean salinity in many parts of the harbor. Such changes in salinity and temperature can produce steric changes on the order of the eustatic signal.

An example showing the changes in mean water levels owing to changes in harbor configuration is provided by Aubrey et al. (1990). Using Chatham Harbor (Massachusetts) as an example, they modeled the tidal mean set-up within the harbor as the configuration of the harbor changed because of natural barrier beach elongation. Using a fully non-linear one-dimensional numerical model, they found that the tidal set-up within the harbor was on the order of 10 cm, primarily because of the extremely shallow water depths and a long confined channel system. As the barrier beach elongated accompanying southerly inlet migration, the tidal mean set-up at any single location within the channel varied. Because this cycle of inlet migration and barrier elongation occurred on a time scale of 100 years, the equivalent rate of relative sea-level rise caused solely by changes in set-up is 1 mm/y., the same order as the hypothesized eustatic sea-level rise. Modifications to harbors of this magnitude can produce changes in mean water levels, leading to rates of change that are comparable to, or even exceeding, those caused by eustatic or tectonic processes.

Shelf Waves and Seiches

An entire class of motions creates impacts on coastal tide-gauge recordings that span the period range from hours to months. These motions are referred to in various ways, but are here termed shelf waves and seiches. Various specific motions such as Kelvin waves are included in this terminology. These waves can be set up by local wind forcing or by ocean-scale events such as the El Niño/Southern Oscillation. Although they appear on all coasts to some extent, these features have been best studied along the eastern Pacific Ocean.

Roden (1960) investigated non-seasonal variations in sea level along the Pacific coast of North America. He found good coherence between anomalies of sea level and of atmospheric pressure at nearly all frequencies between 0 and 6 cycles per year, showing a direct coupling between the two. Coherence between anomalies of sea level and sea-surface temperature was moderate to poor. Roden found some coherence between the sea level and the southerly component of geostrophic wind, and locally near river mouths good correlations between sea level and river discharge.

Brink et al. (1978) studied current, temperature, and tide-gauge data off the Peru coast for 1976. Alongshore currents propagating poleward were not always well correlated with local wind stress. However, temperature and onshore/offshore current fluctuations were correlated with the local wind stress, as expected in this area of severe upwelling. Enfield and Allen (1980) examined the monthly mean sea levels along the eastern Pacific coast, finding strong correlations between sea-level anomalies along the coast and the El Niño/Southern Oscillation. The poleward propagation of these anomalies was documented at a phase speed of 180 ± 100 km/day. Chelton and Davis (1982) also examined the monthly mean sea-level variability along the Pacific coast of North America, finding the signal to be dominated by the inverse barometer effect north of San Francisco, but not south of San Francisco. The interannual variability is correlated with El Niño occurrences, propagating poleward as Kelvin waves. Wind effects contributed to the non-barometric response, particularly in the northern part of the study area. Much of the variability south of San Francisco could not be explained by wind or other known contributions. Mitchum and Wyrtki (1988) examined the sea-level variability in the entire Pacific, attributing the variability variously to El Niño/Southern Oscillation, inverted barometer, local wind forcing, steric changes, and shelf waves such as Kelvin waves. Also examined were equatorial waves, zonal current instabilities, long-period tides, and mesoscale eddies. Clearly, the oceanographic fluctuations are broad in spectral width and caused by a variety of different forcing.

Similar studies have been conducted in the North Atlantic Ocean, but not to the same detail. For instance, Thompson (1986) discussed North Atlantic sea levels and their relationship to circulation. Specifically, he related sea levels to changes in ocean-scale winds and described a strong meteorological dependence for secular sea-level trends. Upon removal of the sea-level signal caused by secular changes in wind stress across the Atlantic, he reduced the standard error of the sea-level trend by half. An investigation by Meade and Emery (1971) of the relationship between sea levels recorded by tide gauges and variations in river runoff along the Atlantic coast of the United States showed that variations in river flow accounted for only 7 to 21 percent of the variation in sea level.

Seiches can be caused by many processes, including submarine processes, tides, and atmospheric pressure fluctuations. The term "seiche" initially described standing oscillations in lakes (Forel 1895), but later it was extended to describe free stationary sea-level oscillations in other water bodies having some type of constraint (enclosure, bathymetric trapping, etc.). Seiches in enclosed or partly enclosed basins have natural periods of free oscillation that depend on their geometry. Whereas the modes of oscillation are well described by present theoretical understanding and numerical techniques, their causes are not nearly as well understood.

For instance, it is well documented that seiches can be caused by abrupt changes in the direction or velocity of the wind, by gradients in atmospheric pressure, or by co-oscillation with an adjoining body of water. Since seiches represent resonance phenomena, even small external forcing can create significant motions in the confined areas. However, other factors also may be involved, such as coupling of baroclinic and barotropic motions (Chapman and Giese 1990).

Natural periods for these oscillations can vary from nearly 1 h to more than 10 h. The Baltic Sea, for instance, has multiple natural periods of 39.4, 22.5, 17.9, 12.9, 9.4, 7.3, and 6.9 h (Lisitzin 1974). Oscillations can reach significant elevations, locally exceeding 3 m (such as at the Balearic Islands in the western Mediterranean Sea). Seiches locally are given different names by which they are reported; this variety of names can cause confusion in the literature. For instance, in Nagasaki, Japan, seiches are known as abikis, but this term is not in general use even in Japan. The increased attention now being paid to seiches probably is overdue, given their broad distribution. However, the early work of Honda et al. (1908) and Platzman (1958) along with that of Munk (1962) and coworkers gave rise to the belief that the causes for seiches were well known, and the field held no future scientific promise. Recent studies (Giese and Hollander 1987; Giese et al., 1990; Chapman and Giese 1990) have challenged the fundamental tenets of what drives coastal seiches, and it is likely to lead to improved understanding of forcing mechanisms, with hope for improved prediction. Work in the Sulu Sea, Palawan Island, Puerto Rico, and elsewhere suggests that deep-sea internal waves may couple to shallow shelves, producing seiches where the geometries are appropriately tuned.

Shelf waves and seiches can affect estimates of mean sea-level rise, particularly for short record lengths. Since the recurrence interval of shelf waves can be on interannual time scales, their presence can impact estimates of sea-level rise. Some shelf waves can have amplitudes of 10 cm; if they occur at intervals of 5 to 10 years, their signals can swamp mean sea levels. Particularly along coasts where shelf waves are abundant, estimates of sea-level rise must be based on records that are long compared with the dominant scale of the shelf waves in order to reduce the bias introduced by shelf waves. Seiches generally are much shorter period features so they have little impact on monthly averaged or yearly averaged sea levels.

Tsunamis

Tsunamis are generated by seismic activity and in turn they can generate seiches on shelves. Tsunamis are also known as seismic sea waves or tidal waves, although these motions have nothing to do with the astronomical tides. Lisitzin (1974) translated the term tsunami to mean "harbor wave" in Japanese. Although marine earthquakes are the most frequent cause of tsunamis, severe volcanic eruptions and submarine landslides also have been implicated. Tsunamis have periods of 10 minutes or longer, although their period is difficult to measure precisely because of the impulsive nature of the wave. Many tsunamis are accompanied by several crests, some from the tsunami itself, and some from seiches generated by the tsunami.

Tsunamis are known to have destroyed coastal human populations and structures for thousands of years in the eastern Mediterranean Sea, as chronicled by Shalem (1956) and Ambrayses (1962). Perhaps the most famous tsunami in the Atlantic Ocean was the one caused by the Lisbon earthquake on 1 November 1755 that reportedly killed 60,000 people, many of them by drowning (Lyell 1850, p. 476–480; Reid 1914). It nearly destroyed Lisbon (which Voltaire later memorialized in his Candide). Waves reached 15 m above sea level at Lisbon, 18 m at Cadiz, and several meters in the West Indies. In the Pacific Ocean the eruptions of Tamboro in 1815 and Krakatoa on 27 August 1883 produced tsunamis nearly 15 m high that destroyed tens of thousands of people and many villages along coasts of the East Indies. Another initiated by a submarine landslide in the Aleutian Islands on 1 April 1946 caused considerable damage along coastal Hawaiian Islands, as recorded in detail by Shepard et al. (1950), who noted that waves rose to 16 m along northern coasts and were low on leeward ones, in long estuaries, and behind coral reefs. Emery could not detect it in the Marshall Islands. Others reach the Hawaiian Islands about once every seven years (Pugh 1987). The Chilean earthquake of 22 May 1960 caused a tsunami that propagated across the Pacific to Japan, killing more than 100 persons. An 8.3-magnitude earthquake near Valdez, Alaska, on 28 March 1964 produced a tsunami whose waves perturbed at least 105 tide-gauge records throughout the Pacific Ocean, and the earthquake shocks through the lithosphere caused seiches that also were recorded by tide gauges in the Gulf of Mexico (Spaeth and Berkman 1967). Tsunamis are especially common in the Pacific Ocean because of the highly faulted and volcanic framework of the Pacific continental margins, "The Ring of Fire."

Just as for seiches, tsunamis have little impact on estimates of long-term sea-level rise. Although tsunamis are energetic features (sometimes attaining heights of tens of meters), their impact on tide gauges is reduced because they last only tens of minutes. With hourly sampling of tide-gauge records and subsequent strong filtering due to averaging, the bias produced by tsunamis is negligible. For instance, a 10-m high tsunami might be averaged over 8766 hourly samples in a year, producing a maximum bias of only 1 mm. The peak of the tsunami would likely not be recorded (if, indeed, the tide gauge survived the tsunami), so the bias would be less. This 1-mm upper limit must be viewed in the light of normal variability in annual mean sea levels, which are on the order of tens of mm or more. Thus, the rare tsunamis are not an important source of error in mean sea-level estimates.

Surface Gravity Waves

Surface gravity waves can impact tide-gauge records by two primary processes: first, the tide-gauge stilling well might rectify the wave motion because of its dynamics and cause a bias in the sea-level measurement; second, if the tide gauge is on the open coast (as many are) it can be influenced more or less by wave set-up depending on the severity of the waves. The importance of the first factor, that of wave rectification in the stilling well, appears to be slight; however, wave set-up and set-down provides a bias of poorly quantified magnitude. For instance, wave set-up can exceed 1.0 m, whereas wave set-down can exceed 0.3 m. Although wave set-up is larger than wave set-down in maximum magnitude, the set-up near a tide gauge generally is smaller than the maximum at the shore, because the gauge is located farther offshore. Simple calculations show that wave set-up and set-down may be similar in magnitude for typical installations, but for other installations wave set-up can bias the tide-gauge recordings, rendering them sensitive to the wave climate for that particular year. Except for areas having waves that are excessively high for a significant percentage of the time, this bias is likely to be small. In general, wave set-up is believed to produce no significant bias to yearly estimates of mean sea levels.

3. Ancient to Modern Changes in Relative Sea Levels

Source Materials

Background

Changes of mean annual sea level are so slow that they scarcely can be observed by humans, especially because they are obscured by much longer and faster but briefer changes caused by waves, tides, and episodic tsunamis and storm surges. Local changes of land level, however, often have been observed in association with uplift or downdrop along faults or folds that are at or near the shore. These movements may span several meters vertically in only a few seconds or minutes, and they can permanently submerge roads and buildings or strand ships and fishing facilities above the surface of the ocean. Slower, but still observable during a lifetime are broad downwarpings associated with removal of groundwater (as at Venice, Italy) and of oil (as at Long Beach, California, and other coastal oil fields) as well as subterranean movements of magma (as at Pozzuoli, Italy). Direct observation also cannot provide direct information on changes of sea level, either real or relative, that occurred before humans were present or before the invention of written records. However, many effects of these changes are recorded in rocks and sediments; study of outcrops and drill-hole samples and indirect geophysical measurements can yield information about dates and extents. Burial and/or erosion of strata have reduced access to some of the geological evidence, especially for the most ancient changes of sea level versus land level, but human ingenuity has found and will continue to find means for obtaining information about past events. Drive for this search is provided by natural human curiosity and striving for economic advantage in discovering new mineral resources such as petroleum and metals.

The many causes of changes in relative sea level (see Chapter 2) can be combined into four groups: tectonic, sedimentation, glacial, and oceanic. The tectonic group is characterized by greatest amplitude and longest time span, sediment deposition causes a slow rise of displaced water, the glacial group is fast but intermittent, and the oceanic group always is present but has low amplitude and shortest time span. As pointed out by Kuenen (1955), Higgins (1965), Donovan and Jones (1979), and many others, tectonism is most important in many ways even though less recognized than glacial causes of relative sea-level change.

A brief review of the general sources of information that can clarify changes of relative sea level and the history of their understanding precedes analysis of the results of study. These information sources (Fig. 30) are listed in descending order of their time spans and in more or less direct order of their precision in date and reliability of changes.

Rocks on Continents

Oldest rocks of the Earth are those on continents, because those on the floors of ancient oceans repeatedly have been lost by subduction beneath the continents. Oldest rocks cannot exceed the age of the Earth itself and of meteorites, about 4.6 b.y. B.P. (Patterson 1956; Badash 1989). Considerable time was required for a crust to accumulate and for water to be expelled from the core and mantle and condense to form an ocean in which marine sediments could be deposited. A reasonable date for the first ocean may be 4.0 b.y. B.P (Fig. 30), and the earliest life forms occurred at least by 3.8 b.y. B.P., according to indirect evidence (Cloud 1976). Marine stromatoliths as calcareous reefs were present by 3.1 b.y. B.P.—during the Archaean Era. With thicknesses of 2.5 to 6 m, their intertidal habitat indicates rather high tide ranges during the Proterozoic Era (Cloud 1968). Marine sediments of the Proterozoic are far more widespread and less metamorphosed than those of the Archaean, so their deposits on the then margins of continents provide more information for at least the larger and longer shifts of relative sea level. Progressively more information about these shifts comes from sedimentary rocks of Paleozoic, Mesozoic, and Cenozoic eras with their well-preserved remains of hard-shelled mollusks and other fossils.

Marine fossils in the strata of mountains had been noted for centuries, but they customarily were explained on the basis of the biblical Flood of 2368 B.C. (Genesis 7)—a tale much like its predecessor flood of about 4000 B.C. described in the Gilgamesh (Gardner and Maier 1984, p. 228-244). In fact, Leonardo da Vinci, who recognized fossil shells in uplifted strata of Italy, probably in the Alps (Mather and Mason 1939, p. 106), may have been confronted with the same argument by religious officials. The use of fossils as well as of compositions and textures of sedimentary rocks to distinguish between marine and non-marine strata has generated many

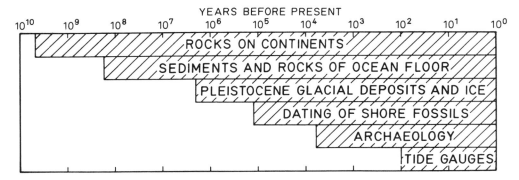

FIGURE 30. Diagrammatic listing of sources of data for estimating past changes of relative sea level; age span of each source is indicated.

estimates of the long-term position of relative sea level with respect to continents by workers such as H. Stille, C. Schuchert, H. Termier and G. Termier, and K. Strakhov. These estimates, compiled and compared by Wise (1974), show general submergence of continents during early Paleozoic and late Mesozoic times, but with many variations that indicate both short-term and local changes in land level versus sea level. Similar results were obtained by Ronov et al. (1980) and by Hallam (1977, 1984). Improved understanding of the shorter fluctuations of relative sea level necessarily is limited by the crudeness of the stratigraphic method and its inherent restrictions in numbers and accuracies of dates, as well as by uncertain vertical movement of strata on the continents after deposition. For example, marine shorelines of mid-Cretaceous in western United States have been uplifted about 2 km by epeirogenesis (Weimer 1983; Sahagian 1987). Similar uplift of Paleogene sediments (Cherven and Jacob 1985; Daly et al. 1985) is recorded. Different amounts of epeirogenic uplift on different continents indicate that a search for a global sea-level curve for the geological past cannot be successful (Chappell 1987).

Sediments and Rocks on the Ocean Floor

A source of more continuous data on relative sea-level changes is the ocean floor, where sediments have been deposited more or less continuously during most of the past 180 million years in an area about three times larger than that of the continents (Fig. 30). This approach was initiated by Wegener in his book *The Origin of Continents and Oceans*, first published in 1915 (Wegener 1929). This attempt failed until the theory was revised and transformed into plate tectonics much later (Dietz 1961; Hess 1962) after more had been learned about the ocean floor. The concept of plate tectonics, or sea-floor spreading, revitalized the field of geology because it provided a framework into which many previously unrelated facts of Earth composition and history could fit and new relationships could be predicted and tested. As a result, a new era of ocean-floor exploration began, using more intensive sampling, continuous deep-sea drilling, and extensive seismic-reflection and other geophysical surveys. This effort was preceded and accompanied by widespread interests in production of offshore petroleum and other mineral resources that were supported by increasingly more intense studies of distributions of foraminiferans and other microfossils, radiometric dating, geochemistry, composition, and processes of modern and ancient marine sediments. Results of these studies have yielded more information about changes of relative sea level during the past 180 million years than even could have been dreamed during Wegener's time. Many of the results are indicated on charts of seismic stratigraphy in which individual sequences are bounded by unconformities and illustrate the use of offlap, downlap, truncation, and down-slope shifts to infer relative changes of sea level along the margins of all continents (Vail et al. 1977c). The pattern is claimed to have had general global synchroneity through the Phanerozoic, but it still must have been affected locally by tectonic movements of the Earth's crust. Inclusion of additional seismic detail, correlation with drill samples, and corrected stratigraphic ages have added much knowledge for the Mesozoic and Cenozoic portions; nevertheless, many questions about geochronology and eustasy versus tectonism remain for future investigation.

Pleistocene Glacial Deposits and Ice

A special and highly important source of information about relative sea levels of the past is continental glaciers (Fig. 30), because large volumes of water become frozen in ice sheets and later are released during their melting. It is hard to realize that as recently as only 150 years ago geologists opposed Louis Agassiz's views (Mather and Mason 1939) of former larger than present glaciers in Switzerland as vigorously as they later opposed Wegener's proposal of continental drift. In fact, the incomplete information from continents later restricted ice sheets to only four advances during the Pleistocene Epoch (Nebraskan, Kansan, Illinoian, and Wisconsinan in North America and Günz, Mindel, Riss, and Würm in Europe). This view remained until Emiliani (1955) measured ^{18}O in planktonic foraminiferans sampled in cores from deep-water sites in the Caribbean Sea, followed by studies of cores from other oceans by other workers (Shackleton and Opdyke 1973, 1976; Pisias et al. 1984) and cores of calcite deposited by groundwaters of the Great Basin in North America (Winograd et al. 1988). Dates were determined and later made more reliable by radiocarbon and other isotopic ratios, magnetic-reversal intervals, and adjustment for the Milankovitch curves of solar insolation based on cyclical changes of the geometry of the Earth with respect to the sun. They also were confirmed by dating of Pleistocene sea levels inferred from raised coral reefs of New Guinea and Barbados (Bloom et al. 1974; Chappell 1974; and others). There were found to be 10 intervals of high concentration of ^{18}O relative to ^{16}O (indicating low air temperature, large glaciers, and low sea levels) during only the last third of the Pleistocene Epoch. Accordingly, there may have been as many as 30 glacial advances, although not all necessarily started from ice-free interglacial conditions. Another source of information about the number of glacial events is seismic stratigraphy and examination of drill cores on the continental shelf and slope. Using these methods, Armentrout (in press) has showed the presence of at least seven sets of facies that mark changes of relative sea level during the Pleistocene and Pliocene in the northern Gulf of Mexico off Texas and now at depths between the present ocean floor and 3500 m. Support is provided by similar studies off Louisiana by Pacht and Bowen (1990). Still more detailed information about minor glacial advances and retreats were provided by radiocarbon datings (between 13,000 and 8000 y. B.P. of uplifted moraines in Scandinavia (Mörner 1969, p. 181–183). Nineteen advances were identified, at an average interval of only 263 years—probably too local and too short to be reflected in sea-level changes. The number of Pleistocene glacial epochs and their geographical extents remain poorly known.

In addition to dates for sediments deposited directly by glaciers or by other agents during times of cold climate or low sea level, useful information about times of growth or melt of glaciers can be obtained from ice cores drilled through them. Cores from ice sheets of Greenland (Dansgaard et al. 1971) and Antarctica (Gow et al. 1973; Lorius et al. 1985, 1988) were analyzed for the heavy oxygen isotope ^{18}O, whose ratio with ^{16}O is low in rain and snow deposited during times of low air temperatures. The record is most detailed during the last 100,000 years; earlier ice is compacted and lost through lateral flow toward the margins of ice sheets. Detailed studies of compaction and ice-flow patterns at an Antarctic drill site and adjustment of chronology according to the Milankovitch curves of solar insolation (Martinson et al. 1987) offer a means for increasing knowledge of the number and ages of reliable inferred low temperatures and thus of low sea levels.

Shoreline Classification

Many generations of college students have been taught the genetic classification of shorelines that was devised by Davis (1896, 1909) and Gulliver (1899) and made more detailed and better illustrated by Johnson (1919). This classification was copied in innumerable textbooks and was considered such an advance over descriptive approaches (Salisbury and Atwood 1908) that it became widely and uncritically accepted. In effect, it has four main classes, all based on the physiography of the coastal belt: 1). emergent shorelines (gently sloping topography, development of offshore barrier islands and lagoons, filling of lagoons by marsh deposits, landward migration of the bar atop the marsh, and later the cutting of seacliffs), 2). submergent shorelines (drowned river and glacial valleys exhibiting progressive closure by baymouth bars, filling by marsh deposits, landward erosion of intervening headlands, and finally erosion of the filled valleys themselves), 3). neutral shorelines (deltas, alluvial fans, outwash plains, volcanoes, coralgal reefs, and faults), and 4). compound shorelines (having characteristics of more than one of the above classes). The modifications of the original shoreline by marine erosion and deposition gave rise to the familiar subclasses of initial, youth, maturity, and old age. Later, Johnson (1925, p. 129–152) pointed out that complications ensue from the presence of evidence that indicates several successive times of emergence and submergence for shores in many or most regions.

Shepard (1937) noted that "all shorelines are potentially shorelines of submergence" because of the considerable but still incomplete return of sea level from its lowest positions during times of maximum continental glaciation. He, accordingly, developed a new classification based on the original (primary—nonmarine) shaping agent with subsequent modification (secondary) by marine agents of erosion and deposition. Both classifications were applied by Emery and Uchupi (1972, p. 16) to more than 400 topographic quadrangles along the United States's shores bordering the Atlantic and Gulf of Mexico. This compilation showed that according to the Johnson classification most shorelines south of New York are emergent with some compound and essentially no submergent ones, whereas according to the Shepard classification almost all of them are secondary (marine deposition). Most eastern Canada shorelines would have been classified submergent by Johnson and primary erosional by Shepard. In reality, nearly all of the southern ones are submergent because of returned glacial meltwater, whereas those of eastern Canada would have been emergent because of crustal rebound after removal of the weight of glaciers—quite the opposite of the movement inferred from the Johnson classification of shorelines.

Even more complication arises through use of radiocarbon dating of late Quaternary and Holocene shore deposits (mentioned in next paragraph and discussed in later chapters), and especially through the use of tide-gauge records. The main conclusion to be reached is that the physiography of the shoreline, or coastal, region indeed provides evidence of previous higher and/or lower than present stands of sea level, but that the dates of those stands rarely are included within the physiographic evidence. For example, the present general low-latitude sinking of the land (rising of relative sea level) is leaving too faint a mark on most shorelines in comparison with the larger marks left by non-marine erosion during times of glacially lowered sea level or with even more abundant raised marine terraces as evidence of land uplift caused by tectonic movements of the land (Cronin 1981). Clearly, identification of changed sea (land) levels through classification of shorelines requires correct identification of dates as well as of physiographic features. Thus, recourse must be made to tectonic history, radiocarbon dating, and tide-gauge records or their equivalents as well as to maps of topography.

Dating of Shore Fossils

Changing sea levels of latest Pleistocene and Holocene have been studied widely by radiocarbon dating of remains of marine animals and plants that lived in sediments of the intertidal belt that subsequently became submerged or emerged mainly by tectonic movements of the land or by real changes of sea level. Among the first such age measurements are by Kulp et al. (1951) and de Vries and Barendsen (1954), and their application by Shepard and Suess (1956) to rising sea level by melting of glaciers. Subsequently, at least a thousand such samples have been dated (Fig. 30), of which 325 for the margins of the Atlantic Ocean were plotted and discussed by Emery and Uchupi (1984, p. 62–67). Differences in direction and rate of relative sea-level change from place to place are ascribed mainly to local vertical tectonic movements. These movements also produced a wide scatter of age-depth data points that increases with age before present. Nevertheless, the mean curve through the data points and the envelope enclosing most of them indicate lowest sea levels of late Pleistocene about 15,000 y. B.P. Earlier dates imply shallower depths as though for an interglacial time about 25,000 y. B.P. in contrast with 75,000 to 125,000 y. B.P. for the Sangamon (Riss-Würm) interglacial event. This difference may be real or simply reflect minor contamination of older samples by later sediments or by weathering, and it indicates the need for datings other than by radiocarbon.

Archaeology

An early stage of investigating Holocene changes of relative sea level is being followed by a few archaeologists and geologists (Fig. 30). Most such work merely indicates the presence of submerged sites formerly occupied by humans (Masters and Flemming 1983). Little regard was given to dating or relationship to adjoining regions—typical for early studies in any field. In contrast, the shores and shallow waters of the Mediterranean Sea are relatively well explored, and knowledge of their archaeology and history is advanced. Using modern diving methods, Flemming and Webb (1986) investigated many archaeological sites (harbors, piscinas, shipways, roads, and buildings) throughout the Mediterranean Sea beginning in 1958. Some (156) sequences show land uplifted (to +8.5 m), 46 sequences identify land submerged (to −11 m), and 204 sequences are at about at the same levels as when the structures were built. Nevertheless, those authors concluded that sea level rose eustatically from −1.1 m 5000 y. B.P. to 0 m 2000 y. B.P. (a mean rate of 0.4 mm/y.).

Perhaps the earliest well documented geo-archaeological study was Babbage's (1847) report on the Roman Temple (or marketplace) of Serapis at Pozzuoli, 7.5 km west of Naples, Italy, and its relationship to nearby coastal sites. The temple was built on shore but sank an unknown amount so that the sea surface reached 4.9 m above the floor of the temple, as indicated by marine mollusk borings in three remaining marble columns. Subsequently the site rose again, bringing the top of the bored part of the columns 5.6 m above sea level. Parts of other nearby temples and two roads still were submerged during Babbage's visit during June 1828, but a stone pier 800 m distant at Pozzuoli had been elevated so that the borings reached about 1.2 m above sea level. Thus, the change clearly was caused by local deformation of the land, not surprising in view of the many volcanoes, active hot springs, and earthquakes in this area near the center of the Phlegrean Fields (Lirer et al. 1987). Further attesting to Charles Babbage's analytical power were his abilities as a cryptographer and as the designer of a powerful computer (Difference Engine) that remained unbuilt 150 years later (AGU Committee on the History of Geophysics 1988; Corcoran 1988). Similar histories of complex vertical tectonic movements are inferred from archaeological sites along the coasts of Israel and Sinai (Neev et al. 1987). Fourteen sites or groups of sites had 4 to 20 indicators of former positions of relative sea level, with inferred movements at different sites indicating different extents and even different directions of vertical motion during many intervals of the past 15,000 years and especially during the past 6000 years when most of the structures were built. The technique was to use archaeological evidence for most dates (mainly pottery and other artifacts, construction methods, occasional contemporary inscriptions, plus radiocarbon) and to use sediment evidence (species of mollusks plus radiocarbon) for positions relative to sea levels and dates. For example, at Akhziv Harbor, a small Phoenician tomb dating from 2700 y. B.P. is at 5.5 m above mean sea level. In itself, this tells nothing about changes of relative sea level, but it was built atop swamp deposits (meaning that the site had been raised above sea level before the tomb was built). Atop the tomb was beach sand containing marine shells (indicating submergence after construction), but its present position well above sea level denotes a subsequent emergence.

In addition to their utility in identifying changes of relative land level (Flemming 1968; Masters and Flemming 1983; Neev et al. 1987), archaeological sites often provide dates at which coastal faults or volcanoes were active and at least capable of displacing shorelines locally. A few examples are provided by studies of deformed graves at Kyparissi and Lamia in Greece (Stiros 1988), collapse of walls at Troy

(Wood 1985, p. 230–241), destruction at Thera/Santorini (Mavor 1969) and Knossos. Many other examples of this kind of indirect information from faults and volcanoes are included in other archaeological studies that mainly provide evidence of submergence or emergence of sites.

Tide Gauges

Finally, we come to tide-gauge records (Fig. 30). There must have been some continuous records of tide levels from calibrated staffs or scales cut in rock wharves of ancient harbors but none are known to us. Records of water levels associated with tsunamis or other violent damaging storms probably were inscribed on building walls or stones then as now, but the earliest one known to us was cut in 1704 at Stockholm (Mörner 1979). Probably more common are inscribed records of river floods; for example, annual floods of the Nile River are indicated on some Roman colonial coins. A bronze drachm of Antoninus Pius, 138–161 A.D., contains a Nilometer reading of 16, a favorable augur for abundant crops and prosperity (see also Strabo, *in* Hamilton and Falconer 1854, v. 3, p. 222). As changes in mean annual relative sea level are based on yearly averages of sea level at one-hour intervals, we cannot expect useful data from tide staffs because of the great time and effort that would have been required for making hourly readings day and night for years. When automatically recording instruments were developed, data for mean sea level computations became available, but few such records exist for earlier than about 1870. Thus, tide-gauge information about either eustatic or tectonic changes of relative sea level are mainly for the 20th century, but they provide the most precise information for our use.

Phanerozoic

General

The most orderly way to discuss changes in relative sea level during the past is to begin with the earliest events (when least is known about dates and details) and proceed to the latest events (when most is known per unit time; Fig. 31). However, much information for the oldest and longest time span (post-Archaean) is extrapolated from relationships to events within the Phanerozoic (Cambrian to present). The latter is better known because much of the record for it comes from sediments and rocks of the ocean floor, whereas the record before that time must depend on evidence available essentially only in sediments and rocks of the continents, because the early contemporaneous ocean floor was later subducted and destroyed. Thus, we must understand the causes of changing sea levels during the Phanerozoic before considering the less clear evidence for earlier sea levels. An alternative would be to begin with latest (modern) sea levels and then work backward through those of the Pleistocene Epoch and the Phanerozoic to late Archaean. We note, though, that Pleistocene and early Holocene sea levels can be well correlated with glacial and interglacial climates, in contrast with domination of tectonism for more ancient sea levels except during episodes of ancient widespread glaciation. Moreover, the next stage of knowledge about sea levels (for the past century) is based on tide-gauge records for which data are so abundant that most of the rest of this book is devoted to this time span, during which we believe that tectonism again dominates.

The opening of the present Atlantic Ocean and the closing of the previous one was described by Wilson (1966), giving rise to the term "Wilson cycle", which implies repeated opening and closing of the ocean in response to alternating divergent and convergent plate movements. Earliest knowledge about the opening of the present Atlantic Ocean was based on outlines of opposing shorelines, shapes of opposing 500-fathom contours, positions of spreading belts (especially the Mid-Atlantic Ridge), hot-spot trails, shapes and dates of magnetic-reversal patterns, and movements of continental crustal rocks relative to magnetic pole positions derived from remnant magnetism of rocks. Later came information from sediment thicknesses and ages, coupled with radiometric ages of underlying oceanic crustal basalts. These data mostly from the ocean floor were confirmed by both older and newer information about the dates and direction of forces that caused folding and uplift of mountain ranges along the margins of continents (plate convergence) and by listric faulting and stretching of continental crust along the margins of the continents (plate divergence).

Seismic Stratigraphy

Still another powerful source of information about changes of ocean basins through inferences derived from relative sea-level changes came from seismic (sequence) stratigraphy, summarized by Vail et al. (1977b). The basis for this method is that global unconformities caused by global changes in sea level can be recognized in the stratigraphic record. Although ignoring tectonic influences (as pointed out by Bally 1981; Hallam 1984; Watts and Thorne 1984; Miall 1986; Summerhayes 1986), the method focuses on stratigraphic relationships that indicate regressions and transgressions and it uses available stratigraphic data to estimate time relationships and thence to reconstruct global sea-level curves. Most dating is from biostratigraphy of offshore wells (Loutit and Kennett 1981; Poag and Schlee 1984; Thorne and Watts 1984) and from the deep-sea drilling project (van Hinte et al. 1985).

Seismic stratigraphic methods rely on reflections of sound from physical surfaces within sediments, such as bedding, stratal surfaces, or unconformities. Where contrasts in acoustical impedance across an unconformity are insufficient to generate a reflection, these surfaces can be mapped using discordance between dips of overlying and underlying reflectors. Primary unconformities used in the analysis include coastal onlap (progressive landward encroachment of deposits of a

58   3. ANCIENT TO MODERN CHANGES IN RELATIVE SEA LEVELS

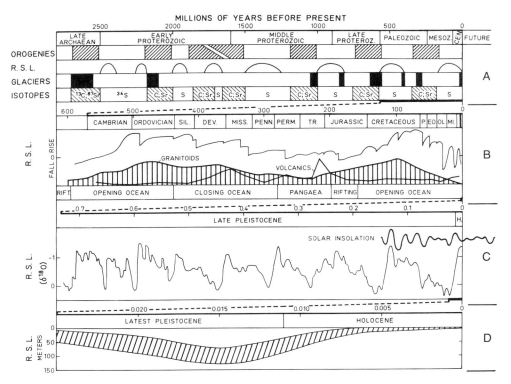

FIGURE 31. Relative sea levels and related phenomena during the past.
A. Longest term (late Archaean to present) changes of relative sea level inferred from times of continental growth (orogeny caused by plate convergence) and times of continental fragmentation (plate divergence) associated with general times of glaciation and with inferred concentrations of the stable isotopes of carbon, strontium, and sulfur in marine sediments. Based largely on data from Nance et al. (1986).
B. Changes in relative sea level (first- and second-order) during the Phanerozoic based on seismic stratigraphy discussed by Vail et al. (1977b) with respect to emplacement of igneous rocks (granitoid plutons–Fischer 1984; volcanic types–Ronov et al. 1980) and to growth and fragmentation of continents (Fischer 1984; Worsley et al. 1984).
C. Composite of changes in relative sea level during the last third of the Pleistocene glacial epoch derived from concentrations of the stable isotope of oxygen in foraminiferal tests of marine sediments (from Emiliani 1978). The curve of relative sea levels inferred from Milankovitch cycles of solar insolation for Latitude 45°N (Broecker and van Donk 1970) supplies a good comparison with the past 150,000 years and provides a projection into the future.
D. Changes of relative sea level during the past 25,000 years based on radiocarbon ages and present water depths of fossils of intertidal animals and plants. The envelope encloses 96% of 307 data points for continental shelves of the Atlantic Ocean (Emery and Uchupi 1984, p. 66); its width is a function of episodes of variable change of real sea level, regional and local tectonism, post-depositional movement of fossils, and errors of age determination.

The presentation in the figure shows more compressed time scales and lesser detail for older time spans, in accordance with more limited knowledge of ancient than of later relative sea levels. If all were at the same time scale as that for the past 25,000 years, the graph would cross 112,000 pages.

given facies) as well as downlap, toplap, and truncation. In early work, Vail and associates assigned these chronostratigraphic indices directly to changes of relative sea levels; now they distinguish between changes of onlap and relative sea levels (to reflect facies change from alluvial to coastal-plain sediments). Onlap curves are used with other data to derive sea-level curves. For instance, the Paleozoic portion of the curve (Fig. 31B) is based on the North American cratonic sequences of Sloss (1963) and on data other than marine sediments (because there are no oceanic sediments of that age in present ocean basins). Vail et al. (1977b) designated long-term sea-level cycles as having a duration of 200 to 400 million years, second-order cycles as 10 to 80 million years, and third-order cycles as 1 to 10 million years (Fig. 31B). Glacial lowerings of sea level during the Pleistocene Epoch are so brief (fourth-order) that they cannot be shown on this sea-level curve. Other brief lowerings of the order of 10 m may be associated with spilling of ocean water into marginal depressions formed during early stages of sea-floor spreading and accompanied by massive deposition of salt (Burke and Sengör 1988) and perhaps spilling into other kinds of coastal depressions (Cercone 1988).

Support for the various Vail curves has been mixed (Hallam 1978; Loutit and Kennett 1981; Tucholke 1981; Hazel et al.

1984; Poag and Schlee 1984; Thorne and Watts 1984; van Hinte et al. 1985; Miall 1986). In general, observations support the concept of most supercycles, but often more unconformities are predicted than are found, some major unconformities are not covered by the model, and some series show an opposite stratigraphic relationship to relative sea levels. Some discrepancies can be attributed to poor resolution of some seismic records (spans of several million years), poor biostratigraphic resolution in some areas, and the spatial averaging procedure used to generate the global curve of Vail and coworkers. Part of the discrepancy is certainly attributable to the influence of tectonics on relative sea levels. Bally (1981), for instance, believed that major plate reorganizations exert the major control on stratigraphical sequences. Hallam (1984) thought that changes in ocean basin volume are more important than glaciation and deglaciation in controlling sea levels prior to the Quaternary. Summerhayes (1986) asserted that plate tectonics control these stratigraphical relationships.

The sea-level curves of Vail et al. (1977b, p. 84, 85) used in Figure 31B reveal one first-order high level back to mid-Triassic about 218 m.y. B.P., with eight second-order peaks. At least 10 second-order peaks of sea level are shown for the time interval between mid-Triassic and Precambrian. Third-order peaks number 31 back to 200 m.y. B.P., omitting others for the Pleistocene Epoch and presenting only vague peaks for most of the Cretaceous Period, because data for that time span had not been released by Exxon Production Research. Since publication of the method and early results in 1977, additional work by Vail and his associates and by others has modified it and provided more detail. A curve prepared by Hallam (1984) from somewhat later stratigraphic information is generally similar and suggests that sea level during Late Ordovician was 600 m higher than at present and during Late Cretaceous, 350 m higher than now. Latest is a publication by Haq et al. (1987), whose diagram for 256 m.y. B.P. (in Late Permian) to Holocene contains 119 third-order sea-level peaks. Many of these peaks are irregular (Fig. 32) as though suggesting that each eventually may be resolved into several peaks. Complications arise with respect to eustatic versus tectonic causes of sequences (Christie-Blick et al. 1988b) and especially with time scale, because of the use of different materials for radiometric dates and the need to correlate faunal zones using measured magnetic reversals (Gradstein et al. 1988).

In summary, stratigraphic relationships clearly record changes in global sea levels, although their interpretation is made difficult by erosional episodes and facies changes. However, local stratigraphic relationships also record land movements on a regional basis; averaging of stratigraphic data around the globe to remove these regional biases is certain to leave a residual bias on proposed sea-level curves. Detailed examination of these changes may allow separation of global versus regional episodes of relative sea-level change, permitting greater resolution of sealevel curves.

Volumes of Igneous Rocks

Accompanying the two times of first-order high sea levels during the Phanerozoic were massive emplacements of igneous rocks (Fig. 31B). Most pertinent would be estimates for volumes of oceanic basalts, but subduction associated with sea-floor spreading has removed essentially all such basalts from floors of oceans prior to the present ones, and andesites on continents have been lost by erosion of mountains. Nevertheless, some evidence remains in the form of granitoid plutons that once formed the cores of the mountains. Their peak volumes at about 470, 370, and 100 m.y. B.P for the Phanerozoic of North America, estimated by A. E. J. Engel and C. G. Engel in 1964 and adapted by Fischer (1984), correspond closely with the times of first-order high sea levels (Fig. 31B). Estimates by Ronov et al. (1980) of the volumes of volcanic rocks emplaced on all continents except Antarctica during the Phanerozoic indicate peaks in volcanism at about 352, 215, and 4 m.y. B.P. The differences between dates for peak emplacement of granitoids and volcanics support expectations that the former correspond roughly with times of mountain building associated with plate convergence, and the latter with times of rifting related to plate divergence. Obviously, estimates for volumes of granitoid and volcanic rocks are to be considered imprecise, especially for the volcanics, because their estimates do not include the vast quantities emplaced on the ocean floor during times of widening oceans.

Sequence of Plate Movements

The information in Figures 31 and 32 is supplemented by a vast array of data from marine geophysical surveys (of bathymetry, seismic reflection and refraction, seismicity, magnetic-reversal patterns, gravity, and heat flow), drill-hole samples (yielding information about ages and depths of sedimentary rocks, ages and kinds of igneous rocks, and distribution patterns and evolution of planktonic and benthonic flora and fauna), geology on land (distribution and ages of mountain-building orogenies, dates and distribution patterns of volcanic rocks associated with continental rifting), and other information that can be coordinated within the framework of plate tectonics. This supplementary information is a basis for inferring causes of the sea-level changes that in turn can serve for estimating times of sea-level changes whose stratigraphic record has been lost by subduction, metamorphism, or other ancient geological processes. This information is too extensive to review here, and it was uncovered, reported, disputed, and confirmed by many workers especially during the past three decades. Much of it was synthesized by Emery and Uchupi (1984). This information can clarify more about the causes for relative sea-level fluctuations of the eustatic kind (fluctuations that depend mainly on changes in volume of ocean basins that

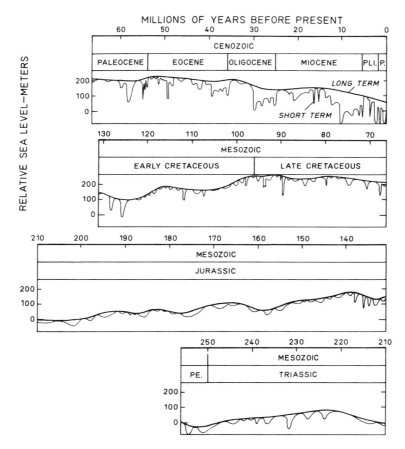

FIGURE 32. Changes of "eustatic" sea level during the past 256 m.y. B.P. (since within Late Permian) according to sequence stratigraphy from seismic profiles, wells, and outcrops, according to Haq et al. (1987). The indicated sea levels range from about 75 m below the present level to 255 m above it, but in our opinion the lowest levels must be about 150 m below the present one, at least during some of the Pleistocene glacial events. Note also the decreased detail in the older parts of the record, which probably reflects lesser knowledge of older strata.

result from plate movements associated with sea-floor spreading). Other eustatic changes of sea level that were caused by changes in volume of water are important only during times of glaciation; best known of these changes are the ones during the Pleistocene Epoch, but those will be discussed separately.

The present pattern of plate movements increases the areas of the Atlantic and Indian oceans at the expense of the Pacific Ocean. This trend is well shown by the magnetic-reversal patterns that denote new ocean floor now being formed along the Mid-Atlantic Ridge with progressively older oceanic crust near the continents that border the Atlantic. In general, the oldest oceanic crust in the North Atlantic (about 180 m.y. B.P.) borders both North America and northwestern Africa, and the oldest in the South Atlantic (about 110 m.y. B.P.) borders both South America and southwestern Africa. Complications ensue from later secondary spreading belts that separated North America from Greenland, Greenland from Iceland, Iceland from Rockall Bank, and Rockall Bank from Europe. Other secondary spreading belts are in the Caribbean Sea and in the southern South Atlantic Ocean. Even some subduction occurs where the Atlantic plate underthrusts the Caribbean plate and the Scotian plate (near the Falkland Islands). Much the same pattern exists in the Indian Ocean, the Arctic Sea, and the Mediterranean Sea. Sea-floor spreading also occurs in the eastern Pacific Ocean, but basically the Pacific is surrounded by trenches into which oceanic crust moves en route to underthrusting the continents along the Pacific perimeter where it melts to form a more silicic magma that rises to form volcanic coastal mountains such as the Andes. Thus, at present the Pacific Ocean is being made smaller by subduction, whereas other oceans and seas beyond the Pacific are being made wider by sea-floor spreading that produces an average widening of about 2 cm/y.

The first new oceanic crust for the present North Atlantic Ocean was emplaced about 180 m.y. B.P., separating North

America from northwestern Africa, but rifting of the continent had begun there about 230 m.y. B.P. Separation of North America from Europe was later—109 m.y. B.P., long after the earliest rifting there about 145 m.y. B.P. The first new oceanic crust in the present South Atlantic was emplaced about 119 m.y. B.P., long after the earliest rifting about 200 m.y. B.P. Further complications are associated with secondary spreading belts where emplacement began later: Gulf of Mexico—150 m.y. B.P.; Saō Paulo/Angola—109 m.y. B.P.; Africa/Australia—80 m.y. B.P.; Greenland/Rockall—56 m.y. B.P.; South America/Antarctica—35+ m.y. B.P.; Drake Passage—27 m.y. B.P.; Cayman—20 m.y. B.P.; East Scotia—7+ m.y. B.P. The rifting that led to formation of the Atlantic Ocean occurred within the supercontinent of Pangaea, which had been produced by closing of a previous Atlantic Ocean that existed during the Paleozoic Era until the Pennsylvanian Period about 300 m.y. B.P. The collision of northwestern South America and then northwestern Africa (part of Gondwana) against eastern North America (part of Laurasia) produced the Appalachian and other late Paleozoic mountain ranges along the collision margins of these continents (Irving 1979; Ziegler 1981).

Shifting positions of continents and oceans involves far more than changing the outlines of the oceans. The average depths also are changed (Sclater et al. 1971, 1977). When continents collide, there is a thickening of continental crust beneath the resulting mountain ranges and a corresponding decrease in total area of the colliding continents. Once a new supercontinent (Gondwana, Laurasia, or Pangaea) has formed, the continental cap with its lower heat flow than that of the ocean floor (Anderson 1982) causes the continental crust and the underlying mantle to increase in temperature, expand in volume, and rise higher above the adjacent ocean floor, thereby changing the shape of the geoid (Chase 1979; Crough and Jurdy 1980). This increase in height of continent increases the volume of the ocean basin and produces a fall in sea level. Moreover, the insulating effect of the continent (low heat conductivity) causes a doming of the underlying mantle, so that the continental crust tends to break into segments that slide radially down the side slopes of the mantle dome. Thus, rifting and crustal stretching are initiated, to be followed by emplacement of oceanic crust between the fragments of the supercontinent. The fragments of continental crust, originally hot, soon begin to cool, contract, and sink (Pitman 1978), displacing water of the ocean and adding to the rise of sea level. Finally, the spreading belts that formed between the pieces of supercontinent grow rapidly at first—vertically, laterally, and with length (Sclater et al. 1971), displacing still more ocean water and raising sea level. With the passage of time, the continental fragments cool, the sea-floor spreading slows, and the oceanic crust cools enough to sink by subduction beneath opposing continents so that sea level no longer rises but begins to fall again. After all of the oceanic crust has been subducted, the ocean is gone and the continents collide to initiate a new supercontinent and a new cycle of sea level.

The plate-movement sequence is illustrated by the lowest strip of Figure 31B. Opening of the Atlantic can perhaps be taken as starting at an average of about 160 m.y. B.P., following about 40 m.y. of rifting (from about 200 m.y. B.P.), following assembly of Pangaea about 80 m.y. previously (at about 280 m.y. B.P.), following closing of the previous Atlantic Ocean that began about 160 m.y. earlier (440 m.y. B.P.), following the opening of that ocean during about 160 m.y. (starting about about 600 m.y. B.P.), following rifting of a previous supercontinent. None of these time spans or dates can be considered precise, but their approximations serve as reasonable estimates based on inexactly known events, none of which were instantaneous but were spread over at least tens of millions of years for different parts of the oceans and continents, as suggested by Nance et al. (1986) and Worsley et al. (1984, 1986). The events appear to be controlled by the time required for heat to accumulate beneath an insulating cap on the mantle provided by assembly of a supercontinent (that produced a swelling of the underlying mantle down the slope of which large rifted fragments of the supercontinent slid) and the time required for oceanic crust to form between rifted pieces of continental crust and then to cool (leading to subduction of oceanic crust beneath continental crust and closing of an ocean to reform a supercontinent). The rates are not unique to the present Atlantic Ocean and its surrounding continents; thus, one can suppose that they have driven crustal movements during most of Earth history and caused first-order changes of sea level prior to the Phanerozoic where the changes are recorded in geological strata and other rocks of the continents.

Late Archaean to Present

Orogenies

The cycles of construction and rifting of supercontinents and of narrowing and widening of ocean basins that were illustrated during the Phanerozoic were determined by the time spans required for heating and cooling of the Earth's crust and mantle. These time spans are largely functions of heat supply and heat conductivity that, in turn, are controlled by the internal composition of the Earth. As the latter can have changed little during the past 3 b.y., one can suppose that the cycles of supercontinents and ocean basins have continued for several billion years. Existing evidence from Precambrian rocks supports the view of repeated Wilsonian cycles beginning at least as long ago as 2.5 b.y. B.P. (Condie 1982; Windley 1983). Much of the record for continental collision and continental rifting during Precambrian time has been lost by subduction beneath the continents and weathering and erosion atop them. Nevertheless, geological evidence does indicate extensive mountain building at distinct intervals during the past: 250, 650, 1100, 1600 to 1800,

2100, and 2600 m.y. B.P. (Condie 1982). These are known and named in the various shield areas of the world (for example, Appalachians of eastern North America, Pan-African of Africa, Grenville of Canada, Eastern Ghats of India, and Karelian of Scandinavia). Each orogeny appears to have been preceded by the onset of collision of previous continental fragments and followed by renewed rifting of supercontinents. According to Worsley et al. (1984) and Nance et al. (1986), an average timing may be about 80 m.y. for assembly followed by about 60 m.y. of stability and heating, followed by 40 m.y. for rifting before fragments of the then current supercontinent became separated. This total life of a supercontinent spans about 180 m.y., as shown by six supercontinental episodes between late Archaean and the present (Fig. 31A). After the separation of continental fragments, new oceanic crusts may widen the ocean floor by sea-floor spreading for about 160 m.y., followed by narrowing of the ocean through subduction for another 160 m.y. Thus, an entire cycle from continental collision to continental collision is about 500 m.y., roughly the average of the five complete cycles between the late Archaean and the late Paleozoic ([2700 − 350]/5 = 470 m.y.). The timing appears to have been complicated by a double collision during the late early Proterozoic (Fig. 31A); nevertheless, the periodicity appears robust, considering the present incomplete state of knowledge of ancient geological history, gained during only the past few decades. We can expect additional details to be filled during coming decades, for example by further application of sequence stratigraphy to Proterozoic rocks (Christie-Blick et al. 1988a). According to these average periodicities the present Atlantic Ocean is near its age limit for widening, and it may begin to close in a few million years, more or less.

Relative Sea Levels

When a supercontinent is being assembled, the continental crust is thickened by the folding of mountains and intrusion of batholiths along the lines of collision, the crust and underlying mantle are uplifted by heating, and the adjacent oceanic crust is old, cool, dense, and underthrusts continental crust. All of these factors enlarge the total ocean basin and produce a sea level that is lowest during the middle age of the supercontinent. When rifting starts, the continental crust is thinned by stretching and listric faulting, and the adjoining oceanic crust is new, hot, shallow, and broad. These rock processes displace ocean water and cause sea level to rise to its maximum level during the time of maximum area of the new Atlantic-type ocean and of minimum area of the remaining Pacific-type ocean. Afterward, the cool oceanic crust subsides and underthrusts the fragments of supercontinent that surround the Atlantic-type ocean, causing sea level gradually to fall. According to calculations by Worsley et al. (1984), sea level during a complete cycle may have ranged about 500 m in a glacier-free world (from above the shelfbreak to below it).

Loss of water to the land during times of extensive glaciations could have depressed sea level a further 200 m. Diagrammatic illustrations of the times of first-order high sea levels back to late Archaean are presented in Figure 31A; these are derived from geological mapping of the incomplete record of sedimentary rocks on continents guided by extrapolation into the past of conclusions derived from the far better known record during the Phanerozoic.

Tides

The presence of tides on the Earth and their control by the Moon and the Sun has long been inferred from ancient beach and mud-flat deposits and from growth rhythms in fossils that lived during the Phanerozoic. A recent study (Williams 1989) of siltstones and fine-grained sandstones deposited in an ebb-tidal environment of Australia extends tidal studies back into the Precambrian, about 650 m.y. B.P. Spacing of laminae record daily tidal cycles and spring tides, from which the lunar month was judged to equal 30.5 ± 1.5 days, the year = 400 ± 20 days or 30.1 ± 0.5 lunar months, the perigean cycle = 9.7 ± 0.3 years, and the lunar cycle = 19.5 ± 0.5 years. These periods are 3 to 10 percent longer than those at present, a difference that is small considering the uncertainties and the antiquity of these rocks. Note also that we do not know the number of hours in the Precambrian day.

Continental Glaciation

Even the now-obsolete concept of four main advances of continental glaciers during the Pleistocene led to many different but mostly inadequate explanations for the cause of ice ages —some terrestrial and others extraterrestrial. The recognition about a century ago of older, even Precambrian, glacial deposits and erosional surfaces complicated still more the search for a suitable cause. However, Schermerhorn (1983) noted that the late Proterozoic glaciation occurred during a time when there was little volcanism but extensive deposition of dolomites and limestones. He attributed this glaciation to low concentrations of carbon dioxide in the atmosphere after loss to marine carbonate sediments—a sort of reverse greenhouse effect. Much other evidence compiled by Holland (1984) indicates that variations in concentrations of carbon dioxide and water vapor in the atmosphere have controlled the climate during the history of the Earth. Correspondence of many dates for continental glaciations (Harland, 1983) with dates for supercontinents (Condie 1982) led Worsley et al. (1984) and Nance et al. (1986) further to the idea that high elevation of supercontinents (low sea levels) caused rapid weathering and erosion of their rocks. Transfer of calcium and magnesium silicates to the ocean by streams could have led to combination with carbon dioxide and precipitation as carbonate sediments. Similarly, the transfer of nutrients from land to ocean would have increased organic productivity and deposition of carbon in marine sediments. Thus, carbon dioxide would have been transferred from the atmosphere to the ocean floor, allowing more than usual longwave thermal radiation to escape from the Earth's atmosphere. Glaciation is not known on the supercontinents of the late early Proterozoic, but perhaps supercontinents did not happen to occupy a polar region at that time. Also glaciations occurred in the absence of supercontinents during early late Proterozoic, Silurian, and Plio-Pleistocene, perhaps because of long-term polar positions of continents (as Antarctica now), local high elevation, or extremely low carbon dioxide concentrations in the atmosphere. Despite these exceptions, the concept of glaciation associated with supercontinents or even just large continents provides a useful framework for investigation of glacial chronology. Clearly, the presently available evidence for Precambrian glaciations is far more vague than for the Pleistocene, and it probably precludes recognition and dating of multiple glacial advances of short duration.

Isotopes in Marine Sediments

When supercontinents are high and sea level low, the high organic productivity in the ocean selectively concentrates the light isotope, ^{12}C, in preference to ^{13}C. The remains of these organisms (organic matter and skeletal carbonate) cause the sediments of that part of the tectonic cycle of continents and oceans to have relatively light carbon, in contrast to sediments of the part of the cycle that has high sea levels (as shown by Nance et al. [1986] for the Phanerozoic). Similarly, the rapid weathering and erosion of high supercontinents should tend to produce a high flux of the heavy strontium isotope, ^{87}Sr, because of its greater concentration in continental crust than in mantle-derived oceanic crust. This heavy strontium becomes deposited in marine carbonates. In contrast, the heavy isotope of sulfur, ^{34}S, tends to be precipitated preferentially in evaporites, causing the sediments of rift valleys (that form in early stages of supercontinent rifting—during times of relatively low sea levels) to contain relatively heavy sulfur. At the same time, heavy sulfur is being refluxed from the connecting open ocean so that sediments there during times of low sea level contain relatively light sulfur in contrast to their intermediately heavy sulfur deposited during times of high sea level—when the rift valleys have been flooded. Measurements of these isotopes of carbon, strontium, and sulfur in Phanerozoic marine sediments and of strontium in late Proterozoic rocks have been summarized by Worsley et al. (1986), but they remain to be investigated in still older rocks. Figure 31A indicates how their maximum concentrations may be expected to vary during the more ancient cycles of continent-ocean tectonism. Note that Hallam (1984) developed alternate explanations for variations of the carbon and sulfur isotopes, but general abundances of them still may be related to sea levels. Some caution for inferring details of sea-level changes from isotopes is sounded by Chappell (1987) and others in view of the influence of rate changes of sea-floor spreading and variations in climate (mainly rainfall) on isotope compositions.

Evolution of Life

Life continued to evolve during the same time spans that continents and ocean basins progressed through their tectonic cycles. In fact, some of the most striking evolutionary steps are more or less contemporaneous with times of supercontinental rifting and flooding by rising sea levels, as though geological changes produced the incentive for organisms to evolve or perish. Some of the major evolutionary steps are indicated here; more complete discussions are provided by Worsley et al. (1984, 1986). The earliest atmosphere of the Earth appears to have been anaerobic, allowing rapid weathering of iron minerals in igneous and metamorphic rocks followed by deposition (possibly aided by bacteria) of banded ferrous-iron sediments during the Archaean (Cloud 1976). Stromatolites (cyanophytes) also appeared about 3.0 b.y. B.P., making calcareous structures akin to those of modern algal mats. By 2.5 b.y. B.P., oxygen-producing organisms had evolved and become common, so that by 2.0 b.y. B.P. the atmosphere contained enough oxygen that ferrous-iron deposition gave way to ferric red-bed sediments. Also blue-green algae (heterocysts) enabled the fixation of nitrogen even in the presence of otherwise poisonous oxygen. About 2.0 b.y. B.P. eucaryotes (oxygen obligates) began, and they became abundant by 1.5 b.y. B.P. (Schopf and Oehler 1976). Primitive metazoa were present after about 1.0 b.y. B.P., as shown by burrows in sediments; they diversified about 0.7 b.y. B.P.—the Ediacarian Period of the late Proterozoic (Cloud and Glaessner 1982). They, in turn, were replaced about 0.6 b.y. B.P. by shelled mollusks having better circulation systems, and they evolved rapidly during the early Paleozoic. The first land plants appeared about 0.4 b.y. B.P., greatly increasing photosynthesis on the Earth and providing a food base for land animals that evolved from amphibians through reptiles to mammals (about 0.2 b.y. B.P.). The mammals had a special role as endotherms in their ability to withstand low air temperatures, and thus they were able to adapt to life on most of the Earth's surface.

The broad change from an anaerobic atmosphere to an aerobic one was produced through photosynthesis by organisms that could use solar energy to combine oxygen, carbon, and hydrogen into organic matter—a more efficient process than organic production by anaerobic methods. The progression of efficiency of organisms, however, was not a continuous and steady one, but was intermittent, as though development of new kinds of organisms had to be promoted by ending of comfortable old environments and initiation of new ones. Most drastic must have been those changes associated with rifting and breakup of supercontinents at about 0.5-b.y. intervals induced by the completely independent rates of heating and cooling of the Earth's mantle and crust. The rifting produced smaller continents that occupied a range of climates in each of which evolution could proceed independently without interference, because the continents were isolated from each other. Thus, the organisms that lived in shallow waters around continents must have evolved at rates different from those in the open ocean because straits, seas, and other avenues for migration developed during times of rifting. During all of Precambrian time, the land was barren, with neither plants nor animals living on it. A great change in habitats occurred during the Phanerozoic, when many plants and animals moved to land and were restricted thereto, whereas those remaining in the ocean were free to move within their temperature and other limits; the former evolved much more rapidly than the latter. Realization of the role of plate tectonics in evolution and extinction was identified by paleontologists (Hays and Pitman 1973; Plumstead 1973; Schopf 1974; and many others) soon after the concept of plate tectonics was initiated by geologists and geophysicists, and the literature on the relation of evolution to plate tectonics has become extensive. Understanding still is proceeding rapidly, with great promise for the future.

Late Pleistocene

Oxygen Isotopes

The heavy stable isotope of oxygen, ^{18}O, tends to become concentrated in seawater when evaporation occurs because the more abundant light isotope, ^{16}O, is more readily removed by evaporation. For this reason, during times of glaciation the ocean has a higher $\delta^{18}O$ (ratio of ^{18}O to ^{16}O) and glacial ice has a lower $\delta^{18}O$ than does the ocean during non-glacial times. In addition, a much larger effect on relative concentrations is produced by the ambient temperature: the higher the temperature, the lower the concentration of ^{18}O in foraminiferal tests and other calcareous skeletal structures of other organisms. Measurements of $\delta^{18}O$ in pelagic foraminiferans in samples from deep-ocean cores (where the rate of sediment deposition is slow and reasonably uniform) exhibit variations with depth that Emiliani (1955) attributed to alternations between glacial and interglacial stages. Similar results obtained by other investigators in other oceans showed the isotope variations to be worldwide and contemporaneous. Improvements in chronology using radiometric dates, Milankovitch solar insolation curves (Broecker and van Donk 1970; Hays et al. 1976), magnetic-reversal measurements (Shackleton and Opdyke 1973, 1976), and spectral analysis (Pisias 1976) have occurred. The curve for the past 730,000 years from Emiliani (1978) reveals 10 epochs of low water temperature (Fig. 31C) that may correspond with glacial advances. Measurements by Shackleton and Opdyke (1973, 1976) on cores from the Pacific Ocean dated by paleomagnetic reversals reveal oxygen-isotope variations back to 2.1

and >3.3 m.y. B.P. The earliest record indicates little variation of climate prior to about 3.0 m.y. B.P., with large and frequent fluctuations between 3.0 and 1.0 m.y. B.P., implying glaciations at least as early as 3.0 m.y. B.P., as summarized by Berger (1982) and Ruddiman and Raymo (1988).

Efforts are being made to extend the use of oxygen isotopes farther into the past to learn about more ancient climates and their possible control of sea levels. Recent examples are those by Moore et al. (1982), Savin (1982), and Moore and Kent (1987) for the entire Cenozoic Era. Results from Deep Sea Drilling Project (DSDP) cores generally show the possible onset of continental glaciations (and low sea levels) beginning in middle Miocene and continuing through the Pleistocene with earlier glaciations during the Oligocene. These dates and others are confirmed by continuous seismic profiles that show the presence of submarine canyons cut in continental slopes during times of low sea levels (Emery and Uchupi 1984, p. 408, 470, 509, 511, 587).

Measurements of $\delta^{18}O$ in glacial ice reveal a trend that is broadly the reverse of that in deep-ocean sediments, with heavier oxygen characterizing interglacial and lighter oxygen for glacial times. Only in part is this the effect of selective loss of the light isotope during evaporation from the ocean and its concentration in precipitation and thus in ice. Most of the preference in ice results from a further concentration of ^{16}O at lowest air temperatures. Comparison of the oxygen isotope distribution with depth in a detailed ice core from Vostok Station in East Antarctica (Martinson et al. 1987; Lorius et al. 1988) shows a close inverse relationship (lighter at low temperatures) to the depth distribution in deep-ocean sediments especially for the period since 160,000 y. B.P. The oxygen isotopes in ice at greater depths, older than 160,000 y. B.P., is far less closely related to the trend in ocean sediments because of flow and thinning of ice at depth. A depth frequency of deuterium (δ^2H) occurs with lighter hydrogen occurring at low temperature (Lorius et al. 1988). Moreover, measurements of carbon dioxide in the ice core show concentrations about 50 percent higher than concentrations during the Holocene and the previous interglacial stage than during the glacial stage between about 115,000 and 15,000 y. B.P., adding to the evidence that glaciation and low sea levels may have been caused by a reverse greenhouse effect. Finally, the temperatures of ice accumulation range from only $-10°C$ during glaciation to $+2°C$ during both interglacial and Holocene times, according to Lorius et al. (1988).

Solar Insolation

Many workers have investigated and used theoretical variations in the amount of solar energy reaching the Earth in order to determine the dating of widespread glaciation and lowerings of sea level. These variations are caused by precession of the Earth's axis of rotation—23,000 years, axial tilt with respect to the ecliptic—41,000 years, eccentricity of Earth's orbit—100,000 years, and latitudes of positions on the Earth's

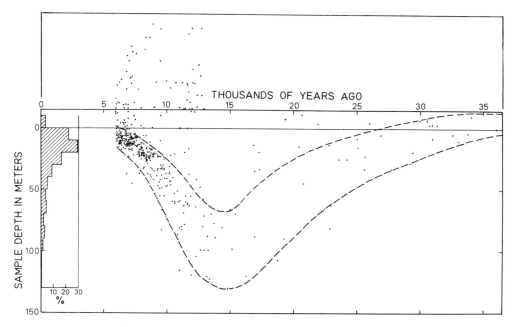

FIGURE 33. Sea levels based on radiocarbon ages and present water depths of calcareous skeletal structures or organic carbon in 325 former intertidal materials. Modified from Emery and Uchupi (1984, figs. 28, 29) plus additional ages above sea level from Bloom (1977).

surface. The resultant curves of insolation with time are credited to the originator of the concept, Milutein Milankovitch. Illustrations of its application to measurements of oxygen isotope concentrations in cores of marine sediments are given by Broecker and van Donk (1970) and Emiliani (1978). As shown by Figure 31C, the Milankovitch curve of solar insolation during the past 150,000 years has peaks and valleys that correspond with dates of interglacial and glacial times. The main objective here, however, is to use its potential for predicting future climates. As interpreted by Broecker and van Donk (1970), the Milankovitch curve reconstructed for the coming 60,000 years implies moderate or low temperatures that may imply onset and continuation of a long but moderate glaciation.

Latest Pleistocene and Holocene

The remains of many animals and plants that formerly lived in the intertidal realm or nearly so have been dated by radiocarbon during the past 40 years to learn the movements of relative sea level during the past 30,000 or so years. The best materials were found to be shells of the common edible oyster and of a few other species that were not only large but restricted in habitat. Intertidal grasses of saltmarshes also served well. Other materials such as freshwater swamp grasses and trees could denote elevations of deposition well above high tide, and corals and calcareous algae could have lived well below low tide. Beachrock, although cemented intertidally, consists of calcareous and other materials originally from either above or below sea level. Oolites are commonly but not exclusively intertidal in origin. Thus, informed decisions must be made of what material to choose for dating. The choice becomes less clear when dates for high-latitude shorelines or for now-deep former beaches are desired, because some common midlatitude mollusks were unable to tolerate low temperatures of ocean waters during late-glacial or early post-glacial times. Reliance then must be placed on identification of sediment structures diagnostic of intertidal levels and containing calcareous or carbonaceous materials derived originally from not much above high tide or below low tide. Alternately, radiocarbon dates can be measured on shells of mollusks that lived intertidally at middle latitudes during glacial times or during postglacial climatic optima, but now live only at high latitudes or low latitudes (Emery et al. 1988b). Efforts to use shells or skeletons of animals that live at depth, perhaps attached to the bottom, are fraught with uncertainty because of inadequate knowledge of the depth range of their habitat, both locally and regionally.

Hundreds of articles have now been published on relative sea-level changes inferred from radiometric ages of the remains of now-submerged former shore-living plants and animals. Depth-age plots from nearly a hundred of the studies were compiled into an atlas (Bloom 1977) for the convenience of all investigators. Subsequently many others have been published, two of the most comprehensive of which are studies by Newman et al. (1980a) for 14 points along the Atlantic coast of the United States in an attempt to adjust for post-glacial rebound, and Marcus and Newman (1983) for different coastal regions of the world. Another summary of previous data was published by Pirazzoli (1976a) for more than 700 data points of the world. His analysis (Pirazzoli

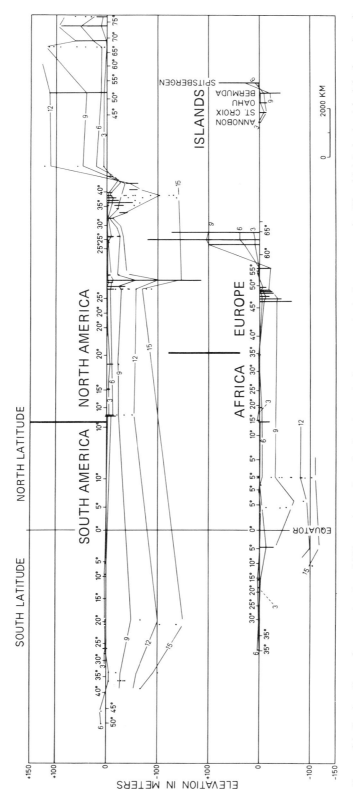

FIGURE 34. Co-time lines of past sea levels at 3000-year intervals at depths relative to present sea level. Dots and lines show positions of radiocarbon dates for intertidal or near-intertidal materials. Comparison for opposite sides of the Atlantic Ocean reveals similar patterns of uplift at high latitudes (glacial rebound) and irregularities at all latitudes attributed to tectonism. From Emery and Uchupi (1984, fig. 28), with permission from Springer-Verlag.

1977) concluded "that no coastal region of the world may *a priori* be considered wholly stable. The consequent lack of an absolute basis makes hazardous, at present, any precise determination of 'eustatic' sea levels since 2000 B.P."

When only a few dates are available, there is a strong tendency to overlook error in depth ranges of the materials or in their measured radiocarbon dates. Some authors drew point-to-point connections between data points and thereby assumed that the lines trace the path of eustatic sea levels (Fairbridge 1961; and many others). Kidson (1982) compared many such point-to-point graphs, showing large differences between them. When depth-age measurements became abundant, they exhibited such a wide scatter of data points that errors in habitat or age must be present or that the data points reveal local vertical tectonic movements of the shore rather than changes of broad eustatic levels. Much of the depth variation of data points older than about 6000 y. B.P. near glaciated areas was caused by post-glacial rebound of the crust and by collapse and migration of a crustal bulge peripheral to the glaciated areas, as well shown by Pardi and Newman (1987). An example of the depth scatter is provided by 325 data points between 36,000 and 6000 y. B.P. for continental shelves of the Atlantic Ocean and 80 data points for previous sea floor but now on land (Fig. 33). The envelope encloses 91 percent of the data points for the present sea floor. The part of it between 25,000 and 6000 y. B.P. below present sea level was transferred and extrapolated to the present in Figure 31D (about an equal number of data points are present in this extrapolation). The value of such a presentation is that it shows a general trend of lowering and rising sea level associated with extensive continental glaciation that apparently reached a maximum lowering about 15,000 y. B.P. Obscured are variations ascribable to changes in eustatic sea levels and those caused by local or regional tectonism and to glacial rebound of the crust (Fig. 34). Radiocarbon measurements of intertidal materials are neither numerous enough nor dense enough in areal coverage generally to identify either short-period eustatism or local tectonism. For the present we can rely more on tide-gauge records to attempt to differentiate between short-period eustatism and local tectonism, acknowledging the complications arising from contributions by other non-geological factors (oceanography, changes in environment at tide gauges, etc.). The implication here is that during the long fall and rise of sea level caused by growth and melt of continental glaciers, loss and gain of ocean waters obscured the effects of local tectonism. When sea level reached about 5 m below its present level about 6000 years ago, the rate of eustatic rise became so slow relative to tectonic movements of the shore region that the tectonic effects on relative sea level have been comparable with or larger than the effect of returned meltwater. In this connection, most plots omit data points for Holocene sea levels higher than present sea level even though such are known, as illustrated by Figure 34. This omission gives the erroneous impression that the scatter of data points is much less during Holocene than late Pleistocene times.

A completely different method was developed by Mörner (1969, 1976, 1980, and many other articles). He identified at least 49 shorelines that date back to 13,700 y. B.P. (dates by varves, radiocarbon, and pollen) and have been raised by post-glacial rebound as much as 280 m above present sea level. The dates reveal a general rise of the land, with complications caused by changes in rate of uplift about 7800 y. B.P., shift of center of uplift, competing eustatic rise of sea level, possible geoidal effects, and local faulting or warping. The earliest uplift appeared to be as a block with no tilt; later, the uplift became domal accompanied by peripheral sinking, with a side slope between the dome peak and the peripheral trough varying in steepness, direction, and time. Nevertheless, Mörner measured profiles up the slope and across dated former strandlines and their tilted terraces. From these by trial-and-error fitting he derived an eustatic curve of sea level for the past 12,700 years and an especially detailed one for the past 8000 years. In his publications (especially Mörner 1971) he compared and found similarities of this curve with other curves made in areas elsewhere in the world claimed by previous workers to be tectonically stable. We are not impressed with these similarities, other than the diminished rate of rise of sea level beginning about 6000 y. B.P., and we are skeptical about tectonically stable areas anywhere. Moreover, we do not fully understand Mörner's method for adjusting eustatic sea levels for post-glacial rebound of the land on which the former shorelines are located and know of no later published record of the method being used by other workers. Peltier and coworkers have used radiocarbon dates from Fennoscandia to calibrate a numerical model describing the Earth's response to deglaciation; this technique, however, is vastly different from that of Mörner.

The maximum depth of sea-level lowering during the latest major advance of continental glaciers as judged from the radiocarbon dates of former intertidal habitats in Figure 32D might be taken as about 130 m. This depth happens to be the same as Shepard's (1963, p. 257) average depth for the point at which the greatest change of slope occurs at the shelf margin. The relationship implies that the shelf edge was established by former shallow-water marine deposition or erosion and that subsequent general deformation has been unimportant. This depth also is near the estimates of sea-level lowering (105–123 m during the classical Wisconsinan Epoch as inferred from the estimated volume of glacial ice; Donn et al. 1962). Values of $\delta^{18}O$ in seawater and in calcareous skeletal materials (especially of foraminiferans) record the concentrations of the heavy isotope left during evaporation of seawater; maximum concentrations occur at times of maximum glaciation. The most useful measurements are in regions where the heavy isotope does not also record water temperature, such as in surface waters of the tropics and bottom waters at high latitudes, in both of which temperature may change little between glacial and interglacial times. New and detailed $\delta^{18}O$ measurements on Pacific deep-ocean drill cores thus yielded values for the same classical sea-level lowering (about 18,000 y. B.P.) of about 125 m (Shackleton

1987). A greater lowering may have occurred in an earlier (about 140,000 y. B.P.) glaciation and a still larger one during an even earlier glaciation. Similar results might be expected from measurements of $\delta^{13}C$, but its concentration probably is more controlled by organic productivity than by evaporation of ocean water. Concentrations of deuterium, ^2H, are too low in sediments to allow it to serve a useful role now in estimating past sea levels.

The date of about 15,000 y. B.P. for the time of maximum sea-level lowering seems reasonable from the data points of Figure 33, even though they are not yet numerous in that part of the time span. The date of this maximum lowering should correspond with the date of maximum glacier volume integrated over the Earth's surface. On the other hand, some workers (CLIMAT Project Members 1976; McIntyre et al. 1976) assumed a lowest sea level 18,000 y. B.P. on the basis of maximum late-glacial advance at that time in Europe and North America. During the past decade, the 18,000 y. B.P. time has become ingrained in the literature, although it may not be correct. In fact, recent measurements of $\delta^{18}O$ have been made in long cores of large diameter from areas of rapidly deposited sediment of the western North Atlantic, thus permitting high temporal resolution (Keigwin and Jones 1989). Cores from the Bermuda Rise and Bahama Outer Ridge (both distant from sources of large quantities of detrital sediments) reveal minimum temperatures at about 14,400 y. B.P. Other cores from Barbados (Fairbanks 1989) reveal that the lowest sea level there was 120 m below the present level 17,100 y. B.P. after correction for an assumed mean rate of uplift of the island. Many additional measurements likely to be obtained in the near future should be able to resolve the question of dating the maximum glaciations and minimum sea levels of the Quaternary.

4. Previous Studies of Relative Sea Level from Tide Gauges

Data Base

The records of the world's tide gauges can provide more precise and more widespread information about changes of relative sea levels during the past century than can be expected from radiocarbon or other kinds of measurements. Annual average levels at each station can be computed and compared with levels at the same station during earlier or later years as well as at other stations. If the year-by-year changes are systematic at a given station, each annual average tends to reinforce the others. Thus, a station record of many years is inherently stronger than a single radiocarbon date on organic or carbonate material that was deposited at a somewhat uncertain elevation above or below mean sea level.

The major problem is that there are only a few hundred tide-gauge stations having long enough records to permit satisfactory separation of the signal of changing relative sea level from the background noise. This noise is produced by variations in climate of both land and ocean, in the method used for recording, in stability of the structure on which the tide gauge is mounted, and on other variable factors (see Chapter 2). The limited number of stations cannot be quickly increased, because the required length of record is usually at least 15 years. Emery (1980; Fig. 35) accepted records from 247 stations having a worldwide distribution. Of these, 25 percent had a span of 54 years or more, 50 percent—39 years or more, and 75 percent—23 years or more. By accepting shorter time spans and/or lower t-confidence levels in statistical treatment the number of station records can be increased to gain better geographical coverage, but that expansion with less reliable records may produce less meaningful conclusions. Nevertheless, the store of information in the tide-gauge records is large enough to yield a significant contribution to the knowledge of change in levels of the ocean surface and of the ocean floor.

Illustrating the problem of length of record versus number of records are world-distribution plots of all tide-gauge records of the PSMSL and other sources used in this book that have time spans of any length beyond 10 years, beyond 20 years, beyond 40 years, beyond 60 years, and beyond 80 years, respectively (Fig. 36). These plots contain records from 588, 395, 151, 69, and 27 stations, respectively. Only those PSMSL stations having datum corrections (to Revised Local Reference—RLR) were plotted therein.

Previous Methods and Interpretations

United States

The earliest plots of change in mean annual relative sea levels from tide-gauge records in the United States are those of Marmer (1927). His book is a manual for measuring sea levels and computing the lunitidal interval (difference between time of meridional transit of moon and time of high tide), the mean sea level for tide-gauge stations (mean of hourly readings of the tidal curve for one station during a year—8766 data points), and other statistical measures at tide-gauge stations. These measures essentially eliminate the effects on the tidal record of irregularities caused by waves, seiches, tsunamis, episodic currents, and episodic strong winds, and they are intended to establish mean sea level to be used for a datum plane to which geodetic and other leveling surveys are referred. He noted that the mean sea level at any given station changes from year to year and that the measurements should span 19 years in order to include a complete lunar cycle, but that 19 years is a long time to devote to finding mean sea level for a given site. Although a 9-year span seemed to be long enough for the averaging and short enough to avoid changes in the volume of ocean water or in the elevation of the land, still plots of 9-year averages showed secular changes of sea level. Marmer did not continue to the next step (causes of regional differences in secular changes of mean sea level), probably because of the then short time span of United States Coast and Geodetic Survey tide-gauge measurements (maximum of 1883 to 1925) and of sparsity of the longer term stations (only 12) for Atlantic, Gulf of Mexico, Pacific, and Alaska coasts of the United States.

In a later study based on only a few of the longer records, Marmer (1949) recognized that "varying ratios of rise and fall of sea level at different places along the shore of a continent obviously point to differential movements of the continental crust." Plots of relative mean annual sea level for tide-gauge stations of the United States were periodically updated by Disney (1955), Hicks and Shofnos (1965), Hicks and Crosby (1974), Hicks (1978), Emery (1980), and Hicks and Hickman (1986). A tabulation of some results (Table 6) shows annual mean sea levels during the total span of years between the establishment of a tide-gauge station and a few years before the date of each summary. All averages here are based

70   4. Previous Studies of Relative Sea Level from Tide Gauges

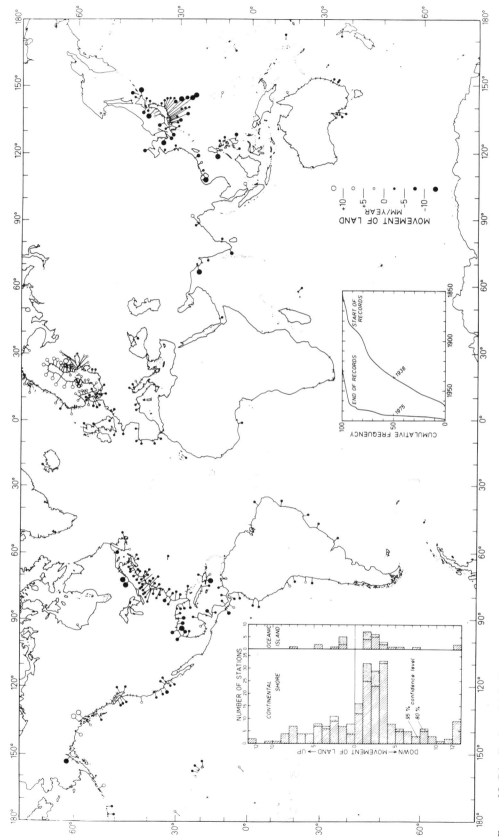

FIGURE 35. Distribution of 247 tide-gauge stations used by Emery (1980), arranged according to direction and rate of relative sea-level movement. Some additional stations having briefer and more erratic records are indicated by short lines across coasts. Insert at lower middle shows frequency of starting and ending dates for station records. Insert at lower left is histogram of direction and rate of movement of *land* level in units of 1 mm/y. for the 247 acceptable station records. With permission of National Academy of Sciences.

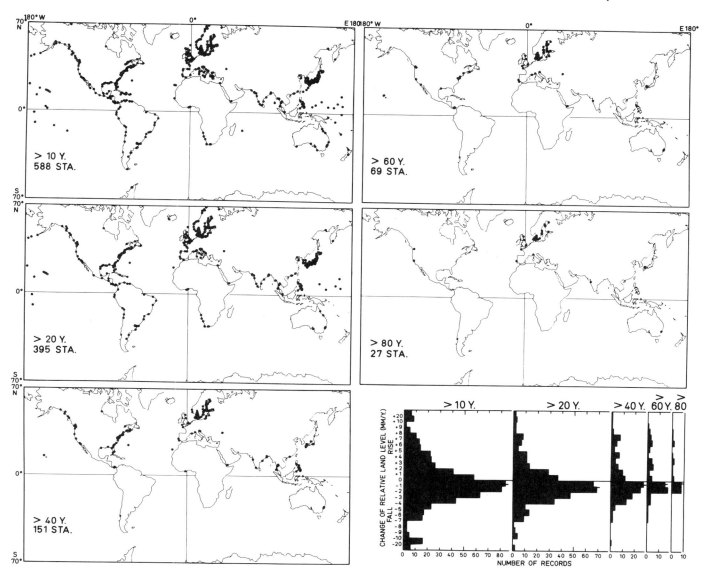

FIGURE 36. World distribution of tide-gauge records accepted for analysis in this book that have time spans exceeding 10, 20, 40, 60, and 80 years. Inset shows histograms for changes of relative land levels for each of the five subsets of station lengths.

TABLE 6. Summaries for relative sea-level change in regions of United States. Numbers indicate relative sea-level change (+ is rise of sea level; − is fall of sea level) in mm/y. In parentheses are number of stations contributing to calculations of change.

	Marmer (1927)	Disney (1955)	Hicks and Schofnos (1965)	Hicks and Crosby (1974)	Hicks (1978); Emery (1980)	Hicks and Hickman (1986)
Up to year:	1925	1955	1965	1972	1975	1986
Northern Atlantic	+1.47 (5)	+3.51 (4)	+3.79 (17)	+3.22 (24)	+3.47 (32)	+2.95 (20)
Southern Atlantic	− (1)	+4.95 (3)	+3.34 (5)	+2.66 (5)	+2.22 (8)	+2.54 (5)
Gulf of Mexico	−1.40 (3)	+3.51 (5)	+4.99 (5)	+4.33 (5)	+5.43 (11)	+2.18 (3)
Pacific	+0.90 (3)	+4.95 (3)	+0.84 (11)	+0.88 (10)	+1.25 (14)	+0.93 (11)
SE Alaska	− (0)	0.00 (1)	−6.34 (4)	−5.27 (4)	−7.71 (5)	−4.81 (4)

on stations whose records covered at least a decade with no or few missing measurements. Differences in averages for the same regions in different tabulations are attributable to inclusion of more stations for later summaries (tending to dilute the effects of a few atypical records). It also illustrates the dilemma between restricting the studies to a few long-term tide-gauge stations versus many short-term ones; the few stations may be unduly influenced by one or two aberrant ones, whereas the many short ones can reflect a decade or so of temporarily changed weather or oceanographic patterns. The high rate of rise in relative mean annual sea level for stations in the Gulf of Mexico is largely the result of subsidence by compaction of sediments in the Mississippi Delta, but in later summaries also by land subsidence caused by pumping of petroleum and water from subsurface reservoirs at Galveston and vicinity. The high rate of fall of relative mean sea level in Alaska is caused by tectonic uplift. Differences between northern and southern Atlantic regions may be attributed to sinking of a periglacial bulge that had developed when the continental crust of eastern Canada was downwarped by the weight of the Pleistocene ice sheet, and to movements of large crustal blocks affected by plate tectonics.

Europe

A prominent early study of tide-gauge records in Europe was that of Gutenberg (1941) for crustal uplift after melting of the Pleistocene ice sheet in Fennoscandia. His findings showed uplift of 100 to 110 cm/century, or 10 to 11 mm/y. at the north end of the Gulf of Bothnia, in general accord with earlier (Celsius 1743; Witting 1918) and later studies of tide-gauge records and with uplift indicated by present elevations of the Littorina I strandline of 7000 y. B.P. (Sauramo 1939).

The earliest plots of changing sea levels in Europe may be those of The Netherlands, where changes are very important because much of the country has been reclaimed from the ocean by construction of dikes and still lies below sea level. After brief service early in World War II and being shot down in his airplane, Kuenen (1945) occupied his time by investigating tide-gauge records of The Netherlands. The results, partly also shown in Kuenen (1950, p. 534), and discussed by Kuenen (1954), are for the period 1832 to 1942, and they represent averages at 5-year intervals of 1 to 11 stations (8–11 after 1866). Because many later workers have referred to the results with approbation, apparently without reading the original articles (most in Dutch), they are worthy of brief review here. A regression curve for Kuenen's averages of 1845 to 1942 reveals a mean annual rise of relative sea level amounting to +1.52 mm/y. He deduced from geological, archaeological, and precision-leveling data a secular sinking of the crust of 0.5 mm/y., concluding that the eustatic rise was +1.02 mm/y. At the same time he noted (Kuenen 1954) that investigations by van Veen (1945) and Saarloos (1951) showed that Dutch tide gauges are "untrustworthy for studying levels." Lennon (1976–1978) apparently also considered them unreliable or poorly documented because they were omitted from his general compendium on mean heights of sea levels of the world. Furthermore, the instability of The Netherlands coast is illustrated by its position atop the Rhine Delta (subject to compaction) and between the area of glacial rebound of the crust in Fennoscandia and the area of crustal sinking of northwestern France (Emery and Aubrey 1985). Further discussion of these gauges is presented in Chapter 5.

Many analyses of tide-gauge records have been made for southern Europe and throughout the Mediterranean Sea. Among the earliest are two by Polli (1947, 1948) showing for 13 Italian stations a land sinking of 2.0 mm/y. with about 3 mm/y. for Trieste even before 1910. In the western Mediterranean he obtained an average land subsidence of 1.9 mm/y. for 10 stations of southern France and 1.3 mm/y. for 6 Mediterranean stations of Africa. In his many articles on the subject, Polli (1962a) realized that the relative rise of sea level in the Mediterranean Sea was progressive for the 37 useful stations mainly along the central and western coasts. After an interval of little interest, several studies on tide-gauge records of the Mediterranean Sea appeared nearly simultaneously (Goldsmith and Gilboa 1987; Pirazzoli 1987; Emery et al. 1988a). These publications point to more than usual tectonic control of relative sea-level movements in this region, as will be discussed in a later section of this book.

World (Through 1980)

Gutenberg (1941) considered the world average rise of relative sea level on the basis of 71 tide-gauge stations, 67 of which are in the northern hemisphere. None were from the Fennoscandian region of glacial rebound. The stations had records spanning 15 to 129 years and averaged 36 years with a standard deviation of 16 years. His work was before the development of electronic calculators, so he chose to avoid the use of tedious manual least-squares regression analysis and used instead the slope between the average of the first 10 years and the last 10 years of record. This, of course, gives misleading results if the long-term record is irregular. In any event, the average slopes computed in this way exhibited a wide range of relative sea-level change: +4.8 to −0.8 mm/y., with changes that are neither regular nor simultaneous. The mean rise of relative sea level was 1.2 mm/y. with a standard deviation of 1.3 mm/y. Concern about spatial concentrations of stations (especially in France, Italy, United States, and Japan) led Gutenberg to group the 71 stations into 22 regions (having one to eight stations) in which the average rise of relative sea level was 1.1 mm/y. with a standard deviation of 0.8 mm/y. He clearly tried to eliminate the effects of obvious tectonism by omitting Fennoscandian tide-gauge stations (that showed crustal rebound after melt of glaciers) and one at Sicily (affected by volcanic movements), but still he included three from Scotland that reflected glacial rebound, one at Galveston that was strongly affected by subsidence caused by pumping of petroleum, and three at Venice that showed 2.3 to 3.1 mm/y. of subsidence caused by pumping of water from wells. Clearly, the tide-gauge records were affected by land movements as well as eustatic rise of sea level, and much of his article dealt with possible causes of these land movements—relative to internal properties of the Earth. However, the main limitation to his work, and one that was unavoidable, was the scarcity of tide-gauge data from the southern hemisphere, which contains 57 percent of the area of the world ocean, and the short time span represented by most available records.

In 1958 Lisitzin during a search for the mean sea level of the world selected six tide-gauge stations: two from Scandinavia, two from western Germany and France, one from southern France, and one from India. For the period 1891 to 1943, she found the average rise of relative sea level for the four stations south of Fennoscandia to be 1.4 mm/y., using both regression slope and Gutenberg's interpolation method. This cannot be considered a very significant figure, because four tide-gauge stations (three in Europe) are not indicative of the entire world ocean, but it has been quoted widely as though it were significant.

By 1936 the need for mean relative sea level as a datum level or base for geophysical problems had reached such a stage that the Association d'Océanographie Physique resolved to collect and publish the then-existing data in a volume, which appeared under the secretaryship of Proudman (1940). In it were listed mean monthly and annual relative sea levels at 391 tide-gauge stations of the world through 1936. Unfortunately, 91 percent of them were from the northern hemisphere, many had time spans shorter than 5 years, and statistical treatment was not included. Six updates through 1964 using the same formats and under secretaryships of Corkan (1950), Doodson (1953), and Rossiter (1958, 1959, 1963, 1968) were published by the same organization. Under Doodson's (1954) secretaryship the mean annual change of relative sea level by regression analysis was tabulated for 176 tide-gauge stations, of which 94 percent are in the northern hemisphere. General results include a fall of relative sea level (rise of the land) in Fennoscandia reaching −9.4 mm/y. For all non-glaciated areas (148 stations) the average rise of relative sea level was 1.9 mm/y., but regional differences in direction and rate probably reflect tectonism directly and indirectly (including deltaic, volcanic, and fluid pumping effects). With the end of the series published by the Association d'Océanographie Physique, the task of compilation was continued by the Permanent Service for Mean Sea Level at the Institute for Oceanographic Sciences also at Bidston Observatory, Merseyside, England. Under the directorship of G. W. Lennon (1976–1978) a three-volume listing of mean monthly and annual relative sea levels was published. It included no statistics on mean changes, but the rigor of data acceptance was tightened. About 720 tide-gauge stations were present in this compilation, which included measurements to 1975. After these volumes were printed, subsequent data were made available at intervals in the form of magnetic tapes by the PSMSL (secretary—D. T. Pugh,

and later P. Woodworth). We have used versions that were updated on a world basis through 1978, 1980, and 1984 for many tide-gauge stations. Pugh and Faull (1983) published a listing of the world's tide-gauge stations operating during 1982 with descriptions of positions, equipment, nature of records, and operating organizations. This was updated through 1986 by Pugh et al. (1987) and supplemented by a listing of worldwide stations having only short-term records (Spencer et al., 1988). By 1988 the PSMSL computer data bank held tide-gauge series from more than 1300 stations, of which 518 span at least 20 years, and 115 have data from before 1900.

Several workers used the new compilations of tide-gauge averages as soon as they became available. One was the German geologist Valentin (1952, p. 80–84), who tabulated the mean annual change of relative *land* level at 253 tide-gauge stations of the world up to the year 1947. He developed no averages for eustatic change but was particularly impressed by the tectonic movements of the land, especially for post-glacial rebound in Fennoscandia and eastern Canada. The French geologist Cailleux (1952) also recognized that vertical movements of the land are important, and he attempted to compute them as the difference between the tide-gauge record of average rate at each of 78 stations and the average for all stations. The worldwide average (omitting stations in Fennoscandia) was 1.3 mm/y. Differences attributed to vertical land movements ranged from −8.7 mm/y. (sinking) to +3.7 mm/y. (emergence). Essentially the same procedure was followed by Polli (1952), who used 110 tide-gauge stations of the world that he averaged by decades between 1871 and 1940. Computed average sea-level changes ranged from −1.4 to +1.9 mm/y. Differences between sea-level changes and tide-gauge records ranged from −4.8 mm/y. (sinking land) to +0.8 mm/y. (rising land). In general, the sinking coasts were between Latitudes 20°S and 45°N, and the rising coasts were north of 45°N. The main limitation for Cailleux's and Polli's methods is the uncertainty in obtaining global eustatic rise of sea level by overall averages of all tide-gauge records.

About the same time Munk and Revelle (1952) used Proudman's (1940) compilation and supplements to investigate annual eustatic rise of sea level in their search for a cause of the retardation of the Earth's rotation by 1.6 msec/century. They found that even a rise of 10 mm/y. was too small to account for the retardation, but noted that the changes in relative sea level were surprisingly variable with regard to time and that "a large part of the recorded changes must be due to crustal movement." Later, Lisitzin (1974) pointed out the many causes of irregularities in tide-gauge records that are related to variations with time and place of atmospheric pressure, wind, precipitation, evaporation, river discharge, seawater density, ocean currents, glacial meltwater, nodal tide ranges (18.6-year period), and land uplift and subsidence. She particularly noted the uplift of Fennoscandia and the sinking of Venice. With these excellent early recognitions of the effects of vertical land movements on tide-gauge records of "sea level," one can hardly understand the widespread later assumption that the records really show sea-level changes (see IAPSO, 1985) and the search for a global eustatic sea-level rise.

An annual cycle of sea-level heights was recognized from lesser tidal information in Japan by Nomitsu and Okamoto (1927), who related it to changes in water density and barometric pressure, and for the Atlantic Ocean by Polli (1942) who assigned no cause. The investigation by Munk and Revelle (1952) on possible effects of eustatic rise of sea level on rotation rate of the Earth led to their next effort (Pattullo et al., 1955), a more detailed study of the seasonal variation of sea level in which they used the same compilation of base data. They accepted records through the year 1946, using 419 stations arranged in 92 station groups in order to avoid bias by the dense groupings within Europe, United States, and Japan. The stations spanned 1 to 133 complete years and averaged 20.6 complete years of record. Results showed annual variations in departure, with lowest sea levels in each hemisphere during the spring and highest ones during the fall. Amplitudes vary from a few centimeters in the tropics to a few decimeters at higher latitudes and more than a meter in the Bay of Bengal. Similar but briefer results appeared the same year in an article by Lisitzin (1955) that also was based on the compilation by Proudman (1940) and its updates.

Another investigation of sea level (Fairbridge and Krebs 1962) was related to the Southern Oscillation, an east-west two- to three-year cycle of high and low atmospheric pressure. The high and low centers are in the Indian Ocean near Java and the southeastern Pacific Ocean near Easter Island. This oscillation largely is responsible for a corresponding periodicity of low and high sea level having an amplitude of 10 to 30 mm. The data base for sea level was that of Doodson (1954), Disney (1955), and other workers. After omission of tide-gauge records from presumably tectonically unstable and other anomalous areas, five-year running averages for an unstated number of stations provided "a residual curve that may approach a eustatic standard probably glacio-eustatic." The mean sea level is inferred from this curve to have fallen about 1.7 mm/y. between 1860 and 1890, followed by a rise of 1.2 mm/y. between 1900 and 1950, but with a rise of 5.5 mm/y. between 1946 and 1956. Without proper statistical treatment these rates must be considered too uncertain to have much significance, although they have been quoted widely. Another curve for the world ocean developed by Klige et al. (1978) for the period 1900 to 1963 was based on methods that are not clear to us. It is rather similar but delays the low point of sea level to about 1923 with a subsequent mean rise of 2.3 mm/y. Between 1956 and 1963 this plot shows sea level falling precipitously at about 7.0 mm/y., countering the sharp rise between 1946 and 1956 that was indicated by Fairbridge and Krebs (1962). Such discrepancies illustrate the broad spectrum of variability recorded by tide gauges.

Considerable study has been made of vertical crustal movements in areas beyond ice sheets but still caused by removal of weight by melt of ice (Walcott 1972b; Chappell 1974; Farrell and Clark 1976; Peltier 1976, 1980, 1986, 1988; Peltier and Andrews 1976; Clark et al. 1978; Peltier et al. 1978; Clark 1980; Chappell et al. 1982; Wu and Peltier 1983). All of these authors confirmed vertical movement of the Earth's surface beyond the limits of former ice sheets. These changes of sea level in turn have isostatic effects (water loading or unloading) that cause vertical movements of the ocean floor long after the water has moved (Bloom 1963; Bloom et al. 1974; Wu and Peltier 1983). The main new global finding was that the Earth's surface responds to glacial loading and unloading at high latitudes in the Northern Hemisphere differently at different distances from the area of ice sheets. For example, Clark et al. (1978, fig. 15) and Peltier et al. (1978, fig. 8) showed that the areas beneath the Fennoscandian, Greenland, and Laurentian ice sheets emerged after partial melting of the ice. A peripheral area (originally the peripheral bulge) became submerged. A third belt should emerge slightly, and the fourth and fifth belts are world-encircling ones with presumed land submergence and emergence, respectively. The last area (the sixth), at Antarctica, should emerge because of adjacent water loading and probably partial ice melt. These changes were initiated by the mass of ice whose gravitational attraction caused a rise of sea level near the ice, and the weight of ice and water downwarped the underlying crust and deformed the geoid. At distance from the ice, the removal of water by evaporation led to deposition of snow at high latitudes and to underloading of the Earth's crust at low latitudes. The amplitude and the deformation and spacing of the deformed belts were controlled by the densities and viscoelastic properties of the crust and aesthenosphere, thus different models were assumed and tested against actual deformation inferred from radiocarbon-dated former intertidal and other sediments. The deformations still continue, so they may serve to understand regional differences in direction and rates of vertical land movements shown by tide-gauge records if these records were controlled by Earth rheology. As will be shown in later text sections and maps, most tide-gauge records are more variable than would be expected from this kind of broad-belt treatment alone, being due to cooling of rifted crustal plates, plate subduction, volcanic activities, weight and compaction of large deltas, faults and folds associated with local mountain building, as well as activities of humans. Clearly, there are many causes of vertical land movements other than glacial loading and unloading, but perhaps the chief value of knowledge about glacial effects at distance from the ice sheets is that when these effects are recognized and subtracted from total movements the residue expresses the relative roles of other geological processes in producing changed relative land levels.

During the late 1970s considerable interest was being expressed in scientific literature and the press about the greenhouse effect of gases from fossil fuels and other sources (Revelle and Suess 1957; Broecker 1975; Bolin et al. 1979; Woodwell et al. 1979) and their roles in warming the atmosphere and melting glacial ice (Clark and Lingle 1977; Mercer 1978; Madden and Ramanathan 1980). Because of a long interest in post-glacial rising sea levels measured by radiometric dating of now submerged intertidal organic remains, Emery considered the use of tide-gauge data to learn whether sea levels might have increased their rate of rise in accordance with increased air temperature (and thence of water temperature) and with melt of glacial ice. The only previous study especially directed toward the world eustatic rise was that of Gutenberg (1941). Marmer (1927, 1949) had come close with his analysis of United States tide-gauge records. Others were more local in scope: Kuenen (1945, 1950, 1954) for 11 stations in the Netherlands, and Mörner (1969, 1976, 1980) for Scandinavia and mainly for Pleistocene, not for just the past few decades. Still others (Munk and Revelle 1952; Valentin 1952; Pattullo et al. 1955; Fairbridge and Krebs 1962; Lisitzin 1974; Klige et al. 1978) had investigated worldwide tide-gauge records, but most had directed their efforts mainly toward learning about oscillations of the ocean surface with only incidental investigations of eustatism. In view of the nearly 40-year accumulation of tide-gauge data since Gutenberg's (1941) publication and the very comprehensive compilation of tide-gauge records by Lennon (1976–1978), updating of the evidence for possible increased rate of eustatic rise of sea level seemed worthwhile.

The data base for Emery's (1980) investigation of world relative sea-level changes was the compilation by Lennon (1976–1978) plus a few previously unpublished records from the People's Republic of China—a total of more than 725 tide-gauge stations. Most were unacceptable because of being too short (less than 11 years), too interrupted, too irregular, or had such scattered data points that their least-squares regression lines had t-confidence that were too low. Regression lines for only 211 stations had correlation coefficients higher than 0.335, r-tests higher than 95 percent, generally 99.9 percent (computed from correlation coefficient and number of data points), standard errors such that the slope of the regression line minus two times the standard error is positive (meaning that the t-confidence is higher than 95 percent and that the slope of the regression line is significantly larger than double the standard error). In order to be able to include some representative stations from South America and Oceania 36 stations were added even though the t-confidence of their regression lines was between 80 and 95 percent. The median ending date for the 247 stations was 1975, but some continued through 1978; their median span was 39 years, and a few spanned more than 100 years. The 247 acceptable tide-gauge stations are poorly distributed around the world ocean (Fig. 35). Most (65 percent) are along the coasts of United States, Japan, and Fennoscandia. Only one station of Africa and four of South America had regression lines with t-confidence higher than 95 percent. Most stations (92 percent) are in the Northern Hemisphere—a general correlation with the standards of living of the coastal nations in which they are located.

Results of the work (histogram at lower left of Fig. 35) showed 64 tide-gauge stations with rising relative sea level and 153 stations with falling sea level along continental coasts. There were 8 stations with rising sea level and 22 with falling sea level along coasts of oceanic islands (a total for all stations of 86 rising and 161 falling relative sea levels). Of this total, 135 stations exhibited a fall of relative mean annual sea level between 0 and 5 mm/y., and the median fall for all 247 stations was 2.0 mm/y. This median is not significant because it includes stations that are located on rising land areas, especially those that are undergoing post-glacial rebound. As shown by the open circles on the world map of Figure 35, most stations on rising land are in Fennoscandia, some in eastern Canada also are due to post-glacial rebound, but still others (in western Canada, Central America, South America, southern Europe, Australia, and eastern Asia) are distant from former continental glaciers and must owe their uplift to tectonism (folding, faulting, volcanic uplift) or to other causes. Clearly, one cannot obtain useful information on eustatic rise of sea level merely by eliminating areas of known post-glacial crustal rebound, because about 20 percent of all tide-gauge stations that exhibit land uplift are outside the areas of Pleistocene continental glaciation. Moreover, tide gauges bordering the areas of glaciation must be undergoing subsidence of the land (local rise of relative sea level) caused by relaxation of a peripheral ridge that was produced by the downbowing of crust directly beneath the ice sheet. In addition, minor uplift of the land at an unknown number of stations that show rise of relative sea level must be reducing the rate of rise of relative sea level. Some workers may consider that all stations that exhibit a fall of relative sea level are dominated by post-glacial rebound or tectonic uplift and all stations that exhibit a rise of sea level are dominated by return of glacial meltwater or warming of ocean water. For the latter condition the median rise of sea level (or subsidence of land) is 2.9 mm/y., as summarized in Table 6.

An attempt to circumvent the uncertainties about rates of relative sea-level change produced by a combination of eustatic and tectonic movements was made by comparing the change for the period 1970 to 1974 with that for the entire period of record in the belief that a change in rate is more likely to be caused by changed rate of melting of glaciers rather than a changed rate of land uplift. The result (Emery 1980, fig. 2) again revealed a bimodal distribution, but with a 14 mm/y. rise rather than the 3 mm/y. rise for the long term, but this crude comparison is contaminated by short-term effects of temporary weather and ocean changes and the results are considered not useful. A similar but longer term change in trends of tide-gauge records from some United States tide-gauge stations occurred for successive dates from rates reported by Hicks and associates, as summarized in Table 6.

The conclusions that were reached by Emery (1980) were that relative sea levels are rising (or the land is sinking) at a *median* rate of about 3 mm/y. in low and middle latitudes, and land is rising at high latitudes because of post-glacial rebound of the crust. Local irregularities in change of relative sea level at all latitudes are due to tectonism (including volcanism) and to other causes (probably mostly compaction of sediments in river mouths, where many tide gauges are installed). Perhaps relative sea level rose less rapidly before 1970 than afterward, but use of tide-gauge data with their problems of reporting and the poor worldwide distribution of useful records limits their usefulness for computations of the sort that was done. Clearly, any signal recorded by tide gauges about an eustatic rise produced by return of meltwater and heating of ocean water is complicated and locally even exceeded by post-glacial rebound and other tectonic movements of the land to which the tide gauges are attached.

World (After 1980)

The value of correct knowledge of the rate of eustatic rise of global sea level is illustrated by calculations by Etkins and Epstein (1982), who attempted to extend the work of Munk and Revelle (1952) to correlate the possible rise of sea level with the slowing of the Earth's rotation. The well-measured increase of carbon dioxide and some other "greenhouse" gases in the Earth's atmosphere (from burning of fossil fuels and manufacture of cement and chlorofluorocarbons) has been paralleled by an increase of air temperature, which, according to Etkins and Epstein, has amounted to between 0.3 and 0.6°C from 1890 to 1940 in the Northern Hemisphere. Similar increases were postulated by Hansen et al. (1981). One would expect that such an increase would have led to a parallel increase of ocean surface temperature and to a transfer of glacial ice to the ocean as meltwater, essentially a movement of mass toward the equator. Etkins and Epstein calculated that an increase of temperature of 0.3°C within the top 70 m of the ocean should have caused a rise of sea level amounting to 24 mm, about half the rise indicated by Fairbridge and Krebs (1962) on the basis of tide-gauge measurements from unstated locations along coasts of the world. Between 1940 and 1970 the air temperature in the Northern Hemisphere appears to have dropped 0.2 to 0.3°C, despite increased production of greenhouse gases. Yet, Etkins and Epstein interpreted Emery's (1980) investigation as showing that global sea level had risen eustatically 3 mm/y. during this same period. They explained this opposition of presumed cooling against presumed eustatic sea-level rise in terms of cooling of the atmosphere by increased melting of ice (latent heat of fusion). Checks on the assumptions could be provided by better than presently available measurements through time of the average ocean surface temperature and of the thickness of ice sheets in Antarctica and Greenland. If 50,000 km³ of glacial meltwater had been added to the ocean during the past 40 years, global sea level would have risen about 125 mm, or 3 mm/y., enough to slow Earth rotation 1.5×10^{-8} sec, about three quarters of the observed slowing since 1940, according to Etkins and Epstein. The problem with these interesting calculations is that we do not yet know with sufficient exactness the rate of increase of water temperature, the rate of melting of glacial ice, or the rate of eustatic rise of sea level.

Articles on rising eustatic sea levels by other climatologists soon began appearing. Gornitz et al. (1982) questioned Emery's (1980) computation of change of relative sea (land) level by averaging all acceptable (high t-confidence for regression line) tide-gauge records of the world. They made three main changes in their study. First was their omission of all records from known regions of glacial rebound (Fennoscandia), seismic activity (Pacific coast of Japan), and rapid subsidence (Galveston, Mississippi River Delta). These omissions plus exclusion of stations having records shorter than 20 years (but apparently no removal of station records whose computed regression had a low t-confidence) left 193 accepted tide-gauge stations. The omission of stations in regions of known vertical crustal movement was well intentioned, but what about stations where crustal movement occurs but is not evident *a priori* or simply not known to Gornitz et al.?

The second change introduced by Gornitz et al. (1982) was an attempt to avoid the bias in average relative sea-level rise caused by the concentrations of tide gauges in Fennoscandia, eastern United States, and Japan that was inherent in Emery's (1980) averaging of all acceptable tide-gauge records of the world. This they did by dividing their 193 acceptable station records into 14 regions "on the basis of geographic proximity and the expected similarity of isostatic or tectonic behavior." In this way, each region contained 1 to 32 stations (omitting 47 in Fennoscandia that were not incorporated in their later averages, anyway). All station records within each region were combined to obtain an average relative sea-level curve for that region. These curves (except the one for Fennoscandia) then were averaged together with equal weighting to obtain a global mean relative sea level curve. The problem here is that each of the 14 regions is given the same weight in construction of a global relative sea-level curve regardless of the number or reliability of its tide-gauge records and regardless of the area of the ocean region for which it serves as an indicator. Moreover, 10 of the 14 regions are in the Northern Hemisphere. Gornitz et al. (1982) presented no map of their 14 regions, but we constructed one from their listing of the distribution of the tide-gauge stations. Areas per station ranged from about 0.008×10^6 km² for the southern Baltic Sea and 0.024×10^6 km² for eastern North America (both of which are affected by glacial rebound anyway) to 33.3×10^6 km² for Africa (with its only two stations, neither of which we consider useful). In any event, a comparison of the Gornitz et al. (1982) global relative sea-level curve for the period 1880 to 1980 with the ones for 12 of their regions showed that the four main high stands and four main low stands of the global curve had only a 43 percent correlation with the curves for individual regions, a 22 percent non-conformity, and a 35 percent omission from the regional curves. The relationship between conformity and non-conformity leaves much to be desired for deciding that the sea-level variations are global eustatic rather than partly regional tectonic in origin.

The third change by Gornitz et al. (1982) is a reduction of the raw mean relative sea-level change by the long-term rise of relative sea level during the past 6000 years. They reported that this long-term correction averages about 0.2 mm/y. among the 14 regions, according to profiles given by Bloom

(1977) in his *Atlas of Sea-Level Curves*, which are based on now-submerged intertidal materials that had been radiocarbon dated. They also apparently assumed (from Clark 1980) that perhaps 90 percent of the long-term trend is residual isostatic downwarp of the continental shelf due to the weight of returned melt water from glaciers. Subtraction of this isostatic factor of 0.2 mm/y. from the raw value of tide-gauge average global rise of sea level (1.2 mm/y.) yields a net rise of 1.0 mm/y. In order to compare the Gornitz et al. (1982) net estimated global rise of sea level with other estimates, one must remove this correction.

Barnett (1982, 1983a, b) pointed out the clustering distribution of tide gauges that affected Emery's (1980) presentation and the fact that the increased rise of 1970 to 1974 probably is an artifact caused by short-term climate change. He also noted for the method used by Gornitz et al. (1982) the dominance of regions that border the Atlantic Ocean, the inhomogeneous time series of tide-gauge records, the use of equal weighting for each region that allows some regions having few tide gauges to dominate the global average, and the uncertainty of their correction for long-term land movement. In order to avoid some of these problems, Barnett selected 10 primary tide-gauge stations and seven secondary ones on the basis of length of record and distribution pattern that was guided largely by estimates of vertical motion of the crust during the period 1000 to 2000 y. B.P. (Newman et al. 1980b). For the period 1903 to 1969 he used the records for four Pacific, three Atlantic, and two Indian Ocean stations. The number was chosen in roughly equal proportion to the area of each ocean, but the choice left eight in the Northern Hemisphere and only one in the larger Southern Hemisphere — a result of the actual poorer distribution and shorter records in the south. For the period 1930 to 1975 (to detect possible change in rate of rise of sea level) he chose three Pacific, three Atlantic, and one Indian Ocean stations (having to omit three primary stations whose records available at that time ended before 1975).

The original selection of 10 primary stations included one (Takoradi in Africa) whose regression showed a fall of relative sea level, and it was omitted from the calculation for the period 1903 to 1969. Two of the secondary stations (Hosojima and Helsinki) also recorded a fall of relative sea levels. The nine accepted primary stations for 1903 to 1969 and the seven for 1930 to 1975 were subjected to eigenanalysis followed by plotting of the principal components through which a regression line was fitted to determine the mean relative sea level rise. This was believed more suitable than merely averaging slopes of regression lines for each station. The result was 1.54 ± 0.15 mm/y. rise of relative sea level. A similar analysis for 1930 to 1975 yielded 1.79 ± 0.22 mm/y., not significantly different and thus not necessarily indicating a recent increase in rate of eustatic rise of sea level. The rate of 1.54 mm/y. for 1903 to 1969 is based on 9 tide-gauge records. The tenth primary station (Takoradi in Ghana) was omitted because it showed a fall of relative sea level. The average annual rise for the nine remaining records ranged from +3.29 (Baltimore) to +0.32 m (Tonoura) from their inception through the latest year of report, 1882–1909 to 1969–1986. Curiously, both Tonoura and Hosojima (a secondary station) are near the boundary between sinking crust and rising crust in Japan, to be described in a later section of this book. We consider nine worldwide tide-gauge stations to estimate the relative rise of global relative sea level to be too small a number, especially when there was no evident attempt to estimate local land deformation by comparison with tide-gauge records from other nearby stations.

Evidently, Barnett soon realized the problems inherent in using only a few tide-gauge records to estimate change in global relative sea level. The next year he published (Barnett 1984) a new method based mainly on records of the Permanent Service for Mean Sea Level supplemented by a half dozen records from the North Pacific Experiment. After eliminating all records shorter than 30 years and stations having obvious problems, the number of suitable stations was 155. Averaging of records in high station-density areas reduced this number to 82. These were grouped into 6 oceanic regions, having the following number of stations for 1880 to 1980 and 1930 to 1980: Europe and western Africa — 10 and 10, western North America — 19 and 19, eastern North America and Caribbean Sea — 33 and 33, eastern Asia — 6 and 6; Indo-Pacific — 10 and 10, and Central America and western South America — 0 and 4. Thus for the longer period (1880–1980) there are five regions and for the shorter one (1930–1980) there are six regions. For each region and time span the data were analyzed by empirical orthogonal function (eigenanalysis), and the percentages of variance captured by eigenmodes 1, 2, and 3 were computed. Averages of relative sea-level rise for each region exhibited somewhat similar trends except that for eastern Asia (where sea level fell) and North America (where sea level rose most rapidly). Estimates of overall averages by year also were plotted for both time spans, and regression lines through the data points yielded average rises of mean relative sea level of 1.43 ± 0.14 mm/y. for 1880 to 1980, and 2.27 ± 0.23 mm/y. for 1930 to 1980. Barnett (1984) concluded that there is no unique way to average sea-level data and that the scatter of data (noise) is so large that a man-induced increase in rate of rise will be extremely difficult to detect during realistic time spans. He also noted that "estimation of 'global' sea-level change does not appear possible with existing data." His method of averaging station records is probably better than any previously published, in view of the poor worldwide distribution of good records. On the other hand, the assumption is made that the changes in relative sea level recorded by tide gauges are solely due to eustatic sea-level change and are not affected by crustal movements except in areas known to have post-glacial rebound. This assumption cannot be supported by geologists.

After the Gornitz et al. (1982) article, its authors realized that averages of regional averages allowed regions having only one or perhaps two poor tide-gauge records to have the same weight in formulating global sea-level change as regions having many excellent records. In a later study (Gornitz and

Lebedeff 1987) two methods were used. The first method was a slight modification of the one used earlier. A regression line was fitted to each of 286 station records that spanned at least 20 years. The arithmetic mean for the regression slopes of all these 286 station records yielded 0.6 ± 0.4 mm/y. rise of relative sea level at a 95 percent t-confidence level. By omitting 54 station records in Fennoscandia, the rise became 1.7 ± 0.3 mm/y. Next, they corrected this rate by the long-term rise of sea level based on an expanded data base of radiocarbon age-depth measurements for 6000 to 7000 years. These long-term trends were grouped among 71 cells distributed in 11 regions (8 in the Northern Hemisphere, and 3 in the Southern Hemisphere). The average long-range trend within each cell was subtracted from all tide-gauge records also within that cell, and the arithmetic mean global rise of sea level for 130 corrected tide-gauge records was computed as 1.2 ± 0.3 mm/y. Their second method of computation was more complex. It began with grouping of tide-gauge records into subcells and weighting the records inversely according to their distance from the center of the subcell. Sea-level curves for each cell were combined into a composite coastal-area weighted sea-level curve for each region. The curves for the 11 regions were averaged with equal weight for each region and also averaged by weighting each region according to the reliability of its data. The results were nearly identical: $1.0 + 0.1$ mm/y. and 0.9 ± 0.1 mm/y., respectively.

The statistical treatments used by Gornitz and Lebedeff (1987), like those used in the previous study by Gornitz et al. (1982), made no allowance for tectonic effects other than those that might be included in their long-term correction. The latter comes from a regression that ignores temporary vertical movements and ones that involve only part of an area in which radiocarbon dates are measured on material originally living in or deposited on the shore zone. There is considerable question about how much of this long-term effect is caused by isostatic sinking of the continental shelf because of increasing weight of overlying water and how much is the slower return of glacial meltwater after 6000 years ago than prior to that time, when glaciers were larger and probably melting more quickly. In view of this uncertainty, perhaps we should be satisfied for the present with the many uncertainties in the gross relative sea-level change than by contending with the additional uncertainties inherent in any long-term tectonic correction. Although Gornitz and Lebedeff (1987) assumed that the long-term tectonic correction is uniform in their regions for purposes of computing global eustatic sea-level rise, they recognized and discussed some of the local tectonic variations in vertical tectonic movements of the crust in northwestern Europe, eastern North America, and western North America—topics that will be treated in more detail within following chapters of this book.

Further discussion of these tectonic factors in a review article by Gornitz (in press) attempted to set general limits on rates and durations of vertical movements of the land caused by tectonism and anthropogenic activities. These movements are estimated from sources other than tide gauges and grouped into wave lengths longer or shorter than 100 km. Estimated upper limits of long wavelength movements are: glacio-isostatic—10 mm/y., lithospheric cooling and loading by sediment and water—0.03 mm/y., and faster loading by continental shelf sediment—5 mm/y. Upper rates for short wavelength movements are: neotectonics—5 mm/y., and local sediment loading (river deltas)—5 mm/y. For land movements caused by anthropogenic activities the upper rates are: water loading behind dams—0.75 mm/y., groundwater mining—0.7 mm/y., and local withdrawal of water, oil, or gas—5 mm/y. For comparison, shifts of sea-level caused by oceanic-atmospheric effects are: geostrophic currents—1000 mm, low-frequency atmospheric forcing—40 mm, and El Niño—500 mm. In following chapters of this book comparisons of these estimates will be made with actual tide-gauge measurements attributed to the same tectonic, anthropogenic, and oceanic-atmospheric causes.

Summary

In reviewing the publications on relative sea-level changes, one realizes that the subject is complex and the changes are far from uniform throughout the world ocean. Clearly, sea level fell and rose in step with the growth and melt of large ice sheets during the latest glacial epoch of the Quaternary. The same was true but remains less well known for earlier glaciations of the Quaternary and especially for still earlier ones back to Archaean times. Relative sea level also has undergone large falls and rises during times of slowing and speeding of sea-floor spreading and associated subduction of oceanic crustal plates beneath continents and collision and rifting of continental crusts. Likewise, changes of relative sea level are produced by shorter period oscillations and other movements of ocean water itself. We can discriminate easily between the general causes on the basis of their differences in time spans, areas, and relations to geology, climate, and oceanography. However, the literature indicates considerable confusion between local tectonism and return of glacial meltwater during the past 6000 or so years. We believe that this confusion came about because most of the easiest melted glaciers were gone by that time, so that the eustatic rise of sea level from returned meltwater and thermal expansion of ocean water slowed about 6000 y. B.P.

Prior to that date, the effects of local land movement were unimportant compared with the larger effects of glacier melt and water expansion, but after that date the effects of meltwater and expansion became relatively smaller than the effects of land movements.

Some confusion can be ascribed to the different means of measurement that can be used for changes in time spans of thousands of years and those for only decades: geology plus radiocarbon ages versus historical records and tide-gauge measurements. The intermittent and widely spaced results obtained from geology and radiocarbon ages are acceptable because the amplitude of sea-level changes during the late Quaternary was large, but they are not so useful during the past 6000 years of lesser change of relative sea level. During the past century tide-gauge records provide continuous and more accurate data, but they cannot discriminate between changes of sea level and land level except within groups of stations, where geological processes can supply supplementary information to help explain differences between individual stations within the groups. In addition, the recognition of atmospheric warming related to gases liberated during the industrial revolution produced an expectation of drastically increased melting of glaciers and a resulting eustatic rise of sea level. This expected rise of sea level might provide confirmation of the greenhouse effect, but the confirmation is limited by the fact that automatic tide gauges that provide records capable of yielding mean annual sea levels were developed and emplaced only during and not before the industrial revolution. In addition, the most avid searchers for eustatic sea-level rise associated with the greenhouse effect are climatologists, many of whom have attempted to substitute statistical treatment of variations in tide-gauge records in place of geological interpretations. Finally, changes in tide-gauge methodology (both hardware and analysis) have led to differences comparable to the magnitude of signals expected.

During our work after 1980, we soon found that tide-gauge records so closely reflect geological processes that they can serve as a means for learning some details of these processes in real time rather than after the passage of long geological time. We started with contours of relative sea-level change in areas having a dense spacing of tide gauges, especially along the coasts of northern Europe and Japan. These contours indicated systematic differences in relative sea-level change that could not possibly be interpreted as caused by eustatic rise. They showed changes of land level in the form of regional warps and/or block movements such as would be expected from post-glacial rebound, subduction of oceanic crust, local faults, volcanism, and sinking of continental crust after rifting. We were able to use results of these detailed studies to interpret better the less densely distributed tide-gauge records of South America, Australia, the Mediterranean Sea, the Caribbean Sea, and southern Asia. In all areas we found that movements of the land far exceeded eustatic rise of sea level during the past century and, in fact, we are led to believe that any eustatic effects merely constitute a bias on the record of tectonic land movements. Examples of these results are given in the next next section of this book.

5. Detailed Mapping of Tide-Gauge Records in Specific Regions

General

The broadest view of changes in relative sea levels or land levels is given by histograms based on all tide-gauge records of 10 or more years' duration. One such histogram (Fig. 37) is an update and modification of the insert at the lower left of Figure 35 and contains more than twice as many tide-gauge records. Note that the modal change of relative sea level is a rise of approximately +1.0 mm/y. and that rises and falls occur in both hemispheres. Hemispheric relationships are clearer in Figure 38A, which shows far more tide-gauge records in the north than the south. This latitudinal distribution is the reverse of the distribution of ocean area, but it accords with the distribution of large seafaring industrial nations of the world. Moreover, the fall of relative sea level (rise of land) is mainly but not solely caused by removal of crustal loads by melting of glaciers. Many stations that show rise of relative sea level are at low latitudes where they probably are the result of tectonic uplift (mainly faults, folds, volcanoes) as well as rise in the ocean level. Probably the chief conclusion to be drawn from Figures 37 and 38 is that the wide range of rises and falls of relative sea level cannot reasonably be ascribed to simple eustatic changes caused by changes in the volume of ocean water. The range and geographical distribution of these simultaneous changes in the world ocean are far more reasonably ascribed to changes in land level mainly caused by tectonism including relief of crustal loads by melt of Pleistocene ice sheets and effects of convergence and divergence of crustal plates by sea-floor spreading. A method of checking such a conclusion is to investigate regions of closely spaced tide-gauge records in order to learn whether variations in land level correspond with those to be expected of tectonism. Although such a correspondence may not be conclusive, this comparison permits formulation of hypotheses that can be tested in future research.

The changes of relative sea level or land level are based on regression analysis (best-fitting least-squares line through the data points). For the 587 PSMSL station records of 10-year or longer spans, a plot of number of years of record versus t-confidence level reveals a fairly close relationship (Fig. 39). The t-test here is a calculation of the confidence level that the true regression slope is within ± 1 mm/y. of the indicated slope. Most records longer than 30 years have confidence levels higher than 0.95. Those between 30 and 20 years in span are intermediate, averaging perhaps 0.95 but ranging down to 0.73, whereas ones shorter than 20 years have t-confidence levels averaging perhaps 0.80 but ranging down to 0.50. Obviously, for comparison of relative changes of land level, we prefer to use only long-term tide-gauge records having t-confidence levels of 0.99 or more. However, these are not abundant enough, especially in some regions, to yield patterns that are useful for interpretation of causes of relative sea-level change, whether eustatic or tectonic. Thus, we must accept some station records having lower quality than ideal. An attempt was made to limit the regional studies to station records longer than 20 years and t-confidence levels higher than 0.80. These limits were not everywhere attainable, as indicated in the ensuing discussions for some regions.

A general way of plotting the changes of sea level or land level for many stations in a region is to show the direction and rate of change as a symbol at the site of each station. This method was followed in the world plot of Figure 35, in which solid dots represent sinking of the land in 5-mm/y. increments and circles represent rising land in 5-mm/y. increments. Greater precision is given by Figures 40 and 41 for the Atlantic and the western Pacific oceans, respectively. The trends of relative sea level are represented by regression lines in rectangular panels for typical longer span stations. Numbers along the coasts denote relative movements of the coastal lands derived from slopes of the same regression lines. Although these methods do show the ranges and patterns of the changing levels, they are too compact to allow the detailed comparisons needed to infer geological controls on the land levels.

A quite different approach to understanding the message of tide gauges is provided by detailed comparison of the directions and rates of relative sea-level movement in areas having numerous tide-gauge stations. These studies generally reveal about the same wide range of direction and rate that is exhibited on a broad world scale, showing that relative sea levels are less controlled by eustatic addition of meltwater and expansion of water volume or by changes in the geoid than by local effects of tectonism. Visualization of the local changes is clearest where there are many tide-gauge stations having long records and where the coast is sinuous enough to permit two-dimensional mapping of mean annual changes (as in northern Europe and Japan). For straight coasts having numerous stations (such as most of those of the United States) the only effective method is by plotting the mean annual

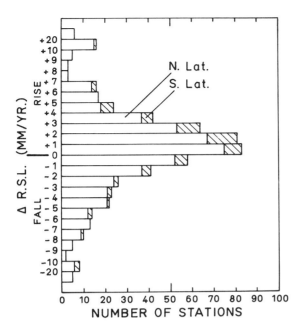

FIGURE 37. Histogram of the mean annual change of relative sea level in the world ocean based on 587 tide-gauge records from PSMSL exceeding 10 years in duration.

of a total of about 150 in the region; Appendix A), but it is widely accepted as containing portions that have been and still are undergoing uplift from crustal rebound after melting of a thick late Pleistocene ice sheet. During the 19th century many workers recognized in Fennoscandia the deposits of former glaciers and subsequent shoreline topography and sediments that record post-glacial uplift. De Geer (1888–1890) contoured raised former shores up to 213 m above what was then sea level. Later more detailed studies revealed 26 shorelines that were dated by varve counts, pollen analysis, lichenometry, archaeology, and finally radiocarbon measurements. Sauramo (1939) showed that his oldest shorelines (9000 y. B.P.) reached 180 m above what was then sea level, mapped several hingelines, and found a succession of brackish, freshwater, brackish, and marine environments. The oldest shoreline would have been uplifted at an average annual rate of about 30 mm/y.—and more considering that the uplift started from a crustal surface depressed below sea level by the weight of ice. Mörner (1979) computed from his detailed studies of elevations of terraces (46) and slopes that total uplift may have been as much as 830 m for his oldest shoreline of 13,700 y. B.P. (an average rebound rate of 60 mm/y.). About the same time, Jamieson (1865) recognized similar but less crustal rebound in Scotland after its smaller ice cap melted. Later, Nansen (1922) and

changes along coastal profiles. Coasts of any configuration that have only a few stations can be studied only by plotting tide-gauge records at station positions because there can be little confidence in either contours or continuous profiles drawn for only a few stations.

Examples of interpretation are arranged within following text sections in geographical order beginning with northern Europe and proceeding generally eastward around each of the continents back to northern Europe. These examples (all that can be given with the existing distribution of tide-gauge stations having reasonably long records) reveal changes of relative sea level usually so abrupt or so continuous that they cannot mean other than vertical movements of the land beneath the tide gauges. Thus, the tide gauges are recording aspects of modern (*actuo*) geology. This means that tide-gauge records serve better for learning present movements of the land than for learning present movements of real sea level. Comparison of these recent land movements with those inferred from the geological record can help establish the continuity of geological forces and the reasonableness of a tectonic hypothesis for tide-gauge records. For these reasons the contours are labeled to express movements of land level with plus indicating uplift and minus denoting subsidence of the land.

Northern Europe

The first area chosen for discussion of detailed contouring of mean annual changes of relative land level is Fennoscandia and adjacent regions (Emery and Aubrey 1985). This area not only contains many useful tide-gauge stations (134 acceptable

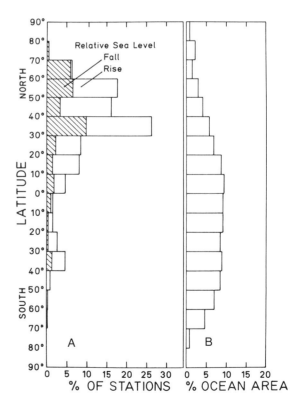

FIGURE 38. Latitudinal distributions of relative sea-level changes at 10°-latitude intervals.
A. Histogram of 587 tide-gauge station records showing mean annual fall or rise of relative sea level (rise or fall of land).
B. Histogram of ocean areas within each 10° latitude cell.

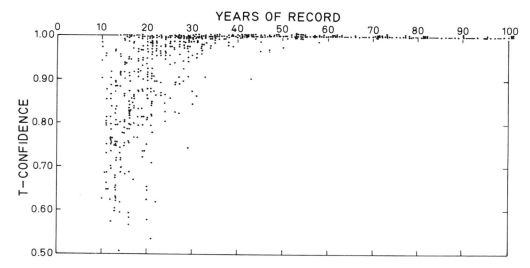

FIGURE 39. Plot of *t*-confidence level against years of record for 587 tide-gauge stations of the world, omitting all data for stations having fewer than 10 years of record. From data compiled by PSMSL.

others postulated a peripheral bulge caused by lateral subcrustal movement away from the area depressed by the weight of an ice mass and its subsequent sinking after the ice melted. Officer et al. (1988) concluded that the adjustment involved the upper mantle rather than the whole mantle or the lithosphere.

Regression analyses of the mean annual relative sea level versus date were computed for each of the tide-gauge records compiled by the Permanent Service for Mean Sea Level supplemented by a few from West Germany and The Netherlands — a total of 134 stations having time spans of at least 10 years with *t*-confidence limits higher than 0.90 for 78 percent of the stations. The resulting mean changes of relative land level plotted in map form were contoured at 2-mm/y. intervals (Fig. 42). The contours reveal an upwarp over the Gulf of Bothnia and the adjacent land areas of Sweden, Norway, and Finland at a rate of more then 8 mm/y. and in a small part of Finland at 10 mm/y. A similar but smaller uplift of about 6 mm/y. occurs in Scotland. Also shown is a long narrow uplift along the Gulf of Finland that possibly is caused by postglacial rebound but more likely is controlled by a fault or other crustal tectonics along the boundary between the Baltic Shield and the Russian Platform (see insert of Fig. 42). More striking is a curvilinear area of sinking land level that extends from western France along the English Channel (La Manche) through the North Sea. It may be part of the peripheral crustal bulge developed during glacial times by crustal flexure and outward lateral displacement of lower crust or upper mantle from beneath the area of crustal depression caused by the weight of ice. After the ice melted, the depressed crust rose and the peripheral bulge sank as the flexure relaxed and the subcrustal material moved back to the area beneath the original ice center. Subsidence of part of the coast, that along The Netherlands, is augmented by compaction of deltaic sediments that had been deposited by the Rhine and other rivers in that region. Of course, the full extent of the peripheral bulge and of present crustal sinking cannot be traced by tide-gauge records in either the deep Norwegian Sea or on land in eastern Europe where tide gauges are absent. These are areas where new technology such as Very Long Baseline Interferometry (VLBI), differential Global Positioning Satellite (GPS), and absolute gravity measurements might yield measurements of such crustal movement.

Essentially the same pattern of modern crustal rise in Fennoscandia and Scotland is revealed by contours of mean annual relative land movement computed from eigenanalysis. Eigenanalysis is a statistical treatment for separating data sets (commonly used for weather, ocean–atmosphere interactions, and water waves) into orthogonal spatial and temporal modes to describe efficiently the variability in data sets. The method permits identification and rejection of aberrant records, allows the use of records having different but overlapping time spans, and adjusts for records having gaps of missing data. Eigenanalysis of the 128 tide-gauge stations in northern Europe that have time spans longer than 15 years between 1924 and 1980 yields three spatial eigenfunctions that contain 58, 13, and 6 percent of the total variation in the original records, respectively. The three spatial functions were combined with the temporal functions to produce the contours of mean annual land movements shown in Figure 43. Differences between the contours of Figures 42 and 43 are due to the usually shorter but uniform time span for records in the eigenanalysis, the incomplete coverage of total variation of tide-gauge records by only three spatial functions of eigenanalysis, the elimination of a few extreme rates included in the regression analyses, the omission of the possible uplift along the Gulf of Finland that was based largely on two Soviet stations having only 11-year time spans that ended in 1938, and approximations associated with the technique.

An interesting confirmation of the contours from tide-gauge records is provided by contours that show vertical

84   5. Detailed Mapping of Tide-Gauge Records in Specific Regions

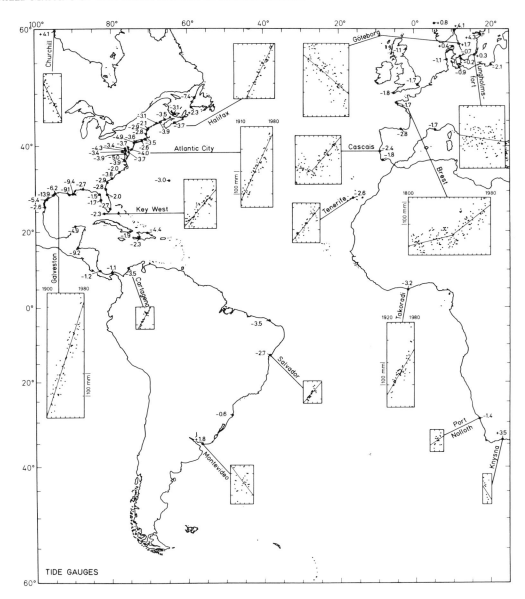

FIGURE 40. Trends of relative sea levels along the coasts of the Atlantic Ocean. The sea-level trend for some stations is expressed in plots that show the regression line through the data points for mean annual relative sea level. Trends for these and for most other stations are indicated as relative movement of land level (+ = rise, − = subsidence) in mm/y. From Emery and Uchupi (1984, fig. 32), with permission from Springer Verlag.

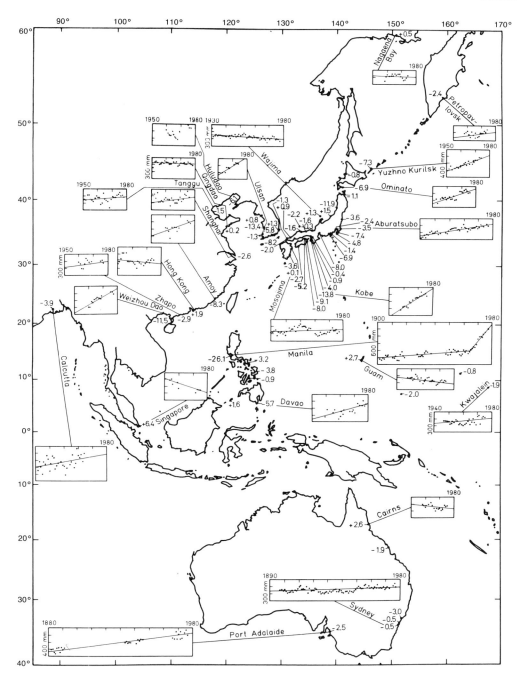

FIGURE 41. Trends of relative sea levels (rectangular plots) and of relative land levels (numbers along coast as in Fig. 39) for the western Pacific Ocean. From Emery and You (1981) with permission from Oceanologia et Limnologia Sinica.

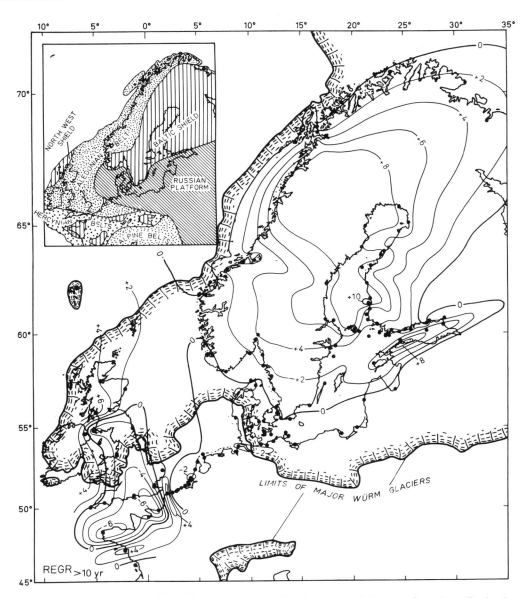

FIGURE 42. Mean annual uplift of Fennoscandia and Scotland and subsidence of southern England, northwestern France, Belgium, Netherlands, Denmark, southern Norway, West Germany, northern East Germany, northern Poland, and westernmost U.S.S.R. Contours in mm/y. are based on least-squares linear regression analyses of tide-gauge records from 134 stations (dots) having records longer than 10 years within the period 1880 to 1980. Also shown are limits of Würm ice sheet and tectonic provinces. The figure is from Emery and Aubrey (1985, fig. 3) with permission from Elsevier Science Publishers.

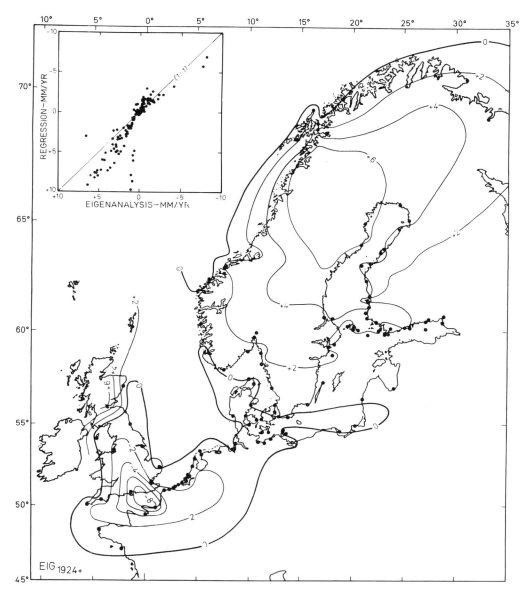

FIGURE 43. Mean annual uplift and subsidence of same area as in Figure 42 based on eigenanalysis of tide gauge records at 128 stations, each having a record longer than 15 years between 1924 and 1980 and having unit variance. Note the similarity of results obtained by regression and eigenanalysis shown by the insert at the upper left. From Emery and Aubrey (1985, fig. 6) with permission from Elsevier Science Publishers.

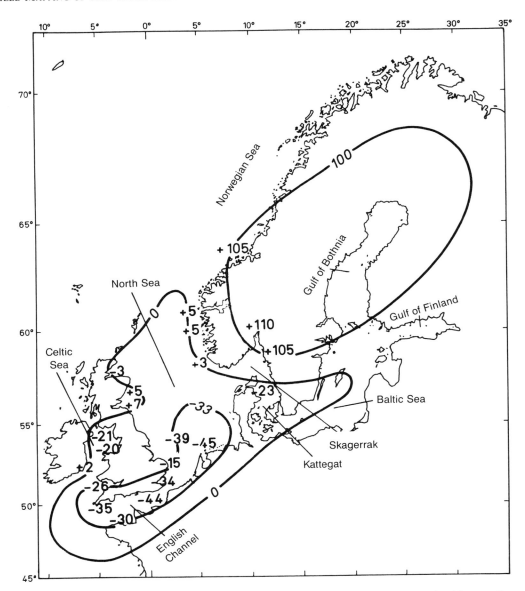

FIGURE 44. Elevation with respect to present sea level of 9000-y. B.P. shore level obtained from radiocarbon dates of shore sediments compiled by Emery and Aubrey (1985, fig. 2) from many local studies in the area of Figure 42. Published with permission from Elsevier Science Publishers.

movement of the 9000-y. B.P. shoreline surface with respect to present sea level (Fig. 44). These contours are based on elevations to shoreline materials dated by radiocarbon at many sites that have been studied by various workers, as discussed by Emery and Aubrey (1985). Rises above +100 m (probably to +280 m) in Fennoscandia and to at least +7 m in Scotland, and sinking to −44 m between England and France within the relaxed original peripheral bulge are portrayed. Additional confirmation for post-glacial rebound on land comes from repeated precise leveling surveys beginning in southern Finland (Sauramo 1939), continuing throughout the rest of Fennoscandia, and summarized by Balling (1980). Balling's map of uplift shows a maximum of 9 mm/y. and contours based on both geodetic leveling and tide-gauge data. Not surprisingly, these contours are closely similar to those of our Figures 42 and 43. The sinking of the floor of the North Sea is augmented by long-term subsidence initiated by crustal stretching caused by rifting during early Mesozoic times, as shown by seismic and drill-hole stratigraphy (Thorne and Watts 1989).

Southwestern Europe

In contrast with northern Europe, southwestern Europe has relatively few tide-gauge stations and mostly ones having only short record spans (Figs. 45, 46) acceptable to PSMSL.

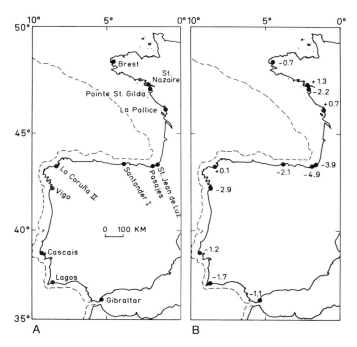

FIGURE 45. Southwestern Europe. Dash line denotes approximate position of shelf break.
A. Tide-gauge stations along Atlantic coasts of France, Spain, Portugal, and Gibraltar having records of 15 years or longer and regression lines with a *t*-confidence level of at least 0.82.
B. Mean annual change expressed as change of land level in mm/y. at each station (+ = rise; − = subsidence).

The record at Brest began in 1807 but ended in 1934; Cascais spans 1882 to 1985; and Lagos spans 1908 to 1985. Notably, Cascais and Lagos are in the small nation of Portugal with its early seafaring history.

The three northernmost stations of Figure 45 are repeated from Figure 42 of northern Europe where two were interpreted as indicating continued sinking of a former peripheral bulge associated with Pleistocene weighting of crust beneath a glacier in Scotland and Ireland. Note that rates of crustal movements at these three stations differ in Figures 42 and 45B, because some additional years of record became available after Figure 42 was compiled and published by Emery and Aubrey (1985). The rise at St. Nazaire and the sinking at Pointe St. Gilda may reflect wave-like movements of the crust having diminished amplitude with distance from the originally glacier-weighted crust. Perhaps more likely, it is an indication of inaccurate estimates limited by the short time spans of the records. The same inaccuracy also may account for the somewhat erratic changes of level farther south. The only generalizations that can be attempted from the poor data are the contrast between the relatively fast sinking of the coastal region between St. Jean de Luz and Vigo (perhaps associated with the western extension of the Pyrénées Mountains along the coast of northern Spain folded during the Alpine orogeny of the late Cenozoic) and the relatively slow sinking at Cascais and Lagos (perhaps reflecting rise of eustatic sea level and hydrostatic weighting of the continental shelf).

Mediterranean Sea

Navigation in the Mediterranean Sea began many thousands of years ago, but tide gauges have been unimportant there for shipping as compared with their greater use in coastal regions of the open ocean where tide ranges are far greater. Nevertheless, 36 tide-gauge records from 31 localities are available plus 6 others in Algeria and Tunisia that were used only in the next text section for the perimeter of Africa. The 36 records

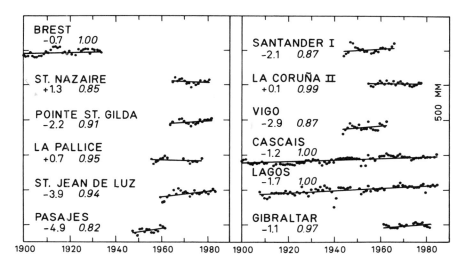

FIGURE 46. Plots of relative vertical movements of land levels at stations in Figure 45. Numbers beneath station names denote mean annual change in relative sea level in mm/y. (vertical numbers) and its *t*-confidence (slant numbers). Records at two stations begin before 1900 (Brest−1807, and Cascais−1882). Data are from PSMSL.

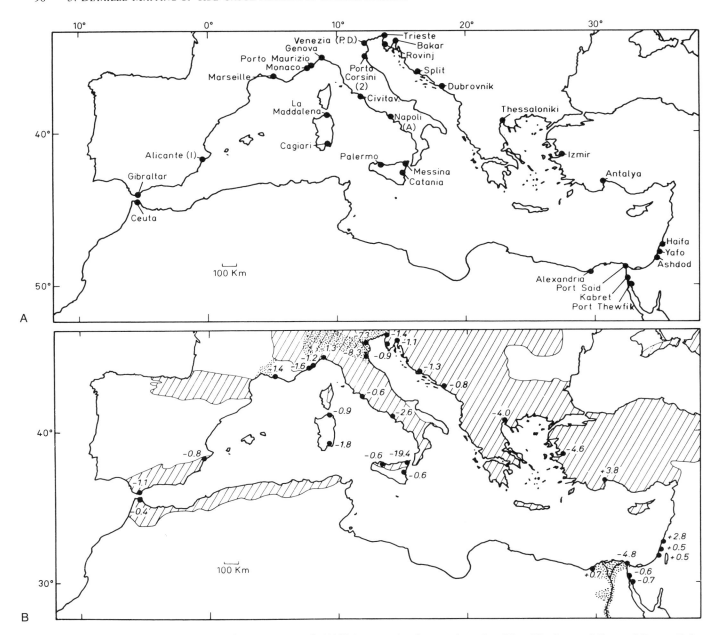

FIGURE 47. Mediterranean Sea redrawn from Emery et al. (1988a). A. Tide-gauge stations having acceptable records. B. Annual vertical movement of relative sea level at each station (mm/y.) from regression lines through plot of relative mean annual sea levels at each station (Fig. 48). Areas of diagonal lines—Alpine folding; blank areas of Europe—Hercynian folding and later sediments; blank areas of Africa and Asia Minor—Paleozoic to Neogene platform sediments; dotted areas—deltas.

were discussed more fully by Emery et al. (1988a), so less detail and documentation are needed here. These records are concentrated along the coasts of Egypt–Israel and of Yugoslavia–Italy, as shown by Figure 47A. Changes of annual relative sea levels (Fig. 48), expressed as slopes of regression lines, are indicated in map form as vertical movements of the land beneath the tide gauges, because the wide range of changes (−19.4 to +3.8 mm/y.) precludes the existence of eustatic sea-level changes as the sole cause. The t-confidence level in the slopes of the regression lines is between 1.00 and 0.95 for 28 of the records and 0.91 to 0.79 for only 5 of them that were retained for the importance of their messages.

Many previous studies of Mediterranean tide-gauge records have been published, but mostly for only small parts of the region (Polli 1947, 1948, 1962a, b; Uziel 1968; Goldsmith and Gilboa 1985, 1987; Pirazzoli 1986; Flemming and Woodworth 1988). Tide gauges in the region yield evidence of changes in sea/land levels during only the past century; for earlier evidence we must consider other kinds of information. An interesting source is provided by archaeological

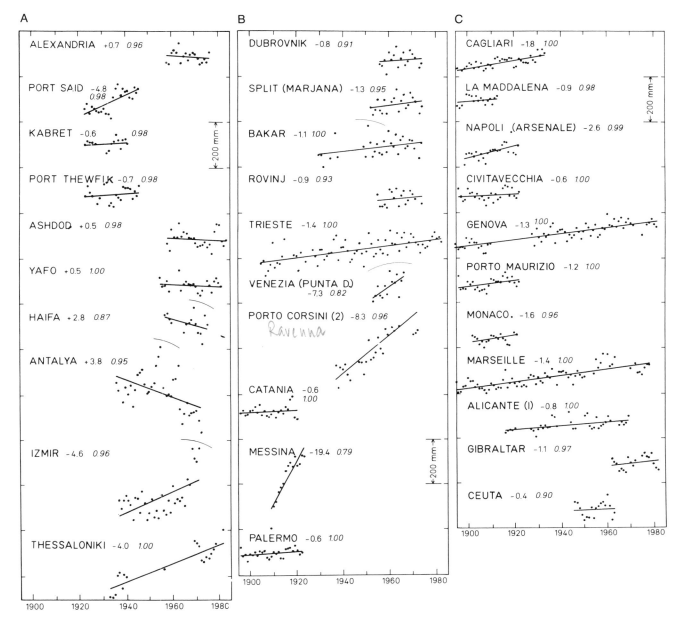

FIGURE 48. Mean annual movements of relative sea levels at tide-gauge stations bordering Mediterranean Sea (dots for each tide-gauge station and linear regression line through data points). Redrawn from Emery et al. (1988a). Vertical numbers—slope in mm/y. of regression line; slant numbers—*t*-confidence level. Data are from PSMSL and other sources listed in text.
A. Eastern Mediterranean Sea
B. Northern Mediterranean Sea
C. Southern Mediterranean Sea.

studies that extend the dates back to 6000 B.P. and exceptionally to 10,000 B.P. Knowledge about changes in level comes from observation of present levels of structures usually of stone and originally built to conform with contemporary water levels—fish ponds, bathhouses, wharves, harbors, slips for ship building and repair, seawalls, and, less precisely, storehouses, temples, homes, and protective walls around coastal towns. Especially precise are levels of fish tanks (Pirazzoli 1979-1980). Investigations during decades of diving have been made by Flemming et al. (1978), Flemming and Webb (1986), and Raban (1983, 1989). On a larger scale are general studies of submerged cities (Dolan and Goodell 1986; Pirazzoli 1982; Raban 1989). More detailed are comparisons of the depth of re-dug ancient wells with present ground-water levels (Nir and Eldar 1987) and stratigraphy in deltaic sediments (Rizzini et al. 1978; Coutellier and Stanley 1987; Stanley 1988a). Dates for these levels are based on construction types, potsherds, and other artifacts as well as on radiocarbon measurements.

Distribution patterns of earthquakes denote regions of relative coastal instability, as shown by Ambraseys (1971),

Poirier and Taber (1980), Ben-Menahem (1979), Espinosa et al. (1981), and Stiros (1988). Many of these earthquakes were accompanied by tsunamis (Strabo in Hamilton and Falconer 1854, v. 3, p. 174; Shalem 1956; Ambraseys 1962, 1971; Striem and Miloh 1975; Sharaf El Din and Moursy 1977; Papadopoulos and Chalkis 1984). The effects of still more ancient earth movements are exhibited by submerged and emerged shores (Polli 1962b; Kafri 1969; Pirazzoli and Thommeret 1973; Pirazzoli 1974, 1976b; Kafri and Karcz 1975; Norman and Chase 1986; Mart 1987; Pirazzoli et al. 1982a, b, 1989; Frihy 1988; Kayan 1988). Dumas et al. (1988) documented from study of 11 raised shorelines in southwestern Italy east of the Strait of Messina reasonably uniform uplift rates. The rates varied from 0.9 to 1.0 mm/y. spatially, a remarkably small range even for a small area (20 km of shoreline). However, the rates across the strait are more variable, with changes through time shown by different rates between strandlines.

Knowledge of geological structures, especially those associated with plate movements, allows inferences about shore stability to be carried even farther back in time. Some geological results have been published for the Mediterranean Sea region (Girdler 1958; Quennell 1958; UNESCO and Bundesanstalt für Bodenforschung 1962–1980; Yanshin 1966; Freund et al. 1970; Ben-Avraham and Hall 1977; Neev and Ben-Avraham 1977; Ross and Uchupi 1977; Ryan 1978; Garfunkel 1981; Mart 1984, 1987; Garfunkel and Almagor 1985; Steininger et al. 1985; Malinverno and Ryan 1986; Mascle et al. 1986; Neev et al. 1987; Steckler et al. 1988).

Many interrelationships between tide-gauge records, archaeology, earthquakes, raised or lowered shorelines, sediment facies and ages, and general geology were summarized by Neev et al. (1987) for the southeastern Mediterranean Sea. Some of them were extended throughout the entire Mediterranean by Emery et al. (1988a). In general, the conclusions from previous publications and Figures 47 and 48 indicate that the Mediterranean region contains many subregions in which vertical crustal movements are so different in direction and rate that one cannot confidently identify and separate the effects of eustatic rise of sea level during the past few decades from the larger effects of crustal instability.

Local subsidence because of compaction of sediments and weighting of the underlying crust is exhibited by deltas in the region. Largest is the Nile Delta where the single tide gauge (at Port Said) indicates sinking at a rate of −4.8 mm/y.; this is supported by radiocarbon ages of sediments sampled in boreholes by Stanley (1988), which show an average rate of −5.0 mm/y. during the past 7500 years. During the past, this sinking was compensated through deposition of sediments brought by the Nile River; these sediments now are trapped far upstream above the Aswan High Dam (operational since 1968). Almost complete retention of the sand fraction at the dam also has caused marked retreat of beaches in Egypt and farther northeast (Inman et al. 1976; Frihy 1988; Stanley 1988a; and others). Sinking of the Po Delta is recorded by six tide gauges (four at Venice and two at Porto Corsini) showing subsidence rates of −7.3 and −8.3 mm/y. at the two cities. Subsidence has produced the well-known flooding of St. Mark's Square and water damage to many prominent buildings and their contents. Previous studies by Polli (1962b) and Pirazzoli (1974, 1982) and results of our present study indicate long-term subsidence because of compaction of deltaic sediments augmented perhaps about 1940 by increased compaction caused by overpumping of water wells (see also Dolan and Goodell 1986). Another delta, but one having only a single tide-gauge record, is Marseille at the eastern edge of the Rhône Delta. This record reveals a relative sinking of only −1.4 mm/y.—showing little compaction (consolidation of sediments), as suggested also by an observation by Pirazzoli and Thommeret (1973) that part of that coast had risen relative to sea level about 30 mm in 2100 years. Last is a tide gauge at Izmir (Turkey) that exhibits a rather erratic relative subsidence of −4.6 mm/y., perhaps caused by compaction of marine or fluviatile sediments at the head of the long embayment. Much more striking than deltaic subsidence, however, is the sinking at a rate of −19.4 mm/y. at Messina on the northeastern flank of Mount Etna of Sicily. This movement, like that at Pozzuoli discussed in an earlier section on archaeology, surely is caused by local volcanism, as may also be the lesser sinking at two tide-gauge stations in Napoli.

Most of the remaining tide-gauge stations are in areas of known tectonism, mainly associated with Alpine mountain building (Fig. 48), as shown by geotectonic maps by Yanshin (1966) and UNESCO−Bundesanstalt für Bodenforschung (1962–1980). Some areas of tectonic subsidence have received thick marine sediments whose compositions, ages, and thicknesses are known from wells drilled during the search for petroleum. These data permit back-stripping to obtain computed effects of sediment loading on subsidence of the basement crust. Difference between the depth of this computed subsidence and the observed depth of basement is considered a result of tectonism through crustal cooling or actual vertical tectonic movements. An example in the Mediterranean region is the Gulf of Suez where tectonic subsidence computed for six wells was as much as 1600 m in 25 m.y., or an average of 0.06 mm/y. in the southern Gulf. Tide-gauge records at Kabret and Port Thewfik (Figs. 47, 48) indicate subsidence of the land there about 10 times as fast, but both stations are in the northern Gulf where sediments probably are thicker than farther south.

In regions of Alpine folding that border the Mediterranean Sea only one site (Antalya) exhibits a rise—of +3.8 mm/y (Kayan 1988). All others reveal subsidence of −0.4 to −1.4 mm/y., possibly less than the rate of eustatic rise of sea level and thus indicating uplift of the land relative to original positions but not uplift relative to sea level. Uplift of land outside the belt of Alpine folding occurs on both sides of the Nile Delta. One of the largest uplifts (+2.8 mm/y.) is at Haifa (Israel) on the northern flank of the Carmel Uplift, a horst or anticline transverse to the coastal trend (Kafri 1969; Ben-Avraham and Hall 1977). Lesser rates of rise farther south in Israel occasion no surprise, because archaeological-geological studies summarized by Neev et al. (1987) at many sites in Israel and Sinai showed clear evidence during the Holocene of movements of blocks between the major bordering longitudinal Gulf of Elat–Arava–Dead Sea–Jordan Valley–Kinneret Fault at the east and the Pelusium Fault (Neev 1977) along the outer continental shelf at the west. Some blocks were uplifted while others sank at about the same time. Moreover, oscillatory movements were inferred, whereby the upthrown side may gradually sink with respect to the downthrown side, as noted also in Japan by Lensen (1974) and Thatcher (1984) from tide-gauge records over a period of about 30 years.

The rise of land shown by the tide-gauge record at Alexandria (+0.7 mm/y.) may be part of the same system, because Alexandria is at the edge of the Nile Delta and atop pre-delta Miocene strata (Said 1962). Ben-Menahem (1979) documented active tectonism in this region that was indicated by several major earthquakes during historical times. One (A.D. 320) caused great damage and casualties in Alexandria; a second (A.D. 796) toppled the Pharos lighthouse; a third (A.D. 1303) damaged city ramparts and battlements and destroyed the remnants of the Pharos lighthouse; and a fourth (1870) caused much damage and generated a tsunami. The latter three earthquakes exceeded 7.2 in magnitude.

Africa

The second largest continent, Africa, has relatively few useful tide-gauge records. Only 21 stations have 15 or more (to 64) years of record and t-confidence levels for regression lines of 0.90 or higher. We included two stations, Gibraltar and Aden, that are not in Africa but are separated by only the narrow straits of Gibraltar and Bab-al-Mandeb and are on continental crusts having the same tectonic history as the nearby African coast (see also Uchupi 1988 for the Gibraltar connection). Data for 15 of these stations are listed by the PSMSL (Lennon 1976–1978; Pugh et al. 1987). Records for six others in Algeria and Tunisia are listed by the Association d'Océanographie Physique (Proudman 1940; Corkan 1950; Doodson 1953; Rossiter 1958, 1959, 1963, 1968; Pirazzoli 1986) but are not accepted by PSMSL because they have not reported periodic re-leveling results that are needed to assure that the base levels have not changed. Still, they provide useful information about changes of mean annual relative sea level or vertical movements of the land. Africa contains other tide-gauge stations whose records are considered too short to be useful or are otherwise impaired. One is at Dakar, Senegal, where the record consists of 11 years distributed throughout a total time span of 25 years, but with a high t-confidence level of 0.99. Another is Takoradi, Ghana, with a continuous record of 56 years and also a t-confidence level of 0.99. The problem at Takoradi is that in 1976 an abrupt drop of relative sea level is recorded, reaching in 10 years a level 300 mm below the previous trend. The cause is unknown, but it is about what might be expected of settling of a pile-mounted tide-gauge, a change in datum, or subsidence due to groundwater withdrawal. If the change is inferred to be a rise of land level, the rate is +1.9 mm/y.; if the points after 1975 are omitted, the mean annual change is −3.9 mm/y. We included the plot of data points for Takoradi in Figure 49 to illustrate the problem, but gave no regression results. Even though the record has been accepted by others (Barnett 1983, 1984; Pirazzoli 1986) probably because of its length and location, we prefer to omit it. Lesser excursions from general trends are present in other tide-gauge station records, notably at Oran and Bône of Algeria.

The 21 records accepted here (Figs. 49, 50) are mostly relatively short ones and are concentrated along the coasts of South Africa and of former French and British colonies. The limited distribution of long-term records precludes very convincing conclusions about the movements of African relative sea levels or vertical coastal movements, but it is all the tide-gauge information that seems to be available.

When the records of tide-gauge stations are viewed in the context of the general structural composition of Africa, differences and similarities of the records appear to be grouped in a sensible way. Most of the area of Africa (Fig. 51) consists of cratons, or shields, of ancient rocks that have remained reasonably stable for 550 ± 100 m.y. (since the Pan-African Orogeny) or even for 1100 ± 200 m.y. (since the Kibaran Orogeny). Only four of the tide-gauge stations (Tema, Walvis Bay, Lüderitz, and Port Nolloth) closely border these cratons, but their records are similar, with relative coastal movement of −1.0 to −2.6 mm/y. The other 17 stations occur along four parts of the coast that are more mobile. Oldest of these is the Cape Foldbelt, at the southernmost tip of Africa where folding of Paleozoic sediments occurred 400 to 180 m.y. ago in response to stages of separation of Africa from South America (Dingle et al. 1983; Emery and Uchupi 1984). Four stations border this foldbelt, and they indicate relative coastal movement of +0.4 to −4.0 mm/y. Another foldbelt of nearly the same age borders the cratons of northwestern Africa, the Mauritanides–Anti-Atlas, but no acceptable stations are present. In contrast are the Atlas Mountains of northernmost Africa, which were folded and uplifted during the Alpine Orogeny of about 40 m.y. to present. The eight stations along this coast exhibit considerable variation in relative coastal movement, +1.7 to −6.1 mm/y., just like the other stations in the area of Alpine folding on the opposite (northern) side of the Mediterranean Sea. Last, eastern Africa is bordered by or contains the various branches of the Red Sea–East African Rift (Gass 1970; Dingle et al. 1983)–a complex region of long graben structures that border or lie within previously stable cratons. The rift generally is believed to indicate an early stage in development of an ocean of the future. Only five stations occur within this rifted province, all at the north, and they reveal relative coastal movements of +0.7 to −4.8 mm/y. The main length of the tectonic province is inland, beyond tide gauges, but the southern end of the rift where it reaches the Indian Ocean at the Zambesi graben of Mozambique unfortunately has not a single station. This area is near

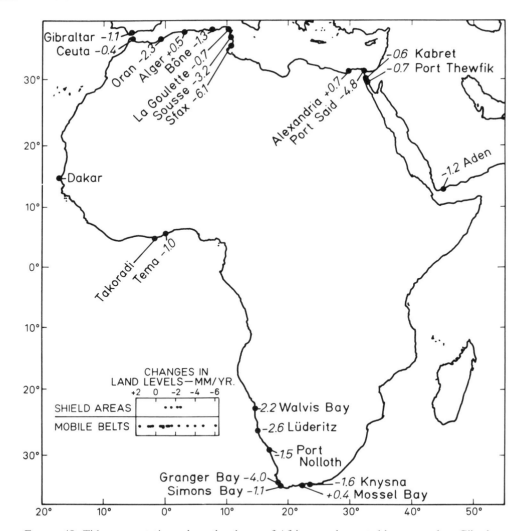

FIGURE 49. Tide-gauge stations along the shores of Africa supplemented by one each at Gibraltar and Aden. Positions of two African stations, at Dakar and Takoradi, are indicated but without trend of mean annual relative land level, because of too few years or apparently too poor quality of data. Insert at lower left shows similarity of mean annual change for stations off shield areas versus diversity of those in folded or rifted belts of the continent.

the perhaps oldest cratonic rock of the Earth but separated from it by an ophiolite belt that was emplaced about 3.5 b.y. B.P. (de Wit et al. 1987). Moreover, the sediment fills in east coast basins have subsided to greater depths than have those of the same age in west coast basins (Dingle 1981), suggesting that present subsidence along parts of the eastern coast may be faster than along the western coast.

Unfortunately, little supporting geodetic evidence is available to refine our tide-gauge estimates. The limited available geodetic evidence appears to be concentrated about the East African Rift System and the Afar (Mohr 1977). Faure and Elouard (1967), Faure and Hebrard (1977), and Faure (1980) discussed some limited radiocarbon evidence for changing land levels in Africa during the Holocene, citing clear differences from one site to another, which they interpreted as resulting from differences in tectonic setting, coastal hydroisostasy, or changes in the geoid. The maximum rate of relative movement cited in these studies is on the order of one meter per thousand years (1 mm/y.) since 5000 y. B.P. Faure (1980) cited vertical movements in the interior of the African continent, calculated from denudation rates of uplifted swells and geomorphological evidence. The Precambrian basement under the Cenozoic volcanic rocks of the Saharan massifs has been elevated with a mean rate of about 100 meters per million years (0.1 mm/y.) since the beginning of the Miocene. In sedimentary basins of Chad, the rate of subsidence since the Pliocene has been estimated at about half the above rate. Thus, Holocene rates of relative sea-level change appear to be one or two orders of magnitude greater than other Cenozoic movements, suggesting a hydroisostatic cause.

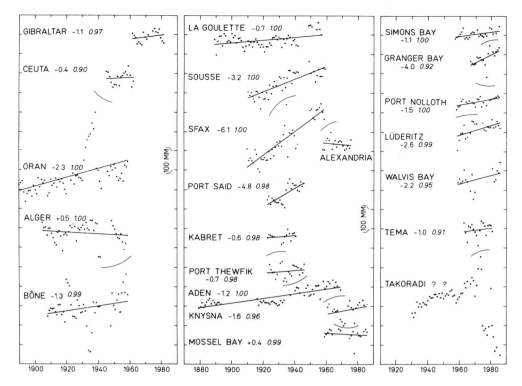

FIGURE 50. Mean annual relative sea levels at stations along shores of Africa plus the nearby stations at Gibraltar and Aden. Relative sea levels for Takoradi are shown, because other investigators have used them despite their poor distribution patterns, although we do not use them. Vertical numbers after station names denote mean annual changes of land level in mm/y. (inverse of mean annual changes of relative sea level), and slant numbers are t-confidence levels. Data are from PSMSL and other sources listed in text.

Other geological evidence for neo-tectonism along the coast comes from Tunisia in northern Africa. Here, early researchers (Flick and Pervinquière 1904) proposed recent tectonic activity along the coast, but later workers (e.g., Coque 1962) rejected that interpretation. More recent work by Richards (1986) indicated that active vertical tectonic uplift is occurring along the entire coast of Tunisia, a finding that is supported by isotopic dating instead of the previously inferred ages based only on paleontology and stratigraphy. Richards assumed that Tunisia is within the area of rebound caused by the Fennoscandian glaciation, and he removed an estimate of the rebound to obtain estimates of vertical tectonic uplift. He found Holocene uplift rates to vary from +1.1 to +2.9 mm/y. Such uplift matches closely the uplift of Holocene shorelines in southern Tunisia (Perthuisot 1979; Rognan et al. 1983) that appears to be related to the convergence of the African and Eurasian plates. Many local faults exhibit considerable vertical tectonic movement; thus, local tectonic activity is shown again to be drastically undersampled by available long-term tide-gauge recordings.

In summary, the scanty information from tide gauges of Africa reveals that relative coastal movements along the mobile belts (whether folded or faulted in origin) are much more varied than those at the edges of long-stable shields, or cratons, of the continent (Fig. 49, insert at lower left). This conclusion accords with results obtained using similar geological associations in both Australia and India.

Antarctica

Not surprisingly, Antarctica is the continent having the fewest tide gauges—only two. The better one is at Argentina's Station Almirante Brown on the Palmer Peninsula (Lat. 64°00′S.; Long. 62°52′W.). Its record spans 16 years (1957–1980) and shows a relative sinking of land level of −0.25 mm/y. from regression analysis but with a t-confidence level of only 0.82. The other station is Japan's Showa Base on the Prince Olav Coast (Lat. 69°00′S.; Long 39°35′E.). It has been at several nearby locations, with shifts caused by both ice and design problems (Oda and Kuramoto 1989). Records began in 1961, but with only a few weeks of observations each year. More continuous records between 1975 and 1980 apparently were intended for tide-prediction purposes rather than determination of mean annual sea levels. The single available long-term tide-gauge record in Antarctica is neither long enough nor reliable enough to allow interpretations of changes of sea level or land level for the continent.

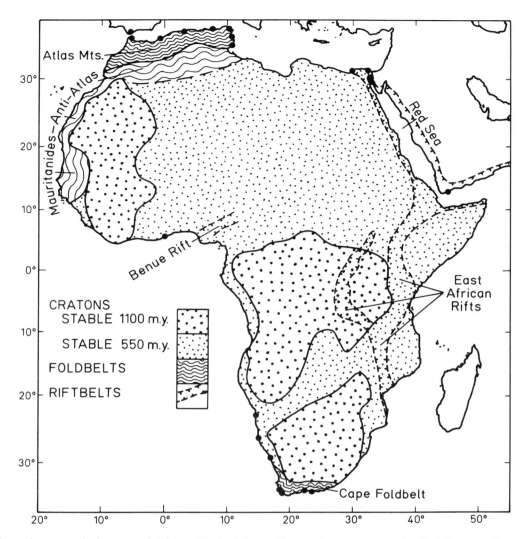

FIGURE 51. General structural elements of Africa. Most of the continental crust consists of cratons, or shields, that were formed of ancient rocks and have remained stable for more than 550 m.y. and even more than 1100 m.y. although covered by patchy later sediments. At the south and northwest are foldbelts formed 400 to 180 m.y. B.P., and the Atlas Mountains folded during the past 40 m.y. At the east are the Red Sea and the East African Rift Valleys whose bordering faults have been active during the past 30 m.y. Compiled from Clifford (1970), Gass (1970), Choubert and Faure-Muret (1981), and Dingle et al. (1983). Large dots along the shore indicate the positions of useful tide-gauge stations from Figures 49 and 50.

India

Previous published studies of tide-gauge records in India (Rossiter 1954; Chugh 1961; Ramaraju and Hariharan 1967; Arur et al. 1979; Arur and Basir 1982; Pirazzoli 1986; Premchand and Harish 1990) considered only a few of the records, or their data sources are not available to us. Lennon (1976–1978) and updates from PSMSL through 1982 listed 16 stations, and they were used in a recent article by Emery and Aubrey (1989). Of these 16 stations, half were established when India was a crown colony before 1948. Four of the later eight stations are shorter than 14 years or have t-confidence levels lower than 0.60 and were not used for our studies. The remaining 12 stations are well distributed (Fig. 52), have t-confidence levels higher than 0.68 (nine are higher than 0.95), and seven have time spans of 44 years or more (Fig. 53). The longest record, for Bombay (Apollo), is 105 years. Long gaps in the records for Kidderpore and Madras, the clusters of data points at Kandla, Bhavnagar, Saugor, and Kidderpore, and the short record at Tuticorin reduce the value of these particular records for inferring geological or other causes. The remaining five stations (Bombay, Cochin, Vishakhapatnam, Diamond Harbor, and Calcutta) include two (Diamond Harbor and Calcutta) that are on the Ganges–Brahmaputra Delta and these probably owe their high rates of land subsidence to compaction and consolida-

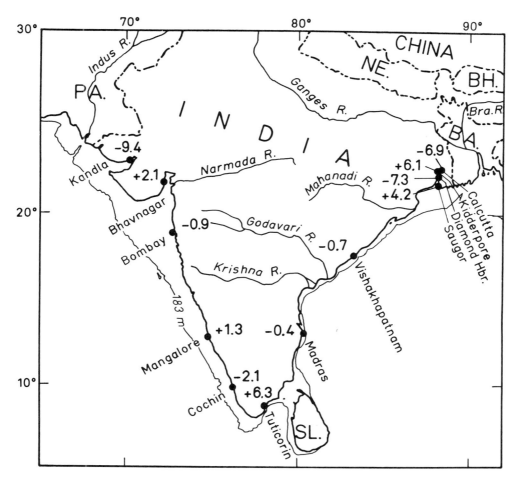

FIGURE 52. Useful tide-gauge stations of India with their mean annual change of relative land level (mm/y.) computed from regression lines of Figure 53. Also indicated are names of rivers, approximate shelfbreak (100 fathoms = 183 m), and abbreviations for adjacent countries.

tion of underlying deltaic sediments. Keeping these limitations in mind, let us determine whether a useful message can be obtained from the tide-gauge records shown in Figures 52 and 53.

All of the tide gauges of India are in the southern region (Peninsular India) south of the sediment-filled Indo-Gangetic Plain and the Himalaya Mountains that underlie the neighboring countries of Nepal, Bhutan, China, and northeastern Pakistan (Fig. 52). Peninsular India is underlain mostly by Precambrian intrusive and metamorphic rocks (Fig. 54A) that comprise a shield rifted from Africa, Antarctica, and Australia. Before rifting, the Precambrian shields of these continents were united in Pangaea (Hurley and Rand 1969; de Wit et al. 1989) where they became peneplained and then broken by grabens (Fig. 54A) during late Paleozoic. The grabens later became filled with Paleozoic and Mesozoic continental (mostly) sediments. In keeping with its thickness and antiquity, the shield of Peninsular India has few earthquakes (especially compared with the many and much deeper ones of the nearby Himalaya Mountains—Fig. 54B and Espinosa et al. [1981]), free-air gravity anomalies that mostly are between −50 and +50 mgals (also in contrast with the belt of much more negative anomalies in the Himalaya Mountains; Bowin et al. [1982]), and low heat flow (Molnar and Tapponier 1975, 1981). These geophysical characteristics as well as the geological composition and absence of Cenozoic folded mountains suggest that Peninsular India acted as a rigid piston that moved northward between the Laccodive and Ninetyeast ridges (Heezen and Tharp 1965, 1966; McKenzie and Sclater 1971; Curray et al. 1982a) and collided with southern Asia about 40 m.y. B.P. Northward penetration of about 2000 km produced underthrusting by the Precambrian shield of northern India, crumpling of the rocks of both India and southern Asia, and isostatic uplift of the Himalaya Mountains and the Tibetan Plateau that still continues (Valdiya 1984; Molnar 1989).

The two other primary geomorphic divisions of India are the Indo-Gangetic Plain and Extra-Peninsular India. The Indo-Gangetic Plain (Fig. 54A) lies between Peninsular India and Extra-Peninsular India, bordered on the west by the Indus River and on the east by the Brahmaputra River, with the

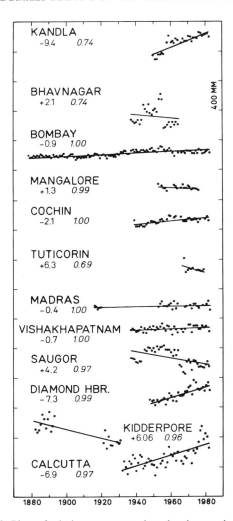

FIGURE 53. Plots of relative mean annual sea levels at each tide-gauge station of India shown in Figure 52 modified from Emery and Aubrey (1989, fig. 2). The mean annual change is expressed as change of relative land level in mm/y. (vertical numbers) with t-confidence of the regression (slant numbers). Data are from PSMSL.

Ganges River lying within it. The Plain contains more than 1-km thickness of terrestrial and littoral Cenozoic sediments derived mainly from the uplifted mountains at the north. Extra-Peninsular India lies along the southern side of the Himalaya Mountains and is contiguous with the Arakan-Yoma Mountains located mostly in Burma. These two mountain chains contain folded largely Mesozoic sediments deposited in the Tethys Ocean prior to collision of India with Asia. In contrast with Peninsular India, both the Indo-Gangetic Plain and Extra-Peninsular India experience appreciable levels of seismicity, both in frequency and magnitude (Fig. 54B).

Comparison of Figures 52 and 54 shows that the less reliable tide-gauge records at Mangalore, Tuticorin, and Madras indicate a wide range of land movement between −0.4 and +6.3 mm/y. Erratic rates at Kandla, Bhavnagar, Saugor, Diamond Harbor, Kidderpore, and Calcutta (between −9.4 and +6.1 mm/y.) may be related to the positions of these tide-gauge stations on deltas, especially that of the Ganges–Brahmaputra rivers, with rates influenced by variations in annual water discharge as well as by compaction of deltaic sediments (note from Figs. 52 and 53 that most stations exhibit high rates of relative land subsidence). These deltas contain very thick sediments: Ganges Cone −15 km (Curray et al. 1982a) and Indus — more than 3.5 km (Udintsev 1975; Kolla and Coumes 1987). Such thicknesses are in keeping with the large sediment discharge of the Ganges–Brahmaputra and Indus rivers (largest and 11th largest of the world in terms of sediment discharge, according to Milliman and Meade [1983]: 1.67 and 0.10 billion tons/y., respectively). The effects of subsidence of the Ganges–Brahmaputra Delta and impacts of relative sea-level rise were discussed by Broadus et al. (1986) and Milliman et al. (1989).

The best tide-gauge records in India from sites that border the Precambrian shield (Bombay, Cochin, and Vishakhapatnam) exhibit subsidence of the land at rates between −0.7 and −2.1 mm/y. Archaeological support for subsidence of the Indian coast is provided by the presence of submerged temples and other structures off Kandla (Wadia 1963, p. 47; Rao 1987) and off Bombay, Mangalore, just north of Sri Lanka, and Vishakhapatnam, and of submerged forests off Bombay, Madras, and the Ganges Delta (Emery and Aubrey 1989), and probably at many other sites yet to be discovered and reported. Most but not all known submerged archaeological sites are underlain by deltaic sediments. A submergence near Kandla during 1819 occurred when 5000 km² of the western side of the Gulf of Kutch suddenly dropped 4 to 5 m, and a smaller nearby area emerged about 1 m (Lyell 1850, p. 442–445; Wadia 1963, p. 47). In contrast with these evidences of Holocene coastal submergence are low raised marine terraces on land along the southeastern coast of India (Brückner 1988), perhaps uplifted in response to sediment loading and downwarping of the adjacent continental shelf.

That vertical tectonic movements appear to be continuing in India is supported by geophysical, geological, and geomorphological evidence presented by Kailasam (1975, 1980). He cited Cenozoic vertical movements including the continued uplift of the Tibetan Plateau following continental collision, uplift of the southern part of Peninsular India as a result of thrust activity in its northern margin, uplift in the Shillong plateau region of northeastern India, and vertical motion in the Deccan Trap areas of west-central and central India. Activity in the Shillong plateau region is supported by gravity data and the horst-like nature of the plateau. Kailasam (1980) hypothesized continued uplift of the region along a series of fractures. Activity in the Deccan Trap region is indicated by pronounced gravity anomalies, precisely where a disastrous earthquake of magnitude 6.5 struck in 1967, with aftershocks continuing to present. The southeastern coastal area is thought to be an area of uplift, primarily on the basis of geomorphological and geophysical evidence. Premchand and Harish (1990) addressed some evidence for submergence of the Kerala coast along the southwestern margin of India. Active vertical tectonism in some limited regions of India is

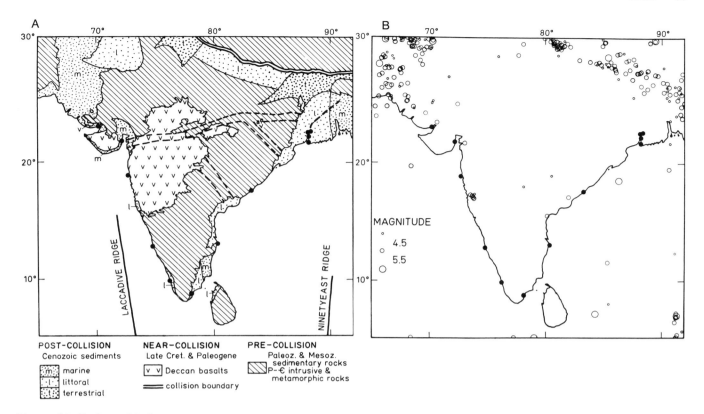

FIGURE 54. Geology of India.
A. Geological map of India showing distribution of pre-collision, near-collision, and post-collision rocks and sediments. Also indicated are collision boundary (parallel continuous lines), northern ends of Laccadive and Ninetyeast ridges (continuous straight lines), late Paleozoic graben structures (parallel dashed lines), and inferred Eocene shelfbreak within Bengal Basin (wide dashed line at east). Compiled by Emery and Aubrey (1989) from Director of Geological Survey of India (1958), Roe (1962), Sengupta (1966), Yanshin (1966), Udintsev (1975, p. 118–119), and Kumar (1985, p. 47).
B. Distribution of earthquake epicenters for India and vicinity between 1961 and 1985 from data of the National Oceanographic and Atmospheric Administration in files of the Woods Hole Oceanographic Institution.

attributed to continental collision and isostatic adjustment following retreat of Himalayan glaciers and amassing of melt water on the Indian continental shelves, as well as deep-seated isostatic adjustment and sediment compaction along continental margins and deltas.

Holocene sea-level changes in Sri Lanka have been investigated recently by Katupotha and Fujiwara (1988), who used radiocarbon dates of submerged and emerged coral to postulate changes in relative sea levels during the past 6000 years. While generally documenting the Holocene transgression and suggesting a high stand of sea level during the past, their data reveal some interesting regional variations that may be attributable to tectonism. Along the southwestern and southern coasts of Sri Lanka, much fossil coral has been dated at about 5800 y. B.P. At some sites this coral is below present sea level by as much as 3.4 m, whereas at sites distant only about 10 km the same kinds of corals are elevated by as much as 1 m with respect to sea level. Since these corals must have lived in the same ecological environment, these data suggest tectonic movements in the area on the order of 4.4 m in 6000 years, or 0.7 mm/y. on average. This differential relative change occurs across small spatial scales, once again illustrating the undersampling of tectonism afforded by the present distribution of tide gauges.

In summary, the best tide-gauge records of India accord with those of southern Africa in showing slow relative subsidence of the land, perhaps in effect being a measure of the present eustatic rise of sea level. Most records that show more rapid subsidence are on thick deltas, indicating that compaction is the probable cause. Five records indicate emergence of the land, but several of them have only short time spans and/or low t-confidence levels. Two (Saugor and Kidderpore) have peculiar data distributions, with several clusters of data points that instill little confidence in the mean slopes of their regression lines; perhaps the clusterings are caused by periodic changes in river discharge or by effects of engineering changes in the stream patterns. Among the important engineering changes is the construction of a barrage (dam) at Farakka in India just across the border from Bangladesh to direct Ganges River water into the Hooghly River during the dry season when it is needed also by Bangladesh. Sand retained by this dam and by others being built in Bangladesh will not reach the subsiding margins of the delta and thus will allow landward intrusion of ocean water.

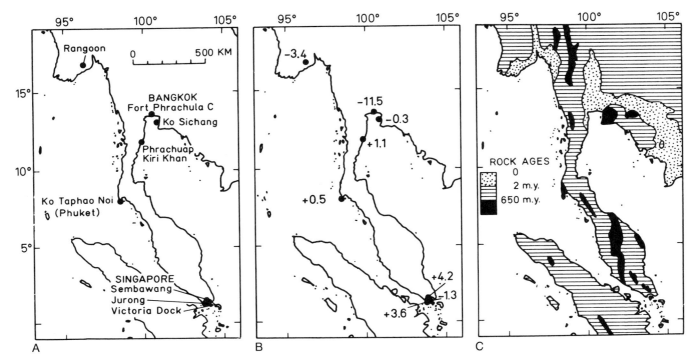

FIGURE 55. Southeastern Asia—between India and the People's Republic of China.
A. Positions of tide-gauge stations having records longer than 15 years and *t*-confidence levels higher than 0.90.
B. Mean annual vertical change of relative land levels (mm/y.) during the span of record at each tide-gauge station.
C. Simplified map of the geology of land areas, from Zshehiv (1966).

Southeastern Asia

As used here, Southeastern Asia is the coastal region between India and the People's Republic of China. Farther west between Africa and India there are only two useful tide-gauge stations. One of them is at Aden, and its record was discussed with those of African stations in association with Figure 49. The other station is at Karachi, Pakistan. Its record consists of 15 years spanning 33 years and thus containing gaps totaling 18 years. The eight stations between India and China (Fig. 55A) have better records, but they are clustered, with three near Bangkok and three close to Singapore. The eight stations span 16 to 47 years with the longest records for Thailand's four stations; their *t*-confidence level for regression analyses is between 0.92 and 1.00 (Fig. 56). As shown by Figures 55B and 56, the regression slopes range from −11.5 to +4.2 mm/y. with fairly regular progressions.

Comparison of the station records with the simple areal geology (distribution of Quaternary sediments, sedimentary rocks of Neogene to Cambrian age, and Precambrian intrusive and metamorphic rocks—Fig. 55C) shows significant relationships. The fastest subsidence of coastal land occurs on thick Quaternary deltaic sediments, as elsewhere in the world, with about −11.5 mm/y. recorded at Fort Phrachula C at the coast of the Chao Phraya delta just south of Bangkok. Subsidence caused by natural compaction of deltaic sediments has increased greatly because of overpumping of groundwater especially beginning about 1955 (Asian Institute of Technology, 1981, 1982a, b; Milliman et al. 1989). In fact, the tide-gauge regression line shown in Figure 56 may be subdivided into three segments in accordance with the groundwater pumping record: 1940 to 1960,—about −3 mm/y.; 1960 to 1981,—about −18 mm/y.; 1981 to 1986,—perhaps −40 mm/y. The station at Rangoon also is on a delta (of the Irrawaddi River) but probably is less affected by pumping of groundwater; its subsidence rate is −3.4 mm/y. One station at Singapore (Victoria Dock) exhibits minor subsidence of −1.3 mm/y. and it appears to be on marshy ground near the entrance to a lagoon.

Four other tide-gauge stations in southeastern Asia (Fig. 55B) are at shores bordered by sedimentary rock or ancient intrusive or metamorphic rocks. Their records indicate land emergence at rates between +0.5 and +4.2 mm/y. Only two sites, Phuket and Phrachuap Kiri Khan, may be distant from influence by local tectonic activity. Although the evidence is not conclusive, the existing data for Southeastern Asia suggest that land subsidence corresponds with compaction of Quaternary sediments of deltas, and land emergence with areas of sedimentary, intrusive, and metamorphic rocks. Clearly, the distribution patterns of relative submergence and emergence of the land do not favor interpretation of tide-gauge data in the region as indicating eustatic changes of sea level alone.

In other parts of Southeastern Asia, measurements have been made of relative sea-level rise using more indirect

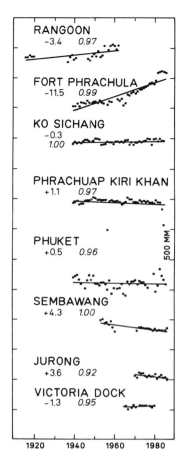

FIGURE 56. Plot of records at each station in southeastern Asia with linear regression line through the data points. Below each station name are listed the mean annual change of relative land level in mm/y. (vertical numbers) and the *t*-confidence of the regression analysis (slant numbers). Data are from PSMSL.

methods. Haile (1971) and Biswas (1973) investigated relative sea levels in West Malaysia using geological indicators, and they showed that most evidence for Quaternary emergence of more than 10 m is doubtful. Samples from borings attest to sediments submerged as far as 80 m, and bathymetry shows the presence of stream channels submerged as much as 165 m, perhaps because of warping. Evidence for the former stream channels now on the floor of the Sunda Shelf came from Molengraaff (1921) and Kuenen (1950, p. 482), whose bathymetry was supported by low-frequency seismic surveys made by Untung (1967) in a detailed search for submerged placer tin deposits near Sumatra, and by Emery et al. (1972) in a more general geophysical study of the entire Sunda Shelf.

Australia

Except for its southeasternmost part, the smallest continent is known geologically for its ancient crust, rarity of earthquakes, lack of Cenozoic mountain ranges, and similarities with southern Africa, India, and Antarctica, which also are rifted parts of Pangaea (Hurley and Rand 1969; de Wit et al. 1989), and all have essentially the same pre-rift geological history. The inferred stability suggests that Australia may be an ideal place for isolating the effects on tide-gauge records of post-glacial eustatic rise of sea level versus active movements of the land caused by local tectonism.

The general geology of Australian geology that bears on tectonic stability of its crust is indicated in Figure 57. Its oldest rocks are intrusives and metamorphics more than 1100 m.y. old and all are within large patches of the western two thirds of the continent. They probably are imbedded in and surrounded by early Paleozoic rocks of the same general kinds but now largely buried under later platform sediments, mostly continental in nature. By contrast, the oldest rocks of the surrounding oceanic crust (Fig. 57, insert) are basalts that are much younger, 153 m.y. B.P. or less. They occur in belts that surround Australia with younger rocks at progressively greater distances from Australia. The ages and distributions indicate rifting of western Australia from Africa and India about 153 m.y. B.P. (McKenzie and Sclater 1971; Veevers and Heirtzler 1974) or perhaps somewhat earlier (Boote and Kirk 1989). Separation from Antarctica at the south was later, 112 to 90 m.y. B.P. (Cande and Mutter 1982). Eastern Australia separated from Lord Howe Rise and New Zealand 80 to 60 m.y. B.P. (Shor et al. 1971) accompanied by volcanism on Australia (Brown et al. 1968) and Lord Howe Rise (Churkin and Packham 1973). These age relationships are well illustrated on the map of bedrock geology of the world by Larson et al. (1985). After rifting, the Pangaean continental crust should have cooled, become denser, and subsided, as shown in general elsewhere by Pitman (1978). This subsidence allowed transgression of marine sediments onto the margins of the crust (Vail et al. 1977a). Such sediments occupy the perimeter of the continent and reach thicknesses exceeding 6 km along parts of the older western and southern margins (Fig. 57). According to Brown et al. (1968), the transgressions began during Cretaceous and accelerated during early Paleogene times. They slowed or ceased during an Oligocene glacial event that lowered world sea levels (Vail et al. 1977b), after which transgression continued with deposition of Neogene sediments. The general depositional sequence of shallow-water marine sediment deposition on subsiding crust is revealed by seismic profiles (Veevers 1974) and drill-hole information around Australia (Veevers 1969; Cande and Mutter 1982; Williams and Corliss 1982). Northern Australia is bordered by presently active plate convergence and underthrusting toward the Indonesian plate along the Timor Trench and its extension to the northeast (Hayes and Taylor 1978).

During Pleistocene glacial and interglacial stages the sea levels off Australia reportedly ranged between −175 m (Veeh and Veevers 1970) and +180 m (Ward and Jessup 1965). Some of this wide range may have been caused by local warping (Phipps 1966; Gill and Hopley 1972). Depths of the shelf-break are poorly known but reportedly range between 110 and 130 m at the northwest (van Andel and Veevers 1965) and

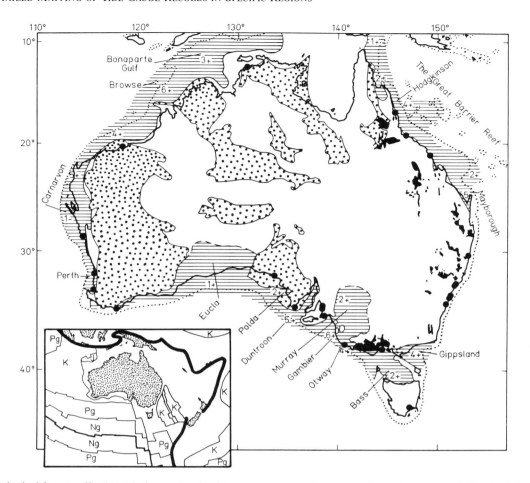

FIGURE 57. Geological factors affecting relative sea levels of Australia. Dotted line—approximate shelf break (200-m isobath); large dots—tide-gauge stations; dotted pattern—distribution of rocks older than 1100 m.y. B.P. exposed in outcrop or only thinly covered by later sediments; irregular black areas—Cenozoic volcanic rocks; lined areas—Cenozoic marine sediments with numbers indicating depths in km to basement within basins (named). Insert at lower left corner shows ages of oceanic crust and belts of subduction and translation (broad lines). Data are from Bureau of Mineral Resources (1965), Brown et al. (1968), Geological Society of Australia (1971), Hayes and Ringis (1972), Wiessel and Hayes (1972), Pitman et al. (1974), and Heirtzler et al. (1978). Modified from Aubrey and Emery (1986b, fig. 1).

between 100 and 160 m at the south (Conolly and von der Borch 1967). Most submarine canyons cut in bedrock off Sydney reach maximum depths less than 120 m (Albani et al. 1988). Uplift of a few meters in an inland belt along the coast during the Holocene is caused by weighting of the shelf by returned meltwater and landward transfer of crustal or subcrustal rock in the manner discussed by Bloom (1967), Thom et al. (1969), Thom and Chappell (1978), Belperio (1979), Chappell et al. (1982), and Chappell (1983).

A previous study of Australian tide-gauge records (Aubrey and Emery, 1986b) indicated some non-uniformity as compared with most records from other continents. This present analysis builds on the previous one and involves a perhaps better selection of tide-gauge records. The previous study was based on 21 records longer than 10 years with t-confidence levels as low as 0.60. The present study accepted only 15 records (Fig. 58B), all longer than 14 years and with a t-confidence level higher than 0.88 except for one (Port Lincoln) that had a t-confidence level of 0.76. This greater selectivity improves the certainty of the data base but reduces the geographical density of coverage. Both factors control the likelihood of correctness of geological conclusions, as pointed out by Bryant et al. (1988).

The change of annual relative sea level at each of the accepted stations is most certain for the records (Fig. 59) that are longest and have the highest t-confidence levels for the regression lines drawn through the data points (Sydney, Fort Denison—1897–1983; Williamstown—1916–1977; Newcastle III—1925–1982; Port Adelaide, Inner—1882–1976, but with two long gaps; and Port Lincoln—1920–1975, but with a long gap after the single oldest data point). The first four have t-confidence levels of 1.00, but Port Lincoln has one of only 0.76. All are in southeastern Australia, presumably the least geologically stable part of the continent. Records in the western area of older crust are shorter and have lowest t-confidence levels. Longest are those at two stations (Thevenard—1934–1970, with a long gap; and Fremantle—1937–1970, also gappy); *t*-confidence levels are 0.99 and 0.98, respectively.

Along the southeastern coast of Australia the four longest span and highest *t*-confidence stations (Sydney, Williamstown, Newcastle III, and Port Adelaide I) have mean annual changes expressed as changes of land level of −2.5, −0.6, −1.8, and −2.5 mm/y. (Fig. 58B). The average change is −1.8 mm/y., about the same as the average for all 11 stations between and including Cairns and Port Lincoln of −1.9 mm/y., but the latter 11 stations have a much wider range of +2.5 to −5.3 mm/y. Note the strong movement during the 7.1-magnitude earthquake of December 1989 at Newcastle. The range for the four stations of western Australia (between and including Thevenard and Fremantle) is +0.5 to −5.1, with an average of −2.4 mm/y. In a broader test, the eight tide-gauge stations distant from rocks older than 1100 m.y. B.P. (Fig. 57) have an average relative land subsidence rate of −2.0 mm/y. and the seven that are on or near those ancient rocks have an average of −2.1 mm/y., essentially identical. Four stations (Fremantle, Bunbury, Albany, and Thevenard) are on or near the ancient rocks and distant from harbor-making faults and volcanoes; their relative land movements range from −0.5 to −5.1 mm/y. Whether this range in rates indicates neo-tectonism or poor gauging is unclear, but testable.

Three quarters of the acceptable tide-gauge stations of Australia are concentrated in the approximately one third of the country's perimeter that may be least stable. This southeastern region contains most of the useful harbors and thus it was first and most densely settled and had greatest need for tide gauges and other aids to shipping. Its harbors at Sydney, Newcastle, Williamstown, Port Adelaide, and Mackay are associated with many minor faults (mapped by the Geological Survey of Australia 1971), whose movements probably helped form the harbors directly by graben faulting or indirectly by weakening the rocks to permit active stream and marine erosion.

An alternate explanation for the uniformity of this set of tide-gauge records relative to those on most other continents is that deformation of the crust in southeastern Australia is small, because the rifting that outlined these coasts occurred long ago (80 m.y. B.P.), and thus presumably much of the crustal margin of Australia is now quite stable. The apparent variation in vertical crustal movement indicated by tide-gauge stations may have contributions from errors in recording and effects of different time spans at tide-gauge stations. Short time spans for the records can yield erratic slopes of regression lines when cyclical variations of relative annual sea levels are produced by variations of climate or oceanic oscillations such as those related to the El Niño (Bryant et al. 1988). As a test, we find that the five long- to moderate-span records having a *t*-confidence level of 1.00 (Figs. 58B, 59) range from +0.1 to −2.5 and average −1.5 mm/y. change of level. In contrast, the 10 shorter term records having lower *t*-confidence levels range from +2.5 to −5.3 and average −2.3 mm/y.

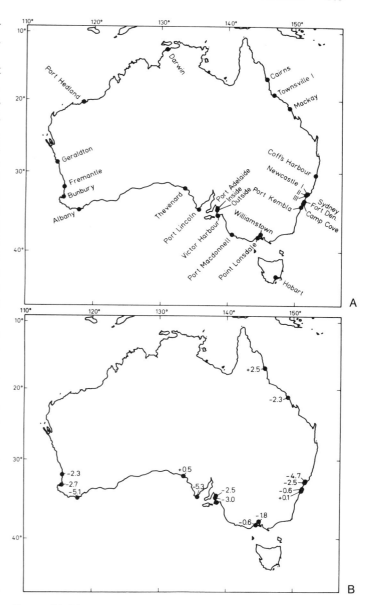

FIGURE 58. Tide-gauge stations of Australia having at least 10 years of record. Modified from Aubrey and Emery (1986b, figs. 2, 4).
A. Geographical distribution.
B. Mean annual changes of relative land level (mm/y.) for tide-gauge stations from linear regression analyses of Figure 59.

Altogether, the evidence from tide-gauge records provides some partial support for a belief in general geological stability compared with other continents that have active folding or other tectonism. This stability implies that the effects of rifting, cooling, and subsidence of the continental crust have largely ceased, but this conclusion is complicated by minor downwarps of the continental shelf and uplifts of the adjacent land caused by weighting of the shelf by returned glacial meltwater. Another decade or two of tide-gauge recording may remove the complications inherent in too few and too short time-span records. Similarly, precise leveling to date appears to reveal differences caused primarily by instru-

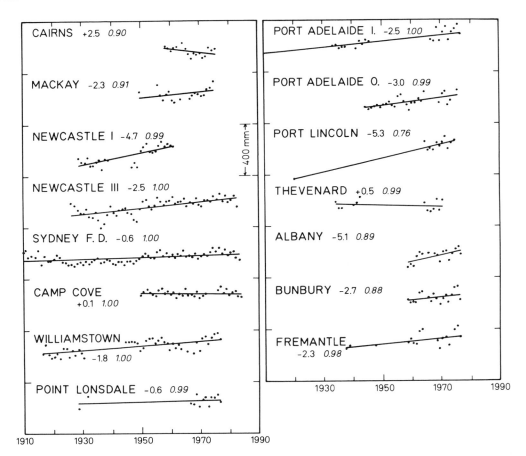

FIGURE 59. Mean annual changes of relative land levels at Australian tide-gauge stations that have records longer than 14 years with linear regression lines through the data points. After each station name is noted the mean annual change of relative land level derived from the regression line (mm/y.—vertical numbers) and t-confidence of the regression (slant numbers). All of these stations have t-confidence levels higher than 0.88, except Port Lincoln. Data are from PSMSL.

mental discrepancies (Aubrey and Emery 1988), hopefully to be eliminated in future surveys. A balanced investigation of Australian coastal regions, combining VLBI, DGPS, absolute gravity measurements, first-order re-leveling, and tide gauging would contribute significantly to resolving the continued debate regarding the absolute stability of Australia relative to sea level. The relative stability of Australia is defended ably by Bryant et al. (1988) in reply to the tectonic instability hypothesis posed by Aubrey and Emery (1986b, 1988) and more locally by Semeniuk and Searle (1986). Resolution of this issue would be helpful, because if Sydney, for instance, were absolutely stable geologically (within a tenth of a mm/y. or so/y.), then the Sydney tide-gauge record might represent a useful measure of global eustatic sea-level change.

New Zealand

Only two acceptable tide-gauge records exist in New Zealand, both on North Island, at Auckland in the north and Wellington Harbor at the south. They have 14 and 16 years of record spread over spans of 60 and 59 years, respectively. Their t-confidence levels are 1.00, but their utility still is reduced by long time gaps between the years for which mean sea level was determined (Fig. 60). Similarities in the two records make them equally reliable, but the one at Auckland indicates a relative sinking of the land (−1.3 mm/y.), whereas the one at Wellington Harbor indicates a relative rise of the land (+2.8 mm/y.). Subsidence at Auckland and elsewhere in New Zealand is reasonable in view of the drowned nature of many large harbors there (Cotton, 1945, p. 438–460). Embayed coasts other than at Auckland and on both North Island and South Island also are present, testifying to past episodes of land submergence or more likely to the effects of post-glacial rise of sea level.

Emergence at Wellington accords with the presence of extensive gravel beaches raised several meters during an earthquake there in 1855. Similar uplift of 2 m occurred in 1931 at Napier (about 270 km northeast of Wellington), where pilings that are bored and barnacle-laden cross the shore and where raised tidal flats now have been converted into farm land (Cotton 1945, p. 164; 1955). Elsewhere in

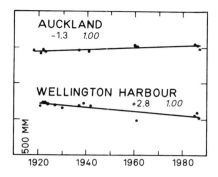

FIGURE 60. Plots of mean annual changes of land level at tide-gauge stations of New Zealand. Beneath the names of the stations are given the mean annual change of land level in mm/y. (vertical numbers) and the *t*-confidence level (slant numbers). Data are from PSMSL.

New Zealand along seacliffs are narrow rocky terraces commonly about 2 m above sea level and variously considered to be eustatic terraces or storm-wave platforms. Many stream valleys contain raised terraces probably matching with marine ones, and supporting the existence of regional tectonic uplifts (Korsch and Wellman 1988). Bull and Cooper (1986) attributed many of the raised (to 1700 m) terraces to eustatic sea levels of the Quaternary, believing that the New Zealand terraces could be correlated with those of New Guinea that were measured and dated by Chappell (1974). Higher elevation of the sets of terraces in New Zealand than in New Guinea served as a measure of higher rates of tectonic uplift in New Zealand. C. M. Ward (1988) questioned the reliability of ridge-crest notches as indicative of marine terraces and considered that tectonic movements along the Alpine fault scarp and extensive glacial erosion would have destroyed the terraces and disrupted the topographic correlation of terraces of eustatic marine origin. As an example of the extreme variability in rate of relative sea-level rise, Ota et al. (1988) identified 11 tectonic subregions within a 500-km coastal area along the eastern coast of North Island. Rates of uplift calculated from transgressive deposits ranged from 0.2 to 4 mm/y.

The difference between the two New Zealand tide-gauge records of Figure 60 is about 4 mm/y., only about half the maximum difference between the 15 Australian tide-gauge records (Fig. 59). A wider difference was expected in New Zealand (and may well have existed if there had been more records) because of the greater seismic activity of New Zealand, the large number of active faults throughout the two islands including especially the 500-km right-lateral strike-slip Alpine Fault, and active volcanism especially in North Island (Healy 1962; Korsch and Wellman 1988). The Alpine Fault has many similarities as well as differences from the San Andreas Fault of California (Blake et al. 1974), but both are systems with nearby faults having differing movements and thus tending to produce vertical as well as horizontal offsets on land and on the adjacent ocean floor. Northeast of New Zealand tectonic activity is even greater along the Kermadec–Tonga Trench, as shown by more numerous earthquakes whose hypocenters reach a 400-km depth to form a westward-dipping Benioff zone (Espinosa et al. 1981). The west side of the trench and the ridges along its west side (Dupont 1988) may be the eastern edge of a continental crust (Katz 1974) that continues in a complex way southward into New Zealand, where at South Island few hypocenters are deeper than 33 km. The interesting suggestion by Bull and Cooper (1986) of faster uplift of marine terraces on New Zealand than on New Guinea, unfortunately, cannot be checked through use of tide-gauge records because there are no acceptable records on New Guinea.

Eastern Asia

The coast of eastern Asia is broadly curved with most tide-gauge stations too widely spaced to provide high resolution of local features (Fig. 61). Only records longer than 12 years were chosen: nine from the People's Republic of China, eight from South Korea, three from Hong Kong, two from the Republic of China (Taiwan), one from Macau, and one from Japan (Izuhara). Of this total of 24 station records, 12 span at least 25 years (Macau is 57 years) and only 4 (from China) are as short as 13 years (Weizhou, Zhapo, Xiamen, and Huludao). Most of the station records were supplied by the PSMSL, but 6 (including the short time-span ones) were obtained in China and published by Emery and You (1981). A more complete analysis of 22 stations (omitting those from Taiwan) was published by Emery and Aubrey (1986a), from which much of this summary was taken. Several years of additional record for some of the stations are included here along with plots of the long-term trends of changing land levels (Fig. 62). Records from perhaps 40 more stations have not been released by China for perceived national security reasons. If they had been available, an analysis much more useful to China could have been made. The two station records from Taiwan (Proudman 1940; Rossiter 1958) were made during the rule there by Japan; they end in 1924 and 1944–before most of the mainland ones began. Although they are not accepted by PSMSL they provide further elements of information. These stations on Taiwan probably also have been continued to the present, but the results are not available from agencies of the Republic of China despite numerous attempts.

Regression analysis of the 24 station records leaves much to be desired, because only 16 of them have *t*-confidence levels higher than 0.90 and nine of these are in South Korea and on Japan's Izuhara Island. Only the records from Macau, Hong Kong's North Point, Qingdao, and Qinhuangdao along the mainland coast of China appear to be highly reliable. When contoured (Fig. 63), however, the results of regression

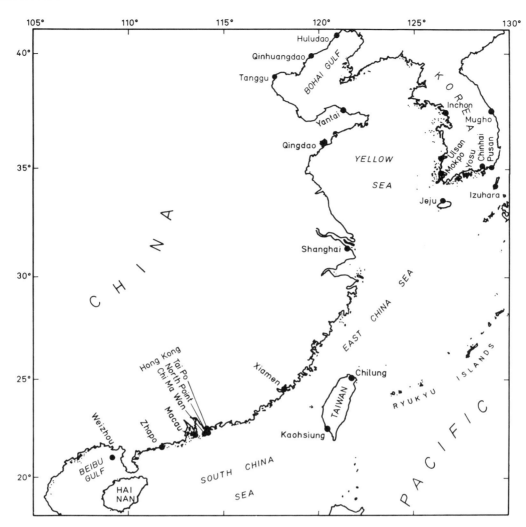

FIGURE 61. Eastern Asia—coastline, water bodies, and positions of tide-gauge stations useful for determining changes of relative sea level. Modified from Emery and Aubrey (1986a, fig. 1) through up-dating of records from some stations and inclusion of two older stations from Taiwan.

analysis for the whole region show a rather simple pattern of general subsidence along the inner shelf between Hong Kong and South Korea with maximum rates of −11.4 and −11.6 mm/y. at Weizhou and Xiamen. Additional areas of lesser subsidence occur off Yantai (on the Shandong Peninsula), and at Tanggu and Huludao (in the Bohai Gulf). These areas of subsidence are separated by ones of relative land emergence. The pattern revealed by contours (Fig. 64) based on the results of eigenanalysis of station records since 1950 are similar to those of regression analysis but are simpler because of fewer stations, especially the omission of the two records from Taiwan that pre-date 1950.

Although the contours of annual change in relative land level directly off China in Figures 63 and 64 have a rather insecure derivation, they do fit well the known or inferred geology of the region. In a general way, the areas of subsidence during the 20th century inferred from tide-gauge records correspond with areas of Cenozoic and late Cretaceous basin sediments (insert of Fig. 63), as though similar subsidence had occurred during at least the Cenozoic and allowed deposition of thick terrestrial and marine sediments. Note also that the fastest subsidence off Xiamen may be related to the thick sediments mapped by seismic methods (Wageman et al. 1970) and drilled (2865 m of Cenozoic near Taiwan; Fletcher and Soeparjadi 1984). Most of the thick Cenozoic sediments as well as present sediments were brought by the large rivers Huanghe (Yellow) and Changjiang (Yangtze) and deposited as deltas and prodeltas. These are the second and fourth largest rivers of the world in terms of total sediment discharge (Milliman and Meade 1983). Perhaps related to sediment thickness also is a large apparently fault-controlled submarine canyon bordering southwestern Taiwan (Yabe and Tayama 1929; Niino and Emery 1961).

The tide-gauge records indicate that the areas of subsidence are separated by ones of relative land uplift. The geology of the continent reveals that the relative uplifts may be associated with Precambrian massifs, early Paleozoic foldbelts, and a Cenozoic foldbelt containing much volcanic rock (insert of Fig. 64). This geological background for China and Korea was derived by Emery (1983) from geological maps published by Yanshin (1966), Alverson et al. (1967), Pushcharovsky and Udintsev (1970), Wageman et al. (1970), Academy of Geological Sciences of China (1975), Hayes and Taylor (1978), Ministry of Economic Affairs of Taiwan (1978), Um and Chun (1983), Zhang (1983), Zhang et al. (1984), and Heezen and Fornari (no date). Some additional discussion was provided by Emery and Aubrey (1986a). The closely spaced contours of vertical land movement along the northwestern side of Bohai Gulf (Figs. 61–63) may be related to the contact of Neogene sediments and Late Cambrian rocks (insert of Fig. 64); it is near this contact that the destructive 8.2-magnitude earthquake of 28 July 1976 occurred at Tangshan.

Local land emergence also is supported by the presence of Holocene and late Pleistocene marine terraces and shores at Hai Nan Island, Hong Kong, Taiwan, between Shanghai and Qingdao, and South Korea (Berry 1961; Yang and Xie 1984; Yang and Chen 1987; Ota 1987; Wang and Aubrey 1987; and others). Ota also compiled evidence for lower than present sea levels during the Holocene and late Pleistocene in the East China Sea based on radiocarbon dates published by many authors including Emery et al. (1971), whose samples continued into Tsushima Strait (between Korea and western Japan). Most of these geological evidences relate to the considerably higher than present and lower than present sea levels associated with Pleistocene glacial and interglacial events of the Earth that exceeded the effects of neo-tectonism recorded by tide gauges.

Northeast of Taiwan there is abundant evidence from earthquake hypocenters (Espinosa et al. 1981) and from seismic profiles that subduction of the Pacific oceanic plate is occurring beneath the Ryukyu Islands along the Kyushu–Ryukyu Trench. At Taiwan the evidence from hypocenters suggests that only shallow hypocenters are present and that they extend southward along the Philippine Islands rather than southwestward along the continental slope off eastern China and southeastern Asia. The two tide-gauge records from Taiwan (Figs. 63, 64) show only slight change of relative land level (uplift of +0.4 to +2.3 mm/y.). This uplift might have been produced by subduction, but it is more likely a result of other tectonic activity or perhaps the records may reflect oceanographic variability unrelated to tectonism. Recent work documents variable rates of uplift across Taiwan. Using radiocarbon dates, Wang and Burnett (1990) found uplift rates of between 4.7 and 5.3 mm/y. along the eastern Coastal Range, 3.3 to 5.3 mm/y. for the Hengchun Peninsula of southern Taiwan, and 1.6 to 2.2 mm/y. for coral terraces on Lanyi and Lutao islands east of Taiwan. These observations,

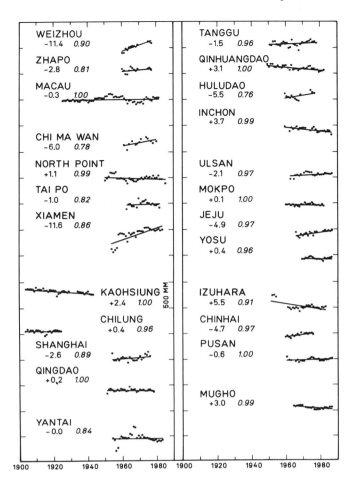

FIGURE 62. Plots of mean annual relative land level at the 24 acceptable tide-gauge stations of Eastern Asia shown in Figure 61. Vertical numbers under station names are mean annual changes in mm/y.; slant numbers are *t*-confidence levels. Data are from PSMSL and Emery and You (1981).

taken over a longer time interval, support the observations from tide gauges (Fig. 63).

Japan

When Aubrey and Emery (1986a) began their work on tide-gauge records of Japan, they knew that Japan's geology is entirely different from that of northern Europe and most other continental regions of the Earth—its much younger rocks, its volcanic activity, and the lack of any great Quaternary glaciation. In fact, they expected contours of relative sea-level (or land-level) change to be complex, reflecting highly localized land movements associated with volcanoes. However, the results turned out to be far more interesting. The PSMSL compilation contains 111 tide-gauge records for Japan and 11 for Korea ending mostly in 1979 or 1980. All stations that spanned fewer than 10 years were eliminated,

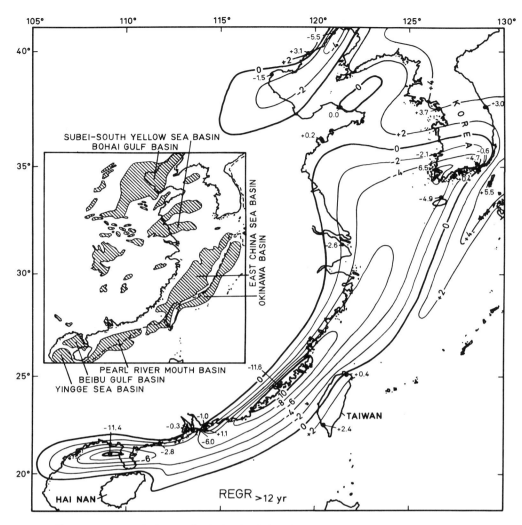

FIGURE 63. Average rate of change of relative land elevation along coastal eastern Asia (2 mm/y. contours) based on linear regression analysis of the 24 indicated tide-gauge stations (dots with rates shown in italics). Insert shows distribution of Cenozoic basins on land and sea floor according to Li (1984) and Wageman et al. (1970). Modified from Emery and Aubrey (1986a, fig. 2).

and the rest were subjected to least-squares regression analysis. A best-fit line having a *t*-confidence level of 0.90 or higher serves as a useful measure of emergence or submergence of the land at a tide-gauge station, but 0.80 was accepted as the lower limit of records in order to increase the areal density of data. The specified criterion was that the expected regression slope be within +2.0 mm/y. of the observed one in order to remove any bias toward large relative change (Aubrey and Emery 1986b). Changes of relative land level at these stations were plotted in map form and contoured at 2-mm/y. intervals (Fig. 65). The contours have a general trend from southwest to northeast (parallel to the coasts of Japan), they show deepening toward the southeast (the Pacific Ocean), and they rise along the coast of the Sea of Japan. The re-entrant along the Inland Sea (Seto) of Japan continues eastward toward the Pacific and southwestward through and beyond the southwestern island of Japan (Kyushu). Another interruption is indicated by the circular contours of land subsidence at Lat. 38°34′N. and Long. 139°33′E. that are caused by slow continued displacement of the land below a tide gauge after an 18-cm drop along a fault in 1964, probably the Tanakura Fault. The contours also suggest that Korea has an entirely different mode of vertical land movement—a separate crustal block from that of Japan.

For eigenanalysis, 51 tide-gauge records of Japan were selected on the basis of more than 15 years duration during the period 1953 to 1980. The first, second, and third spatial eigenfunctions contained 83, 4, and 3 percent of the record variance. These spatial functions were combined with temporal functions to yield estimates of mean annual changes of land level at the station positions. Contours at 2-mm/y. intervals (Fig. 66) are similar to those of Figure 65 from regression

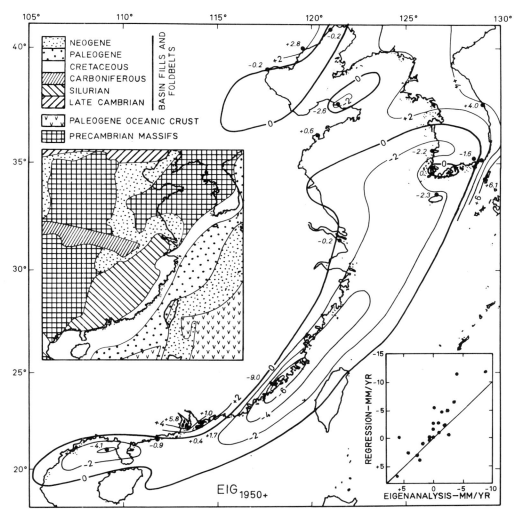

FIGURE 64. Average rate of change of relative land elevations along coastal eastern Asia (2 mm/y. contours) from eigenanalysis of station records after 1950 (individual station rates indicated by italics). The pattern is somewhat smoother than that of Figure 63 because of fewer station records and possibly from uniform time spans. Compare regression and eigenanalyses in plot at bottom right. Insert at upper left outlines massifs and foldbelts discussed in text. From Emery and Aubrey (1986a, fig. 4) with permission from Elsevier Science Publishers.

analyses except for lesser detail permitted by the fewer station records and numerical approximations.

Several previous studies have been made by Japanese scientists who used tide-gauge data in very different ways, but each study led to the conclusion that tectonic movements of the land were important. Hayashi (1969) used only a few tide gauges but had 20,000 km of geodetic leveling made between 1898 and 1950. He assumed that sea level had not changed during the period of geodetic leveling, and his maps of vertical land movement resemble ours. Next, Dambara (1971) used eight tide-gauge records and geodetic surveys of 1895, 1930, 1950, and 1965, computing differences in elevation by averages for 1° squares and attempting to correct for changes in sea level (an uncertain procedure, because his tide-gauge records exhibited variations between +1.3 and −4.9 mm/y., little of which was caused by sea-level change alone). Nevertheless, his 2-mm contour-interval map has patterns similar to ours that he attributed to subsidence along the Pacific side of the northern islands, uplift along the central region, and horizontal plate movements in the southwestern islands. Last, Kato and Tsumura (1979) used monthly tide-gauge records from 96 stations, removing most meteorological and oceanographic components and secular changes of sea level derived from tide-gauge stations within nine regions of Japan. Remaining anomalies were attributed to tectonism caused by underthrusting of the Pacific crustal plate complicated by convergence of the Philippine plate against Japan. Lensen (1974) compared results of repeated precise levelings

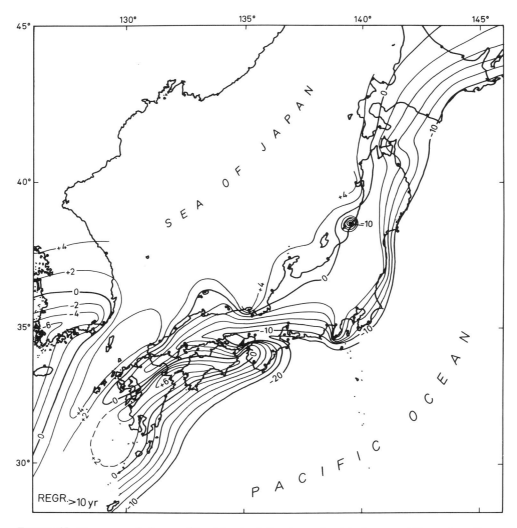

FIGURE 65. Mean annual change of land level relative to sea level in Japan and Korea from linear regression analyses of 97 tide-gauge stations (dots) having records longer than 10 years and within the period 1930 to 1980. Contour interval is 2 mm/y.; note the range between more than −20 mm/y. of subsidence and +6 mm/y. of uplift. Modified from Aubrey and Emery (1986a), GSA Bulletin, v. 97, p. 194–205, fig. 2.

with tide-gauge and other data of the large island of Shikoku in southern Japan. The pattern of his contours of vertical uplift on Shikoku after a large earthquake in 1946 corresponds closely with the contours based on tide-gauge records (Fig. 65). He found many examples of secular vertical land movements interrupted and opposed (opposite movements) by earth shifts during earthquakes and by precursor deformation. Commonly, the vertical land movements associated with earthquakes relaxed after the movement, thereby reducing the total throw of the faults. Some large earth movements even reversed direction later, so that an original normal fault later became a thrust fault, or vice versa. From studies in Japan Thatcher (1984) came to similar conclusions. Lensen had observed similar oscillatory land movements in New Zealand, and Neev et al. (1987) reported them in coastal Israel/Sinai. Such movements complicate any extrapolation of earth movements from the recent past into the future.

At least 500 radiocarbon dates and vertical displacements along the coasts have been reported for samples of peat, shells, wood, and coral by Japanese investigators. Most of the results were summarized by Fujii et al. (1971) and Ota et al. (1982), who reported that sea level rose from below −100 m 18,000 y. B.P. to about +3 m 5000 to 6000 y. B.P., fell to about −2 m 2000 y. B.P., and subsequently rose to its present position. The data exhibit considerable scatter, part of which is attributed to the predominant use of peat that can accumulate above sea level in areas of high ground-water levels, but probably much of the scatter reflects subsequent vertical tectonic movements of the land. Fujii et al. (1971) attempted to obtain the rate of eustatic change of sea level by correcting for an assumed 1.0- to 1.3-mm/y. subsidence of alluvial plains,

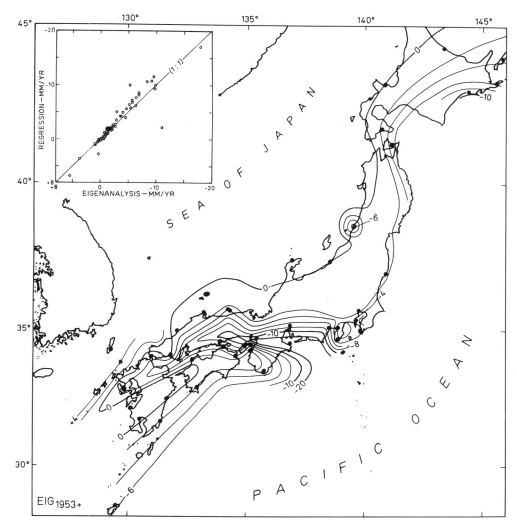

FIGURE 66. Mean annual subsidence and uplift of same area as Figure 65 but based upon eigenanalysis of 51 tide-gauge stations having records longer than 15 years between 1953 and 1980. Modified from Aubrey and Emery (1986a), GSA Bulletin, v. 97, p. 194–205, fig. 3.

which we consider is an unreasonable method because of large variations in vertical movement shown by our Figures 65 and 66. During the much longer Quaternary Era, marine terraces of Japan have been uplifted an average of about 500 m (Sugi et al. 1983), yielding an uplift of the coasts for that 2-m.y. time span that averages about 0.25 mm/y. Although not reported, averages for opposite coasts of Japan would be interesting. A more modest uplift of the marine terrace formed during the last interglacial stage (0.1 m.y. B.P.) has been warped upward to the south as much as 190 m, according to Lensen (1974, Fig. 12). Pirazzoli and Koba (1989) and Kawana and Pirazzoli (1990) studied Holocene levels in the Ryukyu Islands of Japan. They found on two islands a Holocene high stand at about 1.2 m dating to 2800 to 4500 y. B.P. followed by a drop. On other islands there was no evidence of higher stands of sea level.

Results obtained from tide-gauge records (Figs. 65, 66) alone were obtained in a simpler way than the methods used by the Japanese scientists because we did not attempt to isolate tectonic from eustatic processes at this early stage of understanding of the records. The pattern of subsidence and uplift indicated in these figures is explained satisfactorily by the topographic and structural relations mapped in Figure 67 that have been described by both land and marine geologists in many publications cited by Aubrey and Emery (1986b). On land, the Median Line that extends along the axis of the southwestern half of the main island of Japan (Honshu) separates the older continental crust at the north side from the crust at the south. The latter consists of high-pressure metamorphic rocks at depth and at outcrops along the Median Line; elsewhere they are buried beneath sediments of Late Jurassic through Quaternary age and intruded by igneous rocks. The small narrow plate between the Median Line and the Kyushu–Nankai Trough is deformed and ends at the Fossa Magna and the Tanakura Fault. Farther north the rocks consist of late Paleozoic and early Mesozoic metamorphics and

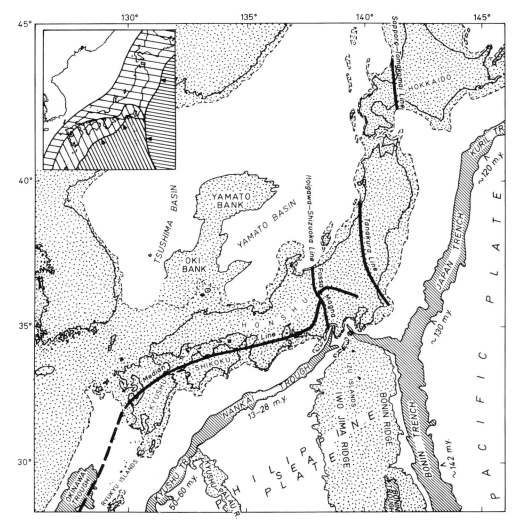

FIGURE 67. Tectonic elements of Japan and vicinity as compiled by Aubrey and Emery (1986a, fig. 7) from previous publications by other workers. Shelf break–dashed line, positive elements–dotted pattern, trenches and troughs–diagonal hatching, plates–named with dates of initial subduction in m.y. Insert indicates crustal plates and their present direction of movement. With permission from GSA Bulletin, v. 97, p. 194–205, fig. 7.

sediments associated with Paleogene granites and volcanics. The volcanic rocks continue northward through Hokkaido and then northeastward along the Kuril Islands.

Dominant features on the ocean floor are two sets of trenches/troughs. The Bonin–Japan–Kuril Trench lies along the boundary between the Pacific Ocean floor and the Bonin–Honshu–Hokkaido–Kuril Islands Ridge. The other set, Ryukyu–Kyushu–Nankai Trough, lies south and southeast of Japan, separating it from the Philippine Sea (Fig. 67). These trenches and troughs separate the Pacific and Philippine oceanic plates from the Japan continental plate (see insert at upper left corner of Fig. 67). The two oceanic plates actively converged on the continental plates beginning prior to 120 m.y. B.P. for the Pacific plate and about 60 m.y. B.P. for the Philippine plate. Present activity is well shown by a westward increase in depth to 700 km of the hypocenters along the Benioff plane related to the Bonin–Japan–Kuril Trench and to depths of 300 km west of the Ryukyu Islands (Fig. 68). The Okinawa Trough west of the Ryukyu Islands may be caused by back-arc spreading that began only 5 m.y. B.P., or perhaps more likely it is a relic of former subduction of the Pacific plate that also is marked by the subsidence embayment along the Inland Sea of Japan.

Westward underthrusting of the Pacific plate in the absence of thick land-derived sediments on the ocean floor is considered responsible for the rapid land subsidence along the eastern and southeastern coast of Japan, probably through tectonic erosion of the bottom of the Japan plate, in the manner suggested by Hilde (1983) or that by von Huene and Lallemand (1990). The subducted material evidently releases magma at depth farther west to form both intrusive and extrusive rocks in the overlying continental plate and uplifting the

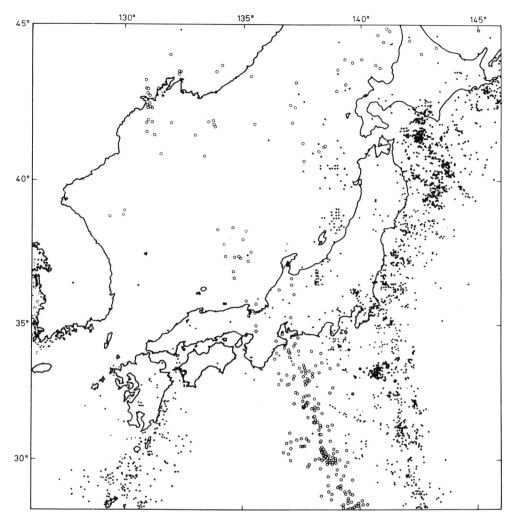

FIGURE 68. Distribution of hypocenters of earthquakes having magnitude > 4.5 between 1961 and 1978. Symbols and depths are: dots < 70 km, circles > 200 km. From Aubrey and Emery (1986a), GSA Bulletin, v. 97, p. 194–205, fig. 8.

coast along the Sea of Japan. A change in direction of plate movement about 45 m.y. B.P. activated the Philippine plate that moved north-northwestward to underthrust western Japan along the Ryukyu–Kyushu–Nankai Trough. This was a general time of broad reorganization of plate movements in the world ocean, including the collision of India with southern Asia, the bend in the Emperor–Hawaiian Seamount Chain, and perhaps initiation of high-latitude spreading belts in the Atlantic Ocean. The subsidence embayment along and beyond the Inland Sea of Japan probably was inherited from an earlier phase of Pacific plate subduction, but it still continues to subside. Less active movement of the Philippine plate appears merely to have pushed northward a sliver of the continental crust of southwestern Japan to leave basins and broad bays near the Median Line as remnants of a former continental slope.

Contours showing rates of vertical relative land movement in Korea differed in the presentations for Japan by Aubrey and Emery (1986a) and for eastern Asia by Emery and Aubrey (1986a), because new tide-gauge data became available between the dates of preparation of these articles. The area of Korea in Figure 63 was restudied and unified in Figure 65, and the pattern for contours of changing land levels in Korea is shown to be similar to that for China and unlike that for Japan, as discussed and illustrated in the preceding text section on Eastern Asia and Figures 62 to 64.

In summary, the contours of Figures 65 and 66 drawn from the tide-gauge records of Japan are supported by vertical movements of the land inferred from topographic and tectonic structures of the region, rock types and distributions, and present seismic activities associated with the margins of the oceanic and crustal plates. These characteristics do not continue into Korea and neither do the patterns of vertical crustal movements; thus, Korea is part of a different tectonic regime. With Japan's differential vertical movements between maximum land subsidence and maximum uplift

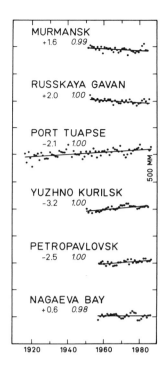

FIGURE 69. Plots of changes in relative land levels at all six reported tide-gauge stations in the U. S. S. R. other than the eight stations that border the Baltic Sea (shown on Fig. 42). Beneath station numbers are mean annual change of relative land level in mm/y. (vertical) and t-confidence of regression analysis (slant). Data are from PSMSL.

totalling about 30 mm/y. one cannot reasonably expect to identify and evaluate a component of eustatic rise of sea level, although that has been attempted by others. At best, eustatic rise of sea level can provide only a bias on the trends of vertical movement of the land. Tectonism of the crust evidently dominates eustatism of the ocean water in this region.

Northern Eurasia

The coast of northern Eurasia (east of Fennoscandia) borders only the Union of Soviet Socialist Republics, an immense area having only six tide-gauge station records beyond its eight in the Baltic Sea (Fig. 42) as listed by the PSMSL (Pugh et al. 1987; and later updates on magnetic tape). Although these six station records are excellent ones that span 31 to 71 years with t-confidence levels of 0.98 to 1.00 for regression analyses (Fig. 69), they are so far apart that they yield only the most general information about changes in relative land levels.

The westernmost station is at Murmansk, where a rise in land level is associated with crustal rebound after melting of the Pleistocene ice sheet mainly farther west (Fig. 42). The next station, on Novaya Zemlya at Ruskaya Gavan about 1300 km to the northeast, also exhibits a rise of land level that averages +2.01 mm/y. The adjacent northeastern half of the island still retains a cover of glacial ice, so presumably this uplift also is related to partial crustal rebound during postglacial times. Novaya Zemlya is the only one of four island groups off Siberia in the Arctic Sea that had Pleistocene ice sheets (Flint 1971) that reports results from a tide gauge. The mainland east of the Ural Mountains had glaciers only on mountain ranges, where they were thin. Absence of widespread ice sheets in Siberia is attributed to lack of enough moisture to make much snow, but the very low temperatures produced extensive areas of permafrost.

Rather different is the record at the third station, Port Tuapse, at the northern coast of the Black Sea. It reveals a relative sinking of the coast at a mean rate of −2.1 mm/y. just east of the edge of Figure 47 — reasonable in view of the belt of Alpine mountain building that extends along the coast at Port Tuapse from farther west where tectonism associated with the mountain building has produced erratic vertical land movements on tide-gauge records.

The three remaining Soviet tide-gauge stations are in the far east of the U. S. S. R. One, Yuzno Kurilsk, is shown near the northeastern corner of Figure 65 just east of the Japanese island of Hokkaido. With its indicated mean annual change of −3.2 mm/y., it indicates that the site is a continuation of the Pacific border of Japan where sinking of the crust is attributed to tectonic erosion of the granitic crust by the subducting adjacent oceanic Pacific plate along the Kuril Trench (Figure 67). The same tectonic setting may well be the reason that the next Soviet tide-gauge station to the north (Petropavlovsk) also indicates land submergence, at a mean rate of −2.5 mm/y. This station is in southeastern Kamchatka adjacent to the northern part of the Kuril Trench. The last station is at Nagaeva Bay near Magadan, a city at the northern coast of the Sea of Okhotsk. The record here indicates a land uplift of +0.6 mm/y. The relative positions of these two last tide-gauge sites are the same as those between Japanese tide-gauge stations that show sinking along the Pacific Ocean coast and those that show uplift along the Japan Sea coast. Possibly the same effects of subcrustal erosion versus upward displacement of the crust by subducted rock continues northeastward from Japan to the U. S. S. R. It is to be hoped that Soviet investigators can obtain additional tide-gauge records and other geodetic information to investigate this apparent northeastward extrapolation of tectonic control of land levels.

Western North America

After viewing the extreme effects of strong oceanic plate subduction on land levels in Japan, it is reasonable to investigate next the effects of lesser subduction along the also-tectonic Pacific coasts of Alaska, Canada, United States, and Mexico. This coast of western North America has subduction that is oblique and is marked by many earthquakes but fewer and at lesser depth of hypocenters than in Japan (Espinosa et al. 1981). The 6200-km coast (Fig. 70) contains 38 tide-gauge stations having records 13 years or longer, of which 36 have t-confidence levels higher than 80 percent for least-squares

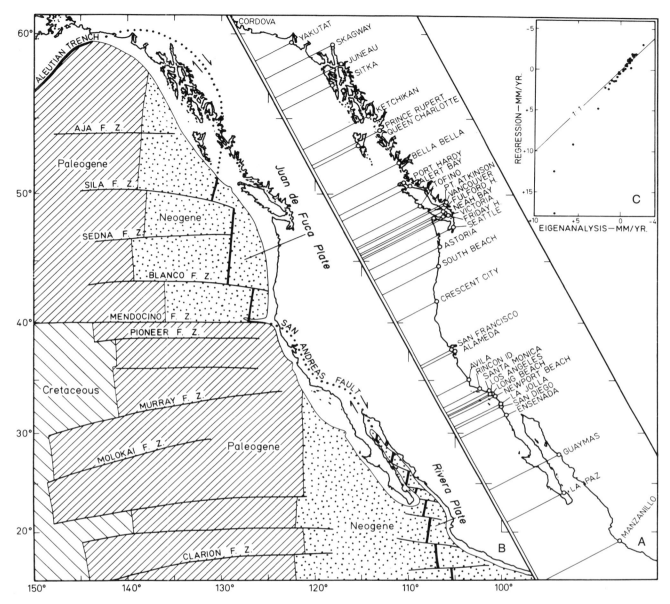

FIGURE 70. Pacific coast of North America. From Emery and Aubrey (1986b), Journal of Geophysical Research, v. 92, p. 13941–13953, fig. 1, copyright by the American Geophysical Union.
A. Positions of tide-gauge stations having useful records and their projection to a roughly parallel straight line.
B. Relation of tide-gauge stations to oceanic plates and their ages.
C. Relation of elevation changes derived from linear regression analysis versus eigenanalysis of tide-gauge records.

regression lines. Five records began before 1911, and most of them continued through 1982. Unusual events at two stations (a 14-m fault uplift at Yakutat in 1899 and about 9-m sinking of land at Long Beach because of excessive withdrawal of petroleum prior to 1963 before unitization of the field; Emery 1960, p. 318) were avoided by using only tide-gauge records made after these dates. Eigenanalysis for 30 records between 1920 and 1982 (Emery and Aubrey 1986b) revealed that the first three spatial eigenfunctions contain 68, 11, and 7 percent of total variation in the data. The first two spatial functions were multiplied by the slope of their respective temporal functions and summed to yield synthetic mean annual changes of land level (Fig. 71).

The results of both regression and eigenanalysis reveal considerable variation along the Pacific coast of North America, far too much variation to be the result of eustatic rise of sea level alone, and the record at nearly half of the stations would require a fall of sea level. Sharp boundaries between regions having different directions and rates of change in level occur at Cordova, Skagway, and Sitka at the far north. Farther south the changes are smaller, but on average relative subsidence increases toward the south. We interpret these changes to be

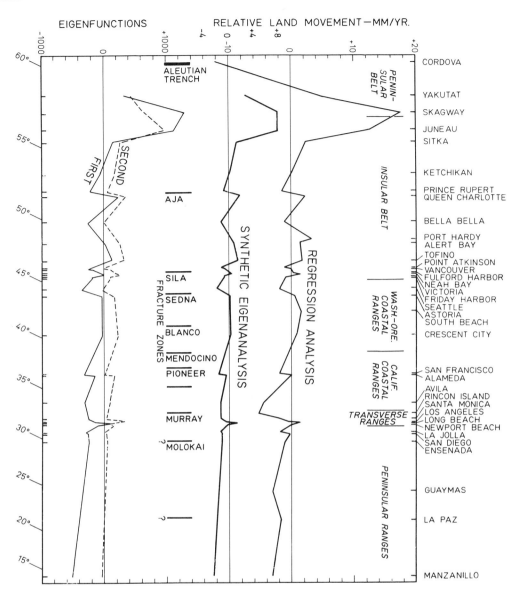

FIGURE 71. Mean annual changes of land level (mm/y.) plotted along straight line of Figure 70A for regression, eigenanalysis, and first and second eigenfunctions. Data are from PSMSL. From Emery and Aubrey (1986b), Journal of Geophysical Research, v. 92, p. 13941–13953, fig. 2, copyright by the American Geophysical Union.

the result of land movements—at the north associated with intersection of the coast by the Aleutian Trench, and farther south by other effects of the movement of oceanic plates. The positions of these plates, fracture zones between them, and directions and rates of plate movements are indicated in Figures 70 and 72.

Beginning at the north, the land level at Cordova is interpreted as sinking (compared with rising land at stations both north and south of Cordova). The sinking is caused by the Aleutian Trench in a region of subduction of the Kula plate that is moving north-northwesterly at 12 cm/y. Depression at the trench occurs along its entire length, but in the east the rate of deposition of river-contributed sediments is faster than the rate of structural depression of the trench. Farther west, the trench receives little sediment from the Aleutian Islands, and there the trench also is older (perhaps having begun only during the Pliocene at the east; von Huene 1972). Uplift of land at Yakutat, Skagway, and Juneau is attributed to resistance of the continental crust with its cover of sediments to being underthrust beneath the crust north of the Aleutian Trench and its easterly extension. This region also is marked by intense faulting. Barnes (1985) measured the gravity in Glacier Bay (between Juneau and Sitka), concluding that uplift there probably is not caused by crustal rebound after

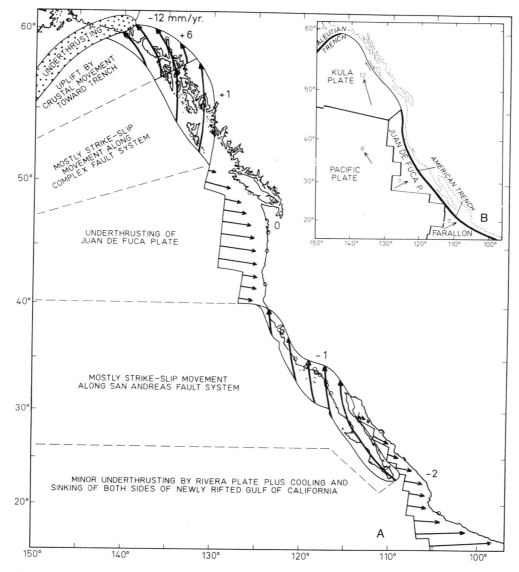

FIGURE 72. Movements of oceanic plates and their inferred effects on vertical movements of coastal land levels of western North America. Numbers for each coastal segment indicate averages of mean land levels in mm/y. from eigenanalysis at tide-gauge stations within those segments. Numbers in insert at upper right show rate of oceanic plate movement in mm/y. From Emery and Aubrey (1986b), Journal of Geophysical Research, v. 92, p. 13941-13953, fig. 4, copyright by the American Geophysical Union.

melting of Pleistocene glaciers. Between Sitka and Crescent City (just north of the Mendocino Fracture Zone), the tide-gauge records exhibit minor differences in mean rates of elevation, but those changes apparently are closely related to fracture zones (or their eastward extensions beyond the spreading belt) that are being subducted by the plate movement. The movement is about 7 cm/y., slower than that of the Kula plate at the north, but movement is directly toward the land (Atwater 1970). Unlike the subduction along Japan, subduction here is in an area of rapid contribution of sediment from land that has filled a possible trench and smoothed the surface of the oceanic plate. Thus, the oceanic plate moves without eroding the bottom of the overlying continental plate, agreeing with the theory outlined by Hilde and Uyeda (1983).

Between the Mendocino Fracture Zone and the Gulf of California, relative land levels are falling slowly but irregularly, in accordance with the findings that the sea-floor spreading belt was subducted by eastward plate movement about 10 m.y. B.P. and that subsequent plate movement of about 6 cm/y. is northward, oblique to the coast and along the San Andreas Fault (Figs. 70, 72). This movement (S. M. Ward 1988) has been well confirmed recently by VLBI measurements of signals from celestial radio sources. Best known of the earthquakes that accompany movements along this right-lateral strike-slip fault is the one at San Francisco on 18–19 April 1906, having a magnitude of 8.3. It was preceded and followed by many other earthquakes along different parts of the fault. The most recent major movement was a right-lateral slip of 1.9 m at a depth of 17.6 km that occurred on 17 October 1989—the Loma Prieta earthquake of 7.1 magnitude (U. S. Geological Survey Staff 1990) that was accompanied by a tsunami and by submarine slides in Monterey Bay (Schwing and Norton 1990). Damage mostly between San Francisco and Monterey Bay has estimated recovery costs of $6 billion. The sharpest change in land level is uplift at tide-gauge stations near the Transverse Ranges that may constitute an eastward extension of the Murray Fracture Zone and that have been shown by precise leveling (Gilluly 1949) to be undergoing uplift of about 5 mm/y. The bend of the San Andreas Fault where it crosses the Transverse Ranges probably also is related to the uplift. The last segment of the Pacific coast in Figure 70 is between the head of the Gulf of California and Manzanillo, Mexico, beyond which it probably continues still farther south. The tide-gauge stations here indicate a relative rise of sea level, with a contribution by sinking of land level caused by crustal cooling after rifting along the spreading belt in the Gulf (Curray et al. 1982b). The oblique movement of the oceanic plate may not much affect the vertical movement of land along the coasts; nevertheless, movement during the Cenozoic has produced large variations in elevations of marine terraces in coastal California and Mexico (Wahrhaftig and Birman 1965; Berryman 1987; and many other workers, including Emery 1960). The same is true for other coastal regions of the world especially in subduction belts.

Since about 1980 much has been learned about continental fragmentation, microcontinents, and microcontinent accretion to continental margins. These continental fragments have been identified in the Pacific Ocean (Ben-Avraham et al. 1981; Nur and Ben-Avraham 1982) and on the North American continent as melanges of exotic terranes (Maxwell 1974; Coney et al. 1980) associated with deep-ocean sediments, oceanic basement rocks, and ophiolites. Uchupi and Aubrey (1988) investigated the possibility that the juxtaposition of exotic and normal terranes may be responsible for variations of land-level changes along the Pacific coast of North America—indicating a need to distinguish between vertical movements of small rigid crustal units and surface effects of subduction at depth. Accretion to the continental margin of exotic (or allochthonous) terranes implies their being welded to the autochthonous craton accompanied by thickening during plate convergence, faulting, and lateral displacement by plate translation, followed by erosion and/or burial under later sediments and volcanics—all giving rise to vertical movements to regain isostatic equilibrium. Rates and changes of vertical land movement (Figs. 70, 72) for the terranes at the coast reveal some relationship to terrane boundaries, but there also occur variations in directions and rates within the identified terranes, indicating that the latter may be composites of varied composition or thickness. Quite clearly, oceanic plate subduction can affect complex exotic terranes as well as simple continental crusts (if such exist), and the complicated variations of land-level changes can be caused by local variations in the terranes as well as by variations in rate and direction of subduction. Because rheological properties may vary from terrane to terrane, the response of a composite continental margin to glacial loading/unloading and other tectonic stresses is expected to be complex and non-uniform.

A further tectonic complication for interpreting land-level changes is isostatic adjustment of continental crust to unloading by glacial melt. Peltier (1986) modeled vertical land movements to be expected from deglaciation, but that pattern is relatively smooth along the Pacific coast (Emery and Aubrey 1986b, fig. 7), quite unlike the pattern of changing land level that is indicated by tide-gauge records. Accordingly, the effect of deglaciation on the land, like that of the resulting rise of sea level by addition of meltwater, may be present but only as a minor bias on the larger-amplitude land movements presumably caused by continued long-term plate tectonics.

Central America

Aspects of tide-gauge records at four stations in northern Mexico were discussed in the previous text section on Western North America (Figs. 70–72). Records at four other stations in southern Mexico and in the rest of Central America were considered by Aubrey et al. (1988) in a general study for South America, Central America, and the Caribbean islands. Subsequently, about half the original 27 records of the 1988 publication were updated by PSMSL, making possible an improvement in the previously used records and the addition of six more stations. In order to take advantage of these new data, the original maps were revised and the three regions are discussed separately beginning with Central America.

For Central America there are now 18 station records longer than 10 years and having t-confidence levels of 0.80 or higher for regression analysis of changes in mean relative land levels. Geographic positions of the 18 stations are shown on Figures 70 and 73, and their plots of change in level on Figure 74. Land subsidences of −2.2 to −3.1 mm/y. are indicated for the three stations in and just south of the Gulf of California (Figs. 70–74), perhaps reflecting crustal cooling of the newly opened Gulf (Pitman 1978; Aubouin et al. 1982). The four station records in southern Mexico are less uniform than those farther north. The ones at Veracruz and Progreso indicate relative land subsidence of −1.5 and −5.2 mm/y., respectively, but between them at Carmen the land is emerging at +3.0 mm/y. Reference to the tectonic map (Fig. 75) shows that the two sites of submergence are located within sediment-filled basins (the Veracruz and Quintana Roo basins) and the area of emergence is on a structural arch between the basins; thus, the tide-gauge records correspond

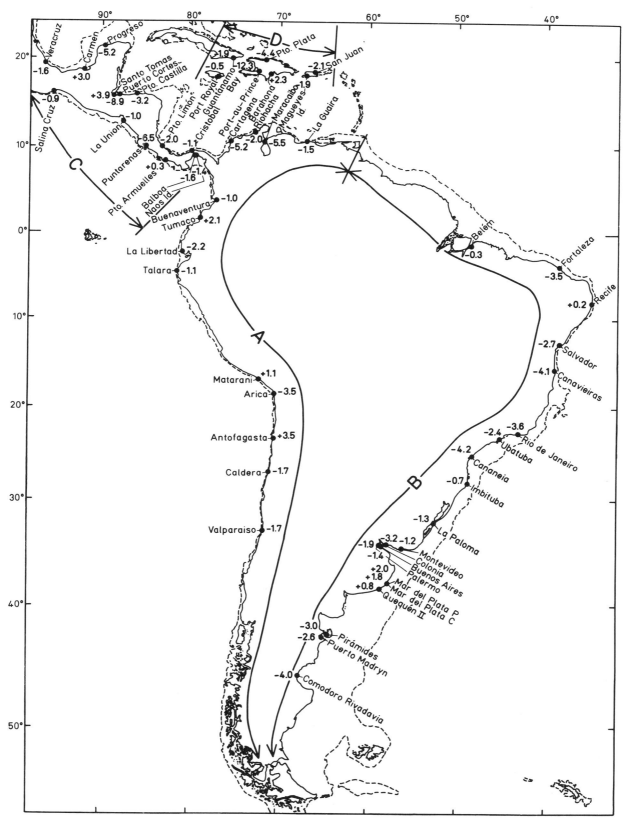

FIGURE 73. Tide-gauge stations of South America, Central America, and the Caribbean islands that span 10 years or more and have 95% confidence levels higher than 0.80 for slopes of linear regression lines. Slopes, indicated by numbers next to station positions, indicate mean annual change of relative land level (mm/y.; − for sinking land, + for rising land). Dashed line is 100-fathom (183-m) contour. Redrawn from Aubrey et al. (1988, fig. 1).

FIGURE 74. Plots of mean annual changes of relative land level at tide-gauge stations of Central America (whose positions are shown on Figs. 70A and 73) and of the Caribbean islands (whose positions are shown in Fig. 73). Beneath the names of the stations are given the mean annual change of land level in mm/y. (vertical numbers) and the *t*-confidence level (slant numbers). Data are from PSMSL.

with longer term tectonic movements. The Veracruz basin borders the south side of the Trans-Mexican Volcanic belt (Fig. 75), a left-lateral megashear between the allochthonous Yaqui and Maya West blocks (Martin and Case 1975; Anderson and Schmidt 1983). This megashear may be an extension of the Tamayo transform fault that was associated with opening of the Gulf of California during the Pliocene (Aubouin et al. 1982). The Quintana Roo basin lies beneath the Yucatan Peninsula (Viniegra 1971; Locker and Sahagian 1984), where basin subsidence, perhaps augmented by cooling of igneous rocks emplaced during Late Cretaceous, may well account for the marked subsidence at Progreso.

The tide-gauge records at Santo Tomas (Guatemala) and Puerto Cortés and Puerto Castilla (Honduras) are even less uniform than those of Mexico, with their relative land movements of +3.9, −8.9, and −3.2 mm/y., respectively. In contrast is the −1.0 mm/y. movement at La Union (El Salvador) on the opposite, or Pacific, coast. Similar slow subsidence may be indicated by two stations at San Jose (Guatemala) about halfway between La Union and Salina Cruz, but these stations have records too short and too low in their *t*-confidence levels to be considered seriously. All three of these Pacific coast stations are bordered on the west by a belt of obducted oceanic crust and on the east by a foldbelt or by Quaternary volcanic rocks. The extreme variation in vertical land movements along the east-west Caribbean coast of the Gulf of Honduras is ascribed to tectonism. The adjacent land area of Honduras and Nicaragua and its northeastward projection as the Nicaragua Rise comprise the Chortis block, an allochthonous terrane probably derived from the Pacific Ocean and that was attached to the Caribbean side of Central America during the Mesozoic (Emery and Uchupi, 1984, p. 343–346). The northwestern side of the Chortis block is the seismically active right-lateral Montagua fault, which extends from southeastern Cuba along the Cayman Trough to the southwestern corner of the Gulf of Honduras and thence onto land in Guatemala. During the Guatemala earthquake of 4 February 1976, the Montagua fault of magnitude 7.5 had vertical displacements down to the north (Plafker 1976). Pliocene and Quaternary strike-slip motion along it and along associated faults developed horsts and grabens (Pinet 1975; Manton 1987). Puerto Cortés and Puerto Castilla are located within two of these grabens, and Santo Tomas may well be on a horst, accounting for their differences in relative vertical land movements.

The southernmost tide-gauge stations of Central America are in Costa Rica and Panama. The four on the Pacific side are atop obducted oceanic crust (Bowin 1975a), and the two on the Caribbean side are on Neogene sediments (Figs. 73, 75) deformed by the obduction (Lu and McMillen 1982). All are south of the Chortis block. With such an obducted terrane one might have expected emergence of the land at all six stations, but only the one at Puerto Armuelles exhibits a rise—of only +0.3 mm/y. (Figs. 73, 74). The largest subsidence rate is at Puntarenas (−6.5 mm/y.) in a graben on the obducted belt (King 1969). All other tide-gauge records, Puerto Limón to Naos Island, have a limited range of coastal subsidence (−2.0 to −1.4 mm/y.), presumably indicating that effects of the obduction have decreased.

In parts of the world that have numerous acceptable tide-gauge records and several more-or-less parallel coasts, one can visualize relationships of land movements to geology and plate tectonics by use of contours showing tide-gauge rates of land movement throughout a region, as for northern Europe (Fig. 42) and Japan (Fig. 65). In other areas where the spacing of acceptable tide-gauge records is much wider than the spacing of major geological structures (inadequate sampling) or where the coast is a simple straight one, recourse must be limited to profiles (as for those along western North America; Figs. 70, 71). For Central America (Fig. 73) there are two coasts, but the geological structure appears to be too complex to be resolved by the available number of tide-gauge records, so the regional relationships of relative land movements can be viewed only by a general profile plot (Fig. 76C). This plot shows large differences in land movements within short distances, especially between Carmen and Progreso, Progreso and Santo Tomas, Santo Tomas and Puerto Cortés, Puerto Cortés and La Union, La Union and Puerto Limón, and Puerto Limón and Puerto Armuelles. Obviously, the

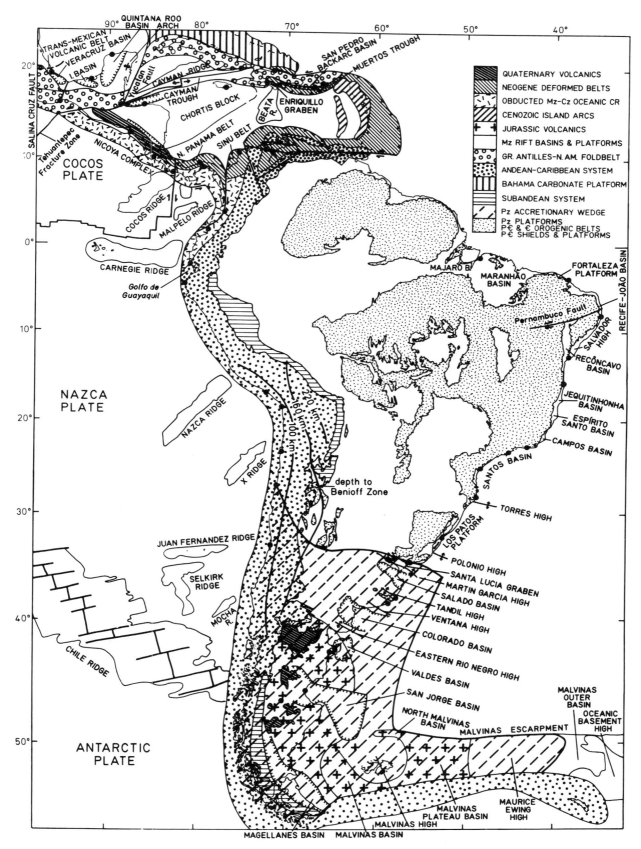

FIGURE 75. Tectonic map of Central America, South America, and Caribbean regions compiled from King (1969), Bowin (1975a), Ponte and Asmus (1976), Herz (1977), De Almeida et al. (1978), Case and Holcombe (1980), Urien et al. (1981), Aubouin et al. (1982), Biju-Duval et al. (1982a), Forsythe (1982), and Bovis and Isacks (1984). Mexico, Yucatan, and the Chortis block are allochthonous terranes attached during the Mesozoic. From Aubrey et al. (1988, fig. 2) with permission of Elsevier Science Publishers.

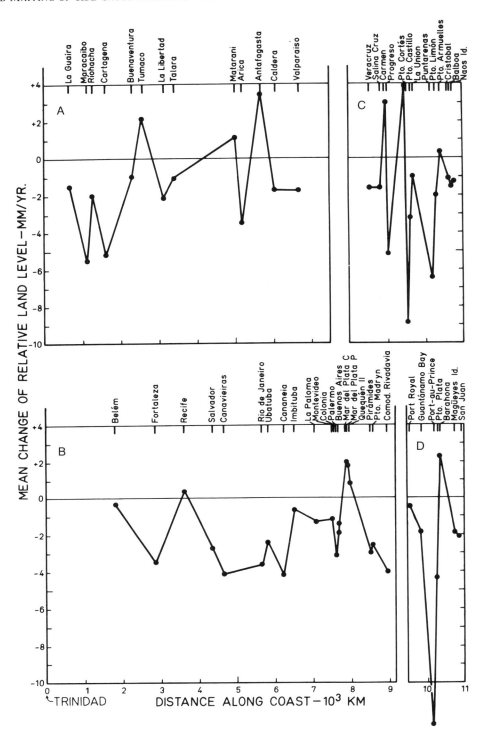

FIGURE 76. Comparison of results from linear regression analyses in the four regions indicated in Figure 73. Modified from Aubrey et al. (1988, fig. 3).
A. Pacific coast.
B. Atlantic coast.
C. Both coasts of Central America.
D. Caribbean Island coasts.

geological structure is more complex relative to spacing of tide-gauge records than in most other parts of the world and especially along the coasts of broad continents. Such complexity is expected *a priori* given the active and varied geology of the region.

South America

As for other large continents and even for Central America, the tide-gauge stations in South America are located along two fairly straight or broadly curved coasts with a broad area of complex geological structure separating them. Thus, one cannot reasonably draw contours across the wide expanses of land on the basis of tide-gauge records on opposite coasts, and we must be content with interpretation of the records along profiles that roughly parallel the Pacific and Atlantic coasts (Figs. 73A and B, 76, 77).

For South America there are nine records from the Pacific coast to which four others from Colombia and Venezuela facing the Caribbean Sea are appended, because all 13 stations are underlain by rocks of related geological origin and tectonic history (Figs. 75A, 76A). Along the rest of the Caribbean coast and all of the South Atlantic coast there are 20 tide-gauge stations having useful records (Fig. 76B). These tide-gauge data are derived primarily from PSMSL, with update of some provided by D. B. Enfield of Oregon State University. All of the 33 station records for South America have t-confidence levels for regression analyses higher than 0.80, and 24 of them are 0.90 or higher. The 13 stations having the highest t-confidence levels (0.99 and 1.00) also have the longest time spans of 16 to 36 years, with time spans for other stations ranging from 12 to 28 years. Comparison shows that the 13 records having highest t-confidence levels and longest time spans include the widest range of mean annual change of relative land level (-5.5 to $+3.5$ mm/y.) in contrast with -4.1 to $+2.1$ mm/y. range for the other station records. Therefore, the wide range of mean annual change of relative land level revealed by Figure 77 cannot be caused only by variations in accuracy of positioning of regression lines through the data points, but more likely they reflect variations in tectonics (uplift or subsidence) of crustal blocks, especially along the volcanic and subducted Andean–Caribbean system (Fig. 75). Details of these crustal movements were discussed by Aubrey et al. (1988), and most of the following summary is based on that discussion.

The four northeastern tide-gauge stations of the Pacific profile (Figs. 73, 77; actually in the western Caribbean Sea) are atop a wedge of Neogene sediments deformed by eastward (relative to South America) movement of obducted Mesozoic–Cenozoic oceanic crust (Fig. 75). This wedge consists of two parts, the Middle Eocene San Jacinto belt on the south and the Pleistocene–Holocene Sinu belt on the north (Duque-Caro 1979). The Sinu belt still is moving, as indicated by deformation of turbidites at the base of the continental slope

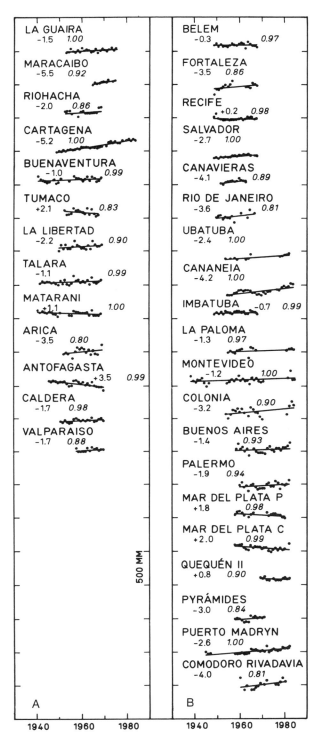

FIGURE 77. Plots of mean annual changes of land level at tide-gauge stations of South America, whose positions are shown on Figure 73. Beneath the names of the stations are given the mean annual change of land level in mm/y. (vertical numbers) and the t-confidence level (slant numbers). Data are from PSMSL augmented by data provided by D. B. Enfield of Oregon State University.
A. Pacific coast.
B. Atlantic coast.

east and west of the Magdalena Fan (Lu and McMillen 1982; Kolla et al. 1984; Vitali et al. 1985). The eastward movement appears to be a continuation of that present along the northeast side of Panama (Wadge and Burke 1983), and it is interrupted by numerous strike-slip faults (Fig. 75) and affected by underthrusting of oceanic crust along Venezuela (Biju-Duval 1982b). All station records exhibit relative land submergence, with fastest submergence at Maracaibo and Cartagena (-5.5 and -5.2 mm/y., respectively), probably associated with vertical movements along faults and with compaction of deltaic sediments at Maracaibo due to pore fluid extraction. Subsidence at Lake Maracaibo, Venezuela, was first reported by Kugler (1933) and discussed in detail by Poland and Davis (1969). Extraction of hydrocarbons from the Bolivar Coastal Fields, discovered in 1917 and expanded during the next decade, caused compaction of shale layers because of increased effective pressure following removal of hydrocarbons. This compaction may be accentuated by some compaction of sands. Maximum subsidence coincides with greatest decline in gas pressure and the thickest Miocene producing zones. The greatest subsidence of 3.4 m occurred between 1926 and 1954 northeast of the Lake; lesser subsidence elsewhere mandated construction of a continuous wall nearly 5 km long to protect upland areas (drained former wetlands) from repeated flooding.

The next four tide-gauge records (Buenaventura to Talara) lie atop or adjacent to the obducted oceanic crust along western Colombia and Ecuador (Figs. 73, 77). They exhibit mean annual changes of relative land level between $+2.1$ and -2.2 mm/y., about as erratic as at stations on similar obducted strata along the western coast of Central America.

Farther south along the coasts of Peru and Chile are six tide-gauge stations having acceptable records—Talara to Valparaiso. Their records reveal changes of land level between $+3.5$ at Antofagasta to -3.5 mm/y. at Arica. At Matarani the change is $+1.1$ mm/y., and another station at Chimbote (Lat. 9°05′S., near Talara) records an emergence of $+5.6$ mm/y. Although the t-confidence level for the latter is only 0.72, the true rate of emergence still likely exceeds 3 mm/y. or so. The remaining three stations have rather similar records (-1.1 to -1.7 mm/y.). Thus, three stations show relative land emergence and four show land submergence. The entire length of the coast of Peru and Chile is a belt of plate convergence with the Pacific oceanic plate being subducted down the Benioff Zone and beneath the continent. Segmentation of the descending plate may occur at tear faults in the plate (Barazangi and Isacks 1979), at flexures in the plate (Bovis and Isacks 1984), or perhaps more likely by ridges carried atop the plate (Nur and Ben-Avraham 1982). We note that the two most reliable sites of land emergence, at Matarani and Antofagasta, are positioned near the eastern ends of the Nazca Ridge and an unnamed ridge (Fig. 75), perhaps the remnants of ridges already partly subducted. Subduction of such ridges causes major subcrustal erosion of the overlying continental plate, according to von Huene and Lallamand (1990). Paskoff (1980) discussed late Cenozoic movements within the South American continent. He reported on the movements in northern Chile, citing the high cliffs along northernmost Chile (Lats. 18° to 27°S), the marine terraces around Coquimbo Bay (Lat. 30°S), and the faulting and uplifting of coastal deposits between Lats. 30° and 33°S. These vertical changes are ascribed to glacio-eustatic variations as well as to non-uniform tectonism. He cited Sillitoe's (1974) hypothesis that the sinking oceanic lithospheric slab is segmented, with individual units that subduct independently and at different rates.

The 20 acceptable tide-gauge records from the Atlantic coast of South America are more uniform in their indications of changes in relative vertical land levels (Figs. 73, 75, 76B, 77B). Only four stations indicate emergence, and three are adjacent stations (Quequen II and Mar del Plata C and P: $+0.8$, $+2.0$, and $+1.8$ mm/y., respectively), indicating a region of uplift probably related to volcanism; the other station (Recife) has only a minor rate of uplift ($+0.2$ mm/y) perhaps associated with the Pernambuco fault (Fig. 75). Two stations, Belém and Comodoro Rivadavia, are in structural basins (Fig. 75) and thus are subject to sediment compaction and other deformation. In contrast, the remaining 14 station records indicate coastal submergence (-0.7 to -4.2 mm/y.) with five of them at rates between -3.2 and -4.2 mm/y. As a result, the profile of land-level change along the Atlantic coast of South America is somewhat more uniform than that along the Pacific-Western Caribbean coast (Fig. 76A and B). The mean annual changes of land level of western South America exhibit a diversity of amplitude nearly as large as that of Central America (Figs. 76A and C); however, the similar number of stations is more widely spaced along the longer eastern South American coast. The greater uniformity along the Atlantic coast of South America in part may be the result of the accretion of an ancient (Middle Devonian to Triassic) thick-crusted wedge from the Pacific plate onto western South America south of about Lat. 30°S (Forsythe 1982). After the wedge was accreted, it became rifted during the separation of South America from Africa during the Mesozoic, and afterward the region experienced magmatism during Middle to Late Jurassic (Güst et al. 1985) and again during the Quaternary (De Almeida et al. 1978). The latter corresponds with the Ventana and Tandil structural highs, where tide-gauge records at Quequen II and Mar del Plata C and P exhibit present emergences of the land (Figs. 73, 75).

The Atlantic tide-gauge stations of South America have comparable records except for those at Comodoro Rivadavia and Belém (both in structural basins), at Quequén and the two at Mar del Plata (in a volcanic region) and at Recife (nearness to a major fault). The other 14 station records are in regions of Paleozoic and older crusts that probably have had time to come to isostatic equilibrium; their change of relative land levels range between -0.7 and -4.2 mm/y. This range is small, considering the length of coast and the potential for minor disturbance by local tectonics and other minor factors. These movements are comparable with those near stable shield areas of Africa, India, and Australia, and they will be considered together in a later section of text. These estimates of Atlantic tide-gauge records and our analyses can be compared with previous studies of tide gauges and relative sea-level change. Brandini et al. (1985) using data from longer time spans (from 1906 to 1986 for Buenos Aires, for example), determined regression slopes significantly different from the present estimates. They determined slopes of −1.3 versus +1.3 mm/y. at Buenos Aires, −1.8 versus +1.8 mm/y. at Mar del Plata Club, −3.1 versus −2.6 for Puerto Madryn, and −7.4 versus −4.0 mm/y. at Comodoro Rivadavia. Lacking the Brandini et al. data to analyze, it is impossible to determine which trends are more indicative of the long-term changes in relative sea level. This disagreement underlines the need for caution when interpreting short-span tide-gauge data from a geological perspective. Fulfaro and Suguio (1980) discussed the neo-tectonics of southern Brazil during the Cenozoic, citing the correlation between tectonism and sedimentation along the southern Brazil coast. The neo-tectonism takes the form of reactivation of ancient Precambrian fault lines. Suguio et al. (1980) cited higher than present Holocene sea levels, with two maxima at 5000 and 3300 y. B.P. A minimum occurred 3800 y. B.P. These extremes are geographically variable with gradients along the coast, suggesting neo-tectonism as well as eustasy as the cause. Vertical deformation of the geoid surface is hypothesized as the cause for this vertical movement. Rates of vertical movement are in the range of 0.1 mm/y.

Caribbean Sea

Seven tide gauges having acceptable records are present on islands in the Caribbean Sea (Figs. 73, 78). Their mean annual change of relative land level ranges from −12.3 to +2.3 mm/y., showing that the islands that contain the tide gauges are rising and sinking independently. In fact, the island of Hispaniola with its three tide gauges has the two fastest sinking stations (Port-au-Prince in Haiti; −12.3 mm/y., and Puerto Plata in the Dominican Republic; −4.4 mm/y.) and the only rising station recorded for the Caribbean Sea (Barahona in the Dominican Republic- +2.3 mm/y.). Stations on the adjacent islands of Jamaica (Port Royal), Cuba (Guantánamo Bay), and Puerto Rico (Magueyes Island and San Juan) show sinking at intermediate rates of −0.5, −1.9, −1.9, and −2.1 mm/y., respectively.

Varying rates and directions of vertical land levels are supported by ages and heights of marine terraces and coral reefs on these and other islands and bank tops of the Caribbean Sea (Zans 1958; Kaye 1959; Monroe 1968; Weaver 1968; Horsfield 1975, 1976; Taylor 1980; Emery 1981; and others). The deformation is attributed to eastward movement of the Caribbean plate and convergence of North America and South America (Ladd et al. 1981; Burke et al. 1978; Biju-Duval et al. 1982a; Sykes et al. 1982). The topographic relief, varied rock types, and structural elements in the Caribbean Sea indicate that the blocks that comprise the region are smaller than the distances between tide-gauge stations. Once again, the geology is under-resolved by available tide-gauge data. This conclusion also is supported by the resemblance between the profiles connecting tide-gauge rates of relative vertical land movement within the Caribbean Sea and within Central America (Fig. 76C and D), both areas being dominated by tectonized belts and associated faulting, folding, and volcanism (Fig. 75).

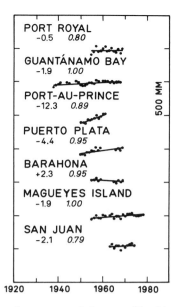

FIGURE 78. Plots of mean annual changes of land level at tide-gauge stations of Caribbean islands, whose positions are shown on Figure 73. Beneath the names of the stations are given the mean annual change of land level in mm/y. (vertical numbers) and the t-confidence level (slant numbers). Data are from PSMSL.

Gulf of Mexico

Shores along the elliptical Gulf of Mexico contain 16 tide-gauge stations (Fig. 79A) having time spans of 16 years or longer. Twelve are longer than 30 years. Fourteen are in the United States and two are in Mexico. Regression analyses of the station records show that the mean annual fall of relative land level is between −1.1 and −2.4 mm/y. at the five easternmost stations (Pensacola to Key West). The −2 mm/y. contour (Figs. 79B and 84B) encloses most of the Florida peninsula (south of Fort Pulaski). Four stations landward of the contour exhibit a sinking of −2.0 mm/y. or less, whereas seven stations seaward of it reveal faster sinking of the land−to −2.9 mm/y. This pattern suggests that the Florida Peninsular Arch (Ocala Arch) of Murray (1961) and Maher (1971) controls the distribution of present rates of relative land subsidence.

Most tide-gauge stations in the western Gulf of Mexico exhibit much faster relative land subsidence, but probably this is local (Fig. 79B). One center of fast sinking is on the

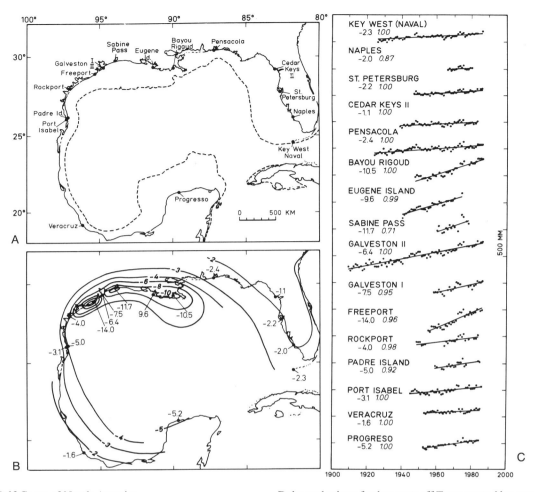

FIGURE 79. Gulf Coast of North America.
A. Positions of tide-gauge stations having useful long-term records.
B. Contours (mm/y.) of mean annual change of land level based upon linear regression analysis of full time spans of all tide-gauge stations. Note the local rapid sinking at two stations on the Mississippi River Delta and others farther west off Texas caused by compaction of deltaic sediments and removal of oil in the sediments.
C. Plots of mean annual changes of land level at the tide-gauge stations; mean annual change of land level in mm/y. (vertical numbers) and t-confidence levels (slant numbers).

Mississippi Delta (Bayou Rigaud and Eugene Island: −10.5 to −9.6 mm/y.) with sinking caused by compaction of very thick deltaic sediments (see, for example, Fisk and McFarlan 1955; McFarlan 1961; Kolb and van Lopik 1966) and by removal of petroleum. Additional sinking that diminishes landward from the coast (Scaife et al. 1983; Deegan et al. 1984) is caused by flood-control measures (levees that prevent flooding and consequent non-deposition of sediment in interfluve areas), and by canal construction for navigation and access to drill sites (that increases dewatering and reduces wetlands by the areas of the canals and adjacent spoil banks). Another center of fast land subsidence is in southeastern Texas (Galveston and Freeport: −6.4 to −14.0 mm/y.), where it is caused by removal of petroleum and groundwater. According to Coates (1983), maximum subsidence in the Houston-Galveston area is 2.75 m with some subsidence throughout an area of 12,000 km² and damage exceeding $10⁹. A third center may be at Sabine Pass (−11.7 mm/y.), but the time span of the record at this station is only 17 years with a regression slope having a t-confidence of only 0.71. Contours and profiles of anomalous local sinking were drawn by Holdahl and Morrison (1974) on the basis of tide-gauge records to 1971 and results of repeated precise levelings at various dates between 1906 and 1973. Intermediate and probably regional sinking rates of −3.1 to −5.0 mm/y. occur farther west in Texas, but there is lesser sinking at Veracruz (−1.6 mm/y.) and faster at Progreso (−5.2 mm/y.); both stations are in areas of coral reefs off Mexico, but the sinking at both sites probably is caused by proximity to faults.

The contours of relative land sinking show that the belt of faster oceanward subsidence that is typical of the Atlantic coast continues around the Florida peninsula into the Gulf of Mexico. Its possible original further continuation past the Mississippi River Delta now is obscured by much faster local sinking caused by pumping of petroleum and groundwater from the coastal strata.

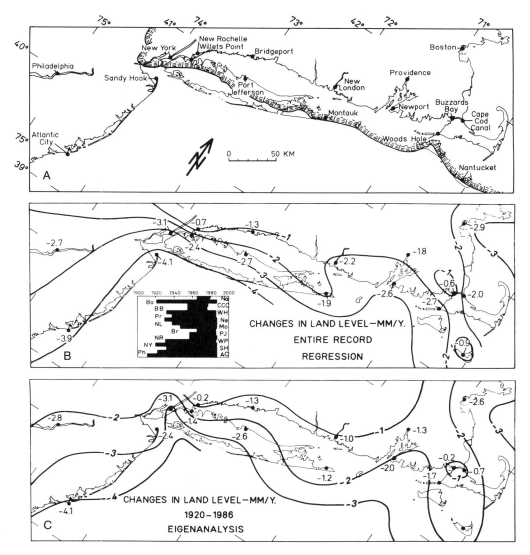

FIGURE 80. Long Island and Long Island Sound, Atlantic coast of United States.
A. Positions of tide-gauge stations having useful long-term records and near or beyond the ends of the Sound. The edge of the latest Wisconsinan (Würm) ice sheet at its maximum extent (about 16,000 y. B.P.) lay along the offshore islands.
B. Mean annual change of relative land level based on linear regression analysis of full time span of records from tide gauges (see insert at lower left for histogram of these time spans). Contours show areal distribution of rates in mm/y.
C. Mean annual change of relative land level based on eigenanalysis of all tide-gauge records of the Atlantic coast of the United States having time spans between 1920 and 1986 and records longer than 22 years. Synthetic rate at each tide-gauge station necessarily differs from that shown in panel B, but the trends of the contours are similar.

Eastern North America

Our first attempt to infer the relative roles of eustatic sea-level change versus tectonic movements of land level for the Atlantic coast of the United States (Aubrey and Emery 1983) was based on tide-gauge records listed by Lennon (1976–1978). A total of 26 stations having 40-year records were chosen for eigenanalysis. The resulting synthetic rates suggested that the coast consists of three segments that exhibit an along-coast change of relative sea level: southern Florida to mouth of Chesapeake Bay—a rise from 1.5 to 3.8 mm/y., mouth of Chesapeake Bay to Portsmouth (northeast of Cape Cod)—a fall from 3.8 to 2.0 mm/y., and Portsmouth to Canada—a rise from 2.0 to 2.7 mm/y. Clearly, such a distribution pattern accords with land movement rather than with simple eustatic rise of sea level, but there is no obvious fault-block mechanism to cause such a pattern of vertical land movement.

A possible explanation is vertical crustal movement caused by melt of the Wisconsinan ice sheet in eastern Canada and northeastern United States. Peltier (1986) developed a world pattern of crustal response to removal of glacial load using

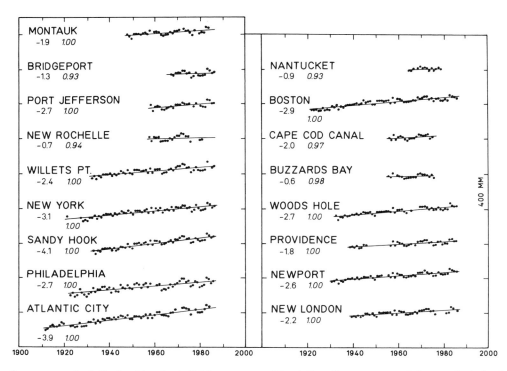

FIGURE 81. Plots of mean annual relative land levels at all tide-gauge stations in Long Island Sound and vicinity that have long-term high-quality records (Fig. 80A). At the left for each record are the name of the station, the mean annual change of relative land level in mm/y. (vertical numbers), and the t-confidence of the regression line (slant numbers). Data are from PSMSL.

radiocarbon-dated samples, tide-gauge records, and theory of crustal response. Using this pattern and some additional tide-gauge records, Braatz and Aubrey (1987) adjusted the gross changes shown by Atlantic coastal tide-gauge records for Peltier's results, finding a residual movement of relative sea level that ranged between about 0 and +2 mm/y., with an average of about +1.0 mm/y. This residual might be a measure of the rate of actual eustatic rise of sea level during the time between about 1940 and 1980, except that Peltier's corrections are based largely on the same locations as the tide-gauge records that are being corrected. Biases arising from local tectonism other than isostatic rebound are calibrated into the model. In addition, the scatter of the corrected rates (about 1 mm/y. on either side of the average value) is so large that there must be one or more factors beside that of distant crustal effects of glacial unloading of the crust. Part of the additional signal emanates from hydroisostatic loading—a process that Peltier's model attempted to include. Other factors include local tectonism, isostatic response of non-radially symmetric Earth rheology (Peltier's model is radially symmetric), and differential response to tectonism from welded terranes of vastly different composition (Uchupi and Aubrey 1988).

Both Pacific and Atlantic coasts of North America are long and broadly sinuous, but the Atlantic coast contains a section of convoluted shores where one can compare tide-gauge records by using contours of change. This section, Long Island and Long Island Sound (Fig. 80A), has eight tide-gauge stations with useful records; nine others are beyond the ends of the Sound but in positions to contribute supporting information. All of these stations except Nantucket (15 years) have records longer than 22 years, and eight are longer than 50 years. Regression lines drawn for all stations exhibit a range of relative land subsidence between −0.6 and −4.1 mm/y. Plots of mean annual relative land level against time reveal overall subsidence with time (Fig. 81). Regression lines through the data points for each tide-gauge station record have t-confidence levels of 1.00 for the 12 station records longer than 30 years, and between 0.98 and 0.93 for the remaining five records. These rates of subsidence calculated from regression were plotted on Figure 80B, where 1-mm/y. interval contours reveal a general slope in rates toward the ocean at the south. An interruption of the trend occurs in the form of a southeasterly projection through western Cape Cod and Nantucket, possibly an artifice related to the presence there of short time-span records.

In an effort to understand the effect of using tide-gauge records of different time spans, Figure 82 was constructed. It shows rates based on time spans for approximately 20-year cumulative portions of the records (1957–1986, 1940–1986, and 1920–1986). The shortest and most recent interval (Fig. 82C) has the most data and has contours that are most similar to those for the entire time span of all station records (Fig. 80B), but the longest and intermediate time intervals (Fig. 82A,B) are less informative because their station records are fewer in number. Nevertheless, the general trends

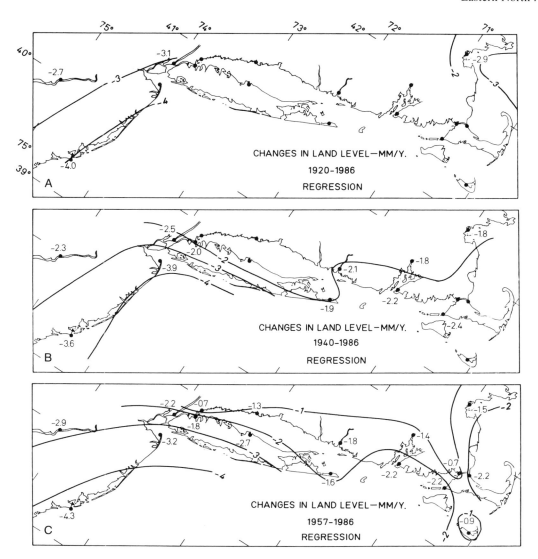

FIGURE 82. Comparison of rates of relative land subsidence (mm/y.) based on linear regression analyses of tide-gauge records in Long Island Sound and vicinity on basis of different parts of the time-span records. Data are from PSMSL.
A. 1920–1986; contours are limited because of few such long time-span records.
B. 1940–1986; information is intermediate in abundance because of the intermediate number of tide-gauge stations for this time span.
C. 1957–1986; has the most stations, but rates of vertical land movements are subject to error because the relatively short time span leads to influence by possible energetic variations in rates recorded by tide gauges.

of the contours of subsidence are similar for all three time spans. These results reinforce our opinion about the utility of eigenanalysis as a means of combining the advantages of the greater precision of long time spans and the greater number of short time spans. Contours at 1-mm/y. drawn through the station positions having synthetic rates determined by eigenanalysis (Fig. 80C) correspond approximately with contours based on regression analysis of all station records of whatever time span. Note that information from Nantucket was omitted, because its time span is only 15 years—much shorter than the chosen limit of 22 years for the eigenanalysis.

Terminal moraines show that the limit of the late Wisconsinan (Würm) ice sheet reached about 16,000 y. B.P. (Prest 1969; and others) lay along the axes of Nantucket, Martha's Vineyard, Block Island, and Long Island, and thence west-northwestward to Lake Erie (Flint et al. 1959; Schafer and Hartshorn 1965; Oldale 1982). The contours parallel those of post-glacial land uplift of the mainland beneath the area of the former ice sheet and subsidence of a peripheral ridge that is somewhat like that of northern Europe shown by Figures 42 and 44. Similar land emergence and ocean-floor submergence for the 12,000 y. B.P. surface through New York is illustrated by Newman et al. (1980a).

Contours of land emergence and submergence rates based on radiocarbon-dated changes in elevation at 41 sites in eastern North America, augmented by considerations of

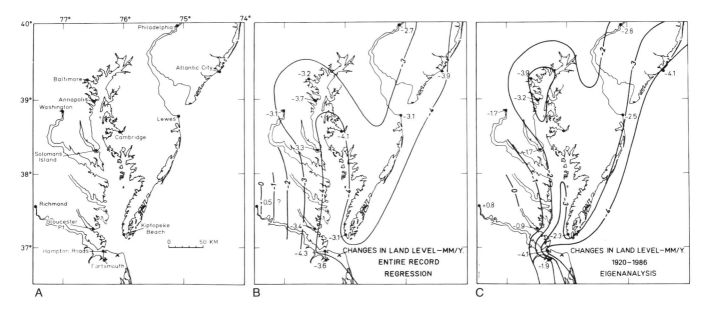

FIGURE 83. Chesapeake Bay and Delaware Bay and vicinity, Atlantic coast of United States. Note that the southeastern end of this figure overlaps the eastern end of Figure 79. Data are from PSMSL.
A. Positions of tide-gauge stations having useful long-term records.
B. Contours of mean annual change of relative land level in mm/y. based on linear regression analysis of full time spans of all tide-gauge records and adjusted to conform with contours from Figure 80B.
C. Contours of mean annual change of relative land level in mm/y. from eigenanalysis of all useful tide-gauge records of the Atlantic coast of the United States having time spans of 22 years or more.

lithospheric thickness and mantle viscosity, were drawn by Peltier (1986). They reveal maximum uplift of land (+11.2 mm/y.) in the southern part of Hudson Bay, and the zero change line is just north of the upper border of Figure 80. He also indicated a belt of land subsidence (of the original peripheral bulge) that amounts to −1.6 to −2.4 mm/y. eastward from Chesapeake Bay (about 150 km southwest of Atlantic City). Note that the rate of land emergence from Hudson Bay to the border of Figure 80 averages 1 mm/150 km/y., about one tenth the average rate expressed by the contours of tide-gauge records in Figure 80B. This difference may indicate a higher rate of post-glacial rebound near the boundary between continental and oceanic crusts, influenced by crustal deformation caused by the weight of returned meltwater atop the continental shelf, as discussed for this region by Bloom (1963, 1971) and Newman et al. (1980a), and for Australia by Chappell et al. (1982). That Peltier's model does not calculate this hydroisostatic response correctly may be because of his use of a radially symmetric Earth representation. The analysis by Bloom is based on radiocarbon-dated changes of relative sea level (land level) at five sites between Massachusetts and Florida. He showed that the changes of level at each of four sites since 4500 y. B.P. is nearly directly proportional to the average present water depth within a radius of 56 km of each site (it also is nearly inversely proportional to the distance of each site from the nearest 100-m contour of water depth). At the fifth site (Plum Island, MA) the great water depth and proximity to the 100-m contour indicates several times greater subsidence than is shown by the depth/radiocarbon-age curve, but this apparent discrepancy clearly is due to concurrent post-glacial rise of the land at that site. Bloom's map of mean water depths within 56-km radii (*isomesabaths*) throughout the continental shelf is strikingly similar to the map of rate of land subsidence from tide-gauge records in Figure 80B, but the indicated rates of subsidence are only 0.3 to 0.8 times those indicated by the tide-gauge records. Both mechanisms for change in land level provide a much more likely explanation for the tide-gauge records of Figure 80 than does simple eustatic rise of sea level.

Another area of the Atlantic coast where convoluted shores permit two-dimensional mapping of rates of relative land subsidence lies just southwest of the Long Island Sound area. It includes 13 tide-gauge stations in the vicinity of Chesapeake Bay and Delaware Bay (Fig. 83A), two of which also were included within Figure 80A for Long Island Sound. Nine of these 13 station records have time spans of 48 years or more, and all except one have t-confidence levels for regression analysis of 1.00 or 0.99. The exception is at Richmond, whose 27-year record is irregular probably because the station is on the James River where fluctuations of level may be caused by annual variations of runoff; thus, the mean annual level there is questioned on Figure 83B. Contours at 1-mm/y. intervals drawn from the rates of change of mean annual land level (Fig. 83B) again reveal a slope toward the ocean, but one that is less steep than that across Long Island Sound. The pattern is deeply indented at Chesapeake Bay perhaps because of compaction of bay sediments under some tide gauges or the presence of a structural graben, but the precision of the shape of this indentation is limited by the relatively few tide-gauge stations and their uneven distribution. Contours based on eigenanalysis (Fig. 83C) may indicate the same deep indentation, but its continuity is interrupted by the omission of the record at Cambridge (only 19 years, less than the minimum limit of 22 years that was chosen for eigenanalysis).

Two previous studies of the Chesapeake Bay/Delaware Bay area (Holdahl and Morrison 1974; Davis 1987) yielded somewhat different subsidence contours from those of Figure 83. They show that pumping of water from aquifers lowered water tables and caused land surface subsidence at Franklin, VA (42 m and 154 mm in 25 y.; a rate of 6.2 mm/y.), West Point, VA (39 m and 118 mm in 39 y.; a rate of 3.0 mm/y.), Dover, DE (57 m and 124 mm in 31 y.; a rate of 4.0 mm/y.), Atlantic City, NJ (16 m and 60 mm in 44 y.; a rate of 1.4 mm/y.), and farther south at Savannah, GA and farther north in Monmouth County, NJ. Fortunately, all except one of these areas are between the distantly spaced tide gauges having the best records that were used for the regional contours of Figure 83. Only the subsidence at Atlantic City obviously influenced the contours, but the rate of subsidence associated with pumping there is less than half the total rate recorded by the tide gauge at Atlantic City. The results of precise leveling indicated by Holdahl and Morrison (1974) represent little conflict with the contours of Figure 83, because the leveling routes do not cross Chesapeake Bay between Kiptopeke Beach and Annapolis.

Two other recent studies of the sediments of Chesapeake Bay are relevant. One, by Colman et al. (1990), is based on many seismic reflection profiles that indicate dissection of the bay floor by three glacial-epoch channels of the Susquehanna River. Differences in positions of the channels (some beneath present land areas) may account for slight station-to-station differences in present rates of relative sea-level change because of differences in thickness of later sediment fills within the channels and of rates of compaction. The other study, by Donoghue (in press), was based on ^{14}C and ^{210}Pb dates along cores, showing a mean sedimentation rate of 1.5 mm/y. for the past 5000 years and a rate of nearly 10 times that for the past century. This increase in sedimentation rate was attributed to an increase in rate of rise of sea level, but it may have been caused by increased discharge of sediment from human land-use practices.

The two areas of detailed studies (Figs. 80, 83) comprise only 27 percent of the general coastal length between the southern tip of Florida and the western tip of Newfoundland, but they contain 64 percent of the useful long-term tide-gauge stations. The 17 additional stations shown on Figure 84 have time spans between 18 and 90 years, with 14 being 40 or more years. The t-confidence levels for regression lines drawn through data points for each station are 1.00 for 15 station records, 0.97 for Port-aux-Basques (20 years), and 0.92 for Pictou (18 years). Although short, the records for the two latter stations were retained, because they are in a region of sparse stations.

Contours were drawn from the mean rates of change in relative land levels obtained by regression analysis of the records. They appear to show lateral extensions of the oceanward increase in relative subsidence of the land revealed by the more detailed contours of Figures 80 and 83 (transferred to Fig. 84B). From Cape Cod southwestward this slope is more uniform except for the naturally greater irregularity to be expected in areas that have more abundant tide-gauge records. There is a bulge at Wilmington that probably is related to the Cape Fear Arch (Maher 1971) that may have influenced the record at Wilmington, and the famous earthquake at Charleston in 1886 (Dutton 1889) indicates tectonic activity there. From Cape Cod northeastward the contours continue in similar fashion along the Gulf of Maine and Bay of Fundy through Nova Scotia and Prince Edward Island with an oceanward increase in rate of land subsidence. In easternmost Maine precise leveling between 1942 and 1966 (Tyler et al. 1979) supports the regression analysis of scanty tide-gauge records in showing greater land subsidence eastward along the coast; in reality, there is greater relative land emergence at the west and north, as shown also by elevations of late-glacial sea level and their radiocarbon dates (Anderson et al. 1984; Emery and Uchupi 1984, p. 65; Belknap et al. 1987). Farther northward and eastward the relative land movement is reversed, exhibiting uplift that reaches a maximum (+6.2 mm/y.) at Churchill near the center of the original crustal depression caused by the weight of the now-melted Pleistocene ice sheet. The gradient of vertical land movement caused by glacial rebound is gentler than the gradient near the coast except between Charlottetown (Prince Edward Island) and Port-aux-Basques (Newfoundland), but this particular steeper gradient may be result from uncertainty at the latter site caused by the relatively short (20-year) record there. All station records between Miami and Eastport having station records longer than 22 years were subjected to eigenanalysis; the resulting synthetic mean annual changes of relative land level are depicted in Figure 84C. Contours from Figures 80C and 83C were transferred and continued beyond the limits of those more detailed charts on the basis of subsidence rates at the other stations. These contours are similar to those obtained from regression analyses for the United States' portion of Figure 84B. Comparison of the results from the two methods of analysis also is given by the plot in the insert at the bottom of Figure 84C, which shows an unusually wide scatter attributed to the limited number of stations and their wide range in time spans, as well as numerical approximations.

The difference in crustal movement southwest of New York (non-glaciated) and northeast of it (glaciated) is reflected both in tide-gauge records of the past century and radiocarbon ages of sea-level indicators of post-glacial time. More than 100 studies of the sea-level (land-level) movements inferred from these radiocarbon ages have been published, and they have been well summarized by Pardi and Newman (1987). Their summary consists of a three-dimensional plot of latitude (Lats. 20°–50°N) versus radiocarbon ages (0–16,000 y. B.P.) versus present elevation of the dated sea-level materials—largely basal peats. For the southwestern coast the plot shows generally faster relative subsidence with age especially before 4000 to 6000 y. B.P. Irregularities in

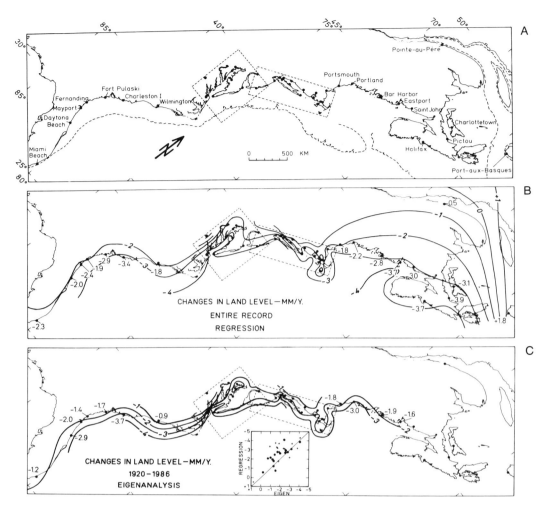

FIGURE 84. Changes in relative land level (mm/y.) for the Atlantic coast of North America. The areas within the rectangles limited by dashed lines contain information transferred from more detailed treatment within Figures 80 and 83. All data are from PSMSL.
A. Positions of tide-gauge stations having useful long-term records. Dashed offshore line denotes the position of the shelf break that ranges from depths of about 70 m at the south to about 160 m at the north.
B. Mean annual change of relative land level based on linear regression analysis of full time spans of all tide-gauge records. Contours in mm/y.
C. Mean annual change of relative land level based on eigenanalysis of all tide-gauge records between Miami Beach, Florida, and Port-aux-Basques, Newfoundland, having time spans between 1940 and 1986. An insert at the bottom shows the relationship of mean annual change of relative land level obtained from linear regression analysis of tide-gauge records during the entire period prior to 1986 compared with those from eigenanalysis of records longer than 22 years and between 1920 and 1986. Solid dots in the insert are for tide-gauge stations in or near Long Island Sound, crosses are for other stations along Atlantic coast of United States.

plotted depths can be produced by errors in measured ages or assumed sea-level environment of deposition, but they equally well may be caused by differences in oceanward distance from the hingeline of warping, and thus are differences in rates of crustal subsidence caused by isostatic weighting by rising sea level. For the northeastern coast the plot shows emergence of the crust above present sea level between 16,000 and 20,000 y. B.P. (isostatic crustal rebound after removal of ice load). This emergence was followed by subsidence below sea level to about 8000 y. B.P. (northward moving and subsiding peripheral bulge). Subsequently, relative emergence continued until 4000 to 6000 y. B.P. This diagram and Figure 34 incorporated and summarized the information from the varied curves of relative sea level (land level) for the Atlantic coast of North America that had been published by many authors during the past few decades, and they set the scene for the tide-gauge results of Figures 80 through 84.

The opportunity to investigate local land movements in two-dimensional map form for Long Island Sound and Chesapeake Bay (Figs. 80, 83) is paralleled by the use of tide-gauges in southeastern Canada. Only six of them were used in Figure 84. Acceptance of shorter time spans (to only 11 years), lower *t*-confidence levels (to only 0.72), and a broader region (through Newfoundland) increased the number of useful Canadian stations to 17, supplemented by four others from northeasternmost United States. In Canada, as elsewhere, tide-gauge stations were established for prediction of tidal currents and water depths for navigation chiefly in harbors rather than for measurement of changing sea levels or land levels. According to G. Carrera of Dartmouth, Nova Scotia (personal communication, November 1989), additional stations were occupied in the Arctic for only a year or so in support of seasonal hydrographic surveys, and their results were not always referred to the same datum; thus, they are not included here. Records even at some of the Canadian stations of Figure 85 have undocumented datum shifts, but all shifts appear to have been before or after the span of the records that were used (Yarmouth in 1966; Boutilier Point in 1971; Pointe au Père in 1983; Halifax in 1987).

The plots of changing relative sea level (land level) for the 21 stations of Figure 85 provided the basis for a new map (Fig. 86) showing changes attributed mainly to removal of the weight on the crust by melt of the Pleistocene ice sheet. Contours of relative vertical land movements at 2-mm/y. intervals are rather different from those of Figure 84B, but they are far closer to those showing the present elevations of the earliest radiocarbon-dated post-glacial marine sediments and shoreline topography compiled by Grant (1980) from many sources. They also are similar to contours from eight Canadian tide-gauge records supported by precise leveling (Vaníček 1976) and several new sets of ^{14}C datings (Carrera and Vaníček 1988). The pattern appears to be influenced by major faults within the region along the Bay of Fundy, St. Lawrence Valley (Logan's Line Fault), and elsewhere (King 1969).

Two additional stations far to the north of Figure 86 are at Churchill, western shore of Hudson Bay (Lat. 58°46′N, Long. 94°11′W), and Resolute much farther north (Lat. 74°41′N, Long. 94°53′W). Their mean annual change of land level are +6.2 and +2.9 mm/y., respectively, but the region has too few marine tide-gauge records to establish contours of land emergence after glacial melt and contours of land submergence in a peripheral belt. Peltier (1986) attempted to remedy their absence by augmenting the marine tide gauges with geophysical calculations based on differences in Great Lakes' tide-gauge records, tilt of terraces around the Great Lakes and their ancestral lakes, radiocarbon dates, lithospheric thicknesses, and mantle viscosities. His resulting contours (Fig. 87) resemble those of the rebound in northern Europe (Fig. 42) that is better defined by raised shorelines and marine tide-gauge records. A network of absolute gravity measurements at 12 proposed stations has been suggested by Lambert et al. (1989) to test the contours.

As shown by Figure 87, crustal rebound of areas previously covered by Wisconsinan ice sheets and sinking of the peripheral bulge essentially eliminate tide-gauge records of eastern and southern North America from serious consideration as

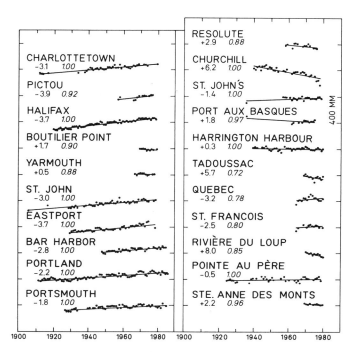

FIGURE 85. Plots of relative vertical movements of land levels shown by tide-gauge records in area of Figure 86. For each station the regression line through the data points, the mean annual change of land level in mm/y. (vertical numbers), and its *t*-confidence (slant numbers) are indicated. Data are from PSMSL.

clean indicators for estimating simple present eustatic rise of sea level. One might accept the possibility that the records for the 10 stations between Pensacola (Fig. 79A) and Fort Pulaski (Fig. 84A) inclusive are least influenced by crustal warping associated with glacial loading and they are in a region of low seismicity and little evidence of modern surface faulting. However, all are in areas of thick Mesozoic and Cenozoic sediments (King 1969; Maher 1971; Emery and Uchupi 1984, p. 434–463) and thus subject to compaction and sinking that may compensate for uplift of the peripheral bulge. The tide-gauge station indications of relative land movement are a composite of sinking by compaction of underlying sediments, rising of the rebounding peripheral bulge, and sinking by rising eustatic sea level caused by return of meltwater and increase in ocean water temperature – a most complex interplay of poorly evaluated rates. For instance, quantitative estimates of sediment compaction are lacking for these areas, although some geological data provide limited insight. By contrast, Peltier (1986) and Peltier and Tushingham (1989) modeled the isostatic rebound process; from these model results, Braatz and Aubrey (1987) and Uchupi and Aubrey (1988) examined the residual signals after subtracting model results of rebound from tide-gauge results. However, the Peltier data are based on a "tuning" procedure that fits the model to radiocarbon dates along the North American coast. Since these radiocarbon dates reflect processes other than just

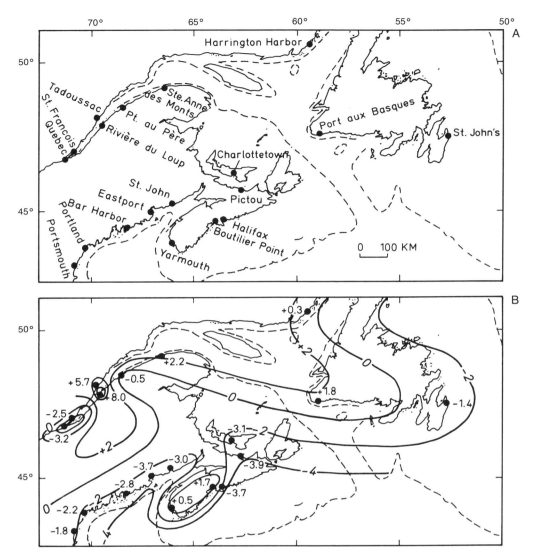

FIGURE 86. Region of post-glacial warping mostly in eastern Canada. Data are from PSMSL.
A. Tide-gauge stations, including 10 from Figure 84, 3 that are beyond the limits of Figure 84, and 6 that have fewer years of record and lower *t*-confidence than those of Figure 84B. Dash line denotes shelf break, 140 to 300 m depending on location.
B. Contours of vertical land movement (2-mm/y. interval) based on linear regression analysis from station records of Figure 85.

glacio-isostatic processes, the tuning procedure introduces inaccuracies of unknown magnitude. The resulting indicated net rates range between −0.9 and −2.4 mm/y., a fairly narrow range. This matter is considered in a later text section together with records from other relatively stable coastal regions of the world.

Islands of the Atlantic Ocean

Two island arcs in the western Atlantic Ocean warrant discussion in terms of present vertical movements of relative land level. One is the South Sandwich Island arc that borders the eastern end of the Scotian Sea off the southern tip of South America, but no tide-gauge records from there are known. The other is represented by the Lesser Antilles and the Greater Antilles at the eastern and northeastern borders of the Caribbean Sea between South America and North America. Although there are no acceptable tide-gauge records for the Lesser Antilles, there are seven along the Greater Antilles. These were briefly discussed in an earlier text section on the Caribbean Sea and illustrated in Figures 73 and 76. Complexities of vertical land movements were attributed to tectonism along a major Mesozoic foldbelt and a Neogene deformed belt (Fig. 75). Study of tide-gauge records in Japan revealed that there is a systematic southeastward (oceanward) increase in subsidence rates of land despite complications caused by secondary continental crustal tectonics.

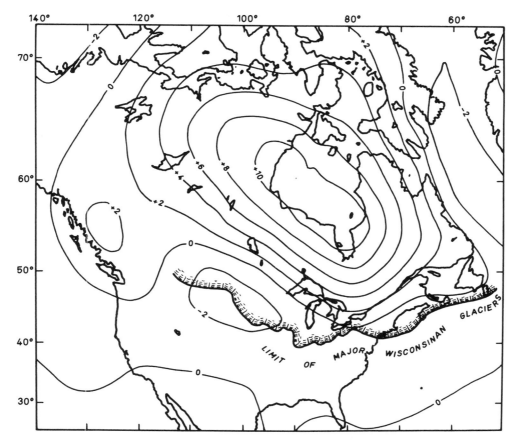

FIGURE 87. Contours of emergence for northern North America in mm/y. calculated from marine tide gauge records augmented by computations from geophysical data of several kinds. Redrawn from Peltier (1986), Journal of Geophysical Research, v. 91, p. 9099–9123, fig. 12a, copyright by the American Geophysical Union.

A subsequent analysis of tide-gauge records in the Philippine Islands indicated a similar eastward increase in subsidence despite local tectonic movements associated with active volcanism (see following section on islands of the Pacific Ocean). Accordingly, the tide-gauge records of the Greater Antilles were re-examined in the light of plate subduction. It is unfortunate that there are no acceptable tide-gauge records in the adjacent Lesser Antilles because this island arc is undergoing rapid subduction accompanied by considerable volcanism (Pelée and La Soufrière in 1902; Agar et al. 1929, p. 299–330) and marine terrace formation.

The Greater Antilles lie within the broad and complex boundary region between the North American and the Caribbean plates. A component of Late Cretaceous to present subduction by the westward-moving North American plate occurs along the Puerto Rico Trench (Molnar and Sykes, 1969); apparently there was also Late Cretaceous to mid-Tertiary compression or subduction along the westward extension of the trench—Old Bahama Channel (between Cuba and the Bahama Islands). Along the south side of the Greater Antilles, subduction is directed northward along the Muertos Trough (south of the Dominican Republic and Puerto Rico) and through the middle of Haiti (Larue 1990). Similarly, there is northward compression along the north side of the Cayman Trough south of Cuba (Emery and Uchupi 1984, p. 542). Greatest seismicity is present in a Benioff zone just south of the Puerto Rico Trench (Espinosa et al. 1981). The many faults in the region and their movements probably have produced numerous variations on the general northward subsidence pattern exhibited by Figure 88. Largest of the variations among the tide-gauge records is that at Port-au-Prince, where mean annual movement of relative land level is −12.3 mm/y. Port-au-Prince lies at the head of the westward-facing gulf of Hispaniola on the Plaine du Cul-de-Sac that is underlain by thick Quaternary sediments deposited presumably in a graben that farther west opens to form the southern part of the Gulf that has a floor depth of more than 3200 m. These depths contrast with higher than 1000-m elevations along the nearly straight east-west peninsula south of the Gulf, the Massif de la Hotte. These structural trends also are represented by roughly east-west faults on the land at either side of Port-au-Prince and by gravity trends on both sea floor and land (Bowin 1975b). The greater subsidence at Port-au-Prince resembles the re-entrant of subsidence contours along

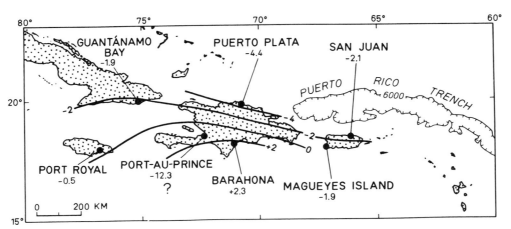

FIGURE 88. Distribution of rates of relative land subsidence (mm/y.) in the Greater Antilles (see Figs. 73 and 76 for data sources). Note the greater northward subsidence rates that are attributed to subduction of the Atlantic oceanic plate moving diagonally along the Puerto Rico Trench (outlined by the −6000-m contour of water depth). The tide-gauge record at Port-au-Prince does not fit the contours of subduction probably because of local tectonism, as discussed in text.

the Inland Sea of Japan (Fig. 66), but there the shape of the re-entrant is well established by numerous Japanese tide-gauge records. For Port-au-Prince the unusually large subsidence is merely questioned here and ignored by subsidence contours drawn on the basis of the remaining tide-gauge records of the region (Fig. 88). Probably less well understood are 28 terraces rising to 600 m above sea level on which Pleistocene reef limestones are present (Horsfield 1975, 1976) and other occur to 300-m elevations in the Dominican Republic (Taylor 1980). Could these terraces have been raised on local horsts and the tide gauges largely restricted to grabens? Or, may the terraces pre-date the present episode of diagonal subduction and subsidence?

Only four tide-gauge stations in the open Atlantic Ocean have acceptable time spans and t-confidence levels for their regression analyses. These are at Barentsburg at the west side of Spitsbergen (Norway), Reykjavik at the southwest side of Iceland, St. Georges at the northeast end of Bermuda, and Santa Cruz at the east end of Tenerife Island (Canary Islands, Spain). Their time spans are 28 to 42 years and their t-confidence levels are 0.97 to 1.00, all excellent records (Fig. 89A).

One station record indicates a rise of relative land level of +1.6 mm/y., at Barentsburg, whereas the other three records reveal subsidences between −2.5 and −3.6 mm/y. The rise at Barentsburg is ascribed to crustal rebound after melt of a Pleistocene ice cap; in fact, 60 percent of Spitsbergen still is covered by glaciers, although they are only a small remnant of the original ice cover (Flint 1971). Underlying rocks consist of early Paleozoic complexes of the Scandinavian craton having structural basins filled with later Paleozoic to Triassic sediments (Yanshin 1966) including coal that is being mined; the considerable ages of these rocks suggest that they are not the cause of present land movements.

At Reykjavik, the land level is subsiding at −3.6 mm/y., which is rather fast (Fig. 89A). This speed may result from crustal weighting by the great thickness of basalt deposited by intense volcanism during the recent past (Palmason and Sigvaldason 1976), probably because Iceland overlies a mantle plume (Bott et al. 1971), and its associated cooling. Continuing tectonic activity is indicated by numerous earthquake epicenters associated with a fracture zone, transform faults, and crustal extension (Einarsson 1979), as well as by active volcanism in 1963 at nearby Surtsey and then Heimaey islands (Thorarinsson et al. 1973). In fact, the entire Reykjanes Peninsula which Reykjavik borders is part of a highly faulted volcanic belt (Sœmundsson 1967).

The remaining two tide-gauge records in the Atlantic Ocean exhibit intermediate rates of land subsidence: −2.7 mm/y. at St. Georges and −2.6 mm/y. at Santa Cruz (Fig. 89A). Bermuda once underwent volcanic activity like that at Iceland, but long ago at 92 and 38 m.y. B.P. (Hyndman et al. 1974; Reynolds and Aumento 1974). Subsequently, thick calcareous sediments were deposited in which are recorded sea levels of the Pleistocene Epoch (Sayles 1931; Bretz 1960; Land et al. 1967). Support for reasonable stability of Bermuda is provided by a series of 14 radiocarbon dates on peat samples probably obtained by A. C. Neumann and tabulated by Redfield (1967). Redfield plotted the ages against the depth below a local datum and obtained a visual fit for an average rise of relative sea level (reported as eustatic rise of sea level during the past 4000 years) of 0.8 mm/y. for 12 of the 14 dates. We measured the rate as 0.7 mm/y. by regression analysis on his 12 records and 0.8 mm/y. using all 14 records. If taken literally, the difference between the tide-gauge and the radiocarbon rates implies a recent three- or four-fold increase in rate of submergence. No human intervention that may affect this rate has come to our attention. Possible confirmation of the general stability of Bermuda may be provided by the similarity of interglacial sea levels 125,000 y. ago relative to present sea levels at Bermuda (+4 to +6 m), New Guinea (+8 m), Barbados (+5 m), and the ocean from stable

Islands of the Atlantic Ocean   137

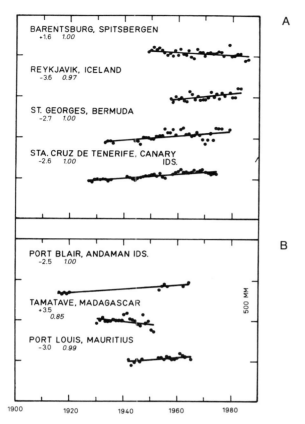

FIGURE 89. Plots of mean annual changes of relative land level at islands of the open Atlantic Ocean (A) and of the open Indian Ocean (B). Beneath the names of the stations are given the mean annual change of land level in mm/y. (vertical numbers) and the *t*-confidence level (slant numbers). Data are from PSMSL.

oxygen isotopes (+11 m), according to Harmon et al. (1981); some doubt arises because both New Guinea and Barbados have many high marine terraces that indicate considerable previous uplift of land.

The fourth station in the open Atlantic Ocean is Santa Cruz, with a relative land subsidence of −2.6 mm/y. (Fig. 89A). Its island, Tenerife, is reported to have been the site of 10 observed volcanic eruptions between 1400 and 1966 (van Pedang et al. 1967). Ash layers in nearby marine sediments of the Canary Island Ridge testify to earlier eruptions in the region, back at least to include Miocene age (Abdel-Monem et al. 1972; Emery and Uchupi 1984, p. 148, 150, 472–473). Thus, land movements associated with volcanism may have been involved in the tide-gauge record at Santa Cruz and the record there cannot be relied on as indicating changes of eustatic sea level alone even though it has a uniform slope.

Islands of the Indian Ocean

The Indian Ocean contains one of the largest island arcs of the world along the Indonesian archipelago that borders the north side of the Java Trench. Unfortunately, there are no accept-

able or available modern tide-gauge records for the region nearer the trench than Thailand and Singapore. Before the founding of Indonesia, the Netherlands East Indies maintained 30 well distributed tide-gauge stations in the archipelago (Proudman 1940). These have records that span 2 to 7 years—too short for effective correlation with tectonics. Although a seven-year time span is insufficient for reliable regression analysis, the rapid submergence that many records indicate may be large enough to provide some confidence in results obtained by their study. Accordingly, the 10 records that span the seven years between 1925 and 1931 were analyzed for mean annual change of relative land level. The data for one station (Benkoelen) are erratic and the regression slope through the annual data points has a large (5.5) standard deviation, so that station record was discarded. The average annual change at Singapore (stations Sembawang and Jurong; Figs. 55, 56) was added to complete the study. The resulting map of the areal distribution of rate of land-level change (Fig. 90) shows a sinking of more than 10 mm/y. at most stations nearest the Java Trench and a rising of 5 mm/y. at two stations farther from the trench. This map is not considered very reliable, but the subsidence trends parallel those at islands along the concave sides of other trenches of the world ocean. They also parallel the trends of the depths of earthquake hypocenters down the Benioff zone that marks the top of the descending northern edge of the Australian plate northward from the Timor Trench (Hayes and Taylor 1978). Jouannic et al. (1988) described the considerable spatial variability associated with island arcs near West Timor. Areas showing little vertical motion since the past interglacial epoch are adjacent to areas that have experienced more rapid emergence. Uplift rates vary from 0.3 mm/y. to much more near the seismic tear zone of West Timor. Somewhat erratic movements can be expected from the dominance of volcanoes and active volcanism along the chain of islands. Notable among the volcanic events are those of Tamboro in 1815 (Stommel and Stommel 1983), and Krakatao on 27 August 1883 (Verbeek 1885; Simkin and Fiske 1983). Both eruptions and caldera formation cast such large quantities of dust and ash into the atmosphere as to reduce the passage of sunlight and produce colder than usual weather around the entire earth (Stommel and Stommel 1983).

There are only three tide-gauge stations on islands in the open Indian Ocean that have acceptable time spans and confidence levels for regression analyses: Port Blair at the east side of the middle of the elongate group of Andaman Islands, Tamatave at the east side of Madagascar, and Port Louis at the northwest side of Mauritius. Their records are for 16 to 21 years (but the shortest number of years, for Port Blair, is spread over a total span of 48 years). The *t*-confidence levels are 0.85 to 1.00 (Fig. 89B).

The record at Port Blair reveals a relative sinking of land (−2.5 mm/y.). This is a common movement for coastal regions of the world despite the Andaman Islands' position on the tectonic belt that extends through Indonesia and Burma where it closely borders the Ninetyeast Ridge that more or

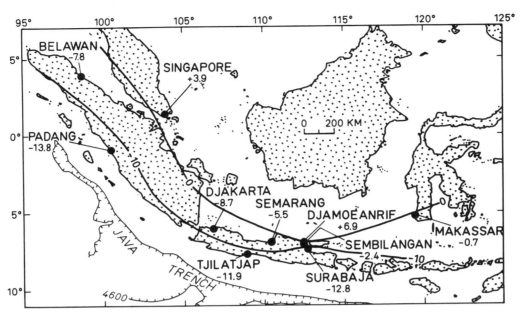

FIGURE 90. Indonesia with mean annual relative land subsidence denoted in mm/y. based on the best available tide-gauge records, ones that span only the years 1925 to 1931, as reported by Proudman (1940). The averages for two later stations at Singapore (Figs. 55, 56) are included. The 5-mm/y. contours of change in relative land level show greater land subsidence toward the Java Trench (outlined by the 4600-m contour below sea level).

less marks the eastern side trail of the northward travel of India about 40 m.y. ago (Heezen and Tharp 1965). However, subsequent geophysical and geological work indicates that the ridge of submerged mountains is not continental but belongs to the oceanic crust in the form of a hot-spot trail that lies near a transform fault (Bowin 1973; Sclater and Fisher 1974; Luyendyk 1977). Earthquakes are common on and near the Andaman Islands, with hypocenters shallower than 70 km, and an active volcano is nearby (Udintsev 1975, p. 111).

Port Louis at Mauritius has a similar rate of relative land sinking (−3.0 mm/y.; Fig. 89B), but Mauritius is in a presently more stable ocean region with few earthquakes, and the nearest reported active volcano is 1300 km distant at Réunion (Fisher et al. 1967; Udintsev 1975, p. 118–119). Latest volcanism at Mauritius is Pleistocene, and much of the island is capped by Cretaceous sediments (in contrast with the Pliocene age at Port Blair in the Andaman Islands; Udintsev 1975, p. 118–119), again implying that tectonic activity at Port Louis is more ancient. On the other hand, the ages of oceanic crust are Paleocene near Mauritius and mid-Cretaceous off the Andaman Islands (Larson et al. 1985). Although Mauritius occupies a somewhat similar position as the Andaman Islands with respect to India, it is on the opposite side of the track followed by the northerly movement of India en route to Asia. Both sides have structural complexities, with the Mauritius side complicated by tectonic movements related to separation of Madagascar and Seychelles Bank from Africa and India through sea-floor spreading along the Carlsbad Ridge (Fisher et al. 1967). One might conclude that Port Louis is situated on a presently inactive volcano that rises above oceanic crust, and Port Blair on a more active volcano belonging to a structural complex of continental rocks strongly influenced by oceanic crustal movements, but this does not explain the similarity in their rates of relative land subsidence.

The tide gauge at Tamatave is on a much larger land mass, Madagascar, than those at Port Blair and Port Louis. Madagascar, unlike the two small islands, is underlain by granitic crust according to outcrops and seismic exploration. The same composition beneath Seychelles Bank supported by observations of ocean-floor structure implies that the Seychelles separated from Madagascar during late Mesozoic time (Fisher et al. 1967). During Late Jurassic or Early Cretaceous the continental block containing the present masses of Madagascar and Seychelles Bank separated from the Somalia–Kenya–Tanzania region (McElhinny et al. 1976; Bunce and Molnar 1977) and moved southward as part of the crust south of the Malvinas–Agulhas Fracture Zone including the Falkland Plateau (Dingle et al. 1983, p. 99; Uchupi 1989). Tamatave, on the east coast of Madagascar, is situated atop a narrow coastal strip of Cretaceous sedimentary rock. Its recorded rise of relative land level (+3.5 mm/y., but with low t-confidence level) appears to be supported by a Flandrian solution nip or visor in limestone at 1.0 to 1.3 m above present sea level near the south end of Madagascar. In confirmation, 100 km north of Tamatave are large slabs of dead corals above sea level. Samples at 1.5 and 0.7 to 1.0 m above present sea level were radiocarbon dated at 3740 and 2930 y. B.P., respectively (Battistini 1970), rates equivalent to 0.4 mm/y. and 0.3 mm/y., respectively.

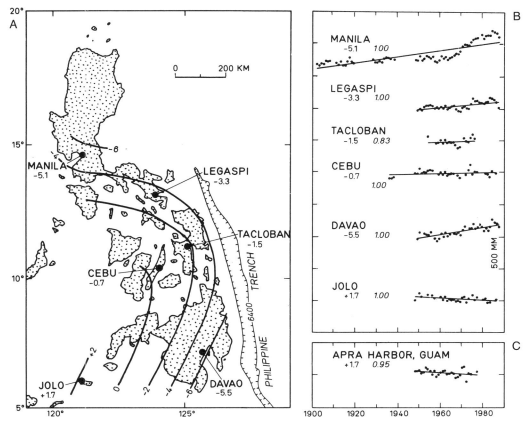

FIGURE 91. Philippine Islands.
A. Positions of Philippine Islands with mean annual change of relative land level in mm/y. indicated below station names and contours of relative land movement indicated by contours at 2-mm/y. intervals. Position of Philippine Trench shown by the 6400-m water-depth contour.
B. Plots of mean annual changes of relative land level at tide gauges of the Philippine Islands. Beneath the names of the stations are given the mean annual change of relative land level in mm/y. (vertical numbers) and the t-confidence level (slant numbers). Data are from PSMSL.
C. Same as B except for the single tide-gauge station of the Mariana Islands at Apra, Guam.

Evidently, tectonic uplift of Madagascar during the past few thousand years continues.

Islands of the Pacific Ocean

Island Arcs

In contrast with the Atlantic and Indian oceans, the Pacific Ocean contains many islands that have acceptable tide-gauge records. Most of these islands are volcanic in origin and belong to either of two groups: islands that border deep-ocean trenches (mainly members of island arcs), and islands distant from trenches and members of hot-spot chains. The island arcs will be considered first.

The Philippine (or Mindanao) Trench is bordered by many islands, but ones that are not as linearly arranged as those of a typical island arc. Six acceptable tide-gauge records are present, and they span 26 to 73 years with t-confidence levels for regression analyses of 1.00 except for the record having the shortest time span for which the level is 0.83. The longest record, at Manila, has a peculiar wave between about 1950 and 1987 that is suggested somewhat also at Legaspi (Fig. 90B). Because both limbs of the wave are present and are on opposite sides of the regression line, the regression probably still approximates the mean annual movement of relative land level at both Manila and Legaspi. Contours drawn on the basis of the areal distribution of mean annual change of relative land levels at the six Philippine Islands' tide-gauge stations (Fig. 91A) show greater land subsidence on the eastern than the western side of the archipelago. This is the same relationship as revealed by rate-of-subsidence contours at Japan (Fig. 65). In both areas the relative land subsidence is largest on the side along which subduction of the oceanic plate from the east is occurring, and there is a slight emergence on the opposite side. At Japan there is a complication from effects of multiple plates that may be paralleled in the Philippine Islands and by active volcanoes, and these complications appear to be supported by the distributions of earthquake hypocenters (see charts by Hayes and Taylor 1978; Espinosa et al. 1981).

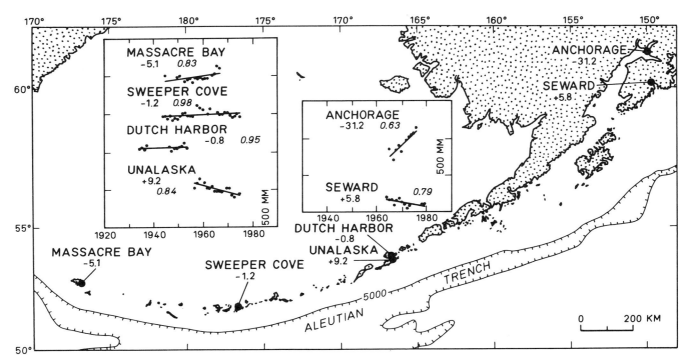

FIGURE 92. Aleutian Islands. Distribution of tide-gauge stations along the Aleutian Islands and the Alaskan Peninsula with inserts that show plots of mean annual changes of relative land level. The Aleutian Trench is indicated by its 5000-m contour of water depth. Insert at left indicates mean annual changes of land level (mm/y.) from data of PSMSL.

The Mariana Islands that border the Mariana Trench about 1700 km east of the Philippine Islands contain a more typical arc of small islands along the concave side of the trench, but there is only a single acceptable tide-gauge record. This record reveals a rise of the relative land level amounting to +1.7 mm/y. (Fig. 91C), but the absence of other tide-gauge records along the arc prevents the making of inferences about the scope of the uplift. Confirmation of uplift at Guam and its northern neighbor islands (to Saipan) comes from studies made there in 1952 by the U. S. Geological Survey. Deep-water Cenozoic sediments were raised to shallow depths by Miocene time and subsequently raised to 400 m above sea level, with about 100 m of uplift during the Pleistocene (Tracey et al. 1964, p. A60–A61). Some of the uplift is recorded by nips (formed by intertidal solution by seawater and biochemical activity) at present elevations to about 100 m above sea level at Guam. Continuing uplift is supported by the presence of erosional terraces along cliffed coasts of both Guam and Saipan.

The Ryukyu Islands bordering the Ryukyu Trench form an island arc between Taiwan (Fig. 63) and Japan (Fig. 65). Acceptable tide-gauge records would have been valuable for inferring present tectonic movements along the island arc, but, unfortunately, none are available. As a substitute, data may be used from emerged intertidal notches, coral reefs, beachrocks, barnacles, and now-submerged limestone quarries (Pirazzoli and Delibrias 1983; Kawana and Pirazzoli 1985). Results of this work were interpreted to show the presence of a general uplift of about 2.5 m centering at Okinawa and vicinity, probably during an earthquake of perhaps about 7.4 magnitude 2400 years ago. Lesser emergence and even subsequent subsidence also is recorded along the arc, implying tectonic relaxation as a prelude to a future large earthquake. Establishment of a tide-gauge station network may prove useful.

The last island arc in the Pacific Ocean having acceptable tide-gauge records is the chain of Aleutian Islands along the north, or concave, side of the Aleutian Trench (Fig. 92). Only four useful tide-gauge records exist along the chain, with time spans of 13 to 33 years—the longest for Sweeper Cove. To these are added station records at Anchorage and Seward to fill the gap between the Aleutian Islands and the Alaskan stations of Figures 70 through 72. Note that the Anchorage record begins in 1965, just after the 8.5-magnitude earthquake there of 27 March 1964. Evidently, post-quake adjustments are responsible for the subsequent fast sinking of land there (averaging −31.2 mm/y.). Many other earthquakes occur along the length of the Aleutian Island chain with hypocenters defining a northward-dipping Benioff Zone (Espinosa et al. 1981). As the islands are volcanic in origin, there have been many dramatic eruptions, among which the one of 11 June 1912 at Katmai was vividly described by Griggs (1922). Another volcano, Bogoslof (Lat. 53°56′N; Long. 168°02′W; Shepard 1948, fig. 73) near Unalaska Island, has had so many changes of form because of eruptions, vertical movements, and wave erosion that the varied descriptions by surveyors raised doubts about the reliability of the surveyors in the minds of their deskbound superiors (Stommel 1984, p. 69). Another, Mount Redoubt, 177 km southwest of Anchorage, was so active in its explosive eruptions beginning on 14 December 1989 that air transports had to be diverted to avoid intake of volcanic dust by their jet engines (Alaska Volcano Observatory Staff 1990; Kienle et al. 1990). Patterns of earthquakes and volcanic eruptions support the evidence of tide-gauge records in showing active tectonism in the form of Pacific plate subduction along the entire length of the chain. The general bedrock geology (Larson et al. 1985) as well as detailed geological and geophysical studies of the region (Scholl et al. 1975; Savage et al. 1986; Stone 1988) confirm the subduction process, and tiltmeter studies by Beavan et al. (1984) showed slow trenchward tilt during about 10 years that was reversed during 1978 to 1980 presumably by an episode of slip believed to be about 80 cm at a depth of about 20 to 70 km.

Pirazzoli and Montaggioni (1988) discussed Holocene sea-level changes in French Polynesia, documenting a stillstand of sea level 0.8 to 1.0 m above the present level between 5000 and 1250 y. B.P. Since then, relative sea level has dropped at a rate of between 0.6 and 0.8 mm/y., in response to possible geoidal changes or hydroisostatic loading. Southwest of French Polynesia, Yonekura et al. (1988) found an analogous maximum stillstand at +1.7 m level dating to 4000 to 3400 y. B.P., with emergence since then. Similar high stands are found throughout the southwestern Pacific Ocean, although rates and timing appear to vary, perhaps systematically (Pirazzoli and Montaggioni 1988).

In summary, tide-gauge records from archipelagoes (Japan, Philippines, Greater Antilles, and Indonesia) serve as links between those of island arcs (Aleutian Islands and the Mariana Islands) and those of subducting continental margins (Pacific coasts of North America, Central America, South America, and the suture between India and Asia). Neither the island arcs nor the continental coasts allow any two-dimensionality of tide-gauge information because of the narrowness of the belt of tide-gauge stations along the latter kinds of coasts. Geological inferences from tide-gauge records along archipelagoes aid the interpretation of tide-gauge records from island arc and subducting mainland coasts.

Hot-Spot Chains

The Pacific Ocean contains many chains of volcanic islands formed by mantle hot spots whose vents penetrated the cover of oceanic crust as the crust migrated over the hot-spot positions during sea-floor spreading. The chains thus are oriented in the direction of plate movement, with the oldest islands far beyond the hot spot and now volcanically inactive, and the islands that are above or just past the hot spot now still active and growing. This arrangement contrasts with that of island arcs, along which volcanic activity is more or less synchronous because underthrusting occurs along their entire length. There are no tide-gauge records at islands that definitely are members of hot-spot chains within the Atlantic and Indian oceans. In contrast, the Pacific Ocean contains many hot-spot islands having acceptable tide-gauge records, but only one chain contains such records on several islands. This chain, the Hawaiian Islands, contains six tide-gauge records including Hawaii (still active with volcanic activity and earthquakes) and Midway Islands (on an old and dead volcano). About 15,000 quakes were recorded on Hawaii during a four-year period, all with hypocenters shallower than 60 km and mostly shallower than 3 km (Koyanagi et al. 1976). They were concentrated primarily at Kilauea and vicinity and secondarily at Loihi, a seamount (Lat 18°56′N; Long. 115°17′W) that was surveyed by Emery (1955) and found to be recent in origin, according to thinness of palagonite on its pillow lavas (Moore and Reed 1963). Elsewhere in the Pacific Ocean, earthquakes recorded between 1905 and 1987 (Walker 1989) probably include ones from incipient or active hot-spot sites within otherwise stable crustal plates. Other means of recording submarine volcanic activity are provided by a large-aperture network of seismic recorders centered on Tahiti (Talandier 1989), and by use of the low-velocity sound (SOFAR) channel at mid-depths in the ocean water (Dietz and Sheehy 1953; Walker and McCreery 1988).

The six tide-gauge records of the Hawaiian Islands are concentrated on the larger and most populated islands at the east-southeast, but there is one at Midway about 25,000 km west-northwest of Loihi (Fig. 93A). Time spans range from 18 to 76 years, the longest being at Honolulu. Regression lines through the data points for mean annual relative land levels have t-confidence levels of 0.86 to 1.00 (Fig. 93B). Comparison shows land emergence at Midway (+1.3 mm/y.) and nearly progressively greater submergence toward the east-southeast to a maximum at Hilo (−3.5 mm/y.). An exception occurs at Mokuoloe Island (same as Coconut Island in Kaneohe Bay), where the record spanning only 18 years shows a rise of relative land level (+0.9 mm/y.). There is no obvious geological rationalization, but the station record appears to have ended with 1974, giving it far less value than the much longer record at Honolulu only 24 km distant. Even if the Mokuoloe Island tide-gauge record were included, the sequence of land-level movements exhibits a sharp drop from Hilo to Oahu followed by gradually lesser sinking from Oahu to Midway (Fig. 93D, circles).

This change is analogous to the subsidence to be expected for cooling and contraction of a transition sequence from younger to older oceanic crust (Menard 1969) and later found to be paralleled for subsidence below sea level of the tops of aseismic seamounts that once had reached above sea level (Sclater et al. 1971). Dates of rocks from the islands and from seamounts were assembled and compared with dates for magnetic anomalies of the oceanic crust near the edifices (Jackson 1976). They were found to differ somewhat, partly because volcanic eruptions had continued for different time spans at different seamounts and islands. Jackson's list of rock

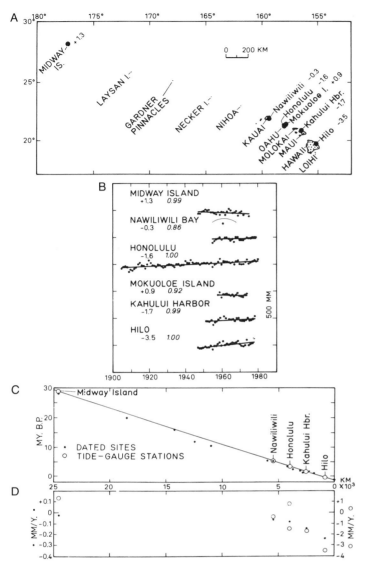

FIGURE 93. Hawaiian Islands.
A. Distribution and names (lower case) of tide-gauge stations having acceptable records. Mean annual changes of land level (from panel B) follow station names. Names in capital letters are for some members of the Hawaiian Islands chain.
B. Plots of mean annual change of relative land levels from linear regression analyses of acceptable tide-gauge records. Below each station name are annual change (mm/y.-vertical numbers) and t-confidence level (slant numbers). Data are from PSMSL.
C. Ages of volcanoes of Hawaiian chain (in m.y. B.P.) plotted against distance along the chain from Loihi, a seamount southeast of Hawaii (see position in panel A). Ages indicated by dots are the ones considered most reliable by Duncan and Clague (1985). Ages of sites of tide-gauge records (circles) are estimated according to the linear regression line computed from the age-distance plots of dated rocks at sites mostly different from those of the tide-gauge stations.
D. Mean annual changes of land level (mm/y.) at tide-gauge stations obtained from slopes of linear regression lines (panel B) are indicated by circles. Dots denote general subsidence rate inferred from ages of volcanoes and seamounts (panel C) according to the method of Detrick et al. (1977) as discussed in text. Horizontal (distance) scale same as in panel C.

ages along the Hawaiian chain was revised by Duncan and Clague (1985) on the basis of better samples and better choices of rock types for dating. Of 30 well-dated rock samples, Duncan and Clague chose 13 as representing the best defined dates. To these, the zero age of Loihi Seamount was added to give a total of 14 dates that were plotted (dots) against distance from 0.0 age at Loihi Seamount along the Hawaiian chain to 24.7 m.y. B.P. at Midway (Fig. 93C). The regression through the distance-age points yielded an average rate of volcanic propagation along the chain of 86 mm/y. From this line the ages of the tide-gauge sites (circles) were determined again according to distance from Loihi Seamount. The sinking rates for seamounts that once had reached sea level were found by Detrick et al. (1977) to be parallel to sinking rates of oceanic crust and a function of age: depth proportional to square root of age—a parabolic curve from the zero age/zero depth point. Depths increase rapidly at the beginning and then more slowly along the distance-age curve. Rates of sinking in the Pacific Ocean range between about 200 and 450 m (m.y.)$^{-1/2}$, and probably it is controlled by differences in temperature of the mantle sources (Cochran 1986). Obviously, a comparison of the sinking rates caused by cooling and contraction of the former volcanic islands is possible only where tide gauges exist—on islands. For each of the sites of tide-gauge records between Hilo and Midway the slope of the subsidence curve caused by cooling and contraction of the rock (depth = square root of age $\times$ k$^{-1/2}$) was computed according to the formula (depth change/year = ½ k^{-1} square root of age^{-1}). The resulting subsidence rates (dots on Fig. 93D) are roughly one tenth those revealed by the tide-gauge records (circles on Fig. 93D). The general relation of sinking rates to ages is similar, but the higher rates for tide-gauge records indicate the presence of another factor, perhaps the effect of weighting of the ocean floor by the increasing thickness of the returned glacial meltwater with time (to be discussed in a later section of text). Contrasting with the tide-gauge records that reveal sinking of land levels in the Hawaiian Islands during the past decades, early studies by Stearns (1935, 1978) revealed both emerged and submerged terraces around the islands of the Hawaiian Chain, and recent studies of raised terraces on Oahu indicate uplift of the island to at least 30 m during the past 0.5 m.y. (Brückner and Radtke 1989).

Many other tide-gauge stations in the open Pacific Ocean besides those of the Hawaiian Islands are situated on hot-spot chains. Records of the PSMSL contain 19 such stations having records longer than 10 years, but regression lines for only 8 of them have t-confidence levels higher than 0.8. These eight stations are widely scattered in the western tropical Pacific Ocean (Fig. 94A). Slopes of the regression lines for the data points of each station record (Fig. 94B) exhibit only a small range of change in land level (-2.0 mm/y. to -0.3 mm/y.) except at Christmas Island, where it is $+4.8$ mm/y. Information to complete an age/land-level plot like that for the Hawaiian Island chain was attempted from ages compiled by others mainly through their efforts to learn the direction and speed that the Pacific plate is moving. The scanty infor-

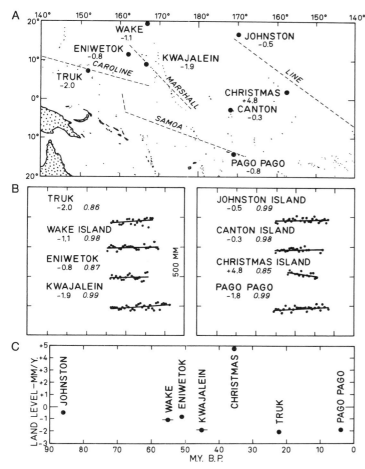

FIGURE 94. Hot-spot chains of Pacific Ocean other than Hawaiian Islands.
A. Geographical distribution of the eight tide-gauge stations having acceptable records and their mean annual changes of land level (mm/y.). Dash lines denote trends of hot-spot trails that contain some of the stations.
B. Plots of mean annual land level. The sequence of stations was chosen as northwest to southeast from Figure 94A. Beneath station names are mean annual changes of land level (mm/y.–vertical numbers) and *t*-confidence levels (slant numbers). Data are from PSMSL.
C. Plot of change in relative land level with geological age of the volcanic islands that contain the tide gauges. See discussion in text. Bars through the data points for Wake and Kwajalein islands indicate more than usual uncertainty in age for these sites.

mation on ages of the volcanic bases for the islands containing the tide gauges was gleaned from several publications. For Truk on Moan Island (Caroline Islands chain) the best information is 22 m.y. B.P. with some subsequent volcanism (Duncan and Clague 1985, p. 100, 114). Clague and Jarrard (1973) reported a date for pyroxene in the basalt at Eniwetok Atoll of 51 m.y. B.P, but no dates for either Wake Island or Kwajalein Atoll, all presumably on the Marshall–Gilbert–Ellice chain (Jackson 1976). Extrapolations from the date at Eniwetok Atoll are uncertain because of complications caused by fracture zones and shifts in direction of the Pacific plate movement. Nevertheless, an average plate movement of 10 cm/y. in the general westerly direction of plate movement and for the distance in this direction between Eniwetok Atoll and Wake Island and between Eniwetok Atoll and Kwajalein Atoll yields rough dates of 55 and 46 m.y. B.P. for Wake and Kwajalein islands, respectively. The uncertainty is expressed by a horizontal bar through the data points for Wake Island and Kwajalein Atoll in Figure 94C. The volcanic rocks of Johnston Island (probably one of the Line Islands chain) are deeply buried beneath calcareous reef rocks (Emery 1956), but Schlanger and Ozima (1977) and Schlanger et al. (1984) reported an argon isotope age of 86 m.y. B.P. (shown on Fig. 94C). They also gave an age of 35.5 m.y. B.P. for Christmas Island (another along the Line Islands chain). Last, the volcanic rocks of Pago Pago (near the active southeastern end of the Samoa Islands chain) have an age of about 4 m.y. B.P. by extrapolation from argon isotope measurements of rocks from nearby seamounts and banks (Duncan and Clague 1985). There is no known reported age for Canton Island (one of the Phoenix Islands chain) so it was omitted from the figure. The plot of annual change in land level exhibits a slow and irregular decrease in subsidence rate with age, not like the curve for the Hawaiian Islands stations (Fig. 93D, 94C), but this difference may be attributable to greater uncertainty in the ages. Most striking is the rise of land level at Christmas Island; it appears to be valid but has an unknown origin.

Long-Term Tide-Gauge Records of the World

The tide-gauge records used for working out relationships between changes of relative land levels and geology necessarily have various time spans in order to achieve maximum practical geographic density of the records. Obviously better would be to use only records exceeding 100 years, but there are only five of these in the entire world. To help make a decision on an optimum time span, all 664 tide-gauge records listed by PSMSL in 1988 were plotted according to their time spans. Note that the number of years of record for any given station may be much less than the total time span because of gaps in the record. The plot (Fig. 95) shows a concentration of records having shorter than 30 years' time span; in fact, 65 percent of the total of 664 records have time spans shorter than 30 years. As shown by the cumulative curve, 106 of them, or about 16 percent (one standard deviation), are longer than 50 years. This may seem to be a reasonable number and time span for use in seeking long-term trends or significant changes in rates and directions of changes in land levels. A further reduction to 98 records was deemed desirable in order to eliminate ones that did not include the entire time span between 1930 and 1980. Lacking only one year of that span were 10, 2 to 5 years—27, and 6 to 15 years—17 records. Nevertheless, all of these records have *t*-confidence levels for the regression lines through the data points of 0.97 or higher, and 89 of them have *t*-confidence levels of 1.00.

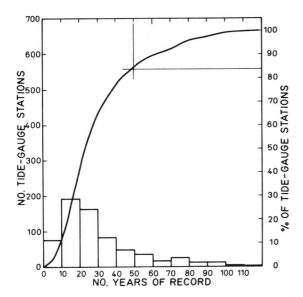

FIGURE 95. Histogram and cumulative curve for number of years of record for 664 tide-gauge records of the world. From these plots came the decision to plot the change of relative land level for the 98 records that span the years between 1930 and 1980 (Figs. 96, 97).

Unfortunately, only two of the 98 records chosen for long-term analysis (Figs. 96, 97) are in the Southern Hemisphere, all in Australia. The high concentration of records in the Northern Hemisphere is only partly due to the lesser area of land or length of inhabited coasts in the Southern Hemisphere (Fig. 38), but it clearly is mainly a function of the lesser and later interest in establishing and maintaining tide-gauge stations by governments of the lesser developed nations that are concentrated in the Southern Hemisphere. For example, Emery (1980) selected 211 records having t-confidence levels higher than 0.95 to which he added 36 records having t-confidence levels between 0.80 and 0.95 in order to increase representation from the Southern Hemisphere, but this increased the Southern Hemisphere records to only 7.7 percent of the total 237 records. For all PSMSL records of 10 years or longer, regardless of their quality, the Southern Hemisphere has only 11.1 percent of the total of 588 records (Fig. 38). For all 664 records (including the ones having less than 10 years of record) the Southern Hemisphere has 14.8 percent. These percentages indicate that the Southern Hemisphere contains a preponderance of short time-span records, yet these records are not just for recently established stations because many station records were terminated decades ago, evidently because of lack of interest or of available funds. In contrast are the records from Fennoscandia where many began during the 19th century. Longest of the records accepted by PSMSL is the one in Brest, France, that began in 1807. The earliest acceptable record in the Western Hemisphere is at San Francisco beginning in 1865. On Figures 96 and 97 the records begin only in 1880, because there are too few earlier acceptable ones to be useful for comparative purposes.

The 98 tide-gauge records shown in Figures 96 and 97 consist of 36 from the United States, 13 from Sweden, 12 from Finland, 8 from Canada, 5 from Japan, 4 from United Kingdom, 3 from India, 2 each from Australia, Italy, Norway, Philippine Islands, Portugal, and the U.S.S.R., and one each from Bermuda, Canary Islands (Spain), France, Macau, and Yugoslavia. This distribution means that there are 43 from mainland sites of North America, 38 from mainland sites of Europe, ten from mainland Asia and Japan, 2 from Australia, and 5 from oceanic islands. There is not one record from mainlands of either Africa or South America, which means that bias is unavoidable because of absence of information from these areas in the Southern Hemisphere. Bias also comes from high density of stations in Fennoscandia compared with other regions of equal area.

The plots of the 98 tide-gauge records having about 51 years of record between 1930 and 1980 (Figs. 96, 97) include 58 that were not so plotted within previous sections of this chapter. They are from northern Europe—32, western United States and Canada—16, eastern United States—10, and Japan—5. All of these particular areas had so many records that only the mean annual changes of relative land level were plotted in map form for the detailed studies of these regions, usually supplemented with contours of change to portray relationships within their geographical regions. The 40 records that appear on Figures 96, 97, and also on the more detailed plots for smaller regions of the world may have different indicated mean annual changes of land level. These differences came about because the mean annual changes of land levels for Figures 96 and 97 are based only on the time spans between 1930 and 1980, whereas earlier in this chapter they are based on the entire lengths of record. The best-fit regression line showing mean annual changes of land level at each station is drawn as a wide line for the data between 1930 and 1980. It merely is extrapolated as a narrow line for earlier and later data points but is not based on these earlier and later data points. Thus, it might serve as a basis for a general estimate of the uniformity of the change of land levels for earlier and later years in comparison with that for the period between 1930 and 1980, inclusive.

Even casual inspection of Figures 96 and 97 reveals a wide range of relative motion from rising relative land levels in Fennoscandia (crustal rebound after melt of ice sheets) and remarkably at Juneau (tectonic uplift associated with plate subduction) to falling land levels for other long-term tide-gauge records elsewhere in the world. The range of mean annual rates of relative land levels is from +12.6 mm/y. for Juneau, through +9.21 to 0.0 mm/y. for most of Fennoscandia, and between +1.4 and −7.0 mm/y. in the rest of the world for the period 1930 to 1980. Twelve records show rises of relative land level faster than +5.0 mm/y., 24 show rises between +5.0 and 0.0 mm/y., 60 show falls of land between 0.0 and −5.0 mm/y., and 2 show falls faster than −5.0 mm/y. This range is not necessarily typical of the much larger number of reliable, but shorter time-span, tide-gauge records of the world.

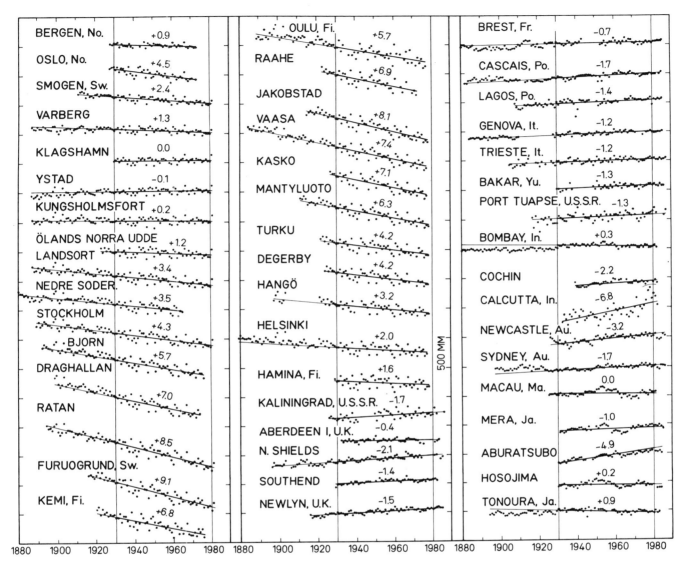

FIGURE 96. Plots of mean annual changes of relative land level during the entire time spans of 49 tide-gauge stations mainly in Europe and Asia that were operated for a period that included the years between 1930 and 1980 (indicated by narrow vertical lines). The linear regression lines and their slopes (mm/y.—vertical numbers) and *t*-confidence levels (slant numbers) are computed only from the data points between 1930 and 1980.

The range may not even be typical of the records of Figures 96 and 97 beyond the time span between 1930 and 1980. Many of these records began long before 1930 or continue for a few years after 1980. The time after 1980 is too short to recognize changes in trends, but the time span of measurements before 1930 may be long enough for 29 of the records to show whether the rates of change in level were different from the rates between 1930 and 1980. Note, again, that the regression lines for data points between 1930 and 1980 are simply extrapolated to earlier and later dates. Departure of these extrapolated lines from the trends of data points can suggest a change in rates before and after 1930. Visual comparison shows that for 15 station records (about half of the 29 records whose pre-1930 data are long enough to be considered significant), there are no obvious changes in rate. Two other records exhibit faster rising of relative land levels before than after 1930 (Draghallan, Helsinki), two show faster sinking (Ystad, Brest), three show slower rising (Varburg, Bombay, Tonoura), and seven show slower sinking (Sydney, Seattle, San Francisco, San Diego, Galveston, Portland, Manila). The slower early sinking at Galveston and perhaps at a few other sites might be ascribed to lesser pumping of fluids from the ground before than after 1930, but tectonic explanations of changes elsewhere are not evident. A faster rise of sea level after than before 1930 (due to increased volume of ocean water: effects of atmospheric heating by the greenhouse effect of the industrial revolution) would have the same effect on tide-gauge records as would increased sinking

146   5. DETAILED MAPPING OF TIDE-GAUGE RECORDS IN SPECIFIC REGIONS

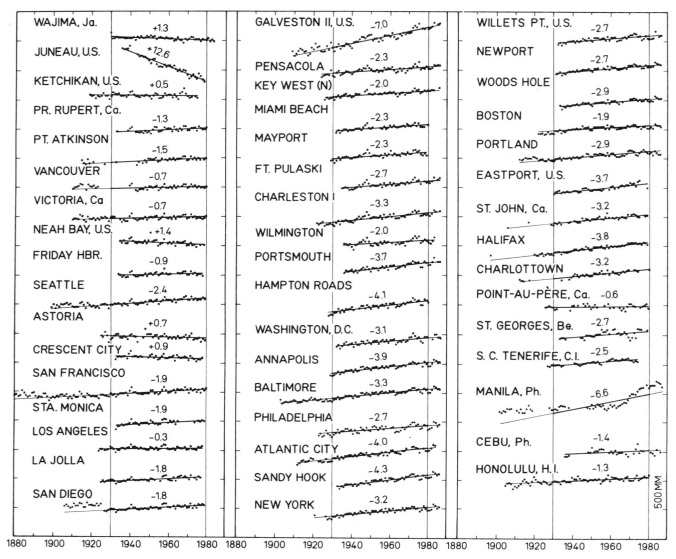

FIGURE 97. Plots of mean annual changes of relative land level (mm/y.) during the entire time spans of 49 tide-gauge stations mainly in North America that were operated for a period that included the years between 1930 and 1980 (indicated by narrow vertical lines). Symbols as in Figure 96.

of land level, but that explanation might apply only to the eight records at Sydney, Seattle, San Francisco, San Diego, Portland, and Manila plus Draghallan and Helsinki. These records constitute only 8 of the 29 long time-span tide-gauge records of the world that include useful data prior to 1930. On the opposite side of the argument is the likelihood that 1930 is not a good date to choose for the onset of sea-level rise caused by the industrial revolution and its greenhouse effect because of the time lag in oceanic response to atmospheric trace gas loading.

Another basis for measurement of trend changes is through eigenanalysis—a more objective method than that of mere visual inspection of Figures 96 and 97. The first step for investigation of trend changes by eigenanalysis is to compare the results obtained by regression and eigenanalysis. The number of stations (98) used for regression analysis in Figures 96 and 97 is large enough to provide a good basis for eigenanalysis for the period 1930 to 1980, but nine records had gaps or missing years of record that we considered to be more serious for eigenanalysis than for regression analysis. Omission of these nine records reduced the number of stations available for eigenanalysis to 89. Results for both methods of analysis are compared in Figure 98. Similar comparisons were made for several particular regions of detailed analysis of tide-gauge records earlier in this chapter (inserts on Figs. 43, 66, 84), but the assembly of records for the entire world is larger and has a longer effective time span than available for the smaller regions. The results show a fairly close

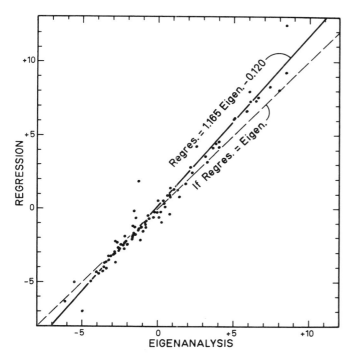

FIGURE 98. Comparison of changes in relative land level obtained by linear regression analysis and eigenanalysis of 89 of the 98 long-term tide-gauge records plotted in Figures 96 and 97. Each of these 89 records has nearly complete mean annual sea level data for all years from 1930 to 1980. The dashed line represents the trend along which the changes in land levels would fall if regression and eigenanalysis provided identical results. The wider continuous line through the data points indicates that in reality regression results show 16.5 percent faster rise or fall than do eigenanalysis results (reflecting the approximations of the eigenanalysis estimates).

(between the years 1930 and 1980) were computed and shown on Figure 99. They contain 45, 15, and 6 percent, respectively, of the total variation in the other records. For comparison, the temporal functions also were computed for the 35 worldwide stations having records longer than 60 years (between the years 1910 and 1980). These temporal functions contain 59, 10, and 6 percent of the total variation (Fig. 100), respectively. Notably, the longer time-span records contain much more of the total variation in the first function than is present in the first function for the 89 records longer than 40 years, a common finding due in part to the greater number of records for the shorter interval. In addition, the 35 records longer than 60 years were divided between the 20 that contain no Scandinavian tide-gauge stations and the 15 that are in Scandinavia. The first temporal function for the Scandinavian stations contains a very high percentage, 81, of the total variation, showing much more uniformity among the Scandinavian stations than among other stations of the world (Fig. 100).

Calculations of the slopes of the first eigenfunctions for different tide-gauge subsets can be used to detect changes in rates of relative sea-level rise during each time interval, although eigenanalysis can produce biases in the estimates of relative sea-level rise that are related to the length of record and the total number of stations reporting for different time intervals (Aubrey and Emery 1986a; Braatz and Aubrey 1987; Solow 1987). However, the present subsets were chosen to minimize gaps in records, and eigenanalysis was performed only on those records that have no significant gaps. Approximately the same number of stations reported

relationship of the trends of relative sea level (or land level) changes during the 1930 to 1980 period. Only 2 of the 89 station records depart more than 1 mm/y. from the best-fit regression line through the data points for regression versus eigenanalysis. However, this regression line has a R^2 confidence of 0.968 (a t-confidence level of 0.99). This means that on the average the change of level at any given station averages a 16.5 percent faster rise or faster fall of sea level or land level by regression than by eigenanalysis, although the significance of this difference is questionable. The cause of this difference is that the "synthetic" slope obtained by eigenanalysis based on the first three temporal functions at any station omits part of the data that are included in the regression analysis; the eigenanalysis results are just estimates having some bias (Solow 1987). On the other hand, the eigenanalysis is capable of comparing groups of station records instead of only a single station record as by regression analysis.

The results of eigenanalysis provide some additional information. En route to the synthetic eigenvalues plotted on Figure 98, the first, second, and third temporal functions for the 89 tide-gauge stations having records longer than 40 years

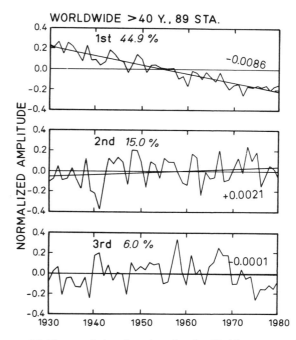

FIGURE 99. Temporal eigenfunctions for the 89 tide-gauge stations of Figure 98.

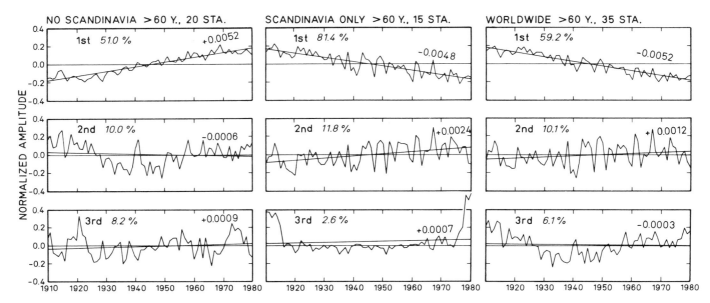

FIGURE 100. Temporal eigenfunctions for the 35 tide-gauge stations having records longer than 60 years with separate plots for those outside of Fennoscandia, within Fennoscandia, and all combined.

for each year of record, although the beginning and the end of the interval have the largest percentage gaps that are nearly equal. However, only about 2 percent of the station data are missing. By 1915, however, all stations reported, and all lasted until at least 1977. Consequently, biases caused by differing numbers of stations reporting are low, as indicated by comparing results of analysis for different station subsets.

To estimate changes in rate of relative sea-level rise as represented by the first eigenfunction, a two-phase linear regression was used (Braatz and Aubrey 1987; Solow 1987). Braatz and Aubrey (1987) used a less rigorous test of changes in slope, but one incorporating differences based on the t-test. Although Solow (1987) described a different two-phase regression model, the present results are not sensitive to which method was used. The worldwide results for 89 stations having a minimum record length of 40 years between 1930 and 1980 (Fig. 99) show little departure from a single regression line; the two-phase regression model reduces residual variance but not significantly better than the single line. However, the longer records (exceeding 60 years between the interval 1910 and 1980; Fig. 100) for the fewer (35) stations have a notable change in the mean slope centered between 1930 and 1935. Although we have tested the changepoint of slope for many records, and these changepoints vary geographically and with statistical model, we use 1930 as an approximate changepoint that is clear in many records. A two-phase linear model is not necessarily ideal nor most appropriate, but it is one test for significant changes in slope testable by error analysis. For longer but far fewer stations records, Gornitz and Solow (in press) found changepoints at different periods that varied geographically. These changes in slope are statistically significant.

When these records are divided into two separate groupings, one excluding Scandinavia, the other consisting solely of Scandinavia, the results (Table 7) show that within each region, the slope prior to 1935 is less than the slope following 1935. For instance, of the 17 stations that meet the length criteria within Scandinavia, only five show significantly different slopes prior to and following 1930. Four of these five show increases in absolute rate of relative change. The fifth, at Ystad, shows a decrease in slope, from a large submergence to a smaller one. This change is consistent with the regional trend of increased emergence (decreased submergence). Half of the remainder of the stations show an increase in rate of change, although they fail the test for statistical significance. For all other countries and their 19 qualifying stations, there are 11 stations having significant differences in pre-1930 and post-1930 slopes. Of these, three show decreasing slopes and eight show increasing slopes. Newlyn is among those showing decreasing slope but that station has only 15 years of data prior to 1931 and barely qualifies for this test. Of the remaining eight stations, six show increases in rate of change.

The data suggest that all regions having long-term tide-gauge records contain a preponderance of stations having an increase in rate of sea-level change since 1930. Rates of land-level rise in Scandinavia appear to be increasing, just as rates of land-level fall appear to be increasing elsewhere. The implications for this finding are curious and not easily explained: if relative sea-level is rising now, it is doing so at an increasing rate compared to pre-1930, and if relative sea-level is falling now, it is falling faster than pre-1930. These findings contrast with what one would expect from a greenhouse warming: as the ocean level rises more quickly, previously submerging areas would submerge more quickly while

TABLE 7. Comparison of pre-1930 and post-1930 rates of relative sea-level rise for all long stations (longer than 60 years between 1910 and 1980) reported by PSMSL.

Country	Station	Pre-1930 slope (mm/y.)	Pre-1930 t-conf.	Post-1930 slope (mm/y.)	Post-1930 t-conf.
Norway	Smogen	2.5	0.820	2.3	0.996 −
	Varberg	0.02	0.988	1.2	0.993 +/*
Sweden	Ystad	−1.6	0.973	−0.3	0.993 −/*
	Kungholmsfort	−0.1	0.972	−0.03	0.983 −
	Landsort	2.2	0.962	3.1	0.969 +
	Nedre S.	2.6	0.995	3.5	0.848 +
	Stockholm	3.5	0.951	4.0	0.963 +
	Bjorn	5.4	0.932	5.7	0.926 +
	Draghallan	7.3	0.860	7.0	0.901 −
	Ratan	7.2	0.918	8.3	0.937 +/*
	Furuogrund	7.7	0.660	8.9	0.933 +/*
Finland	Oulu	4.8	0.864	5.7	0.850 +
	Pietarsaari	6.1	0.656	8.1	0.912 +/*
	Vasa	7.1	0.974	7.4	0.911 +
	Mantyluota	6.3	0.732	6.3	0.926
	Hanko	3.4	0.802	3.2	0.936 −
	Helsinki	3.0	0.978	2.2	0.914 −
U.K.	North Shields	−1.0	0.927	−1.7	0.999 +
	Newlyn	−4.7	0.733	−1.6	1.000 −/*
France	Brest	−0.6	1.000	−0.8	1.000 +
Portugal	Cascais	0.1	0.994	−1.6	1.000 +/*
	Lagos	−2.6	0.814	−1.5	0.989 −/*
Italy	Genova	−1.2	0.999	−1.1	1.000 −
	Trieste	−1.6	0.808	−1.2	0.999 −
India	Bombay	−0.2	1.000	0.3	1.000 +
Japan	Tonoura	−0.6	0.977	0.8	1.000 +/*
Philippines	Manila	−1.5	0.699	−8.3	0.899 +/*
Australia	Sydney	0.8	0.934	−1.4	1.000 +
Hawaii	Honolulu	−2.3	0.802	−1.3	1.000 −/*
U.S.	Seattle	0.0	0.915	−2.4	1.000 +/*
	San Francisco	−0.6	1.000	−1.9	0.999 +/*
	San Diego	2.0	0.768	−1.8	1.000
	Galveston II	5.3	0.712	7.0	0.977 +/*
	Baltimore	−1.4	0.934	−3.3	1.000 +/*
	Atlantic City	−1.1	0.765	−4.0	0.996 +/*
	Portland	2.2	0.745	−2.9	0.999 +

Positive numbers are land emergence; negative, land submergence. T-confidence levels represent the ±1 mm/y. slope sensitivity. Those stations with asterisks indicate the stations having statistically different slopes; a plus means an increase in absolute rate of change; a negative sign indicates a decrease. A minimum of 15 years of data prior to 1931 is required to include the station in this table.

previously emerging areas would emerge more slowly and perhaps even become submergent. The eigenanalysis shows a slowing in rate of change, regardless of sign of change. We are unable to account for this change from a physical or geological standpoint, although one possibility would be global teleconnections through the atmosphere or ocean that caused ocean changes of one sign over most of the northern hemisphere, and of opposite sign restricted largely to Fennoscandia. Sturges (1987) discussed some such large-scale changes, although not global in scope. Conceivably, changes in the shape of the geoid are responsible. Another possibility is human impact, as at Galveston where removal of groundwater and hydrocarbons has accelerated relative subsidence.

The absence of evident physical or geological explanation for the global trend leads one to examine the methods employed to derive these estimates. As shown by Braatz and Aubrey (1987) and in present research, the change in rate of relative rise is a function of neither the number of stations used nor of the data gaps within those stations. This finding is robust in that different subsets of the overall data set show the same results. Thus, neither the numerical methods nor the data sets themselves can explain this global trend. One is left to examine differences in reporting methods or in processing techniques. We have been unable to document any difference in reporting or processing that would have occurred globally since between 1880 and 1930–1935, despite numerous investigations. However, it is possible that such differences, if they had occurred and if the record of changes had been lost to the present generation, could cause the behavior exhibited by the tide-gauge data. For instance, if tide-gauge datums were surveyed to benchmarks on or close to piers or other structures holding the tide gauges, rather than to a network of benchmarks on land surrounding the tide gauges, the measured sea-level changes might exhibit a smaller range and the trend might be dampened. What would be measured, instead of tide-gauge motion relative to land, would be the local fluctuations in sea level caused by imprecise placement of the tide-gauge or imprecise survey of the tide gauge from year to year. Thus, the year-to-year fluctuations would tend to have no trend and might indicate the noise level in the recordings caused by instrumental techniques rather than a trend due to relative sea-level change. If the benchmark networks were expanded following the mid-1930s, or if corrections were made in a different manner, then the subsequent years might record more faithfully the local relative sea-level rise. Certainly, machine automation of some of the tide-gauge processing might contribute to some differences in results. By making processing more automatic, more complete corrections could be made, yielding differences between pre- and post-machine processing eras.

Although we have no evidence or support documenting such a change in method, we have no adequate explanation for the apparent anomalous global increase of absolute rates of change following about 1930–1935. Why sea-level change should be roughly simultaneous with land-level changes (tectonics) is unclear and suspicious, leading one to doubt the wisdom of using the pre-1930 tide-gauge data for estimating either rates of relative sea-level change or changes in those rates. The use of pre-1930 data for this purpose might await resolution of this uncertainty and dichotomy. Thus, we question the utility of the observations of numerous authors (Barnett 1984, 1988; Braatz and Aubrey 1987; Gornitz and Lebedeff 1987; Gornitz and Solow in press) that detect changepoints, when such changepoints may reflect methodological differences or local effects rather than global-scale physical processes. Although Gornitz and Solow use longer records dating to 1801 (some of which are not in the PSMSL data holdings and therefore may not have been corrected adequately for datum), and their conclusions differ somewhat from ours (different times for change points, regional differences, lack of uniformity of change in Fennoscandia), their analysis also may reflect trends that are present in the data series, but perhaps are result from methodological changes or local effects rather than global-scale physical processes.

6. Significance of Tide-Gauge Records

General

In the previous chapter tide-gauge records from different regions of the world were investigated, with particular attention paid to those in regions having many tide gauges with acceptable records in terms of time spans and systematic trends. Both worldwide and regional examinations showed the trends of change in mean annual levels to be too variable for those trends to represent a simple eustatic rise of sea level caused by return of glacial meltwater or by heating of ocean water through climatic warming during post-glacial or even during the span of the industrial revolution (the greenhouse effect). These records reveal rise of relative sea level at some sites simultaneously with fall in relative sea level at other sites. Sites of the opposing trends may be nearby or distant, and the records commonly are supported by others within groups of nearby stations. Short-term changes in the record caused by tides, storms, and tsunamis are filtered out by the hourly averaging process for each year of record. Quite evidently, the long-term records of tide gauges are influenced by something other than gradual change of sea level, namely changes of the level of the land to which the tide gauges are attached as well as longer term events (El Niño and other coherent motions; Sturges 1987). Changes of land level caused by settling of pilings or other structures that support a tide gauge may be revealed by an otherwise inexplicable sudden change in mean annual level, but this is rare because it probably would have been obvious to the tide-gauge attendants. It also would be revealed by the periodic re-leveling required for PSMSL acceptance of records. Moving of a tide gauge from one location to another is indicated by cessation of the earlier record.

If the tide-gauge records that show changes of land level caused by geological processes or human interactions can be identified and mapped, other tide-gauge records that may have been controlled primarily by changes of sea level remain to serve possibly as the basis for estimating the historical rate of rise of sea level. This procedure may permit separation of the signal from the noise in tide-gauge records that have been influenced by changes of both sea level and land level. Even casual consideration of the nature of potential changes of land level indicates the likelihood of three kinds of change that involve considerable lengths of coast: glacial rebound and peripheral crustal sinking, belts of plate subduction and rifting, and stable coasts (Fig. 101). There are many other geological processes that can change the levels along short parts of coasts, but most common of them are volcanic activity, faults and folds, weighting by large deltas and basin fills, and extraction of fluids by pumping. The areas affected by past ice sheets and by plate subduction and rifting will be considered first. They will be followed by mapping of short coasts affected by volcanism (and mantle hot spots), faults and folds, large deltas, and fluid extraction by man. The remaining coasts include those that appear to be stable because of passivity of geological processes and the approximate uniformity of change of relative sea level in all tide-gauge records along such coasts. These coasts and their tide-gauge records may provide the best available clues to the rate of modern eustatic change in sea levels.

Glacial Loading and Unloading

Longest known of the geological changes are those caused by the weight of Pleistocene ice sheets—very large continental glaciers. At the time of glaciation the underlying crust was depressed under the weight of ice, flexing the lithosphere and forcing part of that crust and subcrust radially outward to produce a peripheral bulge. When melting of the ice began 15,000 to 18,000 y. B.P., the progressively less weighted crust began to rise and the peripheral bulge began to collapse. As shown by Figure 101, the process is continuing in Northern Europe where tide gauges are abundant, and in Canada where tide gauges are fewer. In both areas peripheral bulges are known to be sinking wherever tide gauges are present. Greenland and Antarctica still are weighted by ice sheets; rise of their underlying crusts probably has begun but is not documented because of the absence of acceptable tide-gauge records along the shores of these land masses.

The total length of coast (measured in 500-km coastal segments) that has acceptable tide-gauge records and that appears to have been influenced principally by land movements associated with deglaciation is about 20,000 km, roughly 40 percent of the total world length of all such coasts. In other words, the existing geographical distribution of acceptable tide-gauge records verifies vertical changes of land level caused by glacial loading and unloading along only about 40 percent of the world's coasts where glacial effects

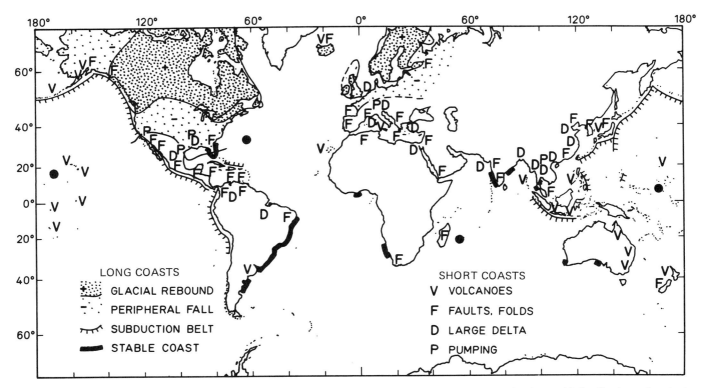

FIGURE 101. Map of the sites of geological processes believed to exert the most important controls over the direction and rate of change of relative land levels at all tide-gauge stations of the entire world that have reliable records. The information was synthesized from the previous chapter and many of its figures. Note that the map is not intended to depict the world distributions of geological processes, but only those parts of the distributions where these processes have affected tide-gauge records. Where no reliable tide-gauge records are present, the map is left blank for the geological processes.

probably are most important (rising of crust beneath melted ice sheets and sinking of peripheral ridges). The areas of Antarctica, Greenland, and most of northernmost Asia probably are undergoing uplift but without documentation by tide-gauge records. In Europe and North America where tide-gauge records and geology indicate major effects of ice loading, there are some local supplementary land movements caused by volcanoes, faults, and folds, but those effects are considered minor compared with the effects of glacial unloading. Where tide-gauge records are in areas affected by glacial loading and unloading but near sites of volcanoes, faults, folds, and deltas, the latter are ignored as overwhelmed by the effects of glacial loading and unloading. Effects of subduction, however, appear to dominate those caused by glacial unloading, so in areas undergoing subduction the effects of subduction dominate those of glaciers in the plots of Figure 102. Support for the belief in their subordinate nature is provided by the range of land movements caused by these same processes in non-glaciated regions of the world (to be considered below). Altogether, there are 179 acceptable tide-gauge records within the areas of northern Europe and North America that were subjected to ice-loading. They exhibit a range of relative land movement between about +15 and −7.5 mm/y. with a median of −0.2 mm/y. (Fig. 102). This low median may be expected because the volume of uplift must equal the volume of downwarp, but precision of measurement is low because of poor distribution of tide gauges in formerly glaciated areas of North America and for ocean floors adjacent to areas of former ice sheets.

Subduction and Rifting

The second geological process that controls shapes and many other aspects of many coasts is sea-floor spreading. Its effects are functions of the directions of relative plate movement at continental margins: divergence or rifting that breaks up continental crusts and initiates emplacement of oceanic crusts, convergence of crustal plates that leads to subduction of denser plates beneath lighter ones or to obduction or collision of more or less equally dense plates, and translation whereby two plates move along each others' shared continental margin that usually started as a rift belt. Examples of rifting are opposing Atlantic continental margins of North America, South America, and Africa; younger ones are along the Red Sea and the Gulf of California. For convergence, we may choose the Himalaya and the Alps ranges or the Pacific continental margins of North America, South America, and Asia

(Fig. 101). For translation, we have the San Andreas fault and its Gulf of California, the Dead Sea fault and its Gulf of Aqaba (Elath), and the Alpine fault of New Zealand.

Most poorly represented by tide-gauge records are coasts of rifting, because youthful rifting in the Red Sea and the Gulf of California occurs in areas bordered by countries that operate no or few tide gauges having acceptable time spans. Initial rifting in the East African Rift Valley is on land (no ocean — no tide gauges). Rifted coasts representative of mature or perhaps old age stages are common, but the tide-gauge indicators of rifting appear to be complicated thoroughly by other processes such as glacial uplift or downwarp along all coasts of the North Atlantic Ocean, and volcanism along the Pacific Ocean coast of Australia. Complications caused by ancient rifting along coasts of southeastern United States, eastern South America, southwestern Africa, India, and southwestern Australia may be so immeasurable that they now may be considered stable ones. Therefore, the few areas of rifting documented by tide gauges are included under effects of faults and folds in Figure 101.

Better represented are coasts of convergence. Mature convergence is represented by mountain ranges such as the Himalayas, Alps, Atlas, Appalachian, and other ranges, all of which are high above the ocean and distant from tide-gauge sites. Much better for study of vertical changes of land level are coasts that represent youthful stages of convergence — where plate subduction is indicated by the presence of deep-ocean trenches, seismic evidence of a Benioff zone of earthquake hypocenters, and by seismic reflection and refraction evidence of a descending oceanic plate. These areas are concentrated around the perimeter of the Pacific Ocean (Fig. 101), where acceptable tide-gauge records are abundant except in southern South America. Elsewhere in the Atlantic and Indian oceans, subduction beneath island arcs is common in the East Indies, West Indies, and South Sandwich Islands, none of which has useful modern tide-gauge records.

Translation movements are best represented along the Pacific coasts of North America, New Zealand, and the Straits of Gibraltar, in all of which the effects are largely obscured by associated convergence, rifting, and active local faulting and folding. Tide gauges there are infrequent. Translation in the Gulf of California has a subduction component and is grouped with subduction effects in Figures 101 and 102.

The best represented of plate movements and sea-floor spreading are youthful convergence or subduction around the perimeter of the Pacific Ocean. The effects of subduction on vertical change of land levels so dominate those caused by glacial loading, volcanoes, faults, folds, and deltas that these latter are ignored in favor of subduction (Fig. 101); most of these volcanoes and faults also are produced by the subduction process. The total length of all coasts affected by subduction and recorded by acceptable tide-gauge records (Fig. 101) is about 36,000 km. Within this belt length are 145 acceptable tide-gauge records whose information indicates a dominance of falling land levels (to −23 mm/y.), but with a few rising ones (to 15 mm/y.) that perhaps indicate the disruptive

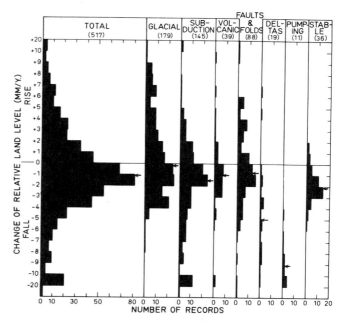

FIGURE 102. Histograms showing distribution of directions and rates of vertical changes of relative land level for 517 acceptable tide-gauge records used in text and figures of Chapter 6. The various station records were grouped according to geological processes that appear to have controlled the vertical movements of land (Fig. 101).

effects of subducting microcontinents of continental crust or oceanic volcanic ridges that were carried atop the oceanic crust. The median rate for subduction coasts (Fig. 102) is −1.5 mm/y. Beyond the concave side of island arcs are areas of uplift that probably originate from the intrusion of subducted masses at depth, raising the overlying continental crusts. The dominance of sinking land level is attributed to erosion of the base of the continental crust just above the subducting oceanic crust, thinning of the continental crust, and sinking of it — as illustrated best along Japan and less so along the Philippine Islands, the Aleutian Islands, the Pacific coasts of North America, and the Indonesian archipelago. Comparison with the changes of relative land levels in regions of glaciation (Fig. 102) reveals a change from dominant uplift to dominant sinking of land levels.

Volcanism

Volcanic activity is well known from historical observations to cause changes of land level along oceanic coasts, because sea level provides a convenient and obvious reference for such changes. Volcanism commonly is associated with crustal plate movements of all kinds and with mountain building that is produced by these crustal movements. Much of the mountain building on continents takes the form of faults and folds, so that changes of land level caused by volcanism generally are more local and more isolated in distribution. Moreover, at high latitudes changes in level produced by volcanism usually are overshadowed by more widespread changes of level caused by glacial loading and unloading of the continent crust (Iceland may be an exception). In contrast, oceanic crust owes its existence to igneous intrusion and extrusion and has been obscured little by glaciation. Thus, changes of land level associated with volcanic activity are more common and more easily recognized on oceanic islands than on continental coasts. On the other hand, changes of land level produced by volcanic activity in oceanic regions can be recorded by tide gauges only where the volcanoes rise above sea level as islands. Few islands have tide gauges, and many of these particular islands consist of volcanic rocks older than a few tens of millions of years and thus volcanism there is now inactive.

As shown by Figure 101, tide-gauge records that are believed to indicate present vertical movements of shores caused by volcanic activity on continents (mainly due to plate rifting or subduction) occur in Italy, Australia, New Zealand, Alaska, and South America. Other sites probably exist but their records were not recognized or were obscured by land movements caused by subduction. Those in oceanic belts of active rifting, subduction, or along hot-spot trails include Bermuda, Iceland, Canary Islands, Mauritius, Andaman Islands, East Indies, Philippines, Japan, and the Aleutian, Hawaiian, Marshall, Mariana, Caroline, and Line islands. Many other small islands of the Pacific Ocean would have been included had their tide-gauge records been long enough or consistent enough to provide acceptable t-confidence levels.

Tabulation of the directions and speeds of change in land level indicated by the 39 acceptable tide-gauge records at sites of volcanic activity indicates a range from about $+10$ to -15 mm/y. and a median of -1.0 mm/y. Total coastal length where volcanic changes dominate is uncertain but probably is only a few thousand kilometers because most coastal volcanoes are associated with subduction—a geological process of greater importance than simple volcanism in causing vertical changes of land level. Where volcanoes are present on island arcs along ocean trenches, their effects on land levels are plotted with subduction on Figures 101 and 102.

Faults and Folds

Mountain building on continental crust produces generally linear ranges that are parallel with the belts along which plate collision or convergence occurs (Appalachians, Andes, Coast Ranges of the United States, for example). Thus, the most recent ranges also parallel continental margins. Some older ranges were segmented by later rifting, so that they now lie at an obtuse angle to the continental margin (Scandinavia, Pyrénées, Atlas, Appenines, for example). The recent mountain ranges are the more active ones, with faulting and folding continuing and causing local-to-regional rising or sinking of the land. Only some of these vertical movements are recorded by tide gauges. Where the movement was sudden and large, and especially where it was near the site of a tide gauge, the record for that station may have been discontinued and a new record begun. Slower bending (strain) of the crust precedes and follows the actual fault break, and this bending as well as warping associated with folds may be the chief aspect of mountain building that is recorded by tide gauges. Doubtlessly, such land movements have been on and still occur on all continents, and some of them are identified on Figure 101. At all indicated sites the irregularities in the record are larger than those in somewhat more distant tide-gauge records, and these sites also are ones where major faults have occurred in historical times or are known from nearby geology according to structural mapping.

The direction and speed of vertical land movements believed to have resulted from faults and folds are less abrupt than the more spectacular ones that are accompanied by earthquakes, but these tide-gauge records are interpreted here as resulting from the local crustal bending that precedes or follows the spectacular breaks of the crust. Conceivably, some of the records contain small vertical movements by faults that reach the Earth's surface. Thus, the rates of change in relative land levels caused by folding and indicated by Figure 102 are far less than those of major crustal breaks that initiated the abandonment of an old record and the start of a new one. The directions and rates of change in the 88 tide-gauge records attributed to folds and faults as slow warping not obviously associated with subduction range from about $+15$ to -25 mm/y. with a median of -0.8 mm/y. The total coastal length where these changes were identified near tide-gauge sites is about 20,000 km, but this length is more poorly defined than those of most other geological processes.

Deltaic Loading

Large deltas cause subsidence of the land surface in two ways: by imposing an extra weight on the underlying crust leading to sinking of the crust to restore isostatic equilibrium, and by compaction and loss of interstitial water from the deeper deltaic sediments through the weight of the shallower ones. Deltaic weighting that causes sinking of the land is not necessarily confined to the land just beneath or near the tide-gauge stations. It can extend farther oceanward beneath the fore-delta where subsidence may affect the crust in a larger area during isostatic equilibrium adjustment of an elastic crust.

Many large deltas have tide gauges that were installed to facilitate the movement of ships bringing and taking cargoes between upstream and overseas sources and destinations. This commerce is the reason why many large cities of the world are located on or near deltas at the mouths of large rivers (for example, London, Amsterdam, Venice, Hyderabad, Calcutta, Rangoon, Bangkok, Shanghai, San Francisco, Galveston, New Orleans, Baltimore, Philadelphia, New York, and Boston). The tide-gauge records reveal faster sinking of the land than exhibited by records at other stations located on either side of the deltas. Using this criterion, we recognize the effects of deltas in 19 tide-gauge records at the deltas of the Rhine, Rhône, Po, at Izmir (probably), Nile, Indus, Ganges/Brahmaputra,

Irrawaddy, Chao Phraya, Changjiang, HuangHe, Amazon, at Maracaibo, and Mississippi rivers (Fig. 101). Other deltas would have been included if they had acceptable tide-gauge records. The direction and rates of land subsidence on deltas differ from those for tide-gauge stations in other geological environments (Fig. 102), exhibiting a range from $+7$ to about -15 mm/y., with a median of -4.9 mm/y.

Extraction of Fluids

Some areas of dense urban population and high industrial demand, areas of high agricultural demand for water, or areas where oil and gas are extracted as fuels have undergone marked sinking caused by the removal of these interstitial materials by man. Best known probably is Venice and vicinity where tourists from many nations have witnessed flooding of Saint Mark's Square at high tide and where buildings are deteriorating from moisture and salt brought upward by capillary movement of water from canals. This sinking is caused by excessive removal of water mainly for industrial purposes. Similar sinking of the Chao Phraya Delta at Bangkok is caused by removal of water for industry and agriculture. In the United States, pumping of petroleum (oil and gas) and water caused considerable sinking of ground levels near Long Beach, at Galveston and Texas City, and at the Mississippi River Delta. In all five areas the subsidence led to intrusion of ocean water and disruption of populations and industries. Some control has occurred through government limitation of pumping rates, but regulation cannot lead to recovery back to original land levels. Many other areas of subsidence are present along coasts having no tide gauges and in inland regions where subsidence is unimportant and perhaps not even recognized by their populations. The five cited examples and their 11 tide-gauge records reveal subsidences at faster rates (about -4 to -15 mm/y., with a median of -9.1 mm/y.) than those produced by natural geological processes (Fig. 101). Pumping has caused local subsidence exceeding 100 mm/y., far surpassing rates resulting from natural geological processes! Note that the natural deltaic sinking adds to the rates of sinking caused by pumping alone.

"Stable" Coasts

Observation

Previous sections of this chapter identified coasts whose vertical movements reflect widespread geological processes such as glacial loading/unloading, subduction, and other plate movements, and more local processes associated with volcanism, faults and folds mainly in areas of recent mountain building, and weighting by deltas, plus man-produced subsidence from pumping of water, oil, and gas. There remain other coasts that have acceptable tide-gauge records, and these coasts may include ones where recent geological activities have been so minor as to allow these coasts to become the best sites for estimating present eustatic changes of sea level. Additional criteria for stability of these coasts would appear to be their great age and thickness of crust, absence of much seismic activity, and considerable uniformity in direction and rates of change of relative sea level/land level. Where records exhibit variability beyond that to be expected from climate and ocean, we assumed that the origin of relative vertical movements was unknown, and those sites were excluded from classification as stable. Elsewhere that acceptable tide-gauge records are lacking, no decision about origin can be substantiated.

Review of Figure 101 reveals such possible "stable" coasts along parts of Africa (4 tide-gauge records), Asia (5), Australia (2), South America (10), and North America (11) plus some islands of the open ocean (4) where the volcanism that produced the islands ceased many tens of millions of years ago. Their coastal length is about 10,000 km. These stable coasts are indicated by 32 tide-gauge stations on continents and by four stations on ancient volcanic islands. The plot of direction and speed of change in relative land level at these 36 tide-gauge stations (Fig. 103) reveals a far narrower range ($+1.1$ to -4.2 mm/y.) than that exhibited by records along coasts affected by presently active geological processes. This range also is far narrower than that (about $+25$ to -25 mm/y.) of the plot (Fig. 36) for all tide-gauge records, for which no attempt was made to separate records that had been influenced by rather different geological processes. One station record is 4.2 mm/y., 18 are between -4 and -2 mm/y., 14 are between -2 and 0 mm/y., and 3 are between 0 and $+2$ mm/y. The median of -2.6 mm/y. for stable coasts also indicates faster subsidence of land levels than for coasts that have been controlled by relatively recent geological processes except deltas. It is from such records as those of Figure 36 that various authors have selected "typical" representatives of different oceanic regions in their attempts to compute the eustatic sea-level rise for the world oceans. The 36 tide-gauge records that may be indicative of stable coasts represent only 6.1 percent of the total of 587 PSMSL records longer than 10 years that were used for Figure 36. They constitute 6.9 percent of many of these same PSMSL stations plus about 50 other records that were used in various coastal regions of the previous chapter but had not been subject to the rigorous re-leveling checks required for acceptance by PSMSL; these are the total of 517 stations indicated under the total heading of Figure 102. Many tide-gauge records tabulated in Figure 36 were omitted from Figure 102 because they have too short time spans or too low t-confidence levels.

An interesting relationship between the 36 tide-gauge stations of stable coasts in Figure 101 emerges when they are plotted on a map of Gondwana (Fig. 104). Thirty of the stations on continental coasts lie along matching coasts that mark the belts where Mesozoic rifting divided the megacontinent of Gondwana into the present continents. The six additional stations are at Phuket and Prachuap Kiri Khan (both in Thailand), and Bermuda, Mauritius, Kwajalein, and Johnston

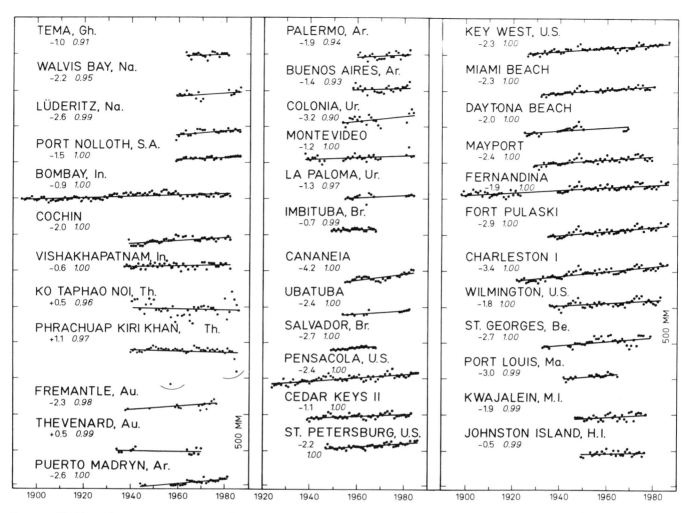

FIGURE 103. Plots of mean annual changes of relative land level (mm/y.) for the 36 tide-gauge records considered to be representative of stable coasts, as indicated on Figures 101 and 102.

Island (all on volcanic islands in the open Atlantic, Indian, and Pacific oceans). None of the coasts for these six tide-gauge stations was part of Gondwana, and at least the island sites did not even exist when Gondwana became rifted to form the present continents and Atlantic Ocean. Stable coasts were not identified around the rest of the perimeter of Gondwana, and they are believed to be absent because of the disturbance produced by subduction and associated tectonism during the radial outward movement of the rifted continents from the opening Atlantic Ocean. Note also that stable coasts are not present in the North Atlantic Ocean, Arctic Sea, nor around Antarctica because the late Pleistocene ice sheets with their loading represented a widespread destabilization that still affects those coasts.

Interpretation

At any particular coastal site a rise of water level relative to the land may be identified as indicating a rise of the level of the entire world ocean. If all of several tide gauges indicate a similar rise of water level, their behavior would seem to support the concept of a rise of the level of the world ocean (some workers even have accepted Kuenen's [1945] listing of eleven Dutch tide-gauge records in that way). The reality, however, is that the records of tide gauges at different sites are rather different. Efforts can and have been made to select "typical" records for different portions of the ocean and to compute sea-level changes of the world ocean on the basis of statistical stratagems without questioning how much the tide-gauge records may differ from each other because of vertical movements of the land to which the tide gauges are attached. In addition, an uninformed observer is unlikely to know that changes in mean annual levels of the ocean may be less than changes in mean annual levels of the land because he sees the large cyclical changes of water level produced in seconds by waves and in hours by tides. On the other hand, a geologist's background includes observations or knowledge that land levels can change far more than water levels—at an extreme reaching kilometers of vertical movement in a few million years.

The more than 500 acceptable tide-gauge records of the world reveal about as many records and sites that show a rise in relative sea level as those that show a fall. Some of these opposing trends are recorded by proximal as well as by distant tide-gauge stations. Part of the overall variation is caused by errors in recording and interpreting the records. Many such errors come about by use of time spans that are too short to filter out the effects of cyclical or episodical changes in winds, currents, salinities, and water temperatures. Time spans of 20 years or longer are desirable but not obtainable in all regions. A plot of all tide-gauge records accepted by PSMSL and longer than 10 years (Fig. 36) reveals that the range in annual changes of sea level at 587 stations exceeds 20 mm/y. rise and 20 mm/y. fall of sea level. Similar ranges are shown by a plot (Fig. 35 insert) of 247 tide-gauge records selected to avoid errors in recording and interpreting by use of long time spans and restriction to statistically high confidence levels in measuring the changes of level that best fit the scatter of mean annual relative sea levels in each tide-gauge record. Similar ranges also are shown by another plot (Fig. 102) of 521 tide-gauge records having a corresponding statistical selection. The three plots and the general similarity of records within coastal regions that are more or less dominated by specific tectonic processes (Fig. 102) indicate that most of the variation in tide-gauge records must be caused by something other than errors in recording and interpreting the information.

Fairly obviously, the contrast in changes of level throughout the world ocean do not favor uniform variations in sea level as much as they favor variations in vertical movements of land. Changes of land level have long been recognized in Scandinavia where shorelines have emerged visibly during the course of only a few decades, and shorelines hundreds of meters above present sea level testify to long-continued rise of the land. These rises of land level in Scandinavia and in northern North America have been attributed to rebound of the Earth's crust that was depressed by the weight of late Pleistocene ice sheets. Moreover, the saucer-shaped depression of the crust beneath Greenland records the presently still unrecovered sinking of land level beneath an existing ice sheet. Other geological processes than ice loading and unloading have been recognized by geologists and engineers as capable of causing vertical changes in land levels. Probably most important are warpings of the crust produced by underthrusting or subduction of continental crusts by denser oceanic crusts during the process of sea-floor spreading, by volcanoes that may rise or sink because of movements of magma at relatively shallow depths and because of excess weight of volcanoes on the underlying Earth's crust, by faults and folds associated with modern mountain building, and by weighting of the crust by thick deltaic sediments deposited by streams at the ocean shore. The pumping by man of water, oil, and gas through wells also is known to cause intense local sinking of ground levels. Note that all of the vertical changes of land level caused by these agents and processes involve movement of rocks and sediments that already were in place; these are not changes in level produced by deposition of new layers of sediment nor by removal of sediment and rock surfaces by erosion.

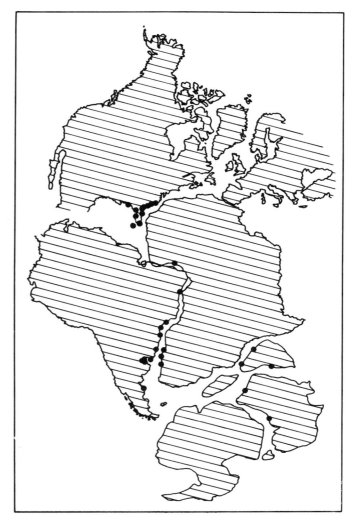

FIGURE 104. Positions of tide-gauge stations sited on continental crusts believed to be stable, compiled from Figure 101 but replotted on a reconstruction of Pangaea developed by Hurley and Rand (1969). There is a close relationship of these most stable coasts to the belts of rifting that broke Pangaea into fragments during the Mesozoic Era. Seven other stations believed to be on stable land are in southeastern Asia and on islands of the open ocean (none being parts of Pangaea).

The total number of acceptable tide-gauge records used in constructing Figure 102 is 517. They have a world distribution but are concentrated in the Northern Hemisphere. Each record was assigned to the geological process that was considered most likely to have caused vertical movement of the land at the tide-gauge site. This decision was based on the geology of the adjacent ocean floor and land, as inferred from geological and structural maps and from publications of many geologists who had studied different coasts. The records believed to have been controlled by glacial loading and unloading exhibit marked rises of land level by crustal rebound after melting of thick ice sheets and corresponding but lesser sinking of peripheral bulges built of crustal and subcrustal materials displaced by sinking of the crust beneath the ice load. Removal of records produced by crustal reaction to loading and unloading by ice sheets eliminated many of the records of rising land and lowered the median level of the total in Figure 102. Removal of records for vertical land movements believed caused primarily by crustal subduction, in contrast, eliminated many of the records of sinking land and raised the median level for the remaining total. Records showing changes of land level attributed to volcanism (beyond those associated with subduction) eliminated some extreme records but made little change in median level of the total. Removal of records for changes related to folding and faulting in regions of coastal mountain building also seems to have changed little the median level but probably intensified the peak of the distribution within the total by reducing the number of records for moderate rises and falls of level. More drastic are the effects of removing from the total the records attributed to sinking by deltaic loading; their removal eliminated some of the records for very rapid sinking and certainly raised the median level of the remaining total. Even more extreme in the same direction is the removal of records that show fall of land level caused by pumping of water, oil, and gas. However, the influence of deltaic loading and of pumping is minimized by the relatively small number of records (30) for both processes.

Eustatic Sea-Level Change

The 517 tide-gauge records for total (Fig. 102) contain 481 records whose changes in level are believed to be determined mainly by geological processes and human interventions. Their removal leaves 36 records attributable to stable coasts, from which we may be able to infer something about the eustatic rise of sea level caused by present rates of melting of glaciers and heating of ocean water. These 36 records have rates that range only from +1.1 to −4.2 mm/y. and they have a median level of −2.6 mm/y, about 1.5 mm/y. faster sinking of land than that of the total for all 517 records. This figure of 2.6 mm/y. by no means should be considered as representing the true rate of eustatic rise for the world ocean. It also includes the sinking or rising of the shore zone (where tide-gauge stations are located) caused by the isostatic adjustment of the ocean floor caused by the weight of returned glacial melt water. For every millimeter thickness of newly returned water one should expect about 0.3 mm of sinking of the ocean floor. This sinking must be accompanied by rise of the adjacent land and of farther offshore ocean floor caused by lateral transfer of crustal and subcrustal material, analogous with the development and later collapse of the peripheral bulge produced by the load imposed by an ice sheet. The main problem here is that the recorded vertical movement of relative land level is controlled by the position of the hinge (boundary between sinking ocean floor and rising land level) with respect to the tide gauge. A tide-gauge station oceanward of the hinge line would record a fall of land level, and another tide gauge landward of it would record a rise. That this factor of position of tide-gauge stations relative to hinge line is important is clearly apparent from the slope of the land subsidence/emergence contours for multishore coasts such as Chesapeake Bay and Delaware Bay (Figs. 80 and 83, respectively). Neither coast is stable, but the warping of ocean floor and land by weight of returned melt or other water must influence all tide-gauge stations within these bay areas regardless of the geological process that dominates the vertical movement at the tide-gauge stations. Only one area of stable coast has a sort of multiple shore with a reasonable number of tide gauges—Florida. As shown by the contours of change in land level given in Figures 79 and 84, the land at or near the shore on both sides of the peninsula is sinking at a rate less than about −2.0 mm/y., whereas about 100 km farther oceanward one can infer sinking at about −3.0 mm/y.

Tide-gauge stations along most coasts occur as a single string of stations along a single shore. Some stations can be landward and some oceanward of the hinge line, with the rate of change of land level being roughly proportional to the distance between the station and the hinge line. Therefore, even though the median change of land level at the 36 stations representative of stable coasts is about −2.6 mm/y., we cannot know just how much of this change is caused by incremental thickness of returned melt water and how much by depression of the crust beneath that water and to elevated crust landward of it. A simple approach to learning about depression of the local crust under the level of meltwater (Bloom 1967) would be to assume sinking of the ocean floor of about 0.3 mm for every 1 mm of returned water (rock density about 3.3 times the density of the water). A tide-gauge indication of 1 mm/y. rise of relative sea level implies a return of 1.3 mm/y. of melt water, of which 0.3 mm/y. occupies the space vacated by sinking ocean floor. Thus, a median of 2.6 mm/y. fall of land level for stable coasts (Fig. 102) might imply a rise of eustatic water level of 2.6 times 1.3, or 3.4 mm/y. If this were the true rate of eustatic rise, an adjustment of this amount should be applied to the median rates for each of the geological processes plotted in Figure 102. Gravitational modeling such as that by Peltier may help resolve some of these questions.

A more realistic approach to estimating coastal deformation caused by load of returned glacial meltwater is that used by Chappell et al. (1982) for northeastern Australia. Their method takes into consideration the changing load, densities of mantle and load, flexural rigidity of the lithosphere, and relaxation-response time to compute the course of deformation. As the investigation was concerned with deformation of a 5500-y. B.P. land and ocean-floor surface, it involved the use of eight new radiocarbon-dated sites plus about as many previously dated sites. The result showed deformation contours of the 5500 y. B.P. surface that range from −1 m on the adjacent ocean floor about 100 km offshore to +3 m about 250 km inland. The average slope of deformation thus is about 10 mm/km for the 5500-y. period. Rate of deformation at the shore is 1 to 2 m for the period, or about 0.2 to 0.4 mm/y. Application of the method to all coasts of the Earth would require enormously more information than is presently available, especially in view of the much wider range of deformation exhibited by the tide gauges in regions where tectonism of many kinds can be inferred from geological relationships.

Broader based theoretical computations for global crustal deformation caused by return of glacial meltwater during long time spans were published by Clark et al. (1978), Peltier et al. (1978), and Peltier (1986). These studies show regional deformation of oceanic and continental crusts radially outward from the areas of late Pleistocene ice sheets. Chronology was based on radiocarbon dates but the results were used to explain variations between tide-gauge records along different coasts of the world. No consideration was given to the role of local tectonism in explaining the great variations between nearby tide-gauge records, and tectonic processes contaminated the model validation and added "noise" to the calibration.

In summary, we believe that tide-gauge records cannot be used for inferring eustatic rise of sea level in a simple way, although many workers have attempted to do so. The reasons are that the records are influenced far more by local tectonism of many kinds (and even by man's activities) and thus they provide no uniform and constant base level for measuring rise of the sea surface. Meltwater was returned to the ocean at a lower rate during the past few thousand years than between 15,000 and 6000 y. B.P. when glaciers underwent rapid post-glacial retreat, but sinking of the ocean floor and rising of the adjacent coastal land belt probably are continuing because of past and present return of meltwater, thus increasing or decreasing the changes of relative levels produced by various tectonic processes. The tide-gauge records also reveal no certain increase in sea-level associated with the supposed rise of atmospheric temperature caused by heating and gas production during the industrial revolution.

In a sense, the attempted use of tide-gauge records to measure a rising eustatic sea level appears to us to be analogous to the use of a rubber tape to measure precisely the freeboard of a barge above the ocean surface during a storm. It may be done, but the result must be considered highly uncertain. The only conclusion that we can reach in this matter is that eustatic change (probably a rise) of sea level must be occurring, but it forms only a bias on larger changes caused by movements of the land beneath the tide gauges.

Even an estimate of average relative change of level at tide-gauge stations (regardless of how much of it is due to change of sea level or of land level) is fraught with difficulties. A simple average is not reliable because of the poor distribution of tide gauges in terms of either geography or tectonics. Moreover, nearly half show a rise of relative land level (fall of sea level) despite a general belief that glacial meltwater and heating must be increasing the volume of the oceans. We cannot merely eliminate all tide-gauge records that show crustal rebound in areas of late Pleistocene ice sheets because this uplift is closely accompanied by crustal downward of peripheral ridges and because many other tide-gauge records that show relative uplift of the land are distant from sites of these ice sheets. Moreover, we cannot estimate satisfactorily average rates of relative change of level from the averages of different categories of movement in Figure 102 again because of the poor distribution of tide-gauge records in terms of coverage through time, geography, and tectonic causes of vertical land movement. With considerable reservation we might take the average change of all tide-gauge records (Total of Figure 102), about 1.0 mm/y. of land sinking plus ocean rising.

During the past 6000 years sea level may have risen an average of 6 m (see Fig. 33). The mean rate might then be considered to be about 1.0 mm/y. but only if we are willing to overlook the absence on this graph (and on others by other authors) of many radiocarbon dates for shore deposits now located above sea level, mainly in glaciated areas (see Fig. 34). Failure to use these elevated radiocarbon-dated materials, but to use tide-gauge records that show rise of land level, makes the averages from the two data sets incomparable even though both may be about 1.0 mm/y.

Slightly more certainty exists in determination of the average rate of rise of sea level between the lowest level about 15,000 to 18,000 years ago (perhaps 130 m) and the 6-m level 6000 years ago (see Fig. 33). A rise of 124 m in 9000 years corresponds with a mean rate of 14 mm/y. This rate is substantially faster than either the rate for the subsequent 6000 years (also by radiocarbon-dated samples) or that for the past 50 years (by tide-gauge records). A high rate for the main time of ice-sheet retreat and the likely maximum rates of return of meltwater and heating of the ocean is reasonable, probably overwhelming the influence of the tectonism that may be responsible only for most of the scatter of data points in Figure 33.

A third comparison can be made with an average rate of subsidence of continental shelves caused by weighting of sediments and by contraction of the crust caused by cooling after rifting. A review of the stratigraphic-stripping calculations that are used for computing this subsidence at seven drill-hole sites off the Atlantic coast of North America is given by Emery and Uchupi (1984, p. 774–777) along with many pertinent references. After that summary was made, similar stratigraphic stripping has been computed for many sites on other continental shelves of the world and with rather similar results. For these seven sites subsidence during the past 50 m.y. ranges from 0.8 to 4.8 km, yielding rates of 0.02 to 0.10 mm/y. These rates are far slower than the unsatisfactory 1.0 mm/y. for tide-gauge records and for radiocarbon dates of the past 6000 years. The implication is that the weighting of continental shelves by sediments and cooling of rifted crust is far slower than the response of the shelves and coastal regions to the rapid and dramatic fall of sea level and subsequent recovery associated with the building and later melting of ice sheets during the late Pleistocene.

The conclusion unavoidably reached by these comparisons is that the total change of level caused by vertical movements of both ocean surface and land underlying tide gauges may be similar to and a continuation of the changes of level recorded by radiocarbon dating of intertidal materials deposited during the past 6000 years. Both sets of data indicate effects by both

160 6. SIGNIFICANCE OF TIDE-GAUGE RECORDS

eustatic sea levels and tectonism but with unevaluated and probably unevaluatable ratios of the roles of sea-level and land-level movements. Thus, we must state that we are unable to assign any reasonable estimate to the present rate of eustatic rise of sea level despite the evident need for such an estimate and the effort by us and many others to determine that rate. In our opinion, the rise can be bracketed only as ranging between 0 and + 3 mm/y.

Statistical Summary

In the previous sections of this chapter we have made connections between geological evidence and rates of relative sea-level change as measured by tide gauges of varying durations and non-uniform spatial distribution. These geological connections have produced several interesting results:

The non-uniform distribution of tide gauges results in severe undersampling of the geological variability, so we cannot resolve unambiguously all appropriate geological scales of change.

The tide-gauge data show remarkable similarity to large-scale geological features and trends and suggest that these data reflect land movements caused by various processes as well as ocean-level change.

Man's impacts on tide-gauge stations are documented in part, showing significant effects at dozens of gauges. Man's influence on records at other stations remain to be quantified, but likely are significant.

Having determined these results, a revisit of the data base appears useful. At the end of Chapter 5 we noted a disparity in tide-gauge recordings that pre-date 1930, compared with those that post-date that time. Apparent increases in rates of land-level rise after 1930 for the long-duration stations of Scandinavia occur at the same time that sea-level rises faster at non-Fennoscandian tide gauges. Although geoidal changes or large-scale water mass re-distribution might produce such contrasting responses, we see no geological or oceanographic evidence that such sizable mass re-distributions are occurring. These findings, along with other considerations, led us to question the utility of the pre-1930 (or so) data for addressing the question of recent accelerations in rates of relative sea-level rise.

Now that the geological and anthropogenic affinities with tide gauges have been described, we investigate the fidelity of various subgroupings of tide gauges and the trends in data density through time. To do so, tide-gauge data for each grouping (spatial or temporal) and each year were averaged together to produce a mean relative sea level for that year. Because the datums at each tide gauge are not necessarily comparable (despite PSMSL's excellent attempt to place all data into a Revised Local Reference), simple averaging would produce artifacts that would depend on which stations were averaged. To minimize that problem, all data were corrected to the mean level for 1950 at each station. The mean level at 1950 was determined by fitting a linear regression line to each data series and calculating the intercept of that line for the year 1950. This value was subtracted from each data point for that station to yield a residual time series. The residual time series for all stations reporting for any particular year were averaged together to produce a time series of mean sea level for each data grouping.

Several problems arise from this methodology. First, small errors in estimating slopes from older station data become magnified when extending the intercept to 1950. These errors may result in anomalously large offsets in residual time series. To estimate the level of error introduced by this approximation, three different intercept dates were selected and anomaly time series were generated for each intercept date and then compared. The resulting mean time series were nearly identical qualitatively, differing slightly in levels but not in trends. Therefore, this error is not deemed significant to the analysis results. A different error is introduced because some data points for the arithmetic mean series are based on averages of 1 station and others on averages of 800 or more stations. Clearly, each carries different credence because the larger series may be more representative of the true average. On the other hand, the averages based on greater numbers of stations may include more outliers, and thus be less representative. We did not quantify the magnitude of this error other than by calculating the variance associated with the estimate of each yearly mean.

The results of this averaging (Figs. 105, 106) illustrate the dangers of calculating long-term trends from either a large number of stations or from pre-selected individual stations (say, the longest ones), and drawing general conclusions about global behavior. The first example (Fig. 105A) provides the results from averaging on a year-by-year basis all PSMSL tide-gauge station data (with some two dozen or so stations added from this study) that exceed 10 years in duration. Plotted on this figure is the number of stations reporting each year. This plot shows few stations reporting prior to 1858, and a gradual increase in number to a peak of more than 800 stations in 1955, and then a precipitous drop-off after 1955. Our data from PSMSL were updated in 1988, so this dropoff represents both a time delay in reporting for stations, but mainly a real decline in number of stations that continue to operate or report data to PSMSL. An example of the latter is the decline in South American stations that reported during the late-1960s. The mean sea-level series shows some interesting behavior. First, for years prior to 1858 when few stations reported, the mean sea levels were lower on average than at other times. The mean linear trend for this total data series was a 0.3-mm/y. rise in relative sea level. Following 1858, more stations began reporting. For the interval 1858 to 1987, the mean trend was − 1.4 mm/y. (fall) in relative sea level. Obviously, this grouping of data includes stations of such short duration that the results presented here cannot be interpreted as representing eustatic sea-level rise or any other global behavior.

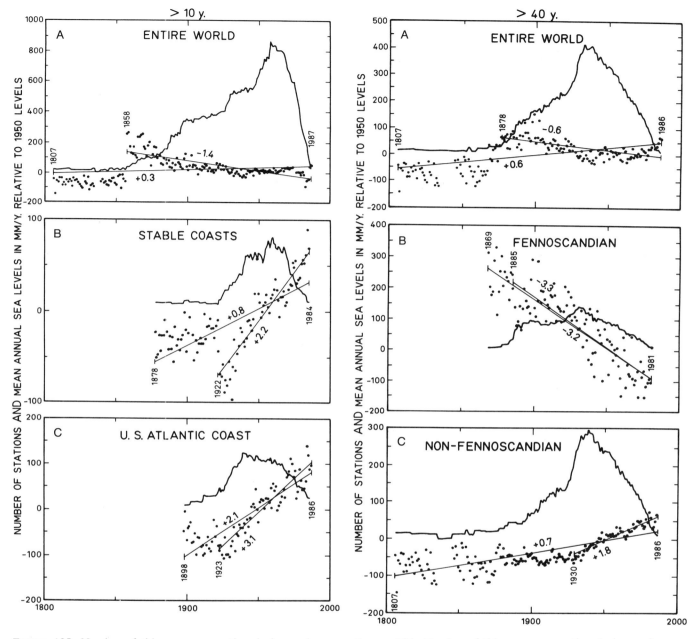

FIGURE 105. Number of tide gauges reporting during each year superimposed on the mean global sea level (relative to 1950 level) calculated from all stations reporting for that year. Numbers indicate mean change of sea level in mm/y.
A. All world tide-gauge stations having lengths exceeding 10 years, from PSMSL RLR data files plus some added from various sources by the authors, showing a falling relative sea level from 1858 to present.
B. Plot for all stations that the authors consider might be on "stable" coasts, showing a distinct difference in mean level prior to 1922 when fewer stations were reporting, and a more consistent rise from 1922 to present.
C. Plot for all U.S. east coast stations, showing a distinct difference in tide-gauge mean levels prior to 1923 compared to the ensuing period, demonstrating either a difference in tide-gauge behavior or incomplete sampling with poorly representative stations prior to 1922.

FIGURE 106. Number of tide gauges reporting during each year superimposed on the mean global sea level (relative to 1950 level) calculated from all stations reporting for that year. Numbers indicate mean change of sea level in mm/y.
A. For all stations having records exceeding 40 years, showing a difference in mean level prior to 1878 when more stations began reporting.
B. For Scandinavian stations only, which show a distinct fall in relative sea level (rise in land level), at a rate that appears to be independent of the number of stations reporting.
C. Plot of all long records excluding Scandinavian records, showing a distinct difference in trend prior to 1930 and following 1930, that is correlative with low number of tide-gauge stations reporting. Such results raise questions about the utility and representativeness of the pre-1930 tide-gauge data.

To improve on this calculation, we examined only those coasts that we identified in the previous section as possible "stable" coasts. Tide gauges for these coasts report only from about 1878 to present. The number of stations reporting each year (Fig. 105B) is low until 1922, following which the number of stations reporting increases until about 1958 and then decreases precipitously. The mean trend in relative sea level for the entire period of recording for "stable" coasts is +0.8 mm/y., whereas for the period following 1922 when more stations reported the trend is +2.2 mm/y. This latter value may represent a more appropriate average for combinations of eustatic sea-level rise and isostatic land sinking. It is also remarkably similar to estimates from Peltier and Tushingham (1989) for global eustatic value. However, we cannot discount significant biases for local tectonism, although we have tried to eliminate stations where tectonism and human impact appeared greatest. This panel illustrates the difficulty of using pre-1922 data for estimating changes in long-term trends. As discussed in the previous chapter, the pre-1920/1930 data appear discontinuous with the data following it, bringing into question the fidelity of longer term tide gauges for representing regional or global averages. The different vertical groupings of the data are directly correlated with the number of stations that reported during the earlier time period; this correlation is not obvious during the period of precipitous drop-off in recording stations following 1958.

For comparison with all-world tide-gauge stations and those of the putative "stable" coasts, we examined only the U.S. east coast stations (Fig. 105C). These data show the same relationship between number of stations reporting and the mean trend in the tide-gauge data. Prior to 1923 few stations reported and there was little mean trend in relative sea level. Following 1923, as more stations reported the trend became more linear and the data were discontinuous with earlier data. The linear regression slope for the entire interval of record is +2.1 mm/y., whereas that for the post-1923 data is +3.1 mm/y. Curiously, once again the precipitous drop-off in number of stations that reported after about 1955 is not accompanied by a decline in the mean sea level. Why the early few stations reported anomalous trends in mean water level in contrast with the later periods when equally few stations reported is not apparent.

Another comparison was made using only those stations having records exceeding 40 years (Fig. 106). If all long stations are examined together (Fig. 106A), the trends are similar to those found in Fig. 105. The mean level of the data is lower for the early interval prior to 1878 when few stations reported. After 1878, more stations reported and the mean level increased discontinuously and dramatically. The mean trend over the duration of the data is +0.6 mm/y. relative sea-level rise. If only the post-1878 data are examined, the mean trend is −0.6 mm/y., showing a fall in relative sea level instead of a rise. Clearly, simple averaging of the entire data set cannot produce good estimates of mean sea-level trends. Since the Scandinavian data are included in these estimates, one might expect poor results to come from simple averaging with no a priori selection process.

We examined Fennoscandian long records separately (Fig. 106B). The mean trend for Fennoscandia after 1869 is a fall in relative sea level of −3.2 mm/y. After eliminating the early part of the record when few stations reported, and using only post-1885 data, the mean trend is a fall of −3.3 mm/y., statistically indistinguishable from the estimate for the entire record. The early period of this abbreviated record, from 1885 to 1920, does show a smaller rate of fall of relative sea level than after 1920, as discussed in Chapter 5.

After deleting the Fennoscandian data, the long records remaining were examined (Fig. 106C). The record shows a nearly flat slope in relative sea level when few records reported, and a larger rise in relative sea level following 1930 when the number of stations reporting increased dramatically. The mean rate of relative sea-level rise for the entire record is +0.7 mm/y., whereas for the post-1930 record it is +1.8 mm/y. However, since the long record contains many known areas of tectonism and human influence, this value cannot be considered as an estimate of the rate of eustatic sea-level rise. The value of this examination is to show that a grouping of long-term records that excludes Fennoscandia and ignores data prior to 1930 provides a linear regression slope close to that for the "stable" coasts. This reasonable proximity between the two estimates is serendipitous and does not reflect any inherent validity to using all tide data to estimate eustatic sea levels. Thus, although the estimates appear close, our analysis in this book indicates that such estimates are severely biased and have enormous errors.

This section is concluded with the following observations:

We can find no justification for using a few long-term tide-gauge stations as representative for regional or global averages. The longer term data appear anomalous compared to the better regional and global estimates derived from more but shorter term data. Although the reasons for this anomaly are only hypothesized here, it appears clear that the long-term station data prior to about 1930 cannot be used to evaluate changes in rates of eustatic sea-level rise.

The trends in relative sea level are strongly dependent on the number of stations reporting, as indicated in the above analysis and by calculations randomly selecting small numbers of tide-gauge stations from the global domain and making trend estimates from these smaller subsets.

The PSMSL data set is the most useful tide-gauge compilation in existence. However, non-uniformity in sampling, both in space and time, renders questionable their use for estimating long-term trends in eustatic sea levels prior to approximately 1930.

7. Future Eustatic Sea-Level Change

Climate Change Scenarios

Scientific consensus on the causes of climate change during the achanges. This uncertainty is illustrated by the debates about the magnitude of global warming during the past century. Whereas few debate that global warming has occurred during the past century, as the Earth emerged from the Little Ice Age, the magnitude of the warming and its possible causes are unclear. The prediction of future climate is fraught with even greater difficulties and uncertainties. Prediction of future climate change can be either empirical or based on numerical models. Each technique makes significant assumptions that cause the predictions to differ substantially for many scenarios of climate forcing. The historical record of climate and the prediction of future climate are discussed separately below.

For purposes of estimating sea-level rise we are concerned with the climate variables of temperature and precipitation. Temperature is important for two primary reasons: in the atmosphere one is concerned with temperature and its effects on ice melting and ocean warming, whereas in the ocean one is concerned with steric expansion of the water column and its contribution to sea level. Precipitation (and its corollaries: glacier distribution, snow/rain patterns, etc.) is important not only for mass balances of glaciers but also for rainfall patterns (both quantity and seasonality) that help control sediment and water inputs into the oceans. The historical record shows that there has been warming of global surface temperatures during the past century and that most mountain glaciers have receded during that time span. Cloudiness appears to have increased in the United States and Australia during the past quarter century, but precipitation has varied greatly depending on location. Beyond these statements, there is little consensus in the scientific community regarding the historical climate record during the past century.

Temperature should be a good measure of climate change because it might indicate global warming; however, the instrumental record over land was fragmentary until the mid- 19th century, when the coverage began to increase. The record for the ocean is even worse in terms of coverage and reliability. The many groups (Hansen and Lebedeff 1987; Jones 1988; Vinnikov et al., in press) that examined land temperature change have produced roughly similar records, but the reasons for this observed temperature change have attracted little consensus. The Northern Hemisphere is best studied because of having better records than the Southern Hemisphere. Even with widespread data, however, interpretation of the rate of warming depends on the length of record, just as for the rise of sea levels from tide gauges. For instance, Jones (1988) showed a rate of warming for the Northern Hemisphere of 0.53°C per century during the interval 1881 to 1989, but the rate is reduced to 0.45°C per century if extended back to 1861. In the Southern Hemisphere, Jones (1988) showed a rate of 0.52°C per century during the period 1881 to 1989, but a rate of 0.45°C per century for 1861 to 1989. As for other climate variables, estimates of global temperature behavior are impacted by poor spatial (Jones et al. 1986a, b) and temporal coverage of the data, instrumentation effects, changes in observational and reporting procedures, and changes in the local environment of observation (the "urban island" effect; Hansen and Lebedeff 1987; Karl et al. 1988; Jones et al. 1989; Karl and Jones 1990; Zhang and Wang 1990). These contributors to uncertainty make difficult the establishment of how much of the thermometric signal, if any, is caused by climatic warming. For instance, there has been intense debate about the magnitude of the urban island effect, with estimates of these impacts in the United States' records ranging from 0.1°C per century, to two to three times as much depending on the size of the city involved.

Understanding of ocean heating is even less certain. Numerous investigators have attempted to compile and analyze ocean sea-surface temperature (SST) records (Farmer et al. 1989; Bottomley et al. 1990), where the number of observations exceeds 80 million. In the Northern Hemisphere there is some evidence of oceanic cooling of 0.1 to 0.2°C during the early 20th century. The overall oceanic warming since the late 19th century appears to be less than over land, which is about 0.3°C. In the Southern Hemisphere, ocean warming during the past century appears to be between 0.3 and 0.5°C. The analyses of ocean temperature are handicapped by poor spatial and temporal coverage and by various types of biases. These biases are related largely to sampling methodology. For instance, water temperature used to be measured by sampling water with a bucket, and then making measurements on board. Since several different types of materials were used for buckets, different biases resulted, and hence different correction procedures. Subsurface temperature variations are even more poorly charted. Levitus (1990) examined subsurface temperatures in the North Atlantic ocean, suggesting cooling in the subtropical gyre and warming beneath the gyre. Roemmich and Wunsch's (1984) data appear consistent with Levitus' findings.

In summary, however, a consensus has developed that the oceans and atmosphere have heated during the past century. Spatial and temporal distributions of such changes are still under examination, as are the absolute magnitudes of such changes. Precipitation changes, by contrast, are less poorly known despite many recent studies (Bradley et al. 1987; Diaz et al. 1989; Vinnikov et al., in press). Considerable spatial variability makes generalizations difficult, although some increase in precipitation during the past several decades appears to have occurred in midlatitudes, along with a decrease in the Northern Hemisphere subtropics and throughout the Southern Hemisphere. The drying of sub-Saharan Africa during past decades is a well-publicized example of regional variability, linked empirically to anomalies in SST patterns (and other factors such as volcanism).

Modeling of climate change has progressed markedly during the past decade, partly in response to the fear of accelerated greenhouse warming. Models are tools that permit assimilation of data and of knowledge from past climatic events, and then prediction of future possible climatic states. Models can have any number of forms, from purely empirical, to statistical, to dynamical, and any mixtures thereof. Because models can be of many different types, and within each type there may be different permutations, output from models can vary considerably. For instance, models may differ depending on the fundamental equations used, by the parameterization of various physical-chemical processes, by resolution of the model grid, and by boundary conditions. Since the physics are so complex, no single model can adequately describe all the physics on all important scales. Because climate dynamics are nonlinear, involving complex feedbacks for which physics are only partly known, predictability is naturally limited.

Several major types of climate models have been used in the greenhouse debate. The first type of model is the palaeoanalogue model, where knowledge of past CO_2 levels (derived from ice cores) and their relationship to temperature are used to predict future changes. The link between CO_2 content and temperature is examined and corrected for changes in the Earth's albedo and solar constant, and then regional patterns of climate are reconstructed for use as analogues to predict future climate, given altered concentrations of atmospheric CO_2 (Budyko and Izrael 1987). Although CO_2 contributes an estimated 60 percent of the anthropogenic greenhouse effect, methane, nitrous oxide, ozone in the troposphere, and chlorofluorocarbons are important and must also be considered (Rodhe 1990).

A second type of model is the atmospheric general circulation model (GCM), which is based on fundamental laws governing flux of mass, momentum, heat, and water vapor in the atmosphere. These equations are described using one of several numerical coding techniques (time or spectral domain), assumptions are made regarding parameterization of sub-grid-scale processes such as mixing and convection, and boundary conditions are prescribed. Models have relatively coarse resolution, typically ranging from 300 to 1000 km horizontally, and using from 2 to 19 levels in the vertical axis. Differences in the models derive from differences in numerical solution schemes, sub-grid-scale processes, land-surface processes (soil processes, especially), and boundary conditions.

Ocean models are similar in construct to the atmospheric GCMs but they use the equations that govern transport of water, salt, and heat in the oceans. These models also have coarse resolution, ranging from 200 to 1000 km in the horizontal, and using 2 to 20 levels in the vertical. Because the finer resolution increases computation time, some models that simulate many of the important physics of the ocean (boundary currents, various long-period waves) cannot be run on existing computers long enough to permit modeling of longer scale climatic changes. In some efforts, coarse ocean models have been combined with atmospheric GCMs to form coupled models of the atmosphere and the ocean.

Important for modeling CO_2 flux is a good model of carbon dioxide exchange between the ocean and atmosphere, necessitating a carbon-cycle model. These models are based on the equations for air-sea flux of carbon dioxide, gas solubility, carbon dioxide chemistry in the oceans, and biological activity. Such models have been coupled to ocean models in the past but their incorporation into full atmospheric-ocean models has yet to be accomplished.

Most model studies to date are equilibrium response studies, where the model is calibrated for present-day forcing and then run for various altered states of the atmosphere. However, some models have been run in a time-dependent state, watching how the newer atmosphere evolves with gradually changing forcing. Cess et al. (1989) evaluated the fidelity of 14 different climate models to equilibrium conditions. For certain aspects of the climate system, all models perform within a narrow range of output. Global mean temperature can be simulated consistently by most models, but there is no assurance that the predictions are correct because of uncertainties in feedback mechanisms (cloudiness, snow-ice albedo effects, water vapor-soil processes, etc.). In general, good skill is exhibited in depicting large-scale distribution of atmospheric pressure, temperature, and wind in both summer and winter. Considerable degradation of skill exists on a more regional scale. Precipitation can be simulated over large scales, but large errors (20–50 percent of the mean annual rainfall) occur regionally. Improved skill has been associated with recent improvements in simulation of convection, cloudiness, and surface processes. Oceanic models describe large-scale ocean temperature and circulation, although correction schemes are needed to account for transport of heat by major currents, and salinity fluxes. Some recent references describing the rapidly evolving field of GCMs and their intercomparison are by Hansen and Takahashi (1984), World Meteorological Organization (1986, 1988), Mitchell et al. (1987), Grotch (1988), Lange (1989), National Academy of Sciences (1989), and Philander et al. (1989).

Sea-Level Change Scenarios

The modeling of climate illustrates that certain aspects of the global ocean/atmosphere system are becoming increasingly predictable, as numerical representation of the driving processes is improved. However, improved prediction of sea-level rise caused by changing climate adds another level of uncertainty. If climate itself is hard to predict, the secondary impacts are even harder to predict. Certainly, available data indicate that major contributors to relative sea-level rise must continue just as they have been active during the past century: tectonism, isostatic rebound, some addition of meltwater to the oceans, and steric effects. Some items can be predicted reasonably well from the historical record and from knowledge of man's activities (groundwater withdrawal, extraction of hydrocarbons, dredging of ports and waterways); however, the addition of meltwater and steric expansion are difficult to predict. Even harder to predict will be those portions of the meltwater addition and steric expansion that are attributable to human activities (greenhouse effect).

Considerable effort has been spent to determine how sea levels might have changed during the past 100 years. A summary of these previous efforts is provided in Table 8. Most estimates are in the range of 10 to 20 cm/century, with few exceptions. That these estimates were made from different groupings of tide gauges covering varied time spans make the observed small spread in estimates remarkable, considering the geological and human controls on relative sea levels described in previous chapters. There is good consensus that sea level has risen during the past century; the rate, however, is at issue. Chapter 2 of this book discussed the causes and estimates of historical rates of sea-level rise. How sea level might change in the future is presented below.

TABLE 8. Estimates of p mm/y.).

Rate (mm/y.)	Con	
>0.5	Cyrologic es	
1.1 ± 0.8	Many statio	
1.2–1.4	Combined	
1.1 ± 0.4	6 stations,	
1.2	Selected st 1900–19	
3.0	Many stat	
1.2	193 static	
1.5	Many sta	
1.5 ± 0.15[a]	9 stations, 1903–1980	
1.4 ± 0.14[a]	155 stations, 1881–1980	Barnett (1984)
2.3 ± 0.23[a]	155 stations, 1930–1980	Barnett (1984)
0–3	Many stations	Aubrey (1985)
1.2 ± 0.3[a]	130 stations, 1880–1982	Gornitz and Lebedeff (1987)
1.0 ± 0.1[a]	130 stations, 1880–1982	Gornitz and Lebedeff (1987)
1.0 to 1.5	44 East Coast U.S. stations, 1920–1983	Braatz and Aubrey (1987)
1.15	155 Stations, 1880–1986	Branett (1988)
0.4–0.6	58 Europe stations, 1920–1980	Pirazzoli (1989)
2.4 ± 0.9[b]	40 stations, 1920–1970	Peltier and Tushingham (1989, in press)
1.2 ± 1.2	Many stations, 1880–1980	Raper et al. (in press)
1.26 ± 0.78	U.S. east coast, 1880–1980	Gornitz and Seeber (1990)

[a]Value plus 95 percent confidence interval.
[b]Mean and standard deviation.

TABLE 9. Estimates of future eustatic sea-level rise (cm).

Author(s)	Thermal Expansion	Alpine Glaciers	Greenland Ice Sheet	Antarctica Ice Sheet	Total
Gornitz et al. (1982)[a,g]	20–30	No est.	20	No est.	40–60
Revelle (1983)[b]	30	12	12	f	71
Hoffman et al. (1983)[c,g]	28–115	28–230 (Melting)			56–345
NAS (1985)[c]	No est.	10–30	10–30	−10–100	10–160
Thomas (1985)[a]				0–220	
Hoffman et al. (1986)[c,g]	28–83	12–37	6–27	12–220	58–367
Robin (1986)[c]					25–165
Wigley and Raper (1987)[e]	4–8	No est.	No est.	No est.	No est.
Jaeger (Villach, 1987)[a]	No est.	No est.	No est.	No est.	−4–140
Oerlemans (1989)[d]	8				33
Oerlemans (1989)[c]					66
Stewart (1989)[c]	3–30	10	±20	±20	20–40
Meier (1989)[a]	20	16	8	−30	34[h]
Raper et al. (in press)[d]	4–12	4–10	1–4	−1–2	8–28

[a]Estimates from 1985 to year 2050; [b]Estimates from 1980 to year 2080; [c]Estimates from 1985 to year 2100; [d]Estimates from 1985 to year 2030; [e]Estimates from 1985 to year 2025; [f]Revelle included 16 cm of change for other causes; [g]Estimates from 1980; [h]Includes 20 cm from groundwater depletion.

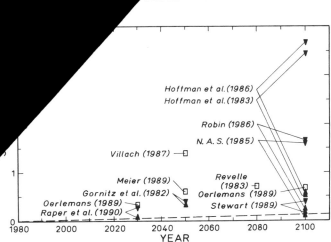

FIGURE 107. Projections of future sea levels by various authors. Hollow rectangles represent spot estimates, solid triangles mark extreme ends of range estimates, and dashed line is trend if present "assumed" rate of eustatic sea level rise continues.

Future sea-level rise must occur because of thermal expansion of the ocean's water and addition of meltwater to the oceans. The thermal expansion depends on the atmospheric warming and how this warming is transferred to the oceans, as well as how the ocean mixes this new heat. Roemmich and Wunsch (1984), Thompson and Tabata (1987), and Levitus (1990) examined how the ocean has been warming, but data are so scarce that global predictions cannot be made. Decades of data will be required to measure the effect of global warming on the ocean steric response. In lieu of suitable measurements, numerical models of steric expansion are used. These models can range from sophisticated three-dimensional models of full ocean circulation (Barnett, 1985) to simpler one-dimensional models. One such model is the box-upwelling-diffusion model (Hoffert and Flannery 1985) that uses simple vertical mixing combined with thermal expansion coefficients to estimate sea-level change. Although such models ignore the details of dynamic mixing and large-scale oceanographic processes, they are useful for estimating trends. For the period 1880 to 1985, Wigley and Raper (1987, in press) estimated thermal expansion to have been between 2 and 5 cm, in general agreement with Gornitz et al. (1982). However, more sophisticated ocean models show trends that are not in full agreement with the Wigley and Raper model (Schlesinger 1986). Estimates for future thermal expansion vary widely. Wigley and Raper (1987) estimated thermal expansion for the period 1985 to 2025 to be 4 to 8 cm (an average rate of 1–2 mm/y.). Oerlemans (1989) projected thermal expansion of 8 cm by 2025 (a rate of 2 mm/y.).

The contributions from small mountain glaciers and the major ice sheets are more problematical, since the present mass balances of these ice sheets are uncertain. For the year 2100, the National Academy of Sciences (1985) projected contributions to sea-level rise from glaciers and small ice caps of between 10 and 30 cm, from the Greenland Ice Sheet of between 10 and 30 cm, and from the Antarctic Ice Sheet of between -10 and 100 cm. Many other estimates of relative sea-level rise range from 2.6 to 11.5 cm for the year 2030 due to mountain glacier melting, 0.3 to 2.8 cm due to melting of the Greenland Ice Sheet, and -1.1 to 0 cm for melting of the Antarctic ice sheet. However, these estimates have huge uncertainties. By combining thermal expansion and ice melt, ranges of sea-level rise by the year 2030 group within the 10- to 32-cm range, for an average rate between 1985 and 2030 of 2.2 to 7.1 mm/y. These increased rates should be measurable, particularly if the upper end of the range occurs.

Estimates of glacier balances to various times in the future, based on various types of data and analyses are summarized in Table 9 and Figure 107. One recent prediction, by Meier (1989), indicated a sea-level rise of only about 34 cm by the year 2050, a considerable drop in projection because of revised estimates of rates of deglaciation. The spread in these values is extreme, making reliable prediction uncertain. Clearly, sea level must continue to rise; uncertain is the rate of rise. This uncertainty must be kept in mind when assessing possible impacts of future sea-level changes.

8. Impact of Sea-Level/Land-Level Change on Society

General

Prediction of the magnitude, timing, and severity of human influence on climate is elusive, but considerable effort has been spent identifying geographic areas and natural processes that are sensitive to climate change. Studies of impacts have been performed for different scenarios of climate change and have focused on the vulnerability of individual regions. Although useful for sensitivity analysis, these approaches are limited because they commonly neglect the full suite of physics associated with climate change. For instance, sea-level rise, the primary concern here, is assumed to inundate coastal regions with little if any consideration for self-maintenance processes such as longshore and cross-shore sediment transport or dynamics of barrier beaches. Essentially, the uncertainties for predicting climate change are compounded by the uncertainties in the modeling of impacts given that climate change. However, a broad range of climate-change impacts can be discussed with some degree of certainty, including some effects of possible increased rate of relative sea-level rise.

Possible increase in rate of sea-level rise caused by the greenhouse effect has attracted considerable public attention for many reasons. Among them are graphic images that arise as one ponders the effects of rising oceans, complete with lost houses, damaged structures, and fleeing populations. Also contributing is the large percentage of population located along the shoreline relative to that farther inland. Based on 1980 census results and projecting into the future, Culliton et al. (1990) determined that approximately 50 percent of the United States population lives and will continue to live in the coastal counties of the United States. Although this value is considerably smaller than the more often quoted 70 percent living within 100 miles (166 km) of the shoreline, a dense population still is at risk to rising seas. This concentration of people along coasts is common throughout the world, not just in the United States, but analogous numbers appear unavailable on a global basis. It is clear that the global populace feels threatened by the possibility of accelerated sea-level rise. Possible impacts of accelerated sea-level rise include coastal inundation, increased erosion, changes in circulation and salinity of estuaries and lagoons, increased storm damage, loss of wetlands, changes in ecotomes and habitats, loss of turtle and bird nesting areas, and increased salinity intrusion into groundwater. These impacts are highly visible and threatening, and much attention has been focused on them.

One measure of the area likely to be impacted by accelerated relative sea-level rise through direct inundation is shown by Figure 108 – a map of the world indicating areas of sinking land depicted by empty squares and areas of rising land (relative to sea level) shown by filled circles, as determined by tide-gauge records of variable lengths. The second panel shows the same delineation between rising land and sinking land (relative to sea level), given an acceleration in rate of sea-level rise of 2 mm/y. (thought by some likely to be reached by the year 2025). A third panel plots the analogous mapping for an increase in rate-of-rise of sea level of 4 mm/y. Comparison of the three panels illustrates those regions where an increase in rate-of-rise of sea level would change the sense of relative land motion from rising to sinking, thus altering the shoreline stability of those regions. Most of the presently emerging land areas would become submerging ones under such a scenario, with the exception of some formerly glaciated areas in Fennoscandia and Canada. Most other coasts would become or remain submergent. It is interesting to compare the distribution of present-day submergence with the distribution of lagoons throughout the globe (Figure 109). Lagoons presently are restricted to areas of relative submergence, according to studies by Nichols (1989) and Nichols and Boon (in press). If there is a cause and effect relationship between presence of lagoons and rates of relative sea-level rise, as one might hypothesize, the changing distribution of submergence and emergence accompanying an increase in rate of relative sea-level rise might increase the distribution of lagoons globally, leading to increased distribution of wetlands rather than fewer wetlands.

Previous discussion in this book indicates that tide-gauge evidence for eustatic rise of sea level is complicated by mostly larger rates of movement of the land beneath the tide gauges. Nearly as many tide-gauge records indicate a rise of land (fall of sea level) as those that indicate a fall of land (rise of sea level). Most of the tide gauges that show a rise of land are in areas formerly occupied by late Pleistocene ice sheets, but many are far beyond those areas. Nevertheless, most coastal areas of the world (Fig. 108) are characterized by fall of land levels at tide-gauge stations. One can question whether a person who owns a home in the coastal belt should be concerned about whether the sea level is rising or the

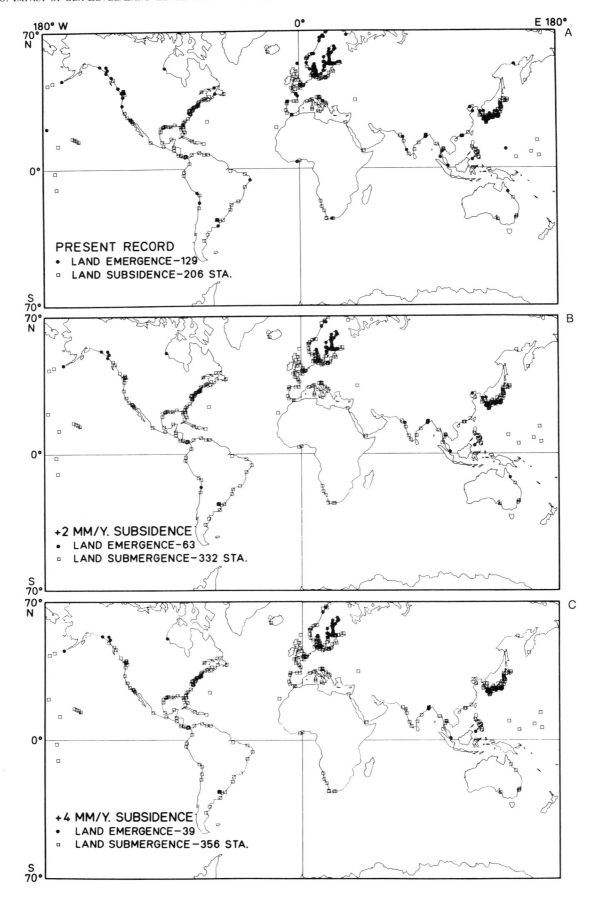

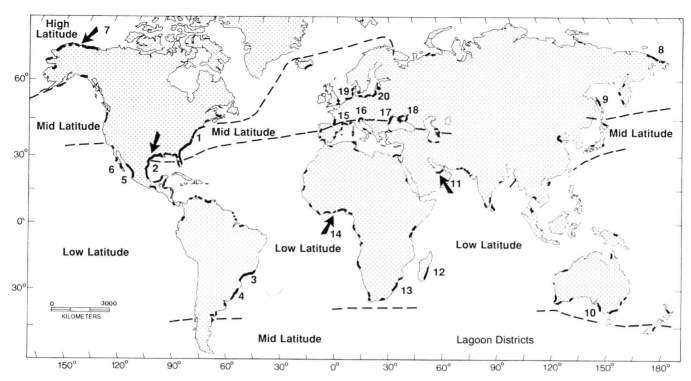

FIGURE 109. Global distribution of lagoons, based on a study by Nichols and Boon (in press). The location of lagoons throughout the globe suggests that lagoons occur only where coastal submergence is presently dominant.

land level is sinking; either process can flood his home and other property. In reality, it does make a difference because if tide-gauge records are the only evidence for relative flooding, this evidence is so poorly distributed that most sites are too far from tide gauges for this source of information to be very definitive for any given site. Moreover, a particular site may behave differently from that of the nearest tide gauge because faults, folds, or other geological boundaries may occur along the coast between the two sites. The sites also can differ because weighting of the ocean floor by returned meltwater can depress the ocean floor and produce a peripheral upwarp along the landward side of the shoreline. The main reason for fear of rising sea level is only the general belief that the atmosphere is warming because of gases liberated by industry, and that this warming is expected to melt glaciers and also expand the volume of the water in the ocean. These effects should cause a rise of sea level even though the rate of rise cannot be measured by tide gauges, because the gauges presently are more influenced by the movements of land levels.

Physical, Chemical, and Biological Impacts of Rising Relative Sea Level

The many potential impacts of increased rates of sea-level rise have been discussed in numerous publications (Barth and Titus 1984; Prins 1986; Titus 1986a; Devoy 1987; National Academy of Sciences 1987; Pugh 1987; Mehta and Cushman 1989). In addition, a series of articles has discussed on a region-by-region basis some of the potential consequences of accelerated sea-level rise (Gable and Aubrey 1989, 1990a, 1990b; Gable et al. 1990). These sources considered the possible impacts of relative sea-level rise for various climate change scenarios. Rather than repeat details of the previously published information, this book discusses only the general processes and illustrates them with a few examples.

Direct inundation by rising water levels and acceleration of storm damage are common concerns in coastal states. Much effort has been spent on examining the consequences. Most studies examine only the passive submergence of land that

◄

FIGURE 108. Coastal areas subject to inundation owing to rising relative sea levels.
A. Present rates of rise determined from tide-gauge data.
B. Present rate of relative rise plus an addition of 2 mm/y. (thought by some researchers to be expected by the year 2025).
C. Present rate of relative rise plus an addition of 4 mm/y.
The difference between A, B, and C depicts those areas where an increase in rate of rise of relative sea level would change the sense of motion from rising of land to sinking of land relative to sea level.

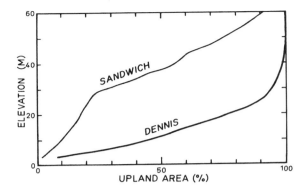

FIGURE 110. Hypsometric curves in towns of Cape Cod, Massachusetts. Curve for Sandwich shows steep gradients associated with a glacial moraine; curve for Dennis shows a low-lying flat outwash plain. Redrawn from Giese and Aubrey (1987).

occurs as sea level rises without any dynamical response of the shoreline. Dynamical processes such as adjustment of the offshore bottom profile, migration of barrier beaches or wetlands, and alteration of longshore or cross-shore sand transport patterns are most commonly neglected. There are many methods to study these processes. One of the simplest is to construct a hypsometric curve for the area in question, showing the percentages of land at or above given contour intervals. These hypsometric curves then show the vulnerability of an area to submergence. For instance, in a study of the Massachusetts coastline, Giese and Aubrey (1987) and Giese et al. (1987) examined hypsometric curves for all the coastal communities in Massachusetts. The curves have distinctly different shapes at nearby towns, related to the complex glacial geology of the region. For instance, the hypsometric curve for a glacial moraine is steep, with most of the land lying at high elevations, whereas the curve for an adjacent region of outwash plain is low-lying (Fig 110). The potential for direct inundation due to altered rates of sea-level rise are greater for the outwash plain than for the glacial moraine because of its flatter profile.

More advanced is the next type of calculation for direct inundation, the "paint-by-number" approach, where the topographic map for a region is interpolated to estimate the potential for various areas to be inundated given specific sea-level rise scenarios. This technique can be applied by hand, using some type of interpolation between contour intervals, or it can be applied in a more sophisticated Geographic Information System (GIS). The GIS advantage is that it can produce color charts of the various scenarios much more rapidly than can the hand approach. However, both presentations rest on the same basis of passive inundation and ignore the physics of how a land mass might actually respond to increasing inundation by changing sedimentation patterns or altering offshore morphology. Although such coloring techniques appear trivial, in fact they are complicated in many instances by absence of good topographic data. In many areas of the world, good topographic control is difficult to obtain, or if

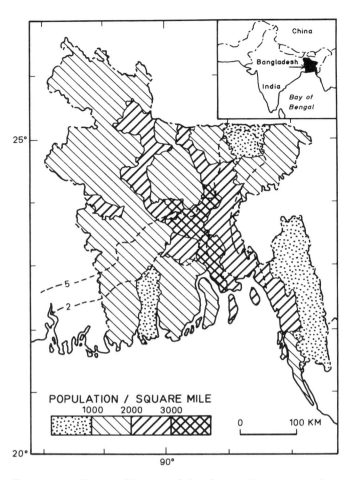

FIGURE 111. Topographic map of the Ganges–Brahmaputra river delta in Bangladesh depicting the low-lying delta region susceptible to inundation. Redrawn from Milliman et al. (1989), Ambio, v. 18, p. 340–345.

obtained, hard to verify. Thus, even the paint-by-number approach has problems in some areas. A good example of the paint-by-number approach is the low-lying Ganges–Brahmaputra river delta, where much of Bangladesh lies at an elevation below 5 m. Here, various sea-level rise scenarios show the susceptibility of Bangladesh to passive flooding (Fig. 111), a condition quantified loosely by economic analysis (Milliman et al. 1989). For the period between 1945 and 1975, total property damages here from tropical storms have been estimated at $7 billion (Murty 1984). During a single tropical storm in 1970, between 200,000 and 300,000 lives were lost as the surge covered approximately 35 percent of the land of Bangladesh.

An analysis by Emery and Aubrey (1989) showed that the maximum number of deaths by cyclonic storm surges in India and Bangladesh comes during May and October, the months of most frequent storm surges. Total deaths average about 4800 per year from these surges, or about 4,500,000 since 1737 when approximate records began to be kept. Floods caused by monsoonal rains dominate during summer months,

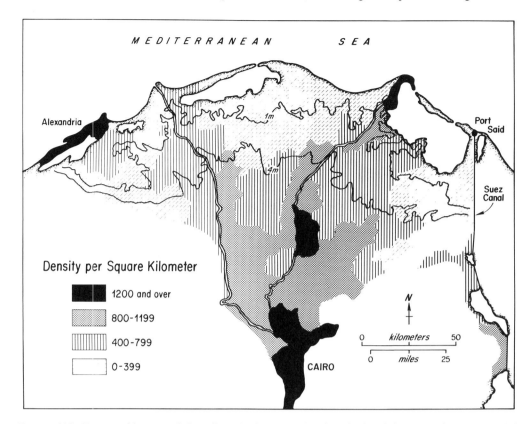

FIGURE 112. Topographic map of the Nile delta in Egypt showing the low-lying coastal areas and their population densities. From Milliman et al. (1989) by permission from Ambio, v. 18, p. 340–345.

drowning or starving additional uncounted millions of inhabitants. If storm surges are increasing because of increasing eustatic rise of sea level, rather than the probably more local sinking of deltaic coastal lands, we can expect the annual number of deaths to increase at least in proportion to the rise of sea level and along more coasts.

Among the areas most susceptible to inundation because of relative sea-level rise are the deltaic areas of the world. Deltas are naturally subsiding accumulations of sediments. Generally located in areas of rapid sediment deposition, delta sediments have large water contents when freshly deposited. As the delta becomes loaded with more sediments, the underlying sediments consolidate and dewater, leading to local compaction. Subsidence rates in some deltas can reach several centimeters per year. For most deltas, local subsidence is more than compensated by sedimentation, so many deltas accrete despite subsidence. Several factors combine to increase the susceptibility of deltas to relative sea-level rise. First, deltas are low plains, lying just above sea level because of the mechanics of deposition of delta sediments. This low profile means that an elevated water level can reach far inland compared to a steeper coast. Second, because deltas are naturally subsiding, any natural or man-made alteration of a river course can allow the subsidence to become evident. For instance, in the Mississippi River delta, the river has altered course many times in response to river sedimentation and buildup of levees (Fisk et al. 1954; Wells 1987; Wells and Coleman 1987). Distal portions of the abandoned deltas already are subsiding at a rapid rate, with consequent rapid erosion of barrier islands (such as the Chandeleur Islands). Man-made changes in river flow can have similarly detrimental impacts. The completion of the Aswan High Dam in 1964 in Egypt, following the early completion of the Aswan Low Dam in 1904, led to nearly complete starvation of sand to the Nile River delta. This starvation is mirrored by rapid erosion of the Nile delta where retreat is as much as 100 m/y., particularly near the former river mouths (Damietta and Rosetta branches; Inman et al. 1976; Stanley 1988a, b; Milliman et al. 1989). Just as for the delta in Bangladesh, the delta in Egypt is low-lying and subject to massive inundation due to rising sea level and land subsidence (Fig. 112). Here, man's activities during the past century have destabilized this section of coast and threaten the primary fishing production within all of Egypt and the adjacent Mediterranean Sea. In the People's Republic of China along the Zhu Jiang (Pearl River) delta of Guangdong province, the Huang He (Yellow River) near Tianjin, and along the Changjiang (Yangtze River) plain near Shanghai, it has been estimated that nearly a hundred million people would be displaced by a rise in sea level of 1 m (Han 1989). Other vulnerable deltas include the Mekong (Indo-China), the Irrawady (Burma), the Indus (Pakistan), the Niger

(Nigeria), the Paraña, the Magdalena, the Orinoco, and the Amazon (the latter four in South America).

Other areas that might be affected include island nations. The Maldives in the Indian Ocean, and Kiribati, Tuvalu, and the Marshall Islands in the Pacific might all be impacted by rising rates of sea level (Lewis 1988; Gable and Aubrey 1989, 1990b). Uncertain, however, is how well these islands may keep pace with rising sea levels. Under natural conditions atolls and marine islands commonly grow upward with sea-level rise through production and accumulation of carbonate sediments. This is the basis for Darwin's (1889) concept of the evolution of fringing reefs, barrier reefs, and atolls atop sinking volcanic islands (his diagrams are repeated in most geology textbooks), and for Daly's (1915) glacial-control theory of coral growth from islands that were beveled during glacial times of low sea levels. Support for upward growth of coral reefs is provided by study of samples from dredgings and drill holes in coral islands throughout the world ocean (Emery et al. 1954, p. 133–141; Schlanger 1963). Nevertheless, relative subsidence of volcanic islands has at times exceeded the ability of calcareous reef-building organisms to keep pace; thus, the flat-topped seamounts known as guyots originated (Hess 1946). Many guyots occur in the Pacific Ocean, and at least some date from the Cretaceous Period, according to fossils dredged from their tops (Hamilton 1956). Since the natural upward growth of coral can exceed 1 cm/y., acceleration in rate of sea-level rise would have to be significant to outstrip a coral island's ability to keep up with the pace. As shown by drill cores off Barbados, *Acropora palmata*, a fast-growing coral that is generally restricted to the upper 5 m of water, was able to keep up with a rise of sea level of 20 m between 17,100 and 12,500 y. B.P. (Fairbanks 1989). This growth corresponds with an average rate of 4.3 mm/y. but probably periods of faster rise are included. Nevertheless, extreme storm conditions at Buck Island in the U. S. Virgin Islands has drowned former reefs of *Acropora palmata*, whose radiocarbon dates show that the reefs grew at an average rate of 5.2 mm/y. 4000 years ago but now maintain a rate of only 0.5 mm/y., and they clearly have been left behind and stranded at a depth of 5.2 m. (Macintyre and Adey 1990). With intense human habitation and interference, it is not clear how present low-lying islands might respond to a considerable acceleration of relative sea-level rise. Even higher islands are not immune; for example, Japan may lose 1700 km^2 in which 4.8 million people live if sea level were to rise 1 m (Mansfield and Nishioka 1989).

Saltwater intrusion into coastal aquifers represents another problem associated with rising sea levels. Specifically, where the coastal plain abuts the coast and the aquifer is an unconfined one within the plain, changes in ocean level can be felt far inland as a contamination of the aquifer by sea water. This effect is certain to be exacerbated by increased use of freshwater from aquifers for industrial and residential purposes. Since the mid-1950s, when many cities switched from surface water to groundwater to satisfy industrial needs, saltwater intrusion has been an increasingly costly problem. For instance, in Japan many coastal cities have reported saltwater intrusion that was caused by excessive mining of groundwater (Fig. 113). Saltwater intrusion also may occur because of increased penetration of seawater into estuaries and rivers. As relative sea level rises, the salt wedge can migrate up many estuaries unless compensated by increased freshwater discharge. This increased intrusion can enable the saltwater to penetrate into aquifers that empty into the river, providing a potentially far-reaching mechanism for salt contamination of freshwater resources. An example of this saltwater intrusion is the Delaware River, where during periods of droughts (analogous to sea-level rise) saltwater intrudes farther into the estuary, contaminating groundwater and surface freshwater resources.

Changes in sea level also can affect circulation in estuaries and lagoons, although just how this circulation would be changed is poorly quantified. With a higher water level, the salt wedge can propagate farther upstream unless compensated by increased freshwater discharge. Minor changes in tidal circulation may result from higher sea levels and significant changes in saltwater distribution could occur. These changes in circulation and salinity distribution could alter the distribution of estuarine organisms. For shallow lagoons and estuaries, the increase in height of sea level could change the very nature of the system and its ability to trap or export sediment from rivers and from the nearshore (Friedrichs et al. 1990).

Considerable concern has been expressed concerning loss of wetlands as sea level rises more quickly (Titus et al. 1984). In order for wetlands to be able to withstand an accelerated rise of sea level, either the sedimentation rate must keep pace or there must be adequate space for encroachment onto the upland. The sedimentation rate is a difficult issue because it concerns both organic and inorganic sedimentation. In some areas (the United States Gulf, south and mid-Atlantic coasts, for instance), sedimentation within marsh systems consists primarily (by weight) of inorganic sediments. Thus, the supply or trapping efficiency of these systems must increase to maintain adequate level for the marshes. In other regions (such as northeastern United States), the primary marsh sediment is organic. In order for marshes to keep up with increased sea-level rise, the rate of production and trapping of organic sediment must increase, a different process than for marshes elsewhere. Where a marsh cannot grow vertically fast enough, it may be forced to migrate horizontally as sea level rises. A successful lateral migration depends on the existence of a suitable adjacent habitat and that the time scale of migration is commensurate with re-colonization times. As many areas of the coast have been engineered to prevent coastal erosion, the ability of many marshes to migrate laterally is not assured, and some loss of wetland is likely to result. Park et al. (1986) estimated that for a high sea-level rise of 1.6 m, a loss of 45 percent of the 4850 km^2 of coastal wetlands in the United States could occur. A lower rise of 0.9 m could result in a loss of 22 percent of the wetlands in the United States.

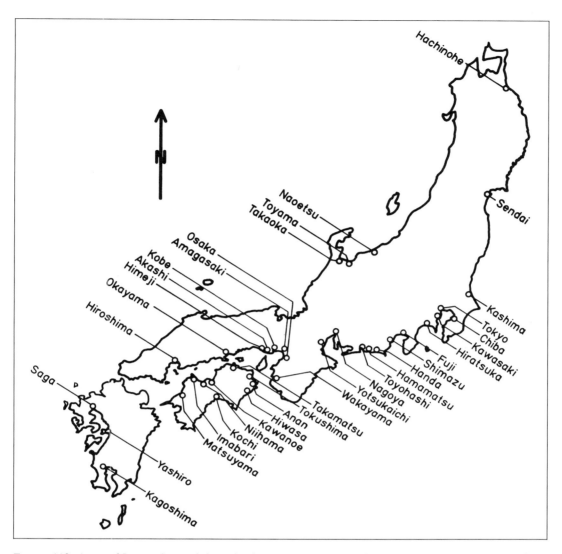

FIGURE 113. Areas of Japan where salt intrusion into groundwater supplies has been reported. Redrawn from Yamamoto and Kobayashi (1984).

Other wetlands, such as mangroves, are critical links in the food chain of coastal ecosystems. These wetlands are critically dependent on both water levels and sediment levels for their health and migration. Since mangroves are beneficial for coastal erosion protection as well as for food-chain purposes, it is important to conserve them. Presently, mangroves are decreasing in density because of human influence (Soegiarto 1985). As the sea level rises, landward migration would result naturally, since vertical migration does not occur in mangroves. However, partly because of their protective function, so many areas landward of mangroves are now used for agriculture (rice fields and fish ponds), that landward migration might not be able to occur naturally (Rosenberg et al. 1989).

Fisheries and other marine resources can be impacted as sea levels continue to change. Waterfowl may undergo changes in distribution as marshes and open-water areas change in density. Birds that rely on marshes and estuarine systems must adapt accordingly. For instance, waterfowl in the Chesapeake Bay and in Louisiana have changed distribution because of the loss of wetlands (Stotts 1985; Meo 1988). Birds that nest near marshes would also be affected, as would birds that feed in intertidal areas. The magnitude of such impact, however, is conjectural. Changes in sedimentation patterns, changes in tidal currents, and changes in water temperature would exert some degree of change on coastal ecosystems in general. As water temperatures increase, the biological and chemical oxygen demand would increase, making worse a critical problem of anoxia and eutrophication in many of our nutrient-loaded coastal water bodies.

Hydrological changes would result if climate changes and sea-level rise accelerates. For instance, the lake levels must change as the water balance adjusts to a warmer atmosphere. As shown by major erosion problems experienced in large lake systems during high stands of lake levels, hydrological changes can have severe economic impact. The rise and fall of lake levels is just another manifestation of sea-level rise, but for different reasons than for the oceans. An interesting example is the Great Lakes in the United States and Canada. The Great Lakes contain more than 8000 km^3 of water covering an area of 246,000 km^2, storing 20 percent of the world's fresh water and 95 percent of North America's surface water. Twenty-nine million Americans and eight million Canadians live along the borders of the lakes. Historical fluctuations of 2 m on time scales of one to two decades have caused significant problems of coastal erosion and related issues. As climate changes the water levels are certain to continue to fluctuate, although they may drop dramatically if the water supply (precipitation) within the drainage basin were to decrease markedly. If the water level were to drop as predicted by some GCM model outputs, hydroelectric power production would be lowered, navigation would be adversely affected (particularly through locks), and water supplies would have to be managed and parceled out differently. Other major water bodies in the world would have analogous impacts.

Biogeochemical changes would be felt as seas continued to rise. Growth of nitrogen and phosphorous concentrations on a regional scale would result from flooding of coastal areas and from soil erosion, particularly in subpolar and midlatitudes such as in the Bering Sea. Many pesticides presently retained in soils could find their way to the ocean, affecting coastal ecosystems.

Even seacliffs may undergo accelerated erosion because of relative rise of sea level; however, most observed erosion of sea cliffs can be ascribed to loss of protecting sand beaches because of blocking of new supplies of sand by construction of dams across tributary streams and of groins, jetties, and other structures across beaches. Their effect is evident for seacliffs (Kuhn and Shepard 1984) just as for low coasts (Pilkey et al. 1975), as pointed out by Emery and Uchupi (1972, p. 15–20) and many others who realize that the shore belongs to the ocean, not to man.

Socio-Economic Impacts

The socio-economic impacts of changing sea levels are difficult to quantify, but during the past decade many economists have attempted to make estimates. Each of the areas described above as having physical, chemical, or biological impacts have some costs associated with them. For instance, inundation would be accompanied by loss of upland and consequent economic damages. As an example, Milliman et al. (1989) estimated that a 1-m rise in sea level would inundate 12 to 15 percent of the arable lands of Egypt and 17 percent of Bangladesh (Commonwealth Secretariat 1989). Estimates in other areas vary, but they include loss of 20,000 km^2 in the United States, worth about $650 billion (Park et al. 1989; Yohe 1990). Recently, Rijkwaterstaat (1990) considered inundation from a 1-m rise in sea level for 181 coastal countries and territories. It estimated that 345,335 km of low coasts, 6400 km of urban waterfronts, 10,725 km of sandy beaches, and 1756 km^2 of harbor would have to be protected. Others (Kana et al. 1984; Leatherman 1985; Titus 1986b, 1990) estimated the loss of oceanfront property from a 30-cm rise in ocean levels to result in tens of millions of dollars per kilometer in the United States.

Losses of natural systems have costs associated with sea-level rise. For instance, one estimate shows coastal marshes have a value to society of $13,600 per hectare. Capitalized values for mangrove wetlands have been estimated as $212,507 (Thorhaug and Miller 1986) and $81,975 per hectare (Logo and Brinson 1979). Values for other wetlands have been even higher. Fisheries associated with these wetlands are difficult to quantify, but clearly they provide economic opportunities that enhance their value. For example, mangrove forests in Thailand have an estimated annual worth of $130 per hectare for fisheries, compared with only $30 for charcoal production (Christiansen 1983). As climate evolves, the occurrence of El Niños must wax and wane, producing large variations in fisheries production off Peru. Since this catch accounts for roughly 20 percent of the value of all protein from fish in the world, such fluctuations could have significant economic consequences, as well as political ones.

The economic impacts on fisheries and habitats are difficult to quantify, not only because of the inherent problems of assigning economic worth to natural systems, but also because the chain of impacts is difficult to project. For instance, an earthquake off Chile on 3 March 1985 caused a coastal uplift of approximately 33 cm. It produced dramatic changes in the rocky intertidal kelp community, including extensive mortality of intertidal organisms and opportunistic re-colonization by barnacles (Castilla 1988). Resultant changes in the biota have economic consequences that are difficult to estimate, given the uncertainties in our ability to predict the rate of change of intertidal communities and the time scales for new communities to develop after such episodic changes. That subtle shifts in the relative sea level can have such dramatic implications makes economic assessment of changes in sea levels even more problematic.

Other consequences would result from changes in biodiversity of the global ecosystems and from changes in the endangered species composition. Further costs would come from changes in recreational use of the shoreline, changes in property values as coastal erosion becomes more prevalent, and increased transportation costs due to threatened coastal highways and transportation links. These all contribute to the economic costs of increasing sea levels.

9. Summary

Major and minor changes of sea level and land level have long been recognized by geologists and archaeologists, but dates, rates, and causes are poorly known. Radiocarbon dating of organic materials deposited in the intertidal region show that some of these materials now are present on the sea floor in water depths of more than 100 m (mainly because of their deposition during late Pleistocene times of low sea levels concurrent with large ice sheets); however, other such dated materials occur at similar elevations above present sea level (due to crustal rebound after melt of glaciers and to other tectonic movements of the land). Archaeological remains also document the shifting levels of land and ocean. Additional and more detailed information about changes of level during the past century comes from records of tide gauges. Most of these gauges originally were installed to facilitate navigation of sailing ships to and from harbors (whose very existence commonly implies local crustal instability and shifting of harbor-floor sediments).

Mean annual sea levels are computed from tide-gauge records as averages of hourly readings throughout each calendar year. Plots of these mean annual sea levels have been accepted by some climatologists and many laymen as measures of the eustatic rise of sea level expected from an increased atmospheric temperature caused by the greenhouse effect of the Industrial Revolution, and in turn, it has been believed to increase the volume of the ocean by addition of meltwater from glaciers and thermal expansion of ocean water. Commonly ignored is variability in tide-gauge records caused by changes in meteorology, ocean dynamics, and tectonism. Because the gauges are located along exposed continental margins where oceanographic variability is greatest (large tides, storm surges, shelf waves) and near the unstable boundary between lighter continental crust and denser oceanic crust (along which tectonic plates commonly collide), these oceanographic and tectonic effects can be magnified beyond their ranges in the interiors of ocean basins and continents.

The present work focuses on contributions to tide-gauge records caused by land-level changes, with less emphasis on contributions by meteorological and oceanographical factors. Omission of tide-gauge records from areas of former ice sheets has been common by those who try to compute an eustatic rise of sea level, but generally ignored is the effect of other land movements: the sinking of ridges peripheral to areas of continental crust that were depressed by weight of ice, sinking of deltas, sinking of rifted and cooling crust, and sinking of ground affected by pumping of water, oil, and gas and either sinking or rising of coasts that border trenches (where oceanic crust is underthrusting continental crust), volcanoes, faults, and folds. Most tide-gauge records reveal vertical movements that can be attributed to changes of land level when regional maps of average rates of change of level derived from tide-gauge records are studied with respect to regional geology. Only a few coastal sites can be considered relatively stable owing to absence of known geological tectonism. Even these few sites probably are being affected by crustal downwarp and peripheral upwarp caused by the weight of returned meltwater onto the underlying ocean floor, especially the continental shelves. These same land movements occur off all coasts as a variable bias affecting changes of level caused by other tectonic or eustatic processes—immensely adding to the complexity of evaluating directions, rates, and causes of vertical changes of land level and sea level.

Further complications come from the poor global distribution of tide-gauge records of sufficient duration (at least 20 years) to yield reliable trends; most of these useful tide gauges border industrial nations of northern Europe, United States, and Japan—all within the Northern Hemisphere. In contrast, the coasts of Africa, South America, Central America, Asia, the Middle East, and oceanic islands have few useful records, a result believed to be due to little industry, shipping, and scientific interest by governments.

Our analysis shows that the signal of a possible eustatic rise of sea level is obscured by "noise" caused mainly by movement of the land beneath tide gauges; thus, study of the "noise" is a potential source of information about modern movements of the Earth's crust—especially of plate tectonics. These data support concepts from geological and geophysical observations of sinking or rising land levels related to crustal movements associated with glaciation, subduction, rifting, sediment loading, volcanism, and general faulting and folding associated with the properties and history of crustal plates and especially their movements during sea-floor spreading. This conclusion should be no surprise to geologists, but it may be unexpected by those climatologists and laymen who have been biased too strongly by the public's perception of the greenhouse effect on the environment. At present, we cannot discover a statistically reliable rate for eustatic rise of sea level alone, but it may not matter to the seafront property owner whether his house becomes flooded because of a rising sea or a sinking coast.

The present rate of relative sinking of land levels along coasts is less than a tenth of that during the time of fastest melt and retreat of late Pleistocene ice sheets, and more than 10 times the average for the past 50 m.y., during most of which the effects of glaciation were absent and geological agents alone were effective. In addition, many present coastal changes (erosion of beaches, collapse of seacliffs, disappearance of salt marshes, sinking of some coastal cities, siltation of harbors, and others) are caused more by indirect and unexpected results of human activities than to changes of sea level or land level. Most coastal instability can be attributed to tectonism and documented human activities without invoking the spectre of greenhouse-warming climate or collapse of continental ice sheets.

Résumé

Les changements du niveau marin et du niveau terrestre, de grande au moindre importance, ont depuis longtemps été reconnus par géologues et archéologues. Mais les époques, les taux de variation et les causes de ces changements sont mal connus.

La datation par carbone radio-actif des matières organiques déposées dans les terrains concernés par l'amplitude des marées, montre que certaines de ces matières sont maintenant présentes sur le sol sous-marin à des profondeurs de plus de 100 mètres (principalement parce qu'elles ont été déposées durant le Pleistocène supérieur, époque de niveau marins abaissés et associés à d'épaisses couches glacières); cependant, d'autres matières de la même époque sont présentes aujourd'hui à des élévations du même ordre au dessus du niveau de la mer (à cause du relèvement de la croûte terrestre, suite à la fonte des glaciers et à d'autres mouvements tectoniques de la terre).

Des restes archéologiques renseignent également sur les variations de niveau terrestre et marin. Des données supplémentaires et plus détaillées sur les changements de niveau au cours des cent dernières années sont fournies par les relevés des echelles de mesure des marées (Limnimètres).

La plupart de ces limnimètres furent â l'origine, installés pour faciliter la navigation à voile à l'entrée et à la sortie des ports (leur existence même, impliquant l'instabilité locale de la croûte terrestre et des mouvements de sédiments sur les fonds des ports). Les moyennes des niveau annuel de la mer sont calculées d'après les moyennes horaires enregistrées chaque jour de chaque année. Les courbes de ces niveaux annuels moyens de la mer ont été acceptées par certains climatologistes et nombre de non-spécialistes comme mesures de la montée effective du niveau de la mer attendue par suite de l'augmentation de la température atmosphérique, causée par l'effet de serre de la révolution industrielle; ceci est de plus supposé augmenter le volume des océans par addition d'eau venant de la fonte des glaciers et l'expansion thermique des mers. La plupart du temps, il ne fut pas tenu compte, dans la lecture des mesures enregistrées par les limnimètres, des variations dûes à la Météorologie, à la dynamique des océans et à l'activité tectonique. Les limnimètres étant placés le long de côtes exposées où les variations océanographiques sont les plus importantes (grandes marées, houles de tempêtes, vagues) et près de la limite instable entre la croûte continentale plus légère et celle de l'océan plus dense (là aussi où les plaques tectoniques se heurtent régulierèment), ces effets océanographiques et tectoniques peuvent être exagérés, par rapport à leur amplitude réelle, à l'intérieur des bassins océaniques et continentaux.

La présente étude des relevés des limnimètres se concentre sur les variations causées par les changements de niveau terrestre et accorde moins d'importance aux facteurs météorologique et océanographique. Les relevés des hauteurs de marées dans les régions autrefois recouvertes par la couche glaciaire ont été généralement omis par ceux qui se sont efforcés de calculer l'augmentations effective du niveau de la mer, et de plus, il n'est pas tenu compte de l'effet d'autres mouvements terrestres: l'enfoncement des plis situés à la périphérie de la croûte continentale qui a été comprimée par la pression de la masse glaciaire, l'abaissement des deltas, l'enfoncement de la croûte lorsqu'elle se fissure et se refroidit, l'affaissement du sol dû au pompage d'eau, de pétrole et de gaz et l'abaissement ou le relèvement des côtes en bordure de fissures (là où la plaque océanique exerce sa poussée sur la plaque continentale), les volcans, les failles et les plissements.

La plupart des relevés des limnimètres revèlent des mouvements verticaux qui peuvent être attribués à des changements du niveau du sol. C'est particulièrement le cas lorsque les cartes régionales donnant les taux moyens de changement de niveau sont étudiées en fonction de la géologie locale. Seuls, quelques sites côtiers peuvent être considérés comme relativement stables par suite de l'absence de tectonisme géologique connu. Même ces quelques sites sont probablement affecté par les deformations de la croûte vers le bas et de sa périphérie vers le haut, causées par le poids de l'eau de fonte revenant sur le fond marin, particulièrement sur les plateaux continentaux. Ces mêmes mouvements du sol constituent le long de toutes côtes, un facteur variable qui affecte les changements de niveau causés par d'autres processus tectoniques ou "eustatiques" - ce qui augmente considérablement la complexité de l'étude des directions, des taux, et des causes des modifications dans le sens vertical du niveau de la mer et de celui de la terre.

L'inégale distribution globale de limnimètres installés depuis suffisamment longtemps (au moins 20 ans) compliquent l'obtention de résultats fiables; la plupart des limnimètres utilisables sout situés sur les côtes des pays industrialisés de l'Europe du nord, des Etats-Unis et du Japon - tous dans l'hémisphère nord. Par contre les côtes d'Afrique, de l'Amérique du Sud, de l'Amérique Centrale, de l'Asie, du Moyen-Orient et des Iles océaniques, ne fournissent que peu de relevés utiles, conséquence peut-on-penser de la faible industrialisation, de l'insuffisance du trafic maritime et du manque d'intérêt dans la recherche scientifique des gouvernements.

Notre étude montre que ce qui pourrait être l'indication d'une montée "eustatique" du niveau de la mer est masquée par le "brouillage" causé principalement par le mouvement de la terre sous les limnimètres; l'étude de ce "brouillage" peut constituer une source d'information sur les mouvements actuels de la croûte terrestre—particulièrement la tectonique des plaques. Ces données corroborent les idées reçues d'observations géologiques et géographiques selon lesquelles le niveau du sol s'abaisse ou s'élève en fonction de mouvements de la croûte associés à la glaciation, la régression, les failles, l'entassement de sédiments, le volcanisme, et la création générale de fissures et de plis propres à la nature et à l'histoire des plaques et plus particulièrement à leurs mouvements lors des épanchements sous-marins.

Cette conclusion ne devrait pas surprendre le géologue, mais peut paraître inattendue aux climatologues et aux non-spécialistes qui ont été trop fortement influencés par l'idée que se fait le public de l'effet de serre sur l'environnement. Nous ne pouvons pas à l'heure actuelle, établir d'une manière statistiquement valable un taux d'élévation intrinsèque du niveau de la mer; mais il importe peu au propriétaire d'une maison au bord de la mer si elle sera inondée parce que le niveau de la mer monte ou parce que la côte senfonce. Le taux actuel d'enfoncement relatif des niveaux terrestres le long des côtes est inférieur au dixième de ce qu'il a été au moment de la fonte et de la régression les plus rapides des couches glaciaires à la fin du Pleistocène. Il est par ailleurs dix fois plus important que la moyenne au cours des dernières 50- millions d'années, pendant lesquelles, la plupart des effets de la glaciation étaient absents et que seuls, agissaient les facteurs géologiques. De plus, plusieurs changements côtiers actuels (érosion des plages, effondrement des falaises, disparition de marécages saumâtres, enfoncement de quelques villes côtières, envasement des ports et autres) sont dûs davantage aux résultats indirects et inattendus d'activités humaines qu' aux changements de niveau marin ou terrestre. La plus grande partie de l'instabilité côtière peut être attribuée au tectonisme et aux activités humaines qui sont répertoriées, sans évoquer le spectre de l'effet de serre, ni celui de la fonte de la calotte glaciaire.

Zusammenfassung

Große und kleine Veränderungen des Meeres- und Landesniveaus sind Geologen und Archäologen seit langem bekannt, aber ihre Dauer, Geschwindigkeiten und Ursachen sind wenig bekannt. Radioaktive Altersbestimmungen (C-14) an organischen Substanzen besagen, daß manche Ablagerungen, die sich heute in einer Meerestiefe von über 100 m befinden, ursprünglich im Gezeitenbereich entstanden waren. Solche Vorkommen entstanden im späten Pleistozän, als der Meeresspiegel wegen der kontinentalen Vereisung wesentlich tiefer lag. Andererseits werden ähnliche Ablagerungen auch bis zu 100 m über dem heutigen Meeresniveau angetroffen. In solchen Fällen sind Entlastung der Erdkruste nach Abschmelzen der Gletscher oder andere tektonische Bewegungen für das Ansteigen verantwortlich. Archäologische Funde bezeugen ebenfalls Veränderungen von Land- und Meeresspiegel. Genauere Information über Niveau-Veränderungen ist in Pegelstand-Meßreihen enthalten. Diese wurden hauptsächlich zur Erleichterung der Schiffahrt in der Nähe von Häfen begonnen. (Die Lage der Häfen selbst ist oft mit örtlichen Schwächen der Erdkruste und ständig sich verschiebenden Sedimenten verbunden.)

Jahresschnitte des Meeresspiegels werden als Mittelwerte von Reihen stündlicher Ablesungen von Pegelständen errechnet. Kurven dieser Jahresmittel werden von manchen Klimatologen und vielen Laien als ein Maß eustatischen Anstiegs des Meeresspiegels angesehen. Ein solcher Anstieg ist als Folge des durch die industrielle Revolution verursachten Treibhauseffekts zu erwarten, der die Temperatur der Atmosphäre, und dadurch indirekt das Volumen des Meeres durch Zufluß von Schmelzwasser und thermische Ausdehnung des Meerwassers, erhöht. Dabei werden oft Schwankungen in den Pegelständen übersehen, die auf meteorologischen, meeresdynamischen oder tektonischen Veränderungen fußen. Da die Pegel sich an offenen Kontinentalrändern befinden, wo ozeanographische Schwankungen (Gezeitenhöhe, Sturmfluten, Schelfwellen) am größten sind, und nahe der labilen Grenze zwischen leichterer kontinentaler und dichterer ozeanischer Kruste (wo tektonische Platten kollidieren), können diese ozeanographischen und tektonischen Auswirkungen weit über die für das Innere der Meeresbecken und Kontinente typischen Masse vergrößert werden.

Dieses Buch handelt hauptsächlich von den Einflüssen der Veränderung des Landesniveaus auf Gezeitenmessungen. Die Einwirkungen meteorologischer und ozeanographischer Faktoren werden weniger betont. Meßreihen von Gezeiten in Gegenden früherer Vereisung werden gewöhnlich in Errechnungen des eustatischen Anstiegs des Meeresspiegels nicht einbezogen, aber die Einwirkungen anderer Landesbewegungen werden vernachlässigt. Bespiele sind das Absinken von Landrücken am Rande kontinentaler Kruste durch Eisbelastung, das Absinken von Deltas, das Absinken driftender und abkühlender Kruste, die Senkung des Terrains wegen Extraktion von Wasser, Erdöl oder Erdgas, und Senkung oder Anstieg von Küsten in der Nähe von Tiefseegräben (wo ozeanische Kruste kontinentale Kruste unterschiebt), Vulkanen, Verwerfungen und Falten. Wenn man Karten, die von Pegelständen abgeleitete Linien gleicher Meeresspiegel-Veränderungen enthalten, mit der örtlichen Geologie vergleicht, so erkennt man, daß die meisten in solchen Meßreihen enthaltenen Anzeichen für vertikale Bewegungen auf Änderungen des Landesniveaus zurückzuführen sind. Nur wenige Küstengebiete können aufgrund fehlender Anzeichen von Tektonismus als relativ stabil angesehen werden. Aber selbst diese Gegenden stehen wahrscheinlich unter dem Einfluß von Absinken oder randlichem Aufbiegen der Erdkruste, die durch die Belastung des Meeresbodens mit Wiederkehr des Schmelzwassers verursacht wurden. Dieses ist besonders in Schelfgebieten der Fall. Die gleichen Landbewegungen finden vor allen Küsten statt und modulieren dadurch zu unterschiedlichem Grade alle Meeresspiegel-Veränderungen aufgrund anderer tektonischer oder eustatischer Vorgänge. Dieser Umstand kompliziert die Auswertung von Vorzeichen, Geschwindigkeit und Ursache der Veränderungen von Landes- und Meeresniveau ungeheuer.

Zusätzliche Komplikationen entstammen der ungleichen weltweiten Verteilung von Pegel-Meßreihen hinreichender Länge (wenigstens 20 Jahre) für zuverlässige Erkennung von Gängen. Die Mehrzahl solch brauchbarer Gezeitenmessungen liegt an den Küsten von Nordeuropa, den Vereinigten Staaten und Japan, also in der nördlichen Halbkugel. Dagegen sind nur wenige brauchbare Meß-Serien vorhanden von den Küsten Afrikas, Süd- und Mittelamerikas, Asiens, des Nahen Ostens und von ozeanischen Inseln. Dieser Umstand beruht wahrscheinlich auf dem weitgehenden Fehlen von Industrie, Schiffahrt und wissenschaftlichem Interesse seitens der Regierungen.

Unsere Untersuchung zeigt, daß eventuelle Anzeichen für Anstieg des Meeresspiegels in den durch Landbewegungen unter Pegeln verursachten „störungen" untergehen. Daher ist Erforschung der „störungen" eine mögliche Auskunftsquelle über gegenwärtige Bewegungen der Erdkruste, besonders der Plattentektonik. Solche Daten unterstützen Konzepte von geologischen und geophysikalischen Beobachtungen absinkender und aufsteigender Landniveaus im Zusammenhang mit Vereisung, Subduktion, Riften, Sediment-Belastung, Vulkanismus sowie Verwerfungen und Faltungen. Diese Vorgänge sind von spezifischen Eigenschaften und Vergangenheit der Krustenplatten abhängig, besonders von deren Bewegungen im Zuge der Ausbreitung des Meeresbodens. Diese Schlußfolgerung dürfte Geologen nicht überraschen, aber sie mag für diejenigen Klimatologen und Laien unerwartet sein, die sich durch populäre Auffassungen über den Treibhauseffekt auf die Umwelt zu stark haben beeinflussen lassen. Wir können zum gegenwärtigen Zeitpunkt keine statistisch zuverlässigen Anzeichen für rein eustatischen Anstieg des Meeresspiegels erkennen. Jedoch mag es für den Besitzer eines an der Küste gelegenen Grundstückes von geringem Interesse sein, ob sein Haus wegen steigendem Meeresspiegel oder absinkender Küste überflutet wird.

Die gegenwärtige Geschwindigkeit des relativen Landabsinkens an Küsten ist unter einem Zehntel von der zur Zeit des maximalen Abschmelzens pleistozäner Eisdecken und über das Zehnfache des Durchschnittes über die letzten 50 Millionen Jahre. Dieser Zeitraum war weitgehend eiszeitenfrei, und nur geologische Auswirkungen machten sich bemerkbar. Außerdem sind viele derzeitigen Küstenveränderungen (Stranderosion, Klippenzerfall, Verschwinden von Marschland, Absinken von Küstenstädten, Hafenversandung und ähnliches) mehr auf indirekte und unvorhergesehene Folgen menschlicher Handlungen zurückzuführen als auf Veränderung des Meeres- oder Landesniveaus. Die meisten Anzeichen für Instabilität von Küsten können durch Tektonismus und erwiesene menschliche Aktionen erklärt werden, ohne daß man dazu das Gespenst der Treibhaus-Klimaerwärmung oder des Abschmelzens der kontinentalen Eisdecken heranzuziehen braucht.

תקציר

שינויים משמעותיים אך גם זעירים, שחלו בעבר במפלסי הימים והיבשות, היו ידועים לגיאולוגים ולארכיאולוגים זה מכבר, אולם גורמיהם, קיצבם ועיתויים עדיין אינם בהירים במדה מספקת. לפיכך תוארכה (בשיטת הפחמן הרדיואקטיבי) תכולת החומר האורגני במספר רב של דגימות של משקעי קרקעית הים אשר נדגמו באתרים רבים בעולם בעומקי מים המגיעים עד לכ-100 מ'. משקעים אלה נוצרו במקורם באזור הכרית (גיאות ושפל). מתוצאות התארוכים הללו ניתן ללמוד כי בתקופת הקרח הגדולה, אשר חלה בפלייסטוקן העליון, היו מפלסי האוקיאנוסים נמוכים בכמאה מטרים בהשוואה למפלס הנוכחי. מאידך נדגמו ותוארכו סלעי משקע ימיים דומים המצויים כיום באתרים רבים על פני היבשות במפלסים הגבוהים במדה נכרת מעל למפלס הים הנוכחי. תוצאותיהם של תארוכים אלה לימדונו כי שטחי יבשה רבים התרוממו באורח טקטוני מאז תקופת הקרח האחרונה וזאת כתגובה של קרום הארץ להתמוססות המסיבית והמהירה של כיפות הקרח אך גם בגלל תנועות טקטוניות אחרות המתרחשות ביבשות. גם מיקומם של שרידי אתרים ארכיאולוגים - בהשוואה למיקומו של קו החוף הנוכחי - עשוי להעיד על תנודות יחסיות במפלסי האוקיאנוס והיבשה. מכשירי הרישום הרציף של הכרית (גיאות ושפל) תרמו מידע מפורט לגבי התהליך של שינויי המפלס אשר התרחשו משך מאת השנים האחרונות. מרביתם של מדי כרית אלה הותקנו במקורם במטרה להקל על ניווטם של כלי שייט בקרבת פתחי הנמלים; עצם הצורך להיעזר במכשיר מסוג זה מרמז על אי יציבות מקומית של הקרום באזורים אלה וכן על הדינמיקה של תנועת המשקעים בקרקעיות הנמלים.

הערכים השנתיים הממוצעים של מפלסי הים חושבו ע"י מיצוע הנתונים שתועדו מדי שעה בעזרת מדי הכרית, עבור כל שנת-לוח בנפרד. העקומות שנתקבלו ע"י הצבתם של חישובים אלה פורשו ע"י כמה מהקלימטולוגים כביטוי וכמדד לעליה האיאוסטטית של מפלס הים וזאת כתוצאה מהעליה שחלה במידות החום של האטמוספירה. עליה תרמית זו נגרמת כתוצאה מאפקט החממה שנוצר בעקבות המהפכה התעשייתית והעשרת האטמוספירה בדו תחמוצת הפחמן. מהאמור לעיל מסתבר כי הגידול שחל בנפחו של האוקיאנוס נגרם כתוצאה של התוספות מי המסת הקרחונים וכן בגלל ההתפשטות התרמית של מי האוקיאנוס. בדרך כלל נהוג לנפות מהרישומים של מדי הכרית את החריגות הנגרמות בגלל שינויים מטיאורולוגיים, כוחות אוקיאניים ותהליכים טקטוניים. מטבע הדברים ממוקמים מדי הכרית לאורך החופים, משמע באזור בו מגיעים לשיאם השינויים האוקיאנוגרפיים (כגון, גלי

הגיאות, מישברי הסערות והזרמים החופיים). בדרך כלל מהווה קו החוף גם את התחום הגבולי הבלתי יציב המפריד בין הקרום היבשתי הקל לקרום האוקיאני הכבד יותר (שלאורכו בדרך כלל מתנגשים הלוחות הטקטוניים). אי לכך מתוגברות באזור רגיש זה עוצמותיהן וטווחייהן של הפעילויות האוקיאנוגרפיות והטקטוניות השונות - זאת בהשוואה לאלו המתרחשות באזורים הפנימיים של האגנים האוקיאניים ושל היבשות.

המחקר הנוכחי מתמקד בתרומתם של מדי הכרית לתיעוד השינויים החלים במפלסי הקרקע ולפיכך עוסק במדה פחותה יותר בהיבטים המטאורולוגיים והאוקיאנוגרפיים של הרישומים. אותם מדענים שניסו לחשב את העליה האיאוסטטית של מפלס הים, נטו להשמיט מחישוביהם את הנתונים שרשמו המכשירים הממוקמים באזורים שהיו מכוסים עד לא מכבר בקרח-עד. כמו כן התעלמו, מדענים אלה, בדרך כלל, מהשפעתן של תנודות קרקע אחרות כגון: התרוממותם של הרכסים המצויים בפריפריה של שטחי קרום יבשתי ששקע בגלל עומס הקרח, שקיעתן של דלתות, שקיעתם של קטעי קרום שנסדק ולפיכך איבד חום והתקרר, שקיעת קרקע הנגרמת בגלל שאיבת מים נפט וגז משכבות מאגר תת קרקעיות, שקיעה או התרוממות של חופים הגובלים בתעלות טקטוניות (באזורי הפחתה שם קרום אוקיאני נדחס אל מתחתיו של קרום יבשתי) וכן בהרי געש, בהעתקים ובקמטים. באזורים בהם ניתן היה להשוות את הנתונים מהמפות הגיאולוגיות למפות המבטאות את קצב שינויי המפלס (ואשר הוכנו מנתוניהם של מדי הכרית) הובחן קיומן של תנועות אנכיות אשר אפשר ליחסן לשינויים שחלו במפלס הקרקע תוך כדי תקופת המדידה. ניתן לסווג אתרי חוף ספורים בלבד כיציבים יחסית וזאת מאחר ולא אותרה שם פעילות טקטונית. אולם נראה כי אף אתרים ספורים אלה מושפעים מתהליכי שקיעת הקרום והתרוממות ההיקפית החלה במקביל. תהליכים אלה נגרמים כתוצאה מהתווספות משקלם של מי הקרחונים, שהומסו והתנקזו לים, לעמוד המים הלוחץ על קרקעית האוקיאנוס. תנועות קרקע תת ימיות אלו, המתרחשות כאמור לעיל באינטנסיביות רבה יותר לאורך החופים, ממסכות את שינויי המפלס הנגרמים ע״י תהליכים טקטוניים או איאוסטטיים אחרים ומגבירות ע״י כך במדה רבה את הקושי לפענח ולהעריך את כיווניהם, קצבם וסיבותיהם של השינויים האנכיים החלים במפלס הקרקע ובמפלס הים.

קשיים נוספים לקביעה מהימנה של המגמות של השינויים הנרשמים באותם מדי הכרית המתעדים נתונים לאורך משכי זמן מספיקים (לפחות 20 שנה), נגרמים בגלל פיזורם הגרוע; מרבית המכשירים הנ״ל ממוקמים לאורך חופיהן של המדינות המתועשות של צפון אירופה, ארה״ב ויפן, אשר כולן

מצויות במחצית הכדור הצפוני. מאידך מתקבלים אך מעט נתונים מועילים מחופי אפריקה, דרום אמריקה, מרכז אמריקה, אסיה והמזרח התיכון וזאת כנראה בגלל התיעוש המצומצם של המדינות שם. זו גם הסיבה לדלילות התנועה הימים אליהן ומהן ומכאן גם מיעוט עניין של הממשלות בעידוד המחקר המדעי.

ניתוח ממצאינו מורה כי הסימנים המעידים על אפשרות התרחשותה של עליה איאוסטטית של מפלסי הים ממוסכים ע"י "רעש" שמוצאו בתנועת הקרקע שמתחת למד הגאות; אי לכך מהווה ה"רעש" הנ"ל מקור מידע פוטנציאלי להתרחשותן של תנועות עכשויות בקרום כדור הארץ – במיוחד טקטוניקת הלוחות. נתונים אלה תומכים בתפישות וגישות אחרות המתבססות על תצפיות גיאולוגיות וגיאופיסיות וזאת בהקשר לתופעות כגון שקיעה או התרוממות של מפלסי קרקע המיוחסים לתנועות הקרום הכרוכות בתהליכי ההתקרחות, קיצור הקרום ע"י בליעתו והתכנסותו, סדיקת הקרום והתרחבותו, מעמס עמוד המישקעים, געשיות וכן שבירה וקימוט הקשורים באורח כללי בתכונות ובהיסטוריה של הלוחות ובמיוחד בתנועת הלוחות המתרחשת תוך כדי תהליך ההתפשטות של קרקעית הים. המסקנה הנ"ל אינה צריכה להפתיע את הגיאולוגים אולם ייתכן ותפתיע את אותם מומחי האקלים וקהל המתעניינים אשר נשבו בקסמי התפישה הפשטנית מדי של השפעת אפקט החממה על התנאים הסביבתיים. עד כה טרם עלה בידינו להגיע לאומדן מהימן מבחינה סטטיסטית של העליה האיאוסטטית נטו של מפלס הים. לבעל הקרקע בחוף הים אין זה משנה אם ביתו הוצף בעקבות תהליך של עלית פני הים או עקב שקיעת הקרקעית.

בתקופה הנוכחית מגיע הקצב היחסי של שקיעת מפלסי הקרקע לאורך החופים לפחות מעשירית מהערך המקביל של תהליך זה בתקופת השיא של המסת מרבדי הקרח ונסיגתם משך הפלייסטוקן המאוחר. מאידך נמצא כי הקצב הנ"ל הנו גדול פי עשרה מהקצב הממוצע של השקיעה משך 50 מליון השנים האחרונות; במרבית התקופה הזו לא התרחשה התקרחות כל שהיא ולפיכך נמשכה השפעתם של התהליכים הגיאולוגיים האחרים בלבד. יתר על כן רבים מהשינויים המתרחשים כיום לאורך החופים (כגון, סחף חופים, מפולת מצוקי חוף, העלמותן של מליחות, שקיעתן של כמה מערי החוף, הצטברות משקעים בנמלים וכו') נגרמים בעיקר כתוצאה עקיפה ובלתי צפויה של פעילויות האדם ורק בחלקן הקטן בגלל שינויי מפלס הים או הקרקע. אי לכך ניתן ליחס את מרבית התופעות של אי-היציבות לאורך החופים לטקטוניזם ולפעילויות אנוש ניתן לתעדן וזאת מבלי לערב בכך צילו המאיים של אקלים החממה המתחמם או את העלמותם המהירה של מרבדי הקרח.

要　約

　海面水準がさまざまの規模で変化していることが地質、考古学者などによって昔から観察、実証されてきた。しかし、その変化の速度、理由についてはなお不明の点がおおく、また説明できないことが多い。潮間帯に堆積した有機物の放射性炭素の年代決定にもとずいた研究によると、現在１００メートル以上も深い海底に堆積している物質が、現在の海面水準面よりもなお高い位置から発見されている。これは最新世後期の大陸氷河の発達にともなった海面の下降期に堆積した地層が氷塊の溶解にともなって拡張し、地面が海面に比較して上昇したことによって説明できる。また、考古学的な遺跡の現在の海岸線からの相対的な距離からも地面と海面の相互位置関係の変化を知ることができる。これらに加えて過去約一世紀にわたって潮位計が海水面の詳細な変化の記録を残してきた。潮位計の大部分は本来帆船の航行の便宜のため、その地方の地殻変動や堆積物の移動にともなう港底の変化を知るために設置されたものである。

　海準面の経年変化は一時間ごとの潮位記録を暦年ごとに平均としてしめされている。多くの気候学者や一般大衆はこれらの海準面の経年変化が産業革命がもたらした温室効果によって極地や高地の氷塊の溶解や温度の上昇によって海水の総体積が増大した結果によると考えてきた。しかしながら潮位計の記録にみられる変化は気候、海洋の変化のみならず地殻変動もその重要な原因となっていることが忘れられていることが多い。潮位計は潮位、嵐　あるいは棚波など海洋からの応力をもっとも受けやすく、また、比重の小さい大陸クラストと密度の高い海洋クラストが衝突する大陸周辺部に位置していることが多い。潮位計のおかれている大陸周辺部は海盆内部などにくらべて海洋学的な影響をはるかに受けやすいのである。

　この研究は気候、海洋学的影響がもたらす効果よりも主として地準面自身の変化によってもたらされる潮位の動向についての知見、論議をまとめたものである。海準面の上昇量を計算するために以前氷床に覆われていた地方に設置された潮位計の記録を無視することがある。氷床の重みで押さえ付けられていた大陸クラストの周辺部、冷却されつつある地殻、地下水や石油の掘削による沈降、海溝周辺での地層の逆押し被せによる上昇、下降、火山、断層そして終局作用などをいつも考慮にいれなければならない。潮位計の記録の大部分は地質学的に検討してみると地面の変化にもとずく上下運動を示している。地殻変動の影響のない比較的安定した海岸とゆうのは世界的にみて非常にすくないと考えられる。この例外的な地点があったとしても、間氷期になって氷が溶解し、ふたたび上昇に転じた海水の自身の重みによってもたらされる地殻のたわみと、それにともなう周辺部の反りあがりから逃れることができない。大陸棚は特にこの傾向が強い。海面の上昇下降が地殻変動によるのか、それとも海水面の変化によるのか、その動向、変化量、そしてこれらをもたらす原因の数々は極めて複雑に交錯している。

　問題をより難しくしているのは、高い信頼度で傾向を推測するのに望ましい期間（少なくとも２０年以上）わたって、続けられた潮位の長期観測データが比較的少なく、また、このような良質のデータが地

理的に片寄っていることである。研究に有用な潮位計の記録は北ヨーロッパ、アメリカ合衆国と日本など北半球の諸国からもたらされることがおおい。これにたいし、アフリカ、中南米、アジア、中東諸国、それに大洋に位置する島島にはそれらの国国の産業の規模、海上輸送の必要性、それに政府の科学行政を反映してか、よい記録が少ない。

　この本の分析が示しているように、しかるべき海面の上昇は潮位計が設置されている位置での地面、地殻の変動がもたらすノイズによって不明確にされている。逆に最近の地殻の変動についての情報がこのノイズに隠されていると考えられる。事実、潮位計の解析データは海底拡張に伴うプレートの機構と歴史に伴って氷河作用、地殻沈降、上昇、堆積物による地殻の加重、火山作用、断層や終局の形成など地質、地球物理学的観察からもたらされた地面の上昇、下降についての理解と一致する。この結論は地質学者にとっては驚くべきことではない。しかしながら温室効果に過敏になりすぎた一部の気候学者や大衆にとっては意外なことであるかも知れない。現在のところ統計的に正しい海面上昇率を算出することは可能ではない。しかし、海面の上昇か、あるいは海岸線の沈降かのどちらの作用でビーチフロントの家、家敷が水浸するのか、などという問題は、オーナーにとってはどちらでも良いことであろう。

　現在の沿岸地方における地面水準の下降速度は氷床の溶解に伴う後退の速度が最も早かった最新世後期１０分の１以下にすぎない。またこの速度は過去５千万年平均の（このうちの大部分の期間は氷河の影響をうけていなっかたのであるが）１０倍以上である。加えて、現在の海岸線の変化、例えば浜岸の侵食、海に沿った崖の破壊、海沿沼（マーシュ）の消滅、沿岸都市の地盤低下、港湾の泥ねい堆積による浅化などは海面、地面の変化よりも、間接的で最初に予測できなかった人工作用によることが多い。海岸地形の不安定な変遷の大部分は地殻変動とすでに記録されている人工作用によるものであって温室効果や大陸氷床の破壊などが強く影響しているという考えはまったく取るに足りない憶測に過ぎない。

КРАТКОЕ СОДЕРЖАНИЕ

Геологи и археологи уже давно признают крупные и незначительные изменения уровня моря и уровня суши. Однако о времени, скоростях и причинах этих изменений известно очень немного. Радиоуглеродное датирование органического вещества осадков, отложившихся в заливавшейся приливами береговой полосе, свидетельствует о том, что некоторые из этих осадков сегодня находятся на глубинах больше 100 метров (главным образом, вследствие их отложения в периоды позднеплейстоценовых понижений уровня моря, совпадающие по времени с периодами обширных оледенений). В то же время другие аналогичные осадки, наоборот, встречаются сегодня выше современного уровня моря на 100 и более метров (вследствие восстановления уровня коры после таяния ледников, а также вследствие других тектонических движений суши). Об изменениях уровня суши и океана свидетельствуют также находки археологов. Дополнительную и более детальную информацию об изменениях уровня моря за последнее столетие дают данные непрерывных наблюдений за приливами и отливами при помощи мореографов. Вначале большая часть этих приборов устанавливалась для обеспечения судоходства в гаванях (само существование которых обычно подразумевает локальную нестабильность коры, а также перемещение в них донных осадков).

Среднегодовые уровни моря рассчитываются по данным измерений мореографов как средние ежечасных показаний в течение каждого календарного года. Построенные на основании этих среднегодовых уровней моря графики рассматриваются некоторыми климатологами и многими непрофессионалами как показатели эвстатического повышения уровня моря, происходящего за счет увеличения температуры атмосферы, вызываемого усилением парникового эффекта - следствием промышленной революции. Последний, как предполагалось, в свою очередь способствует увеличению объема океана вследствие дополнительного таяния ледников и температурного расширения океанских вод. Изменения в данных наблюдений за приливами и отливами, обусловленные сменой метеорологической обстановки, динамики океанских вод, а также тектонических движений, обычно при этом не принимаются во внимание. Измерительная аппаратура устанавливается вдоль открытых континентальных окраин, где изменчивость океанографических характеристик наибольшая (большие приливы и отливы, штормовые волны и зыбь, шельфовые волны), и вблизи нестабильной границы между более легкой континентальной корой и более плотной океанической корой (вдоль этой границы обычно происходит столкновение литосферных плит). Поэтому здесь может быть преувеличена роль указанных океанографических и тектонических воздействий по сравнению с диапазоном таких воздействий во внутренних частях океанских бассейнов и континентов.

В предлагаемой книге основное внимание сосредоточено на работах по наблюдениям приливов и отливов, обусловленных изменениями уровня стояния суши; меньшее значение придается роли метеорологического и океанографического факторов. В тех работах, где делались попытки расчета эвстатического повышения уровня моря, обычными были пропуски данных по приливам и отливам для районов бывших ледяных покровов. В них обычно не учитывалось влияние других движений суши: опусканий хребтов, периферийных к областям континентальной коры, которые происходили за счет давления льда, опусканий дельт, опусканий расколотой и остывающей коры, опусканий поверхности грунта в результате выкачивания воды, нефти и газа. и, кроме того, либо опусканий, либо поднятий побережий, которые граничат с глубоководными желобами (где океаническая кора поддвигается под континентальную), а также либо опусканий, либо поднятий вулканов, разломов и складок. При анализе большинства результатов наблюдений за приливами и отливами можно обнаружить вертикальные движения, которые могут быть отнесены за счет изменений уровня суши. В то же время региональные карты средних скоростей изменения уровня, строящиеся по данным мореографических наблюдений, составляются в интересах региональной геологии. Относительно стабильными могут считаться лишь немногие участки побережий, где отсутствуют тектонические движения. И даже эти несколько участков, по всей видимости, затрагиваются при погружении коры и периферическом вспучивании. Их причиной является давление возвратившихся талых вод на подстилающее океанское дно, в особенности на дно континентальных шельфов. Такие же самые движения наблюдаются в океанах вблизи всех побережий, оказывая непостоянное косвенное влияние на изменения уровня, вызываемые другими тектоническими или эвстатическими процессами. Это очень усложняет оценку направлений, скоростей и причин вертикальных изменений уровня суши и уровня моря.

Получение надежных трендов усложняется также тем, что имеющаяся глобальная сеть достаточно продолжительных (по крайней мере 20 лет) постоянных наблюдений за приливами и отливами недостаточна. Большинство измерителей приливов и отливов этой сети располагается вокруг промышленно развитых государств северной Европы, Соединенных Штатов и Японии - все эти государства находятся в северном полушарии. Что касается побережий Африки, Южной Америки, Центральной Америки, Азии, Среднего Востока, а также океанских островов, то для них, напротив, имеется совсем немного необходимых наблюдений. По-видимому, это результат слабой индустриализации, малого развития судоходства и небольшого интереса к научным исследованиям со стороны правительств.

Наш анализ показал, что выделению сигнала возможного эвстатического повышения уровня моря мешает "шум", обусловленный, главным образом, движением твердого грунта под мореографами. Таким образом, исследование этого "шума" является потенциальным источником информации о современных движениях земной коры - особенно о тектонике плит. Эти данные подтверждают общие представления, возникающие на основании геологических и геофизических наблюдений понижающихся и повышающихся уровней суши, связанных с движениями коры, происходящими в результате оледенений, субдукции, рифтогенеза, перемещения осадочного материала, вулканизма, а также общего разломо- и складкообразования, связанных со свойствами и историей литосферных плит и особенно их движениями в процессе спрединга океанского дна. Этот вывод не должен быть удивительным для геологов, однако он, по-видимому, является неожиданным для тех климатологов и неспециалистов, на которых оказало очень сильное влияние представление широкой общественности о воздействии парникового эффекта на окружающую среду. В настоящее время мы не можем статистически надежно определить скорость только эвстатического повышения уровня моря. Однако собственнику земли на побережье, по-видимому, все равно, затопляется ли его дом вследствие повышения уровня моря или вследствие опускания его берега.

Современная скорость относительного опускания уровня суши вдоль побережий составляет менее одной десятой от скорости во время наибольшего таяния и отступания позднеплейстоценовых ледниковых щитов и более чем в десять раз превосходит среднюю скорость для последних 50 млн.лет. В течение большей части этого времени влияние оледенений отсутствовало, действовали только геологические факторы. Кроме того, многие изменения морских берегов, происходящие в настоящее время (эрозия пляжей, обрушение клиффов, исчезновение соленых маршей, опускание некоторых прибрежных городов, заиление гаваней и др.), в большей мере обусловлены косвенными и непредсказуемыми результатами деятельности человека, а не изменениями уровня моря или уровня суши. В наибольшей степени неустойчивость побережий может быть обусловлена тектоническими движениями и человеческой деятельностью без привлечения свидетельств как потепление климата за счет парникового эффекта, так и разрушения континентальных ледниковых щитов.

Resumen

Los grandes y pequeños cambios de nivel del mar y de la tierra han sido conocidos durante mucho tiempo por geólogos y arqueólogos, pero las fechas, velocidad y causas son pobremente conocidas. La datación por radiocarbono de los materiales orgánicos depositados en las regiones intertidales, muestran que algunos de estos materiales están presentes ahora en el fondo del mar en profundidades de agua de más de 100 m. (principalmente porque su deposición durante el Pleistoceno superior coincidió con bajos niveles de la mar debido a la presencia de grandes mantos de hielo); sin embargo, otros materiales similares aparecen en zonas elevadas a una altura similar por encima des nivel marino actual (debido a la reacción de la corteza, posteriore al deshielo glaciar y otros movimientos tectónicos en tierra). Los restos arqueológicos demuestran también la variación de los niveles del mar y de la tierra. Información adiciónal y más detallada acerca de cambios de nivel durante el siglo pasado, se obtiene de los registros de marea. La mayoría de estos registros fueron instalados originalmente para facilitar la navegación de los buques de vela de y hacia los puertos (cuya propia existencia implicaba comunmente una inestabilidad local de la corteza y variaciones en los sedimentos del fondo del puerto).

Los niveles marinos medios anuales son calculados a partir de los registros de marea como lecturas medias horarias a los largo del calendario anual. La gráfica des estos niveles marinos medios anuales ha sido aceptada por algunos climatólogos y muchos profanos como medidas del ascenso eustático del nivel del mar esperado por un aumento de temperatura atmosférica causado por el efecto invernadero de la revolución industrial, y como consecuencia, se ha creido en un aumento del volumen oceánico por la adición de agua del deshielo de los glaciares y la expansión térmica del agua oceánica. Normalmente ignorada es la variabilidad de los registros de marea causada por cambios meteorológicos, dinámica oceánica y tectonicismo. Debido a que los mareógrafos están situados a los largo de márgenes continentales abiertos donde se da la mayor variabilidad oceanográfica (grandes mareas, oleajes de tormentas, olas de plataforma) y cerca de los inestables límites entre la ligera corteza continental y la más densa corteza oceánica (a lo largo de los caules las placas continentales colisionan normalmente), estos efectos oceanográficos y tectónicos pueden ser magnificados más allá de sus magnitudes normales en el interior de las cuencas oceánicas y de los continentes.

El presente trabajo se centra en la aportación de las variaciones de nivel de la tierra a los registros de marea y con menos énfasis en la aportación de los factores meteorológicos y oceanográficos a dichos mareogramas. La omisión de los registros de marea provenientes de áreas inicialmente cubiertas por mantos de hielo ha sido frecuente por los que intentan calcular una elevación eustática del nivel del mar, así como es generalmente ignorado el efecto de otros movimientos de Tierra: el hundimiento de las crestas perféricas a áreas de corteza continental que ha sido hundida por el peso del hielo, hundimiento de deltas, hundimiento de corteza expandida y enfriada, y hundimiento de terranos afectados por bombeo de agua, petróleo y gas, y en todo caso hundimientos o elevaciones de costas que bordean fosas (donde la corteza oceánica se hunde bajo la corteza continental), volcanes, fallas y plieques. La mayoría de los registros de marea revelan movimientos verticales que se pueden atribuir a cambios de nivel de la tierra cuando se han realizado mapas regionales de los promedios de los cambios de nivel derivados de los registros mareográficos y se relacionan con la geología regional. Solamente unos pocos lugares costeros pueden ser considerados relativamente estables debido a la ascencia de tectonicismo geológico conocido. Incluso estos escasos lugares están afectados probablemente por el peso del agua de deshielo sobre el fondo marino infrayacente, especialmente en las plataformas continentales. Estos mismos movimientos del terreno aparecen frente a todas las costas como un factor variable que afecta a los cambios de nivel causados por otros procesos tectónicos o eustáeticos, sumando complejidad de forma ilimitada, el evaluar direcciones, medias y causas de los cambios verticales del nivel del mar y de tierra.

Complicaciones adicionales provienen de la pobre distribución global de los registros de marea con suficiente duración (al menos 20 años) que proporcionen tendencias fidedignas; la mayoría de estos registros de marea útiles bordean naciones industrializadas del norte de Europa, Estados Unidos y Japón — todos en el hemisferio norte. Como contraste, las costas de

Africa, América del Sur, América Central, Asia, Oriente Medio y las Islas Oceánicas tienen pocos registros útiles, se cree que debido a su escasa industrialización, marina mercante, y interés científico de los gobiernos.

Nuestro análisis muestra que la señal de una posible subida del nivel del mar está enmascarada por el "ruido" causado fundamentalmente por movimientos de la tierra bajo los mareógrafos; por lo tanto, el estudio del "ruido" es una fuente potencial de información acerca de los movimientos recientas de la corteza terrestre — especialmente de las placas tectónicas. Estos datos apoyan conceptos provenientes de observaciones geológicas y geofísicas de hundimientos o elevaciones del nivel de la tierra relacionado con los movimientos corticales asociados con las glaciaciones, subducción, rifting, carga sedimentaria, vulcanismo, y en general, plegamientos y fracturaciones asociados con las propiedades e historia de las placas corticales y especialmente sus movimientos durante la expansión del fondo oceánico. Esta conclusión puede no sorpender a los geólogos, pero puede ser inesperada para aquellos climatólogos y profanos que han influido demasiado fuertemente en la opinión pública acerca del efecto invernadero en el medio ambiente. Actualmente, no podemos presentar una estadística fiable de la ascensión del nivel del mar, pero esto puede no importarle mucho al dueño de propiedades en la costa, si su casa se inunda debido a un mar que asciende o a una costa que se hunda.

La actual magnitud del hundimiento relativo de los niveles de tierra a lo largo de las costas es menor de una décima parte de la que se dió durante los tiempos del Pleistoceno superior, cuando se derretían y retrocedían más de 10 veces la media de los últimos 50 m.y. durante la mayoría de los cuales, los efectos de la glaciación estuvo ausente y solamente fueron efectivos los agentes geológicos. Además, bastantes cambios costeros actuales (erosión de playas, colapso de acantilados costeros, desaparición de marismas, hundimiento de algunas ciudades costeras, situación de puertos, y otros) son debidos más a resultados indirectos y inesperados de la actividad humana que cambios de nivel marino o terrestre. La mayoría de la inestabilidad costera se puede atribuir a tectonicismo y a actividades humanas documentadas sin invocar al espectro del invernadero-calentamiento del clima, ni del colapso de los casquetes de hielo continentales.

Appendix I

Primary Tide-Gauge Stations

This is the list of tide-gauge stations having records longer than 10 years used in this book. Most of them, 586, are PSMSL (RLR) records, of which 57 were not used for the regional charts (Figs. 35 to 104) because they are too short in length or had low t-confidence level but they were used for the statistical summaries of Figures 36, 105, 106, and 108. Other records (termed METRIC here) were not published by PSMSL, but 49 of them were used in the regional charts where PSMSL records were scarce. Thus, the regional charts are based on a total of 578 tide-gauge records and the statistical summaries are based on 586 records. Place names are as listed in PSMSL or other primary references.

Country	Station	Latitude	Longitude	No. years of Record	Start year	End year	Regression slope mm/y.	t-confidence	Source
Antarctic	Almirante Brown	64°54′N	62°52′W	16	1957	1980	0.3	0.82	PSMSL
Iceland	Reykjavik	64°09′N	21°56′W	28	1956	1983	3.6	0.97	PSMSL
Spitsbergen	Barentsburg	78°04′N	15°14′E	40	1948	1987	−1.6	1.00	PSMSL
USSR (Arctic)	Russkaya Gavan	76°14′N	62°39′E	34	1953	1986	−2.0	1.00	PSMSL
USSR (Arctic)	Murmansk	68°58′N	33°03′E	34	1952	1985	−1.6	0.99	PSMSL
Norway	Vardo	70°20′N	31°06′E	19	1947	1965	−2.9	0.90	PSMSL
Norway	Honningsvag	70°59′N	25°59′E	15	1970	1984	0.3	0.68	PSMSL
Norway	Hammerfest	70°40′N	23°40′E	19	1955	1973	−0.3	0.72	PSMSL
Norway	Tromso	69°39′N	18°58′E	34	1952	1985	−0.1	0.98	PSMSL
Norway	Andenes	69°19′N	16°09′E	18	1938	1955	−1.8	0.77	PSMSL
Norway	Harstad	68°48′N	16°33′E	32	1953	1985	−0.1	0.98	PSMSL
Norway	Evenskjaer	68°35′N	16°33′E	24	1947	1970	−5.0	0.87	PSMSL
Norway	Narvik	68°26′N	17°25′E	38	1928	1973	−3.4	1.00	PSMSL
Norway	Kabelvag	68°13′N	14°29′E	38	1948	1985	−1.2	0.98	PSMSL
Norway	Bodo	67°17′N	14°23′E	24	1949	1973	−7.7	0.80	PSMSL
Norway	Bronnoysund	65°29′N	12°13′E	12	1956	1967	6.8	0.62	PSMSL
Norway	Rorvik	64°52′N	11°15′E	14	1972	1985	−113.7	0.51	PSMSL
Norway	Trondheim	63°26′N	10°26′E	39	1945	1985	−0.9	0.98	PSMSL
Norway	Heimsjo	63°26′N	09°07′E	38	1935	1973	−2.0	1.00	PSMSL
Norway	Kristiansund	63°07′N	07°44′E	34	1952	1985	0.3	0.97	PSMSL
Norway	Alesund	62°28′N	06°09′E	35	1945	1985	1.1	0.98	PSMSL
Norway	Kjolsdal	61°55′N	05°38′E	35	1935	1973	0.0	0.99	PSMSL
Norway	Maloy	61°56′N	05°07′E	37	1945	1985	0.7	0.99	PSMSL
Norway	Bergen	60°24′N	05°18′E	49	1883	1973	−0.8	1.00	PSMSL
Norway	Stavanger	58°58′N	05°44′E	42	1881	1973	0.3	1.00	PSMSL
Norway	Tregde	58°00′N	07°34′E	37	1935	1973	−0.1	1.00	PSMSL
Norway	Nevlunghavn	58°58′N	09°53′E	30	1935	1965	−1.7	0.99	PSMSL
Norway	Oslo	59°54′N	10°45′E	52	1885	1973	−4.1	1.00	PSMSL
Norway	Oscarsborg	59°40′N	10°37′E	29	1872	1973	−2.5	1.00	PSMSL
Sweden	Stromstad	58°57′N	11°11′E	60	1900	1965	−2.1	1.00	PSMSL
Sweden	Smogen	58°22′N	11°13′E	71	1911	1981	−2.5	1.00	PSMSL
Sweden	Goteborg-Klipp	57°43′N	11°57′E	81	1887	1968	−1.4	1.00	PSMSL
Sweden	Goteborg-Torsh	57°41′N	11°48′E	13	1969	1981	−0.3	0.81	PSMSL
Sweden	Varberg	57°06′N	12°13′E	95	1887	1981	−0.9	1.00	PSMSL
Sweden	Klagshamn	55°31′N	12°54′E	53	1929	1981	0.2	1.00	PSMSL
Sweden	Ystad	55°25′N	13°49′E	95	1887	1981	0.6	1.00	PSMSL
Sweden	Kungholmsfort	56°06′N	15°35′E	95	1887	1981	−0.3	1.00	PSMSL

Country	Station	Latitude	Longitude	No. years of Record	Start year	End year	Regression slope mm/y.	t-confidence	Source
Sweden	Olands Norra Udd	57°22′N	17°06′E	59	1923	1981	−1.3	1.00	PSMSL
Sweden	Landsort	58°45′N	17°52′E	95	1887	1981	−3.2	1.00	PSMSL
Sweden	Nedre Sodertalje	59°12′N	17°37′E	97	1869	1965	−3.4	1.00	PSMSL
Sweden	Stockholm	59°19′N	18°05′E	93	1889	1981	−4.2	1.00	PSMSL
Sweden	Bjorn	60°38′N	17°58′E	85	1892	1976	−6.1	1.00	PSMSL
Sweden	Nedre Gavle	60°40′N	17°10′E	70	1896	1965	−6.4	1.00	PSMSL
Sweden	Draghallan	62°20′N	17°28′E	77	1898	1974	−8.0	1.00	PSMSL
Sweden	Spikarna	62°22′N	17°32′E	14	1968	1981	−6.1	0.70	PSMSL
Sweden	Ratan	64°00′N	20°55′E	90	1892	1981	−8.0	1.00	PSMSL
Sweden	Furuogrund	64°55′N	21°14′E	66	1916	1981	−9.2	1.00	PSMSL
Finland	Kemi	65°44′N	24°33′E	58	1920	1977	−7.3	1.00	PSMSL
Finland	Oulu/Uleaborg	65°02′N	25°26′E	89	1889	1977	−6.1	1.00	PSMSL
Finland	Raahe/Brahestad	64°42′N	24°30′E	54	1922	1976	−7.5	0.99	PSMSL
Finland	Ykspihlaja	63°50′N	23°02′E	36	1889	1924	−6.6	0.99	PSMSL
Finland	Pietarsaari/Jako	63°42′N	22°42′E	65	1914	1978	−8.3	1.00	PSMSL
Finland	Vaasa/Vasa	63°06′N	21°34′E	96	1883	1978	−7.6	1.00	PSMSL
Finland	Ronnskar	63°04′N	20°48′E	61	1867	1936	−7.0	1.00	PSMSL
Finland	Kaskinen/Kasko	62°23′N	21°13′E	53	1926	1978	−7.3	1.00	PSMSL
Finland	Reposaari	61°37′N	21°27′E	14	1913	1926	−4.7	0.73	PSMSL
Finland	Mantyluoto	61°36′N	21°29′E	69	1910	1978	−6.6	1.00	PSMSL
Finland	Sappi	61°29′N	21°20′E	18	1919	1936	−10.4	0.81	PSMSL
Finland	Rauma/Raumo	61°08′N	21°29′E	44	1935	1978	−5.2	0.99	PSMSL
Finland	Lyokki	60°51′N	21°11′E	79	1858	1936	−5.4	1.00	PSMSL
Finland	Lypyrtti	60°36′N	21°14′E	78	1858	1936	−5.0	1.00	PSMSL
Finland	Turku/Abo	60°25′N	22°06′E	57	1922	1978	−4.6	1.00	PSMSL
Finland	Heligman	60°12′N	19°18′E	17	1920	1936	−8.9	0.83	PSMSL
Finland	Kobbaklintar	60°02′N	19°53′E	43	1885	1936	−7.1	0.90	PSMSL
Finland	Lemstrom	60°06′N	20°01′E	48	1889	1936	−4.5	1.00	PSMSL
Finland	Degerby	60°02′N	20°23′E	53	1924	1978	−4.4	1.00	PSMSL
Finland	Uto	59°47′N	21°22′E	71	1866	1936	−2.7	1.00	PSMSL
Finland	Tvarminne	59°51′N	23°15′E	16	1921	1936	−8.7	0.81	PSMSL
Finland	Jungfrusund	59°57′N	22°22′E	77	1858	1934	−3.0	1.00	PSMSL
Finland	Russaro	59°46′N	22°57′E	70	1866	1936	−2.9	1.00	PSMSL
Finland	Hanko/Hango	59°49′N	22°58′E	62	1897	1978	−3.6	1.00	PSMSL
Finland	Skuro	60°06′N	23°33′E	37	1900	1936	−0.9	0.99	PSMSL
Finland	Helsinki	60°09′N	24°58′E	99	1879	1977	−2.9	1.00	PSMSL
Finland	Soderskar	60°07′N	25°25′E	71	1866	1936	−1.8	1.00	PSMSL
Finland	Kotka	60°27′N	26°57′E	19	1908	1927	0.0	0.81	PSMSL
Finland	Hamina	60°34′N	27°11′E	51	1928	1978	−1.7	1.00	PSMSL
USSR (Baltic)	Sur Sari	60°05′N	26°59′E	18	1919	1936	−6.3	0.79	PSMSL
USSR (Baltic)	Vyborg	60°42′N	28°44′E	53	1889	1944	−1.3	1.00	PSMSL
USSR (Baltic)	Primorsk	60°21′N	28°37′E	19	1921	1939	−5.5	0.73	PSMSL
USSR (Baltic)	Tallinn	59°27′N	24°48′E	11	1928	1938	−5.6	0.66	PSMSL
USSR (Baltic)	Vilsandi	58°23′N	21°49′E	11	1928	1938	−4.9	0.65	PSMSL
USSR (Baltic)	Daugavgriva	57°03′N	24°02′E	63	1872	1938	1.1	1.00	PSMSL
USSR (Baltic)	Liepaja	56°32′N	20°59′E	63	1865	1936	0.8	1.00	PSMSL
USSR (Baltic)	Kaliningrad/Pion	54°57′N	20°13′E	54	1926	1987	1.7	1.00	PSMSL
Poland	Gdansk/Nowy Port	54°24′N	18°41′E	32	1951	1982	2.6	0.98	PSMSL
Poland	Gdynia	54°32′N	18°33′E	10	1931	1966	1.7	0.94	PSMSL
Poland	Hel	54°36′N	18°48′E	25	1901	1966	0.5	1.00	PSMSL
Poland	Wladyslawowo	54°48′N	18°25′E	32	1951	1982	1.2	0.96	PSMSL
Poland	Ustka	54°35′N	16°52′E	32	1951	1982	2.0	0.97	PSMSL
Poland	Kolobrzeg	54°11′N	15°03′E	32	1951	1982	0.4	0.98	PSMSL
Poland	Swinoujscie	53°55′N	14°14′E	32	1951	1982	1.7	0.98	PSMSL
Denmark	Gedser	54°34′N	11°58′E	72	1898	1969	0.9	1.00	PSMSL
Denmark	Rodbyhavn	54°39′N	11°21′E	15	1955	1969	0.2	0.86	PSMSL
Denmark	Kobenhavn	55°41′N	12°36′E	81	1889	1969	0.2	1.00	PSMSL
Denmark	Hornbaek	56°06′N	12°28′E	72	1898	1969	−0.1	1.00	PSMSL
Denmark	Korsor	55°20′N	11°08′E	73	1897	1969	0.7	1.00	PSMSL
Denmark	Slipshavn	55°17′N	10°50′E	74	1896	1969	0.9	1.00	PSMSL
Denmark	Mommark	54°56′N	10°04′E	22	1925	1967	0.2	1.00	PSMSL
Denmark	Fredericia	55°34′N	09°46′E	81	1889	1969	1.0	1.00	PSMSL
Denmark	Aarhus	56°09′N	10°13′E	82	1888	1969	0.5	1.00	PSMSL

Country	Station	Latitude	Longitude	No. years of Record	Start year	End year	Regression slope mm/y.	t-confidence	Source
Denmark	Frederikshavn	57°26′N	10°34′E	75	1894	1969	0.4	1.00	PSMSL
Denmark	Hirtshals	57°36′N	09°57′E	78	1892	1969	−0.5	1.00	PSMSL
Denmark	Esbjerg	55°28′N	08°27′E	81	1889	1969	1.1	1.00	PSMSL
East Germany	Sassnitz	54°32′N	13°40′E	35	1936	1980	0.2	1.00	METRIC
East Germany	Arkona	54°41′N	13°26′E	54	1882	1935	0.4	1.00	METRIC
East Germany	Warnemunde	54°11′N	12°05′E	62	1882	1943	1.4	1.00	METRIC
East Germany	Warnemunde II	54°11′N	12°05′E	32	1944	1980	1.0	0.99	METRIC
East Germany	Wismar I	53°54′N	11°28′E	58	1882	1943	1.8	1.00	METRIC
East Germany	Wismar II	53°54′N	11°28′E	31	1944	1980	1.4	1.00	METRIC
West Germany	Travemunde	53°58′N	10°53′E	27	1885	1881	1.1	0.99	METRIC
West Germany	Travemunde II	53°58′N	10°53′E	60	1882	1943	−0.1	1.00	METRIC
West Germany	Marienleuchte	54°30′N	11°15′E	58	1882	1943	1.4	1.00	METRIC
West Germany	Cuxhaven	53°52′N	08°43′E	21	1938	1959	2.9	0.87	METRIC
West Germany	Bremerhaven	53°33′N	08°34′E	46	1898	1943	0.2	1.00	METRIC
Netherlands	Delfzijl	53°20′N	06°56′E	117	1865	1981	1.5	1.00	METRIC
Netherlands	Terschelling	53°22′N	05°13′E	61	1921	1981	0.4	1.00	METRIC
Belgium	Zeebrugge	51°21′N	03°12′E	21	1942	1980	1.7	0.96	PSMSL
Belgium	Oostende	51°14′N	02°55′E	40	1937	1980	0.7	1.00	PSMSL
Belgium	Nieuwpoort	51°09′N	02°43′E	16	1943	1980	0.3	1.00	PSMSL
United Kingdom	Lerwick	60°09′N	01°08′W	31	1957	1987	−1.3	1.00	PSMSL
United Kingdom	Invergordon	57°41′N	04°10′W	13	1959	1971	−3.9	0.79	PSMSL
United Kingdom	Aberdeen I	57°09′N	02°05′W	49	1931	1983	0.2	1.00	PSMSL
United Kingdom	Aberdeen II	57°09′N	02°05′W	104	1862	1965	0.6	1.00	PSMSL
United Kingdom	Rosyth	56°01′N	03°27′W	24	1964	1987	1.5	0.95	PSMSL
United Kingdom	North Shields	55°00′N	01°27′W	88	1895	1985	2.0	1.00	PSMSL
United Kingdom	Immingham	53°37′N	00°11′W	24	1959	1982	1.7	0.96	PSMSL
United Kingdom	Lowestoft	52°28′N	01°19′E	30	1955	1984	0.7	1.00	PSMSL
United Kingdom	Southend	51°31′N	00°44′E	53	1929	1982	1.5	1.00	PSMSL
United Kingdom	Sheerness	51°27′N	00°45′E	86	1832	1984	1.5	1.00	PSMSL
United Kingdom	Dover	51°07′N	01°19′E	24	1961	1985	2.4	0.99	PSMSL
United Kingdom	Portsmouth	50°48′N	01°07′W	27	1961	1987	3.4	1.00	PSMSL
United Kingdom	Devonport	50°22′N	04°11′W	27	1961	1987	1.2	0.89	PSMSL
United Kingdom	Newlyn	50°06′N	05°33′W	71	1915	1985	1.8	1.00	PSMSL
United Kingdom	Milford Haven	51°42′N	05°01′W	19	1963	1983	−5.8	0.73	PSMSL
United Kingdom	Birkenhead	53°24′N	03°01′W	18	1955	1972	2.1	0.91	PSMSL
United Kingdom	Liverpool Prince	53°25′N	03°00′W	25	1958	1982	2.3	0.98	PSMSL
United Kingdom	Douglas	54°09′N	04°28′W	35	1938	1977	0.2	1.00	PSMSL
United Kingdom	Portpatrick	54°51′N	05°07′W	16	1968	1983	9.1	0.90	PSMSL
United Kingdom	Millport	55°45′N	04°56′W	12	1968	1985	3.1	0.78	PSMSL
United Kingdom	Stornoway	58°12′N	06°23′W	17	1967	1985	2.8	0.86	PSMSL
United Kingdom	Belfast	54°36′N	05°55′W	16	1957	1979	−0.9	0.81	PSMSL
France (Atlantic)	Dunkerque	51°03′N	02°22′E	28	1942	1983	2.7	1.00	PSMSL
France (Atlantic)	Calais	50°58′N	01°51′E	21	1941	1983	−3.8	0.85	PSMSL
France (Atlantic)	Dieppe	49°56′N	01°05′E	29	1954	1983	6.7	0.97	PSMSL
France (Atlantic)	Le Havre	49°26′N	00°06′E	19	1941	1983	3.1	0.93	PSMSL
France (Atlantic)	Cherbourg	49°39′N	01°37′W	12	1974	1985	0.7	0.77	PSMSL
France (Atlantic)	Le Conquet	48°22′N	04°47′W	14	1971	1984	4.9	0.91	PSMSL
France (Atlantic)	Brest	48°23′N	04°30′W	114	1807	1985	0.7	1.00	PSMSL
France (Atlantic)	St. Nazaire	47°16′N	02°12′W	21	1941	1986	−1.3	0.85	PSMSL
France (Atlantic)	Pte. St. Gilda	47°08′N	02°15′W	21	1964	1986	2.2	0.91	PSMSL
France (Atlantic)	La Pallice	46°10′N	01°13′W	21	1956	1979	−0.7	0.95	PSMSL
France (Atlantic)	St. Jean-De-Luz	43°24′N	01°41′W	28	1942	1984	3.9	0.94	PSMSL
Spain (Atlantic)	Pasajes	43°19′N	01°55′W	16	1948	1963	4.9	0.82	PSMSL
Spain (Atlantic)	Santander I	43°28′N	03°48′W	23	1944	1966	2.1	0.87	PSMSL
Spain (Atlantic)	Santander II	43°28′N	03°46′W	12	1963	1974	−3.3	0.70	PSMSL
Spain (Atlantic)	La Coruña I	43°22′N	08°24′W	26	1943	1968	5.4	0.89	PSMSL
Spain (Atlantic)	La Coruña II	43°22′N	08°24′W	24	1955	1978	−0.1	0.99	PSMSL
Spain (Atlantic)	Vigo	42°19′N	08°44′W	21	1943	1963	2.9	0.87	PSMSL
Portugal	Cascais	38°41′N	09°25′W	101	1882	1985	1.2	1.00	PSMSL
Portugal	Lagos	37°06′N	08°40′W	73	1908	1985	1.7	1.00	PSMSL
Gibraltar	Gibraltar	36°07′N	05°21′W	27	1961	1987	−1.7	0.93	PSMSL
Spain (Med.)	Tarifa	36°00′N	05°36′W	16	1944	1965	−11.3	0.83	PSMSL
Spain (Med.)	Alicante I	38°20′N	00°29′W	17	1952	1969	1.6	0.98	PSMSL

Country	Station	Latitude	Longitude	No. years of Record	Start year	End year	Regression slope mm/y.	t-confidence	Source
Spain (Med.)	Alicante II	38°20′N	00°29′W	19	1960	1978	−1.6	1.00	PSMSL
France (Med.)	Marseille	43°18′N	05°21′E	77	1885	1963	1.7	1.00	PSMSL
Italy (Sardinia)	La Maddalena	41°14′N	09°22′E	17	1896	1913	0.9	0.98	PSMSL
Italy (Sardinia)	Cagliari	39°12′N	09°10′E	39	1896	1934	1.8	1.00	PSMSL
Italy (Med.)	Porto Maurizio	43°52′N	08°01′E	27	1896	1922	1.2	1.00	PSMSL
Italy (Med.)	Genova	44°24′N	08°54′E	82	1884	1985	1.3	1.00	PSMSL
Italy (Med.)	Civitavecchia	42°03′N	11°49′E	27	1896	1922	0.6	1.00	PSMSL
Italy (Med.)	Napoli (Arsenale)	40°52′N	14°16′E	24	1899	1922	2.6	0.99	PSMSL
Italy (Med.)	Napoli (Mandracci)	40°52′N	14°16′E	27	1896	1922	2.4	1.00	PSMSL
Italy (Sicily)	Palermo	38°08′N	13°20′E	27	1896	1922	0.6	1.00	PSMSL
Italy (Sicily)	Messina	38°12′N	15°34′E	26	1897	1923	19.4	0.79	PSMSL
Italy (Sicily)	Capo Passero	36°40′N	15°18′E	12	1957	1969	7.1	0.74	PSMSL
Italy (Sicily)	Catania	37°30′N	15°08′E	12	1960	1971	−2.9	0.76	PSMSL
Italy (Adriatic)	Venezia (Arsenale)	45°25′N	12°21′E	25	1889	1913	2.6	0.98	PSMSL
Italy (Adriatic)	Venezia (S. Stefano)	45°25′N	12°20′E	25	1896	1920	3.7	0.96	PSMSL
Italy (Adriatic)	Venezia (Punta)	45°26′N	12°20′E	15	1953	1967	4.7	0.79	PSMSL
Italy (Adriatic)	Trieste	45°39′N	13°45′E	77	1905	1987	1.3	1.00	PSMSL
Yugoslavia	Koper	45°33′N	13°44′E	24	1962	1985	0.6	0.99	PSMSL
Yugoslavia	Rovinj	45°05′N	13°38′E	31	1955	1985	0.2	1.00	PSMSL
Yugoslavia	Bakar	45°18′N	14°32′E	48	1930	1986	1.2	1.00	PSMSL
Yugoslavia	Split Rt. Marjana	43°30′N	16°23′E	34	1952	1985	0.3	1.00	PSMSL
Yugoslavia	Split Harbour	43°30′N	16°26′E	32	1954	1985	−0.2	1.00	PSMSL
Yugoslavia	Dubrovnik	42°40′N	18°04′E	30	1956	1985	0.3	1.00	PSMSL
Yugoslavia	Bar	42°05′N	19°05′E	22	1964	1985	3.8	0.89	PSMSL
Greece	Patrai	38°14′N	21°44′E	14	1969	1984	22.9	0.78	PSMSL
Greece	Katakolon	37°38′N	21°19′E	16	1969	1984	0.2	0.81	PSMSL
Greece	Piraievs	37°56′N	23°37′E	14	1969	1984	0.7	0.73	PSMSL
Greece	Thessaloniki	40°37′N	23°02′E	15	1969	1984	2.2	0.72	PSMSL
Greece	Kavalla	40°55′N	24°25′E	16	1969	1984	−3.5	0.61	PSMSL
Greece	Siros	37°26′N	24°55′E	14	1969	1984	−6.5	0.74	PSMSL
Greece	Soudhas	35°30′N	24°03′E	16	1969	1984	−3.8	0.75	PSMSL
USSR (Black Sea)	Port Tuapse	44°06′N	39°04′E	71	1917	1987	2.1	1.00	PSMSL
Turkey	Izmir	38°24′N	27°10′E	35	1937	1971	4.6	0.96	METRIC
Turkey	Antalya	36°53′N	30°42′E	36	1936	1972	−3.8	0.95	METRIC
Israel	Haifa	32°30′N	34°35′E	19	1957	1975	−2.8	0.87	METRIC
Israel	Jaffa	32°03′N	34°28′E	27	1955	1981	−0.5	1.00	METRIC
Israel	Ashdod	31°29′N	34°23′E	26	1958	1983	−0.5	0.98	METRIC
Egypt	Port Said	31°15′N	32°18′E	24	1923	1946	−4.8	0.98	PSMSL
Egypt	Alexandria	30°51′N	29°53′E	19	1958	1976	−0.7	0.96	METRIC
Egypt	Kabret	30°16′N	32°30′E	15	1923	1941	0.6	0.98	METRIC
Egypt	Port Thewfik	29°57′N	32°34′E	23	1923	1946	0.7	0.98	METRIC
Tunesia	Sfax	33°44′N	10°46′E	41	1910	1959	6.1	1.00	METRIC
Tunesia	Sousse	35°49′N	10°37′E	42	1910	1959	3.2	1.00	METRIC
Tunesia	La Goulette	36°49′N	10°18′E	69	1889	1956	0.7	1.00	METRIC
Algeria	Bône	36°54′N	07°46′E	40	1908	1959	1.3	0.99	METRIC
Algeria	Alger	36°47′N	03°04′E	38	1905	1959	−0.4	1.00	METRIC
Algeria	Oran	35°43′N	03°09′W	39	1889	1956	2.3	1.00	METRIC
Spanish N. Africa	Ceuta	35°54′N	05°19′W	21	1944	1964	0.4	0.90	PSMSL
Spain (Canary Is)	Santa Cruz De Tenerife	28°41′N	17°45′W	12	1949	1960	−24.3	0.73	PSMSL
Spain (Canary Is)	Santa Cruz De Tenerife	28°29′N	16°14′W	41	1927	1974	2.6	1.00	PSMSL
Senegal	Dakar	14°40′N	17°25′W	15	1942	1966	1.4	0.99	PSMSL
Ghana	Takoradi	45°30′N	01°45′W	58	1929	1986	−1.9	0.99	PSMSL
Ghana	Tema	05°37′N	00°00′W	20	1963	1982	1.0	0.91	PSMSL
Namibia	Walvis Bay	22°57′S	14°30′E	22	1958	1986	2.2	0.95	PSMSL
Namibia	Lüderitz	26°38′S	15°09′E	29	1958	1986	2.6	0.99	PSMSL
South Africa	Port Nolloth	29°15′S	16°52′E	26	1959	1986	1.5	1.00	PSMSL
South Africa	Granger Bay	33°54′S	18°25′E	20	1967	1986	4.0	0.92	PSMSL
South Africa	Table Bay Harbour	33°55′S	18°26′E	15	1967	1972	0.1	0.75	PSMSL
South Africa	Simons Bay	34°11′S	18°26′E	28	1957	1986	1.1	1.00	PSMSL
South Africa	Mossel Bay	34°11′S	22°09′E	27	1958	1986	−0.4	0.99	PSMSL
South Africa	Knysna	34°05′S	23°03′E	27	1960	1986	1.6	0.96	PSMSL
Mauritius	Port Louis	20°09′S	57°29′E	19	1942	1965	3.0	0.99	PSMSL
Madagascar	Tamatave	18°09′S	49°26′E	22	1931	1957	−3.5	0.85	METRIC

Country	Station	Latitude	Longitude	No. years of Record	Start year	End year	Regression slope mm/y.	t-confidence	Source
Aden	Aden	12°47′N	44°59′E	66	1879	1969	1.2	1.00	PSMSL
Pakistan	Karachi	24°48′N	66°58′E	17	1916	1948	0.3	1.00	PSMSL
India	Kandla	23°00′N	70°14′E	32	1950	1982	9.4	0.97	PSMSL
India	Bhaunagar	21°45′N	72°14′E	29	1937	1965	−2.1	0.74	PSMSL
India	Bombay	18°55′N	72°50′E	104	1878	1982	0.9	1.00	PSMSL
India	Mangalore	12°51′N	74°50′E	21	1953	1976	−1.3	0.99	PSMSL
India	Cochin	09°58′N	76°15′E	43	1939	1982	2.0	1.00	PSMSL
India	Tuticorin	08°45′N	78°11′E	16	1964	1980	−6.3	0.69	PSMSL
India	Madras	13°06′N	80°18′E	35	1916	1982	0.4	1.00	PSMSL
India	Vishakhapatnam	17°41′N	83°17′E	44	1937	1982	0.6	1.00	PSMSL
India	Haldia	22°02′N	88°06′E	13	1970	1982	−3.9	0.71	PSMSL
India	Saugor/Sagar	21°39′N	88°03′E	45	1937	1982	−4.2	0.97	PSMSL
India	Diamond Harbour	22°12′N	88°10′E	35	1948	1982	7.3	0.99	PSMSL
India	Calcutta	22°33′N	88°18′E	50	1932	1982	6.9	0.97	PSMSL
India	Kidderpore	22°32′N	88°20′E	24	1881	1931	−6.1	0.96	PSMSL
India	Tribeni	22°59′N	88°24′E	20	1962	1982	48.3	0.58	PSMSL
Burma	Rangoon	16°46′N	96°10′E	25	1916	1962	3.4	0.97	PSMSL
Burma	Moulmein	16°29′N	97°37′E	11	1954	1964	−1.0	0.65	PSMSL
Burma	Amherst	16°05′N	97°34′E	11	1954	1964	6.4	0.69	PSMSL
Andaman Islands	Port Blair	11°41′N	92°46′E	16	1916	1964	2.5	1.00	PSMSL
Indonesia	Belawan	03°55′N	98°43′E	7	1925	1931	7.8	<0.8	METRIC
Indonesia	Padang	00°58′N	100°20′E	7	1925	1931	13.8	<0.8	METRIC
Indonesia	Djakarta	06°06′N	106°54′E	7	1925	1931	8.7	<0.8	METRIC
Indonesia	Tjilatjap	07°44′N	109°00′E	7	1925	1931	11.9	<0.8	METRIC
Indonesia	Semarang	07°00′N	110°24′E	7	1925	1931	5.5	<0.8	METRIC
Indonesia	Djamoeanrif	06°58′N	112°45′E	7	1925	1931	−6.9	<0.8	METRIC
Indonesia	Sembilangan	07°06′N	112°42′E	7	1925	1931	2.4	<0.8	METRIC
Indonesia	Soerabaja	07°12′N	112°36′E	7	1925	1931	12.8	<0.8	METRIC
Indonesia	Makassar	05°06′N	119°24′E	7	1925	1931	0.7	<0.8	METRIC
Singapore	Victoria Dock	01°16′N	103°51′E	16	1966	1981	1.3	0.95	PSMSL
Singapore	Sultan Shoal	01°14′N	103°39′E	19	1969	1987	2.0	0.83	PSMSL
Singapore	Jurong	01°18′N	103°43′E	18	1970	1987	−3.6	0.92	PSMSL
Singapore	Sembawang	01°28′N	103°50′E	24	1954	1987	−4.3	1.00	PSMSL
Thailand	Ko Taphao Noi	07°50′N	98°26′E	47	1940	1986	−0.5	0.96	PSMSL
Thailand	Phrachuap Kirikh	11°48′N	99°49′E	47	1940	1986	−1.1	0.97	PSMSL
Thailand	Fort Phrachula	13°33′N	100°35′E	45	1940	1986	11.5	0.99	PSMSL
Thailand	Ko Sichang	13°09′N	100°49′E	47	1940	1986	0.3	1.00	PSMSL
Macau	Macau	22°12′N	113°33′E	57	1925	1981	0.3	1.00	PSMSL
China	Xiamen	24°27′N	118°04′E	30	1954	1983	11.6	0.86	PSMSL
China	Yantai	37°32′N	121°23′E	30	1954	1983	0.0	0.84	PSMSL
China	Qinhuangdao	39°54′N	119°36′E	34	1950	1983	−3.1	1.00	PSMSL
China	Huludao	40°44′N	121°00′E	13	1960	1977	5.5	0.76	METRIC
China	Hai-Ling Tao	21°35′N	111°49′E	13	1960	1977	2.9	0.81	METRIC
China	Wei Chou Tao	21°03′N	109°08′E	13	1960	1977	11.4	0.90	METRIC
China	Shanghai	31°21′N	121°30′E	23	1954	1976	2.6	0.89	METRIC
China	Tsingdao	36°05′N	120°23′E	28	1951	1978	−0.2	1.00	METRIC
China	Tang-Ku	38°57′N	117°43′E	28	1951	1978	1.5	0.96	METRIC
Macau	Macau	22°12′N	113°33′E	31	1937	1967	0.1	0.97	METRIC
Hong Kong	Tai Po	22°27′N	114°11′E	20	1963	1982	1.0	0.82	METRIC
Hong Kong	Chi Ma Wan, Lantai	22°14′N	114°00′E	21	1961	1984	6.0	0.78	PSMSL
Hong Kong	North Point	22°18′N	114°12′E	38	1929	1985	−1.1	0.99	PSMSL
Taiwan	Chilung	25°09′N	120°16′E	21	1904	1924	0.4	0.96	METRIC
Taiwan	Kaohsiung	22°37′N	114°12′E	36	1904	1943	−2.4	1.00	METRIC
Korea	Inchon	37°28′N	126°37′E	28	1960	1987	−3.7	0.99	PSMSL
Korea	Mokpo	34°47′N	126°23′E	28	1960	1987	6.5	0.91	PSMSL
Korea	Jeju	33°31′N	126°32″E	19	1966	1987	4.9	0.97	PSMSL
Korea	Yosu	34°45′N	127°46′E	18	1970	1987	−0.4	0.96	PSMSL
Korea	Chungmu	34°49′N	128°26′E	10	1978	1987	4.3	0.97	PSMSL
Korea	Chinhae	35°09′N	128°39′E	17	1960	1976	4.7	0.97	PSMSL
Korea	Pusan	35°06′N	129°02′E	26	1960	1987	0.6	1.00	PSMSL
Korea	Ulsan	35°30′N	126°23′E	26	1962	1987	2.1	0.97	PSMSL
Korea	Pohang	36°01′N	129°24′E	10	1978	1987	2.5	0.98	PSMSL
Korea	Mugho	37°33′N	129°07′E	23	1965	1987	−3.0	0.99	PSMSL

APPENDIX I

Country	Station	Latitude	Longitude	No. years of Record	Start year	End year	Regression slope mm/y.	t-confidence	Source
Korea	Sogcho	38°12′N	128°36′E	10	1978	1987	3.9	0.90	PSMSL
USSR (Pacific)	Yuzhno Kurilsk	44°10′N	145°52′E	38	1948	1985	3.2	1.00	PSMSL
USSR (Pacific)	Nagaeva Bay	59°44′N	150°42′E	31	1957	1987	−0.6	0.98	PSMSL
USSR (Pacific)	Petropavlovsk	53°01′N	158°38′E	31	1957	1987	2.5	1.00	PSMSL
Japan (Hokkaido)	Monbetu	44°21′N	143°22′E	24	1955	1978	1.3	0.98	PSMSL
Japan (Hokkaido)	Abashiri	44°01′N	144°17′E	16	1965	1980	−3.5	0.91	PSMSL
Japan (Hokkaido)	Hanasaki	43°17′N	145°35′E	21	1957	1977	10.0	0.92	PSMSL
Japan (Hokkaido)	Kushiro	42°58′N	144°23′E	35	1947	1982	10.2	1.00	PSMSL
Japan (Hokkaido)	Hakodate I	41°47′N	140°44′E	25	1961	1985	1.6	1.00	PSMSL
Japan (Hokkaido)	Hakodate II	41°46′N	140°43′E	16	1951	1966	−6.4	0.96	PSMSL
Japan (Hokkaido)	Oshoro I	43°13′N	140°52′E	32	1930	1962	0.1	1.00	PSMSL
Japan (Hokkaido)	Oshoro II	43°12′N	140°52′E	22	1963	1984	0.2	0.99	PSMSL
Japan (Honshu)	Hachinohe I	40°32′N	141°32′E	30	1941	1970	2.2	1.00	PSMSL
Japan (Honshu)	Hachinohe II	40°32′N	141°32′E	14	1970	1983	−3.6	0.91	PSMSL
Japan (Honshu)	Miyako I	39°38′N	141°58′E	26	1941	1966	5.8	1.00	PSMSL
Japan (Honshu)	Miyako II	39°39′N	141°58′E	20	1966	1985	1.0	0.96	PSMSL
Japan (Honshu)	Kamaisi II	39°16′N	141°54′E	12	1973	1984	15.7	0.80	PSMSL
Japan (Honshu)	Ofunato II	39°01′N	141°46′E	13	1973	1985	5.5	0.88	PSMSL
Japan (Honshu)	Ayukawa	38°18′N	141°31′E	27	1958	1985	5.3	1.00	PSMSL
Japan (Honshu)	Soma	37°50′N	140°58′E	12	1973	1984	2.3	0.87	PSMSL
Japan (Honshu)	Onahama	36°56′N	140°55′E	35	1951	1985	0.7	0.99	PSMSL
Japan (Honshu)	Katsuura	35°08′N	140°15′E	18	1967	1984	0.4	1.00	PSMSL
Japan (Honshu)	Mera	34°55′N	139°50′E	53	1931	1985	1.1	1.00	PSMSL
Japan (Honshu)	Tokyo II	35°39′N	139°46′E	18	1968	1985	−0.1	0.95	PSMSL
Japan (Honshu)	Sibaura	35°38′N	139°45′E	25	1960	1984	−0.4	0.99	PSMSL
Japan (Honshu)	Yokosuka	35°17′N	139°39′E	30	1954	1984	5.2	0.99	PSMSL
Japan (Honshu)	Aburatsubo	35°09′N	139°37′E	54	1930	1984	4.4	0.99	PSMSL
Japan (Honshu)	Miyake Shima	34°04′N	139°29′E	20	1964	1984	−4.7	0.75	PSMSL
Japan (Honshu)	Kozu Shima	34°12′N	139°08′E	20	1964	1984	13.9	0.64	PSMSL
Japan (Honshu)	Okada	34°47′N	139°24′E	21	1965	1985	0.8	0.96	PSMSL
Japan (Honshu)	Ito II	34°54′N	139°08′E	12	1973	1984	−30.5	0.86	PSMSL
Japan (Honshu)	Minami Izu	34°37′N	138°53′E	21	1964	1984	−0.2	0.96	PSMSL
Japan (Honshu)	Uchiura	35°01′N	138°54′E	42	1944	1985	−0.9	1.00	PSMSL
Japan (Honshu)	Shimizu-Minato	35°01′N	138°30′E	29	1957	1985	5.5	0.97	PSMSL
Japan (Honshu)	Omaezaki II	34°36′N	138°14′E	16	1970	1985	6.0	0.79	PSMSL
Japan (Honshu)	Maisaka	34°41′N	137°37′E	21	1965	1985	−4.0	0.91	PSMSL
Japan (Honshu)	Onisaki	34°54′N	136°50′E	22	1963	1984	6.7	0.90	PSMSL
Japan (Honshu)	Nagoya	35°05′N	136°53′E	23	1957	1979	8.2	0.86	PSMSL
Japan (Honshu)	Toba	34°28′N	136°51′E	28	1950	1977	24.3	0.95	PSMSL
Japan (Honshu)	Owase	34°04′N	136°13′E	17	1966	1985	−2.2	0.83	PSMSL
Japan (Honshu)	Uragami	33°33′N	135°54′E	21	1965	1985	3.4	0.87	PSMSL
Japan (Honshu)	Kushimoto	33°28′N	135°46′E	29	1957	1985	0.5	0.99	PSMSL
Japan (Honshu)	Shirahama	33°41′N	135°23′E	20	1966	1985	−5.0	0.91	PSMSL
Japan (Honshu)	Kainan	34°09′N	135°12′E	32	1953	1984	1.4	1.00	PSMSL
Japan (Honshu)	Wakayama	34°13′N	135°09′E	24	1957	1980	2.2	0.95	PSMSL
Japan (Honshu)	Tan-Nowa	34°20′N	135°11′E	29	1957	1985	0.9	0.96	PSMSL
Japan (Honshu)	Osaka	34°39′N	135°26′E	21	1965	1985	6.6	0.82	PSMSL
Japan (Honshu)	Kobe	34°41′N	135°11′E	27	1957	1983	6.4	0.90	PSMSL
Japan (Honshu)	Sumoto	34°20′N	134°55′E	21	1965	1985	1.3	0.88	PSMSL
Japan (Honshu)	Uno	34°30′N	133°57′E	29	1957	1985	5.8	0.97	PSMSL
Japan (Honshu)	Kure IV	34°14′N	132°33′E	17	1968	1984	0.5	0.92	PSMSL
Japan (Honshu)	Hirosima	34°21′N	132°28′E	22	1963	1984	3.6	0.97	PSMSL
Japan (Honshu)	Tokuyama I	34°02′N	131°48′E	17	1951	1967	−2.9	0.80	PSMSL
Japan (Honshu)	Tokuyama II	34°02′N	131°48′E	17	1968	1984	−2.1	0.86	PSMSL
Japan (Shikoku)	Kochi II	33°30′N	133°35′E	15	1966	1980	2.8	0.80	PSMSL
Japan (Shikoku)	Kure I	33°20′N	133°15′E	13	1972	1984	−15.8	0.82	PSMSL
Japan (Shikoku)	Tosa Shimizu	32°47′N	132°58′E	28	1957	1985	−0.4	0.99	PSMSL
Japah (Shikoku)	Uwajima I	33°14′N	132°33′E	12	1957	1968	−1.5	0.76	PSMSL
Japan (Shikoku)	Uwajima II	33°14′N	132°33′E	18	1968	1985	−5.5	0.94	PSMSL
Japan (Shikoku)	Matsuyama II	33°52′N	132°42′E	22	1964	1985	1.4	0.97	PSMSL
Japan (Shikoku)	Takamatsu	34°21′N	134°03′E	23	1957	1979	6.4	0.95	PSMSL
Japan (Shikoku)	Komatsushima	34°00′N	134°35′E	28	1958	1985	1.3	0.98	PSMSL
Japan (Shikoku)	Murotomisaki	33°16′N	134°10′E	11	1975	1985	−6.8	0.79	PSMSL

Country	Station	Latitude	Longitude	No. years of Record	Start year	End year	Regression slope mm/y.	t-confidence	Source
Japan (Kyushu)	Mozi	33°57′N	130°58′E	27	1958	1984	0.5	1.00	PSMSL
Japan (Kyushu)	Oita II	33°16′N	131°41′E	14	1971	1984	−7.8	0.91	PSMSL
Japan (Kyushu)	Hosojima	32°26′N	131°40′E	55	1930	1984	−0.6	1.00	PSMSL
Japan (Kyushu)	Aburatsu	31°35′N	131°25′E	26	1960	1985	−0.2	0.99	PSMSL
Japan (Kyushu)	Nisinoomote	30°44′N	131°00′E	20	1965	1984	0.6	0.98	PSMSL
Japan (Kyushu)	Odomari	31°01′N	130°41′E	20	1965	1984	−2.5	0.94	PSMSL
Japan (Kyushu)	Kagoshima I	31°36′N	130°34′E	16	1957	1972	−2.6	0.83	PSMSL
Japan (Kyushu)	Akune	32°01′N	130°12′E	15	1970	1984	−5.7	0.94	PSMSL
Japan (Kyushu)	Misumi	32°37′N	130°27′E	29	1957	1985	−0.2	1.00	PSMSL
Japan (Kyushu)	Oura	32°58′N	130°13′E	21	1965	1985	0.7	0.97	PSMSL
Japan (Kyushu)	Nagasaki	32°44′N	129°52′E	21	1965	1985	0.9	0.99	PSMSL
Japan (Kyushu)	Fukue	32°42′N	128°51′E	21	1965	1985	−87.9	0.53	PSMSL
Japan (Kyushu)	Sasebo I	33°10′N	129°43′E	16	1951	1966	−0.5	0.91	PSMSL
Japan (Kyushu)	Sasebo II	33°09′N	129°44′E	19	1966	1984	−0.1	0.94	PSMSL
Japan (Kyushu)	Kariya	33°28′N	129°51′E	13	1972	1984	−4.2	0.89	PSMSL
Japan (Kyushu)	Izuhara	34°12′N	129°18′E	33	1951	1983	−5.5	0.91	PSMSL
Japan (Kyushu)	Hakata	33°36′N	130°24′E	20	1965	1984	−1.6	0.98	PSMSL
Japan (Amami G.)	Nase II	28°23′N	129°30′E	19	1962	1980	6.8	0.88	PSMSL
Japan (Amami G.)	Naha	26°13′N	127°40′E	11	1975	1985	−5.6	0.81	PSMSL
Japan (Amami G.)	Ishigaki	24°20′N	124°09′E	11	1975	1985	−4.0	0.80	PSMSL
Japan (Honshu)	Shimonoseki I	33°58′N	130°57′E	29	1957	1985	−0.4	0.99	PSMSL
Japan (Honshu)	Hagi	34°26′N	131°25′E	14	1971	1984	−2.3	0.95	PSMSL
Japan (Honshu)	Tonoura	34°54′N	132°04′E	90	1894	1984	0.3	1.00	PSMSL
Japan (Honshu)	Sakai	35°33′N	133°15′E	29	1957	1985	−0.4	1.00	PSMSL
Japan (Honshu)	Saigo	36°12′N	133°20′E	21	1965	1985	−0.5	1.00	PSMSL
Japan (Honshu)	Tajiri	35°35′N	134°19′E	19	1966	1984	3.4	0.98	PSMSL
Japan (Honshu)	Miyazu	35°32′N	135°11′E	13	1957	1969	−4.2	0.84	PSMSL
Japan (Honshu)	Maizuru I	35°29′N	135°24′E	31	1951	1981	0.1	1.00	PSMSL
Japan (Honshu)	Maizuru II	35°28′N	135°23′E	11	1975	1985	−0.4	0.88	PSMSL
Japan (Honshu)	Mikuni	36°15′N	136°09′E	18	1967	1984	−1.3	0.97	PSMSL
Japan (Honshu)	Wajima	37°24′N	136°54′E	55	1930	1984	−1.3	1.00	PSMSL
Japan (Honshu)	Kashiwazaki	37°21′N	138°31′E	30	1955	1984	−1.0	1.00	PSMSL
Japan (Honshu)	Ogi	37°49′N	138°17′E	12	1973	1984	−0.4	0.81	PSMSL
Japan (Honshu)	Awa Sima	38°28′N	139°15′E	20	1965	1984	1.9	0.86	PSMSL
Japan (Honshu)	Nezugaseki	38°34′N	139°33′E	29	1955	1984	1.0	0.95	PSMSL
Japan (Honshu)	Oga	39°56′N	139°42′E	15	1970	1984	−5.0	0.94	PSMSL
Japan (Honshu)	Iwasaki	40°35′N	139°55′E	13	1958	1970	−3.8	0.87	PSMSL
Japan (Honshu)	Ominato	41°15′N	141°09′E	33	1952	1984	4.5	1.00	PSMSL
Japan (Honshu)	Asamushi	40°54′N	140°52′E	31	1954	1984	1.6	1.00	PSMSL
Philippines	Manila	14°35′N	120°58′E	73	1901	1987	5.1	1.00	PSMSL
Philippines	Legaspi	13°09′N	123°45′E	41	1947	1987	3.3	1.00	PSMSL
Philippines	Tacloban, Leyte	11°15′N	125°00′E	26	1951	1976	1.5	0.83	PSMSL
Philippines	Cebu	10°18′N	123°54′E	44	1935	1987	0.7	1.00	PSMSL
Philippines	Davao	07°05′N	125°38′E	40	1948	1987	5.5	1.00	PSMSL
Philippines	Jolo	06°04′N	121°00′E	40	1947	1987	−26.2	0.55	PSMSL
Papua N. Guinea	Rabaul	04°12′S	152°11′E	13	1974	1986	−10.2	0.65	PSMSL
Australasia	Cairns	16°55′S	145°47′E	20	1957	1976	−2.5	0.90	PSMSL
Australasia	Townsville I	19°15′S	146°50′E	12	1966	1977	2.6	0.76	PSMSL
Australasia	Mackay	21°07′S	149°14′E	20	1949	1975	1.9	0.93	PSMSL
Australasia	Coffs Harbor I	30°19′S	153°05′E	15	1956	1970	0.7	0.75	PSMSL
Australasia	Newcastle I	32°55′S	151°48′E	27	1928	1961	4.7	0.99	PSMSL
Australasia	Newcastle II	32°55′S	151°48′E	15	1972	1986	−5.0	0.78	PSMSL
Australasia	Newcastle III	32°55′S	151°48′E	62	1925	1986	2.2	1.00	PSMSL
Australasia	Sydney, Fort Denison	33°51′S	151°14′E	90	1897	1986	0.6	1.00	PSMSL
Australasia	Camp Cove	33°50′S	151°17′E	39	1948	1986	0.2	1.00	PSMSL
Australasia	Port Kembla	34°30′S	150°54′E	13	1958	1970	−0.8	0.60	PSMSL
Australasia	MacDonnell	38°00′S	140°40′E	14	1957	1970	1.6	0.73	PSMSL
Australasia	Port Adelaide (I)	34°51′S	138°30′E	31	1882	1976	2.5	1.00	PSMSL
Australasia	Port Adelaide (O)	34°47′S	138°29′E	27	1944	1970	3.1	0.98	PSMSL
Australasia	Thevenard	32°09′S	133°39′E	15	1934	1970	−0.5	0.99	PSMSL
Australasia	Albany	35°02′S	117°53′E	11	1966	1976	4.9	0.68	PSMSL
New Zealand	Auckland	36°51′S	174°49′E	14	1917	1987	1.3	1.00	PSMSL
New Zealand	Wellington Harbour	41°17′S	174°47′E	16	1918	1987	−2.8	1.00	PSMSL

APPENDIX I

Country	Station	Latitude	Longitude	No. years of Record	Start year	End year	Regression slope mm/y.	t-confidence	Source
Marianas Islands	Apra Harbor, Guam	13°26′N	144°39′E	30	1948	1977	−1.7	0.95	PSMSL
Caroline Islands	Truk, Moen Island	07°27′N	151°51′E	31	1947	1977	2.0	0.86	PSMSL
Caroline Islands	Yap	09°31′N	138°08′E	13	1974	1986	1.7	0.65	PSMSL
Caroline Islands	Malakal	07°29′N	134°28′E	13	1974	1986	−0.3	0.62	PSMSL
Caroline Islands	Ponape	06°59′N	158°14′E	13	1974	1986	3.5	0.68	PSMSL
Nauru	Naura	00°32′S	166°54′E	13	1974	1986	−1.3	0.63	PSMSL
Marshall Islands	Eniwetok	11°22′N	162°21′E	22	1951	1972	0.8	0.87	PSMSL
Marshall Islands	Kwajalein	08°44′N	167°44′E	32	1946	1977	1.9	0.99	PSMSL
Marshall Islands	Wake Island	19°17′N	166°37′E	27	1950	1977	1.1	0.98	PSMSL
Solomon Islands	Honiara II	09°26′S	159°54′E	13	1974	1986	−7.1	0.60	PSMSL
New Caledonia	Noumea	22°18′S	166°26′E	11	1970	1982	−3.5	0.73	PSMSL
New Caledonia	Noumea II	22°18′S	166°26′E	12	1975	1986	−4.8	0.69	PSMSL
American Samoa	Pago Pago	14°17′S	170°41′W	30	1948	1977	1.8	0.99	PSMSL
Phoenix Islands	Canton Island	02°48′S	171°43′W	22	1949	1974	0.3	0.97	PSMSL
Hawaiian Islands	Midway Island	28°13′N	177°22′W	30	1947	1977	−1.3	0.99	PSMSL
Hawaiian Islands	Johnston Island	16°45′N	169°31′W	28	1950	1977	0.5	0.99	PSMSL
Hawaiian Islands	Tern Island	23°52′N	166°17′W	13	1974	1986	−4.3	0.64	PSMSL
Hawaiian Islands	Nawiliwili Bay	21°58′N	159°21′W	25	1955	1979	0.3	0.86	PSMSL
Hawaiian Islands	Honolulu	21°19′N	157°52′W	76	1905	1980	1.6	1.00	PSMSL
Hawaiian Islands	Mokuoloe Island	21°26′N	157°47′W	18	1957	1974	−0.9	0.92	PSMSL
Hawaiian Islands	Kahului Harbor	20°45′N	156°28′W	32	1947	1979	1.7	0.99	PSMSL
Hawaiian Islands	Hilo	19°44′N	155°04′W	33	1946	1978	3.5	1.00	PSMSL
Line Islands	Fanning Island	03°54′N	149°23′W	14	1972	1985	−0.5	0.62	PSMSL
Line Islands	Christmas Island	01°59′N	157°28′W	17	1956	1972	−4.8	0.85	PSMSL
Iles De La Societe	Papeete	17°32′S	149°34′W	12	1975	1986	6.8	0.77	PSMSL
Gambier Island	Rikitea	23°08′S	134°57′W	12	1975	1986	−0.3	0.75	PSMSL
USA (Aleutian Is)	Massacre Bay	52°50′N	173°39′W	24	1943	1966	5.1	0.83	PSMSL
USA (Aleutian Is)	Sweeper Cove	51°51′N	176°11′W	33	1943	1975	1.2	0.98	PSMSL
USA (Aleutian Is)	Dutch Harbor	53°54′N	166°32′W	16	1934	1955	0.8	0.95	PSMSL
USA (Aleutian Is)	Unalaska	53°53′N	166°32′W	21	1955	1975	−9.2	0.84	PSMSL
USA (Alaska)	Anchorage	61°14′N	149°54′W	13	1964	1976	31.2	0.63	PSMSL
USA (Alaska)	Seldovia	59°26′N	151°43′W	17	1964	1980	−4.6	0.67	PSMSL
USA (Alaska)	Seward	60°06′N	149°27′W	15	1964	1979	−5.8	0.79	PSMSL
USA (Alaska)	Cordova	60°33′N	145°46′W	15	1964	1979	12.2	0.87	PSMSL
USA (Alaska)	Yakutat	59°33′N	139°44′W	40	1940	1979	−4.7	1.00	PSMSL
USA (Alaska)	Sitka	57°03′N	135°20′W	42	1938	1979	−2.4	1.00	PSMSL
USA (Alaska)	Skagway	59°27′N	135°19′W	31	1944	1974	−17.6	0.95	PSMSL
USA (Alaska)	Juneau	58°18′N	134°25′W	44	1936	1979	−12.6	1.00	PSMSL
USA (Alaska)	Ketchikan	55°20′N	131°38′W	57	1919	1975	−0.1	1.00	PSMSL
Canada (Pacific)	Prince Rupert	54°19′N	130°20′W	42	1933	1980	1.3	1.00	PSMSL
Canada (Pacific)	Queen Charlotte	53°15′N	132°04′W	19	1957	1980	−3.4	0.97	PSMSL
Canada (Pacific)	Bella Bella	52°10′N	128°08′W	19	1962	1980	0.6	0.87	PSMSL
Canada (Pacific)	Port Hardy	50°43′N	127°39′W	15	1965	1980	−4.9	0.92	PSMSL
Canada (Pacific)	Alert Bay	50°35′N	126°57′W	31	1948	1978	−1.8	1.00	PSMSL
Canada (Pacific)	Point Atkinson	49°20′N	123°15′W	47	1914	1980	1.1	1.00	PSMSL
Canada (Pacific)	Vancouver	49°17′N	123°07′W	54	1910	1980	0.0	1.00	PSMSL
Canada (Pacific)	New Westminster	49°12′N	122°55′W	12	1969	1980	−14.3	0.57	PSMSL
Canada (Pacific)	Steveston	49°07′N	123°11′W	12	1969	1980	−0.1	0.75	PSMSL
Canada (Pacific)	Fulford Harbour	48°46′N	123°27′W	21	1960	1980	−0.4	0.98	PSMSL
Canada (Pacific)	Victoria	48°25′N	123°22′W	72	1909	1980	0.5	1.00	PSMSL
Canada (Pacific)	Tofino	49°09′N	125°55′W	19	1962	1980	−1.4	0.95	PSMSL
USA (Pacific)	Neah Bay	48°22′N	124°37′W	46	1934	1979	−1.4	1.00	PSMSL
USA (Pacific)	Friday Harbor (O)	48°33′N	123°00′W	44	1934	1978	0.9	1.00	PSMSL
USA (Pacific)	Seattle	47°36′N	122°20′W	81	1899	1979	1.9	1.00	PSMSL
USA (Pacific)	Astoria	46°13′N	123°46′W	55	1925	1979	−0.6	1.00	PSMSL
USA (Pacific)	South Beach	44°38′N	124°03′W	13	1967	1979	−1.6	0.83	PSMSL
USA (Pacific)	Crescent City	41°45′N	124°12′W	45	1933	1979	−0.9	1.00	PSMSL
USA (Pacific)	San Francisco	37°48′N	122°28′W	114	1854	1980	1.1	1.00	PSMSL
USA (Pacific)	Alameda	37°46′N	122°18′W	41	1939	1979	0.0	1.00	PSMSL
USA (Pacific)	Avila	35°10′N	120°44′W	26	1945	1979	2.9	0.97	PSMSL
USA (Pacific)	Rincon Island	34°21′N	119°26′W	18	1962	1980	5.2	0.83	PSMSL
USA (Pacific)	Santa Monica	34°01′N	118°30′W	39	1933	1979	1.9	1.00	PSMSL
USA (Pacific)	Los Angeles	33°43′N	118°16′W	57	1923	1979	0.6	1.00	PSMSL

Appendix I

Country	Station	Latitude	Longitude	No. years of Record	Start year	End year	Regression slope mm/y.	t-confidence	Source
USA (Pacific)	Long Beach	33°47′N	118°15′W	17	1963	1979	−1.4	0.84	PSMSL
USA (Pacific)	Alamitos Bay	33°45′N	118°07′W	13	1953	1965	8.1	0.75	PSMSL
USA (Pacific)	Newport Bay	33°36′N	117°53′W	25	1955	1979	−0.1	0.97	PSMSL
USA (Pacific)	La Jolla	32°52′N	117°15′W	54	1924	1979	1.7	1.00	PSMSL
USA (Pacific)	San Diego	32°43′N	117°10′W	74	1906	1979	0.2	1.00	PSMSL
Mexico (Pacific)	Ensenada	31°51′N	116°38′W	30	1956	1985	1.9	0.99	PSMSL
Mexico (Pacific)	La Paz	24°10′N	110°21′W	15	1952	1966	2.2	0.80	PSMSL
Mexico (Pacific)	Guaymas	27°55′N	110°54′W	14	1952	1965	3.0	0.80	PSMSL
Mexico (Pacific)	Mazatlan	23°12′N	106°25′W	14	1953	1966	1.3	0.74	PSMSL
Mexico (Pacific)	Manzanillo	19°03′N	104°20′W	29	1954	1982	3.1	0.94	PSMSL
Mexico (Pacific)	Acapulco	16°50′N	99°55′W	19	1967	1985	4.9	0.79	PSMSL
Mexico (Pacific)	Salina Cruz	16°10′N	95°12′W	28	1952	1979	1.6	0.92	PSMSL
Guatemala (Pac.)	San Jose	13°55′N	90°50′W	10	1960	1969	0.9	0.63	PSMSL
Guatemala (Pac.)	San Jose II	13°55′N	90°49′W	13	1963	1975	4.7	0.68	PSMSL
El Salvador	La Union	13°20′N	87°49′W	21	1948	1968	1.0	0.92	PSMSL
Costa Rica (Pac.)	Puntarenas	09°58′N	84°50′W	26	1941	1966	6.5	0.89	PSMSL
Costa Rica (Pac.)	Quepos	09°24′N	84°10′W	13	1957	1969	−1.6	0.76	PSMSL
Panama (Pacific)	Puerto Armuelles	08°16′N	82°52′W	18	1951	1968	−0.3	0.88	PSMSL
Panama (Pacific)	Balboa	08°58′N	79°34′W	62	1908	1969	1.6	1.00	PSMSL
Panama (Pacific)	Naos Island	08°55′N	79°32′W	20	1949	1968	1.4	0.95	PSMSL
Colombia (Pacific)	Buenaventura	03°54′N	77°60′W	28	1941	1969	1.0	0.99	PSMSL
Colombia (Pacific)	Tumaco	01°50′N	78°44′W	16	1953	1968	−2.1	0.83	PSMSL
Ecuador	La Libertad	02°12′S	80°55′W	37	1948	1984	2.2	0.90	PSMSL
Peru	Callao	12°03′S	79°09′W	43	1942	1984	2.4	0.85	METRIC
Peru	Talara	04°37′S	81°17′W	32	1942	1984	1.1	0.99	PSMSL
Peru	Chimbote	09°05′S	78°38′W	14	1955	1968	−5.6	0.72	PSMSL
Peru	Matarani	17°00′S	72°07′W	28	1941	1969	−1.1	1.00	PSMSL
Chile	Arica	18°28′S	70°20′W	21	1950	1970	3.5	0.80	PSMSL
Chile	Antofagasta	23°39′S	70°25′W	25	1946	1970	−3.4	0.99	PSMSL
Chile	Caldera	27°04′S	70°50′W	21	1950	1970	1.7	0.98	PSMSL
Chile	Valparaiso	33°02′S	71°38′W	13	1958	1970	1.7	0.88	PSMSL
Argentina	Ushuaia I	54°49′S	68°13′W	13	1957	1969	6.2	0.76	PSMSL
Argentina	Comodoro Rivadavia	45°52′S	67°29′W	20	1959	1980	4.0	0.81	PSMSL
Argentina	Puerto Madryn	42°46′S	65°02′W	30	1944	1982	2.6	1.00	PSMSL
Argentina	Piramide	42°35′S	64°17′W	16	1957	1972	3.0	0.84	PSMSL
Argentina	Quequen II	38°35′S	58°42′W	15	1968	1982	−0.8	0.90	PSMSL
Argentina	Mar Del Plata (P)	38°02′S	57°32′W	21	1957	1980	−1.8	0.98	PSMSL
Argentina	Mar Del Plata (C)	38°00′S	57°33′W	26	1957	1982	−2.0	0.99	PSMSL
Argentina	Buenos Aires	34°36′S	58°22′W	25	1957	1982	1.4	0.93	PSMSL
Argentina	Palermo	34°34′S	58°24′W	26	1957	1982	1.9	0.94	PSMSL
Argentina	Isla Martin Garcia	34°11′S	58°15′W	20	1957	1976	−10.0	0.65	PSMSL
Uruguay	Colonia	34°28′S	57°51′W	28	1954	1984	3.2	0.90	PSMSL
Uruguay	Montevideo	34°55′S	56°13′W	36	1938	1984	1.2	1.00	PSMSL
Uruguay	La Paloma	34°39′S	54°09′W	23	1955	1984	1.3	0.97	PSMSL
Brazil	Imbituba	28°14′S	48°39′W	21	1948	1968	0.7	0.99	PSMSL
Brazil	Cananeia	25°10′S	47°56′W	31	1954	1984	4.2	1.00	PSMSL
Brazil	Ubatuba	23°30′S	45°07′W	16	1954	1983	2.4	1.00	PSMSL
Brazil	Rio De Janeiro	22°56′S	43°08′W	20	1949	1968	3.6	0.81	PSMSL
Brazil	Ihla Fiscal	22°54′S	43°10′W	16	1965	1982	16.0	0.78	PSMSL
Brazil	Canavieiras	15°40′S	38°58′W	12	1952	1963	4.1	0.89	PSMSL
Brazil	Salvador	12°58′S	38°31′W	20	1949	1968	2.7	1.00	PSMSL
Brazil	Recife	08°03′S	34°52′W	21	1948	1968	−0.2	0.98	PSMSL
Brazil	Fortaleza	03°43′S	38°29′W	20	1948	1968	3.5	0.86	PSMSL
Brazil	Belém	01°27′S	48°30′W	20	1949	1968	0.3	0.97	PSMSL
Venezuela	Cumana	10°28′N	64°12′W	22	1954	1975	12.0	0.62	PSMSL
Venezuela	La Guaira	10°28′N	66°56′W	23	1953	1975	1.5	1.00	PSMSL
Venezuela	Amuay	11°45′N	70°13′W	20	1954	1975	14.0	0.62	PSMSL
Venezuela	Maracaibo	10°41′N	71°35′W	12	1964	1975	5.5	0.92	PSMSL
Colombia (Carib.)	Riohacha	11°33′N	72°55′W	17	1953	1969	2.0	0.86	PSMSL
Colombia (Carib.)	Cartagena	10°24′N	75°33′W	36	1949	1984	5.2	1.00	PSMSL
Panama (Carib.)	Cristobal	09°21′N	79°55′W	61	1909	1969	1.1	1.00	PSMSL
Costa Rica (Carib.)	Puerto Limón	10°00′N	83°02′W	21	1948	1968	2.0	0.99	PSMSL
Honduras	Puerto Castilla	16°01′N	86°02′W	14	1955	1968	3.2	0.96	PSMSL

APPENDIX I

Country	Station	Latitude	Longitude	No. years of Record	Start year	End year	Regression slope mm/y.	t-confidence	Source
Honduras	Puerto Cortés	15°50'N	87°57'W	21	1948	1968	8.9	0.99	PSMSL
Guatemala (Carib.)	Santo Tomas	15°42'N	88°37'W	18	1964	1983	−3.9	0.87	PSMSL
Mexico (Gulf)	Progreso	21°18'N	89°40'W	32	1952	1984	5.2	1.00	PSMSL
Mexico (Gulf)	Ciudad Del Carmen	18°38'N	91°51'W	11	1956	1966	−3.0	0.80	PSMSL
Mexico (Gulf)	Alvarado	18°46'N	95°46'W	12	1955	1966	−2.8	0.75	PSMSL
Mexico (Gulf)	Veracruz	19°11'N	96°07'W	33	1953	1985	1.6	1.00	PSMSL
Cuba	Guantánamo Bay	19°54'N	75°09'W	31	1937	1968	1.9	1.00	PSMSL
Jamaica	Port Royal	17°56'N	76°51'W	16	1954	1969	0.5	0.80	PSMSL
Haiti	Port-Au-Prince	18°34'N	72°21'W	13	1949	1961	12.3	0.89	PSMSL
Dominican Repub.	Barahona	18°12'N	71°05'W	11	1954	1969	−2.3	0.95	PSMSL
Dominican Repub.	Puerto Plata	19°49'N	70°42'W	16	1949	1969	4.4	0.95	PSMSL
Puerto Rico	Magüeyes Island	17°58'N	67°03'W	23	1955	1978	1.9	1.00	PSMSL
Puerto Rico	San Juan	18°27'N	66°05'W	13	1962	1974	2.1	0.79	PSMSL
USA (Gulf)	Port Isabel	26°04'N	97°13'W	34	1944	1979	3.2	1.00	PSMSL
USA (Gulf)	Padre Island	26°04'N	97°09'W	21	1958	1978	5.5	0.83	PSMSL
USA (Gulf)	Rockport	28°01'N	97°03'W	23	1948	1978	4.1	0.96	PSMSL
USA (Gulf)	Freeport	28°57'N	95°19'W	24	1954	1977	15.9	0.85	PSMSL
USA (Gulf)	Galveston I	29°19'N	94°48'W	17	1957	1973	12.2	0.77	PSMSL
USA (Gulf)	Galveston II	29°19'N	94°48'W	71	1908	1978	6.3	1.00	PSMSL
USA (Gulf)	Sabine Pass	29°42'N	93°51'W	21	1958	1978	11.7	0.71	PSMSL
USA (Gulf)	Eugene Island	29°22'N	91°23'W	34	1939	1974	9.6	0.99	PSMSL
USA (Gulf)	Bayou Rigaud	29°16'N	89°58'W	31	1947	1978	9.4	0.96	PSMSL
USA (Gulf)	Dauphin Island	30°15'N	88°04'W	13	1966	1979	4.2	0.75	PSMSL
USA (Gulf)	Pensacola	30°24'N	87°13'W	58	1923	1980	2.4	1.00	PSMSL
USA (Gulf)	Cedar Keys I	29°08'N	83°02'W	12	1914	1925	1.0	0.84	PSMSL
USA (Gulf)	Cedar Keys II	29°08'N	83°02'W	42	1938	1979	1.3	1.00	PSMSL
USA (Gulf)	St. Petersburg	27°46'N	82°37'W	31	1947	1977	1.7	1.00	PSMSL
USA (Gulf)	Naples	26°08'N	81°48'W	16	1965	1980	2.0	0.87	PSMSL
USA (Gulf)	Key West (Naval)	24°33'N	81°48'W	54	1926	1979	2.2	1.00	PSMSL
USA (Gulf)	Key West	24°34'N	81°48'W	13	1913	1925	−1.3	0.96	PSMSL
Bermuda	St. Georges	32°22'N	64°42'W	42	1932	1979	2.7	1.00	PSMSL
USA (Atlantic)	Miami Beach	25°46'N	80°08'W	47	1931	1980	2.3	1.00	PSMSL
USA (Atlantic)	Daytona Beach	29°14'N	81°00'W	25	1925	1969	2.0	1.00	PSMSL
USA (Atlantic)	Mayport	30°24'N	81°26'W	60	1928	1987	2.4	1.00	PSMSL
USA (Atlantic)	Jacksonville	30°21'N	81°37'W	16	1953	1968	2.1	0.76	PSMSL
USA (Atlantic)	Fernandina	30°41'N	81°28'W	90	1897	1986	1.9	1.00	PSMSL
USA (Atlantic)	Fort Pulaski	32°02'N	80°54'W	53	1935	1987	2.9	1.00	PSMSL
USA (Atlantic)	Charleston I	32°47'N	79°56'W	67	1921	1987	3.4	1.00	PSMSL
USA (Atlantic)	Myrtle Beach	33°41'N	78°53'W	16	1957	1977	5.1	0.85	PSMSL
USA (Atlantic)	Wilmington	34°14'N	77°57'W	53	1935	1987	1.8	1.00	PSMSL
USA (Atlantic)	Portsmouth	36°49'N	76°18'W	53	1935	1987	3.6	1.00	PSMSL
USA (Atlantic)	Richmond	37°34'N	77°27'W	27	1941	1967	−0.5	0.82	PSMSL
USA (Atlantic)	Hampton Roads	36°57'N	76°20'W	54	1927	1986	4.3	1.00	PSMSL
USA (Atlantic)	Gloucester Point	37°15'N	76°30'W	30	1950	1979	3.4	0.99	PSMSL
USA (Atlantic)	Cambridge	38°34'N	76°04'W	19	1942	1979	4.1	1.00	PSMSL
USA (Atlantic)	Washington DC	38°52'N	77°01'W	57	1931	1987	3.1	1.00	PSMSL
USA (Atlantic)	Piney Point	38°08'N	76°32'W	13	1960	1973	7.1	0.74	PSMSL
USA (Atlantic)	Solomons Island	38°19'N	76°27'W	51	1937	1987	3.3	1.00	PSMSL
USA (Atlantic)	Annapolis (Naval)	38°59'N	76°29'W	58	1928	1986	3.7	1.00	PSMSL
USA (Atlantic)	Baltimore	39°16'N	76°35'W	86	1902	1987	3.2	1.00	PSMSL
USA (Atlantic)	Kiptopeke Beach	37°10'N	75°59'W	36	1951	1986	3.1	1.00	PSMSL
USA (Atlantic)	Lewes	38°47'N	75°06'W	48	1919	1986	3.1	1.00	PSMSL
USA (Atlantic)	Philadelphia	39°57'N	75°08'W	66	1922	1987	2.7	1.00	PSMSL
USA (Atlantic)	Atlantic City	39°21'N	74°25'W	75	1911	1987	3.9	1.00	PSMSL
USA (Atlantic)	Sandy Hook	40°28'N	74°01'W	56	1932	1987	4.1	1.00	PSMSL
USA (Atlantic)	New York	40°42'N	74°01'W	68	1920	1987	3.1	1.00	PSMSL
USA (Atlantic)	Montauk	41°03'N	71°58'W	41	1947	1987	1.9	1.00	PSMSL
USA (Atlantic)	Port Jefferson	40°57'N	73°05'W	30	1957	1986	2.7	1.00	PSMSL
USA (Atlantic)	Willets Point	40°48'N	73°47'W	56	1931	1986	2.4	1.00	PSMSL
USA (Atlantic)	New Rochelle	40°54'N	73°47'W	31	1957	1987	0.7	0.94	PSMSL
USA (Atlantic)	Bridgeport	41°10'N	73°11'W	22	1965	1986	1.3	0.93	PSMSL
USA (Atlantic)	New London	41°22'N	72°06'W	50	1938	1987	2.2	1.00	PSMSL
USA (Atlantic)	Providence	41°48'N	71°24'W	41	1938	1986	1.8	1.00	PSMSL

Country	Station	Latitude	Longitude	No. years of Record	Start year	End year	Regression slope mm/y.	t-confidence	Source
USA (Atlantic)	Newport	41°30′N	71°20′W	58	1930	1987	2.6	1.00	PSMSL
USA (Atlantic)	Buzzards Bay	41°44′N	70°37′W	22	1955	1976	0.6	0.98	PSMSL
USA (Atlantic)	Woods Hole	41°32′N	70°40′W	54	1932	1986	2.7	1.00	PSMSL
USA (Atlantic)	Nantucket	41°17′N	70°06′W	15	1965	1979	0.9	0.93	PSMSL
USA (Atlantic)	Cape Cod Canal East	41°46′N	70°30′W	23	1955	1977	2.0	0.97	PSMSL
USA (Atlantic)	Boston	42°21′N	71°03′W	67	1921	1987	2.2	1.00	PSMSL
USA (Atlantic)	Portsmouth	43°05′N	70°45′W	56	1926	1987	1.8	1.00	PSMSL
USA (Atlantic)	Portland	43°40′N	70°15′W	75	1912	1986	2.2	1.00	PSMSL
USA (Atlantic)	Bar Harbor	44°23′N	68°12′W	40	1947	1987	2.8	1.00	PSMSL
USA (Atlantic)	Cutler	44°39′N	67°13′W	14	1964	1977	6.7	0.69	PSMSL
USA (Atlantic)	Eastport	44°54′N	66°59′W	51	1929	1979	3.7	1.00	PSMSL
Canada (Atlantic)	St. John N.B.	45°16′N	66°04′W	56	1906	1980	3.0	1.00	PSMSL
Canada (Atlantic)	Yarmouth	43°50′N	66°08′W	14	1967	1980	−0.5	0.88	PSMSL
Canada (Atlantic)	Boutilier Point	44°40′N	63°58′W	12	1969	1980	−1.7	0.90	PSMSL
Canada (Atlantic)	Halifax	44°40′N	63°35′W	62	1897	1980	3.7	1.00	PSMSL
Canada (Atlantic)	Pictou	45°41′N	62°42′W	18	1957	1980	3.9	0.92	PSMSL
Canada (Atlantic)	Charlottetown	46°14′N	63°07′W	48	1912	1980	3.1	1.00	PSMSL
Canada (Atlantic)	Ste-Anne-Des-Monts	49°08′N	66°29′W	11	1970	1980	−2.2	0.96	PSMSL
Canada (Atlantic)	Pointe-Au-Père	48°31′N	68°28′W	43	1925	1980	0.5	1.00	PSMSL
Canada (Atlantic)	Rivière-Du-Loup	47°51′N	69°34′W	11	1970	1980	−8.0	0.85	PSMSL
Canada (Atlantic)	St.-Francois	47°00′N	70°49′W	16	1965	1980	2.5	0.80	PSMSL
Canada (Atlantic)	Quebec City	46°50′N	71°10′W	16	1965	1980	3.2	0.79	PSMSL
Canada (Atlantic)	Cap A La Roche	46°34′N	72°06′W	16	1965	1980	21.7	0.58	PSMSL
Canada (Atlantic)	Champlain	46°26′N	72°24′W	16	1965	1980	24.5	0.57	PSMSL
Canada (Atlantic)	Grondines	46°35′N	72°02′W	16	1965	1980	18.8	0.59	PSMSL
Canada (Atlantic)	Neuville	46°42′N	71°34′W	14	1965	1980	13.1	0.59	PSMSL
Canada (Atlantic)	Tadoussac	48°08′N	69°43′W	13	1968	1980	−5.7	0.72	PSMSL
Canada (Atlantic)	Harrington Hbr.	50°30′N	59°29′W	38	1940	1980	−0.3	1.00	PSMSL
Canada (Atlantic)	Port-Aux-Basques	47°34′N	59°08′W	20	1935	1980	−1.8	0.97	PSMSL
Canada (Atlantic)	St. John's, Nfld	47°34′N	52°43′W	27	1935	1980	1.4	1.00	PSMSL
Canada (Atlantic)	Churchill	58°46′N	94°11′W	41	1940	1980	−6.2	1.00	PSMSL
Canada (Atlantic)	Resolute	74°41′N	94°53′W	20	1957	1977	−2.9	0.88	PSMSL

Appendix II

Eigenanalysis

Two specific types of information are contained in sea-level records: first, the relative rates of sea-level rise including acceleration or deceleration of this rise; second, the spatial and temporal structure of relative sea-level rise. Linear and non-linear regression techniques that previously were used do not provide the second type of information nor do they determine coherent modes of sea-level change to minimize local aberrations in sea level. Instead, they impose a subjective shape to sea-level curves and optimally fit a local record to that shape. We use here an objective method to determine uncorrelated modes of sea-level change: eigenanalysis.

Eigenanalysis is a well-known widely-used objective technique for determining dominant modes of variation in data sets. The purpose of eigenanalysis is to separate a data set into orthogonal spatial and temporal modes that most efficiently describe the variability of that data set (no other orthogonal functions can more efficiently represent that same data set). The result is a concise description of the spatial and temporal structure of variability in that data set. These patterns may or may not be related to physical processes. However, they are efficient at suppressing "noise."

Mathematically, a data set $\eta(x,t)$ can be decomposed into spatial and temporal functions

$$\eta(x,t) = \sum_{k=1}^{N} C_k(t)e_k(x)(\lambda_k n_x n_t)^{\frac{1}{2}}, \quad (1)$$

where $\eta(x,t)$ is a spatial grid sampled through time with its mean removed, the $C_k(t)$ represent temporal eigenfunctions, $e_k(x)$ represent spatial eigenfunctions, λ_k are the eigenvalues, n_x is the number of spatial points, n_t is the number of temporal points, and N is the lesser of n_x and n_t. Spatial eigenfunctions are determined from the following matrix operations:

$$(A - \lambda I)e = 0,$$

where $\quad (2)$

$$A = \frac{1}{n_x n_t} \eta \eta^T$$

The superscript T refers to the matrix transpose operator, I is the identity matrix, η is an (n_x, n_t) matrix of data, and A has size (n_x, n_x). A is a covariance matrix as shown in (2); if each station variance is set to unity, then A is a correlation matrix. Temporal eigenfunctions are determined from the matrix equation:

$$(B - \lambda I)C = 0.$$

where $\quad (3)$

$$B = \frac{1}{n_x n_t} \eta^T \eta.$$

B is a (n_t, n_t) matrix. The λ's are identical for equally ranked spatial or temporal eigenfunctions. Eigenfunctions are ranked according to their eigenvalues; the first eigenfunction has the largest eigenvalue, the second eigenfunction has the next largest eigenvalue, and so on. The eigenfunctions obey a least-squares criterion: the first eigenfunction best describes the data in a least-square sense; the second eigenfunction best describes the residual of the data in a least-squares sense, and so on. All eigenfunctions are orthonormal (or uncorrelated):

$$CC^T = I$$
$$ee^T = I.$$

Eigenfunction equations (2) and (3) are solved numerically (see, for example, Wilkinson and Reinsch, 1971), after matrices A and B are calculated. Operationally, once either set of eigenfunctions is determined numerically, the second set can be determined from the inner product of the eigenfunction and the data matrix. Once eigenfunctions are calculated the original time series (data set) can be reconstructed completely using equation (1). If the first "m" eigenvalues dominate all other eigenvalues (where m < N), a filtered data set ($\eta'(x,t)$) can then be used to represent the "predictable" or information-rich part of the original signal. In the following analysis, we generally obtain a useful $\eta'(x,t)$ by letting m = 2. A more complete discussion of statistical methods for selecting information-rich eigenvectors is presented in Preisendorfer et al. (1981).

Since either the $C_k(t)$ or $e_k(x)$ eigenfunctions can be viewed as weighting functions, we use the convention of presenting data as normalized $C_k(t)$ for temporal variability, and physical quantities $e_k(x)b_k$ with units of meters for spatial variability, where

$$b_k = (\lambda_k n_x n_t)^{\frac{1}{2}}.$$

Modified eigenanalysis permits use of stations with unequal data lengths (gappy data). The matrix A is formed with elements a_{jk} summed only over the overlapping segments of station j and station k. As a result, mean product elements a_{jk} will be formed over unequal numbers of samples. Although the interpretation of resulting eigenvectors must be done with care, for data with few gaps the method greatly extends the data available for analysis without resorting to interpolation or extrapolation which leads to unacceptable smoothing. Error analysis for this work is discussed by Solow (1987a).

Computationally, the (time) mean is removed from each station for the entire period during which it recorded. Mean product elements are calculated from these series from which means have been removed. Since mean product elements are formed of overlapping segments only, the are in general calculated over dissimilar lengths of time. If the series are stationary during these various time intervals, the resulting covariance (or correlation) matrix is not affected by gaps. If the individual series are *not* stationary during these various intervals, errors arise. To reduce the errors to acceptable levels, we used stations with more than 15 years of record (most of which recorded only between 1953 and 1980) and assured that gaps occupied less than 10 percent of the entire record.

When each station variance is set to unity resulting spatial functions are not scaled correctly to provide true sea-level rise estimates. In this case the weighting function, b_k, is defined:

$$b_k = (\lambda_k n_x n_t V_k)^{\frac{1}{2}}.$$

where V_k is the sample station variance.

Additional techniques are available to discriminate changes in tide-gauge records and to examine for change points. Pertinent literature includes Solow (1987a, b; 1990), Aubrey and Emery (1986a), and Braatz and Aubrey (1987).

Bibliography

Abdel-Monem, A., N. D. Watkins, and P. W. Gast, 1972, Potassium-argon ages, volcanic stratigraphy, and geomagnetic polarity history of the Canary Islands; Tenerife, La Palma and Hierro: American Journal of Science, v. 272, p. 805-825.

Academy of Geological Sciences of China, 1975, Geological Map of Asia, scale 1:5,000,000: Beijing, 20 sheets.

Adams, S. C., 1982, A chronological chart of ancient, modern and biblical history: Jacksonville, Oregon, Southern Oregon Historical Society, 4 chart sections.

Agar, W. M., R. F. Flint, and C. R. Longwell, 1929, Geology from Original Sources: New York, Henry Holt & Co., 527 p.

AGU Committee on the History of Geophysics, 1988, Babbage: American Geophysical Union, Newsletter, v. 3, p. 86-88.

Alaska Volcano Observatory Staff, 1990, The 1989-1990 eruption of Redoubt Volcano: EOS, American Geophysical Union, Transactions, v. 71, p. 265, 272-273, 275.

Albani, A. D., J. W. Tayton, P. C. Rickwood, A. D. Gordon, and J. G. Hoffman, 1988, Cainozoic morphology of the inner continental shelf near Sydney, N.S.W.: Royal Society of New South Wales, Journal and Proceedings, v. 121, p. 11-28.

Aleem, A. A., 1980, On the history of Arab navigation: *in* M. Sears and D. Merriman, eds., Oceanography, The Past: New York, Springer-Verlag, p. 582-595.

Allen, A. S., 1984, Types of land subsidence: *in* J. F. Poland, ed., Guidebook to Studies of Land Subsidence Due to Ground-water Withdrawal: International Hydrological Programme, Working Group 8.4: Paris, United Nations Educational, Scientific and Cultural Organization, p. 133-142.

Allen, D. R., and M. N. Mayuga, 1969, The mechanics of compaction and rebound, Wilmington Oil Field, Long Beach, California, U.S.A.: *in* L. J. Tison, ed., International Symposium on Land Subsidence, Tokyo, Japan: International Association of Hydrological Sciences, Publication 89, v. 2, p. 410-422.

Alverson, D. C., D. P. Cox, A. J. Woloshin, M. J. Terman, and C. C. Woo, 1967, Atlas of Asia and Eastern Europe, 2. Tectonics, scale 1:5,000,000: U. S. Geological Survey, 7 sheets.

Ambach, W., 1979, Zur nettoeisablation in einem höhenprofil am Grönlandischen inlandeis: Polarforschung, v. 49, p. 55-62.

Ambraseys, N. N., 1962, Data for the investigation of the seismic sea-waves in the eastern Mediterranean: Seismological Society of America, Bulletin, v. 52, p. 895-913.

Ambraseys, N. N., 1971, Value of historical records of earthquakes: Nature, v. 232, p. 375-379.

Anderson, D. L., 1982, Hotspots, polar wander, Mesozoic convection and the geoid: Nature, v. 297, p. 391-393.

Anderson, T. H., and V. A. Schmidt, 1983, Evolution of Middle America and the Gulf of Mexico-Caribbean region during Mesozoic time: Geological Society of America, Bulletin, v. 94, p. 941-966.

Anderson, W. A., J. T. Kelley, W. B. Thompson, H. W. Borns, Jr., D. Sanger, D. C. Smith, D. A. Tyler, R. S. Anderson, A. E. Bridges, K. J. Crossen, J. W. Ladd, B. G. Andersen, and F. T. Lee, 1984, Crustal warping in coastal Maine: Geology, v. 12, p. 677-680.

Armentrout, J. M., in press, Paleontologic constraints on depositional modeling — Examples of integration of biostratigraphy and seismic stratigraphy, Pliocene-Pleistocene, Gulf of Mexico: *in* P. Weimer and M. H. Link, eds., Seismic Facies and Sedimentary Processes of Submarine Fans and Turbidite Systems: Frontiers in Sedimentary Geology Series, New York, Springer-Verlag.

Arur, M. G., and F. Basir, 1982?, Yearly mean sea level trends along the Indian coast: Papers and Proceedings of the Seminar on Hydrography in Exclusive Economic Zones, Demarcation and Survey of its Wealth Potential: Calcutta, Hugli River Survey Service, p. 54-61.

Arur, M. G., R. Sivaramakrishnan, and H. R. Ghildyal, 1979, Rise in mean sea levels of Arabian Sea and Bay of Bengal Regions: Indian Surveyor, July.

Asian Institute of Technology, 1981, Investigation of land subsidence caused by deep well pumping in the Bangkok area: Bangkok, Thailand, Office of National Environment Board, Research Report 91, 353 p.

Asian Institute of Technology, 1982a, Phase IV extension of subsidence observation network: Bangkok, Thailand, Office of the National Environment Board, 93 p.

Asian Institute of Technology, 1982b, Groundwater resources in Bangkok area — Development and management study: Bangkok, Thailand, Office of the National Environment Board, 120 p. + app.

Atwater, T., 1970, Implications of plate tectonics for the Cenozoic tectonic evolution of western North America: Geological Society of America, Bulletin, v. 81, p. 3513-3536.

Aubouin, J., J. Azema, J.-Ch. Carfantan, C. Rangin, M. Tardy, and J. Tourmon, 1982, The Middle America Trench in the geological framework of Central America: *in* J. Aubouin and R. von Huene, eds., Internal Reports of the Deep Sea Drilling Project, v. 67, U. S. Government Printing Office, Washington, D. C., p. 747-755.

Aubrey, D. G., 1985, Recent sea levels from tide gauges: problems and prognosis: *in* National Academy of Sciences, Glaciers, Ice Sheets and Sea Level — Effect of a CO_2-induced climatic change: Washington, D. C., U. S. Department of Energy, Carbon Dioxide Research Division, DOE/ER/60235-1, p. 73-91.

Aubrey, D. G., and K. O. Emery, 1983, Eigenanalysis of recent United States sea levels: Continental Shelf Research, v. 2, p. 21-33.

Aubrey, D. G., and K. O. Emery, 1986a, Relative sea levels of Japan from tide gauge records: Geological Society of America, Bulletin, v. 97, p. 194-205.

Aubrey, D. G., and K. O. Emery, 1986b, Australia — A stable platform for tide gauge measurements of changing sea levels?: Journal of Geology, v. 94, p. 699-712.

Aubrey, D. G., and K. O. Emery, 1988, Australia — An unstable platform for tide-gauge measurements of changing sea levels — A reply: Journal of Geology, v. 96, p. 640-643.

Aubrey, D. G., and P. E. Speer, 1985, A study of non-linear tidal propagation in shallow inlet/estuarine systems, Part I. Observations: Estuarine, Coastal and Shelf Science, v. 1, p. 185-205.

Aubrey, D. G., K. O. Emery, and E. Uchupi, 1988, Changing coastal levels of South America and the Caribbean region from tide-gauge records: Tectonophysics, v. 154, p. 269-284.

Aubrey, D. G., C. A. Friedrichs, G. G. Giese, and P. E. Speer, in preparation, Causes of inlet formation at an Atlantic barrier beach: in D. G. Aubrey, and G. S. Giese, eds., Hydrodynamics of Multiple Inlet Systems: Springer-Verlag.

Aveni, A. F., 1981, Old and new world naked-eye astronomy: *in* K. Brecher and M. Feirtag, eds., Astronomy of the Ancients: Cambridge, Massachusetts, Massachusetts Institute of Technology Press, p. 61-69.

Babbage, C., 1847, The temple of Serapis: Geological Society of London, Quarterly Journal, v. 3, p. 186-217.

Badash, L., 1989, The age-of-the-Earth debate: Scientific American, v. 261, no. 2, p. 90-94, 96.

Balling, N., 1980, The land uplift in Fennoscandia, gravity field anomalies and isostasy: *in* N.-A. Mörner, ed., Earth Rheology, Isostasy and Eustasy: New York, John Wiley & Sons, p. 297-321.

Bally, A. W., 1981, Geology of Passive Continental Margins—History, Structure, and Sediment Record (with Special Emphasis on the Atlantic Margin): American Association of Petroleum Geologists, Education Course Note Series 19, 324.

Barazangi, M., and B. L. Isacks, 1979, Subduction of the Nazca plate beneath Peru—Evidence from spatial distribution of earthquakes: Royal Astronomical Society, Geophysical Journal, v. 57, p. 537-555.

Barber, R. T., and F. P. Chavez, 1983, Biological consequences of El Niño: Science, v. 222, p. 1203-1210.

Barnes, D. F., 1985, No measurable gravity change at Glacier Bay regional uplift area: U. S. Geological Survey, Circular 967, p. 88-90.

Barnett, T. P., 1982, On possible changes in global sea level and their potential causes: Scripps Institution of Oceanography, SIO Ref. Series 82-10, 34 p.

Barnett, T. P., 1983a, Recent changes in sea level and their possible causes: Climatic Change, v. 5, p. 15-38.

Barnett, T., 1983b, Long-term changes in dynamic height: Journal of Geophysical Research, v. 88, p. 9547-9552.

Barnett, T. P., 1984, The estimation of "global" sea level change—A problem of uniqueness: Journal of Geophysical Research, v. 89, p. 7980-7988.

Barnett, T. P., 1985, Long term climatic change in observed physical properties of the oceans: *in* M. C. MacCracken and F. M. Luther, eds., Projecting the Climatic Effects of Increasing Carbon Dioxide: Washington, D. C., U. S. Department of Energy, Carbon Dioxide Research Division, p. 91-107.

Barnett, T. P., 1988, Global sea level change: in Climate Variations Over the Past Century and the Greenhouse Effect, A report on the First Climate Trends Workshop, 7-9 September 1988: Washington, D. C., National Climate Program Office; Rockville, Maryland, National Oceanic and Atmospheric Administration.

Barnett, T., N. Graham, M. Cane, S. Zebiak, S. Dolan, J. O'Brien, and D. Legler, 1988, On the prediction of the El Niño of 1986-1987: Science, v. 241, p. 192-196.

Barry, R. G., 1985, Snow cover, sea ice, and permafrost: *in* M. Meier et al., eds., Glaciers, Ice Sheets and Sea Level—Effect of a CO_2-Induced Change: Washington, D. C., National Research Council, p. 241-247.

Barth, M. C., and J. G. Titus, eds., 1984, Greenhouse Effect and Sea Level Rise, A Challenge for This Generation: New York, Van Nostrand Reinhold Company, 325 p.

Battistini, R., 1970, État des connaissances sur les variations du niveau marin a Madagascar depuis 10.000 ans: Semaine Géologique Madagascar, Comptes Rendus 1970, p. 13-15.

Beavan, J., R. Bilham, and K. Hurst, 1984, Coherent tilt signals observed in the Shumagin Seismic Gap—Detection of time-dependent subduction at depth?: Journal of Geophysical Research, v. 89, p. 4478-4492.

Belknap, D. F., B. G. Andersen, R. S. Anderson, W. A. Anderson, H. W. Borns, Jr., G. L. Jacobson, J. T. Kelley, R. C. Shipp, D. C. Smith, R. Stuckenrath, Jr., W. B. Thompson, and D. A. Tyler, 1987, Late Quaternary sea-level changes in Maine: *in* D. Nummedal, O. H. Pilkey, and J. D. Howard, eds., Sea-Level Fluctuation and Coastal Evolution: Society of Economic Paleontologists and Mineralogists, Special Publication 41, p. 71-85.

Belperio, A. P., 1979, Negative evidence for a mid-Holocene high sea level along the coastal plain of the Great Barrier Reef province: Marine Geology, v. 32, p. M1-M9.

Ben-Avraham, Z., and J. K. Hall, 1977, Geophysical survey of Mount Carmel structure and its extension into the eastern Mediterranean: Journal of Geophysical Research, v. 82, p. 793-802.

Ben-Avraham, Z., A. Nur, D. Jones, and A. Cox, 1981, Continental accretion—From oceanic plateaus to allochthonous terranes: Science, v. 213, p. 47-54.

Ben-Menahem, A., 1979, Earthquake catalogue for the Middle East (92 B.C.-1980 A.D.): Bollettino di Geofizica Teorica ed Applicata, v. 21, p. 245-313.

Bennema, J., E. C. W. A. Geuze, H. Smits, and A. J. Wiggers, 1954, Soil compaction in relation to Quaternary movements of sea-level and subsidence of the land especially in the Netherlands: Geologie en Mijnbouw, new series, v. 16, p. 173-178.

Berger, W. H., 1982, Climate steps in ocean history—Lessons from the Pleistocene—Climate in Earth History: Washington, D. C., National Academy Press, p. 43-54.

Bergsten, F., 1954, The land uplift in Sweden from the evidence of the old water marks: Geografiska Annaler, v. 36, p. 81-111.

Berry, L., 1961, Erosion surfaces and emerged beaches in Hong Kong: Geological Society of America, Bulletin, v. 72, p. 1383-1394.

Berryman, K., 1987, Tectonic processes and their impact on the recording of relative sea-level changes: *in* R. J. N. Devoy, ed., Sea Surface Studies—A Global View: London, Croom Helm, p. 127-161.

Biju-Duval, B., G. Bizon, A. Mascle, and C. Muller, 1982a, Active margin processes—Field observations in southern Hispaniola: *in* J. S. Watkins and C. L. Drake, eds., Studies in Continental Margin Geology: American Association of Petroleum Geologists, Memoir 34, p. 325-344.

Biju-Duval, B., A. Mascle, H. Rosales, and G. Young, 1982b, Episutural Oligo-Miocene basin along the north Venezuelan margin: in J. S. Watkins and C. L. Drake, eds., Studies in Continental Margin Geology: American Association of Petroleum Geologists, Memoir 34, p. 347-358.

Biswas, B., 1973, Quaternary changes in sea level in the South China Sea: Geological Society of Malaysia, Bulletin 6, p. 220-256.

Bjerknes, J., 1961, "El Niño" study based on analysis of ocean surface temperatures, 1935-1957: Inter-American Tropical Tuna Commission, Bulletin, v. 5, p. 219-307.

Blaha, J., and R. Reed, 1982, Fluctuations of sea level in the western North Pacific and inferred flow of the Kuroshio: Journal of Physical Oceanography, v. 12, p. 669-698.

Blake, M. C, Jr., D. L. Jones, and C. A. Landis, 1974, Active continental margins—Contrasts between California and New Zealand: *in* C. A. Burk and C. L. Drake, eds., The Geology of Continental Margins: New York, Springer-Verlag, p. 853-872.

Bloom, A. L., 1963, Late-Pleistocene changes of sealevel and postglacial crustal rebound in coastal Maine: American Journal of Science, v. 261, p. 862-879.

Bloom, A. L., 1967, Pleistocene shorelines—A new test of isostasy: Geological Society of America, Bulletin, v. 78, p. 1477-1493.

Bloom, A. L., 1971, Glacial-eustatic and isostatic controls of sea level since the last glaciation: *in* K. K. Turekian, ed., The Late Cenozoic Glacial Ages: New Haven, Connecticut, Yale University Press, p. 355-379.

Bloom, A. L., 1977, Atlas of Sea-Level Curves: International Geological Correlation Programme, Project 61, Sea-Level Project, 123 p (multilithed).

Bloom, A. L., W. S. Broecker, J. M. A. Chappell, R. K. Mathews, and K. J. Mesolella, 1974, Quaternary sea level fluctuations on a tectonic crust—New $^{230}Th/^{234}U$ dates from the Huon Peninsula, New Guinea: Quaternary Research, v. 4, p. 185-205.

Bolin, B., E. T. Degens, P. Duvigneaud, and S. Kempe, 1979, The global biogeochemical carbon cycle: *in* B. Bolin, E. T. Degens, S. Kempe, and D. Ketner, eds., The Global Carbon Cycle: The Scientific Committee on Problems of the Environment (SCOPE-13), International Council of Scientific Unions (ICSU): New York, Wiley, p. 1-56.

Boote, D. R. D., and R. B. Kirk, 1989, Depositional wedge cycles on evolving plate margin, western and northwestern Australia: American Association of Petroleum Geologists, Bulletin, v. 73, p. 216–243.

Born, M. A., 1989, Tidelog, Massachusetts Coast Edition: Bolinas, California, 110 p.

Bostock, John, and H. T. Riley, 1855, Natural History of Pliny: London, H. G. Bohn, 6 vols.

Bott, M. H. P., C. W. A. Browitt, and A. P. Stacey, 1971, The deep structure of the Iceland–Faeroe Ridge: Marine Geophysical Research, v. 1, p. 328–351.

Bottomley, M., C. K. Folland, J. Hsiung, R. E. Newell, and D. E. Parker, 1990, Global Ocean Surface Temperature Atlas (GOSTA): Joint Meteorological Office, Massachusetts Institute of Technology Project; Washington, D. C., Department of Energy, 400 p.

Bovis, M., and B. L. Isacks, 1984, Hypocentral trend surface analysis—Probing the geometry of Benioff zones: Journal of Geophysical Research, v. 89, p. 6153–6170.

Bowin, C., 1973, Origin of the Ninety East Ridge from studies near the equator: Journal of Geophysical Research, v. 78, p. 6029–6043.

Bowin, C. O., 1975a, Caribbean gravity field and plate tectonics: Geological Society of America, Special Paper 169, 79 p.

Bowin, C., 1975b, The geology of Hispaniola: in A. E. M. Nairn and F. G. Stehli, eds., The Ocean Basins and Margins–v. 3, The Gulf of Mexico and the Caribbean: New York, Plenum Press, p. 501–552.

Bowin, C., W. Warsi, and J. Milligan, 1982, Free-Air Gravity Anomaly Atlas of the World: Geological Society of America, Map and Chart series, No. MC-46, 86 charts (folio).

Braatz, B. V., and D. G. Aubrey, 1987, Recent relative sea-level change in eastern North America: in D. Nummedal, O. H. Pilkey, and J. D. Howard, eds., Sea-Level Fluctuation and Coastal Evolution: Society of Economic Paleontologists and Mineralogists, Special Publication 41, p. 29–46.

Bradley, R. S., H. F. Diaz, J. K. Eischeid, P. D. Jones, P. M. Kelly, and C. M. Goodess, 1987, Precipitation fluctuations over Northern Hemisphere land areas since the mid-19th century: Science, v. 237, p. 171–175.

Brandini, A. A., E. D'Onofrio, and E. Schnack, 1985, Comparative analysis of historical mean sea level changes along the Argentine coast: in J. Rabessa, ed., Quaternary of South America and Antarctic Peninsula: Rotterdam, A. A. Balkema, p. 187–195.

Brandt, J. C., 1981, Pictographs and petroglyphs of the Southwest Indians: in K. Brecher and M. Feirtag, eds., Astronomy of the Ancients: Cambridge, Massachusetts Institute of Technology Press, p. 25–38.

Breaker, L., 1989, El Niño and related variability in sea-surface temperature along the central California coast: in D. H. Peterson, ed., Aspects of Climate Variability in the Pacific and the Western Americas: American Geophysical Union, Geophysical Monograph 55, p. 133–140.

Bretz, J. H., 1960, Bermuda—A partially drowned, late mature, Pleistocene karst: Geological Society of America, Bulletin, v. 71, p. 1729–1754.

Brink, K. H., J. S. Allen, and R. L. Smith, 1978, A study of low-frequency fluctuations near the Peru Coast: Journal of Physical Oceanography, v. 8, p. 1025–1041.

Broadus, J. M., J. D. Milliman, S. Edwards, D. G. Aubrey, and F. Gable, 1986, Rising sea level and damming of rivers—Possible effects in Egypt and Bangladesh: in J. Titus, ed., Effects of Changes in Stratospheric Ozone and Global Climate, v. 4—Sea-level Rise: Washington, D. C., United Nations Environment Programme and U. S. Environmental Protection Agency, p. 165–189.

Broecker, W. S., 1975, Climatic changes—Are we on the brink of a pronounced glacial warming?: Science, v. 189, p. 460–463.

Broecker, W. S., and J. van Donk, 1970, Insolation changes, ice volumes, and the O^{18} record in deep-sea cores: Review of Geophysics and Space Physics, v. 8, p. 169–198.

Broecker, W. S., D. Peteet, and D. Rind, 1985, Does the ocean–atmosphere system have more than one stable mode of operation?: Nature, v. 315, p. 21–26.

Brown, D. A., K. S. W. Campbell, and K. A. W. Crook, 1968, The geological evolution of Australia and New Zealand: New York, Pergamon Press, 409 p.

Brückner, H., 1988, Indicators for formerly higher sea levels along the east coast of India and on the Andaman Islands: Hamburger Geographische Studien, v. 44, p. 47–72.

Brückner, H., and U. Radtke, 1989, Fossile strände und korallenbänke auf Oahu, Hawaii: Essener Geographische Arbeiten, v. 178, p. 291–308.

Bryant, E. A., P. S. Roy, and B. G. Thom, 1988, Australia—An unstable platform for tide-gauge measurements of changing sea levels—A discussion: Journal of Geology, v. 96, p. 635–640.

Budyko, M. I., and Y. Izrael, 1987, Anthropogenic climate change: Gidrometeorizdat, 406 p.

Bull, W. B., and A. F. Cooper, 1986, Uplifted marine terraces along the Alpine Fault, New Zealand: Science, v. 234, p. 1225–1228.

Bunbury, E. H., 1959, A History of Ancient Geography among the Greeks and Romans from the Earliest Ages Till the Fall of the Roman Empire: New York, Dover Publications, v 1., 666 p.; v. 2, 743 p.

Bunce, E. T., and P. Molnar, 1977, Seismic reflection profiling and basement topography in the Somali Basin—Possible fracture zones between Madagascar and Africa: Journal of Geophysical Research, v. 892, p. 5305–5311.

Bureau of Mineral Resources, 1965, Geological map of the World—Australia and Oceania, scale 1:5,000,000: Canberra, Australia, sheets 6, 7, 11, and 12.

Burke, K., and A. M. Celal Sengör, 1988, Ten metre global sea-level change associated with South Atlantic Aptian salt deposition: Marine Geology, v. 83, p. 309–312.

Burke, K., P. J. Fox, and A. M. C. Sengör, 1978, Buoyant ocean floor and evolution of the Caribbean: Journal of Geophysical Research, v. 83, p. 3949–3954.

Butcher, S. H., and Andrew Lang, 1930, The Odyssey of Homer Rendered into English Prose: Boston, Hale, Cushman & Flint, 331 p.

Cailleux, A., 1952, Récentes variations du niveau des mers et des terrres: Société Géologique de France, 6th series, Bulletin, v. 2, p. 135–144.

Cande, S. C., and J. C. Mutter, 1982, A revised identification of the oldest sea-floor spreading anomalies between Australia and Antarctica: Earth and Planetary Science Letters, v. 58, p. 151–160.

Cane, M. A., 1983, Oceanographic events during El Niño: Science, v. 222, p. 1189–1210.

Cane, M. A., 1984, Modeling sea level during El Niño: Journal of Physical Oceanography, v. 14, p. 1864–1874.

Cane, M. A., and S. E. Zebiak, 1985, A theory for El Niño and the Southern Oscillation: Science, v. 228, p. 1085–1087.

Carbognin, L., P. Gatto, and F. Marabini, 1986, Correlations between shoreline variations and subsidence in the Po River delta, Italy: Third International Symposium on Land Subsidence, Anaheim, CA, U. S. A.: International Association of Hydrological Sciences, no. 151, p. 367–372.

Carbognin, L., P. Gatto, and G. Mozzi, 1984a, Case History no. 9.15. Ravenna, Italy: in J. F. Poland, ed., Guidebook to Studies of Land Subsidence Due to Ground-water Withdrawal: International Hydrological Programme, Working Group 8.4: Paris, United Nations Educational, Scientific and Cultural Organization, p. 291–305.

Carbognin, L., P. Gatto, G. Mozzi, G. Gambolati, and G. Ricceri, 1984b, Case History no. 9.3. Venice, Italy: in J. F. Poland, ed., Guidebook to Studies of Land Subsidence Due to Ground-water Withdrawal: International Hydrological Programme, Working Group 8.4: Paris, United Nations Educational, Scientific and Cultural Organization, p. 161–174.

Carrera, G., and P. Vaníček, 1988, A comparison of present sea level linear trends from tide gauge data and radiocarbon curves in eastern Canada: Palaeogeography, Palaeoclimatology, Palaeoecology, v. 68, p. 127–134.

Carter, H. 1958, The Histories of Herodotus of Halicarnassus: New York, Heritage Press, 615 p.

Cartwright, D. E, and G. A. Alcock, 1981, On the precision of sea surface elevations and slopes from SEASAT altimetry of the northeast Atlantic Ocean: in J. F. R. Gower, ed., Oceanography from Space: New York, Plenum Press, p. 885–895.

Case, J. E., and T. L. Holcombe, 1980, Geologic-tectonic map of the Caribbean region, scale 1: 2,500,000: U. S. Geological Survey, Miscellaneous Investigation Series Map 1-100, 3 sheets.

Casey, R. E., A. L. Weinheimer, and C. O. Nelson, 1989, California El Niños and related changes of the California Current system from Recent and fossil radiolarian records: in D. H. Peterson, ed., Aspects of Climate Variability in the Pacific and the Western Americas: American Geophysical Union, Geophysical Monograph 55, p. 85-92.

Castilla, J. C., 1988, Earthquake-caused coastal uplift and its effects on rocky intertidal kelp communities: Science, v. 242, p. 440-443.

Cayan, D. R., and D. H. Peterson, 1989, The influence of North Pacific atmospheric circulation on streamflow in the west: in D. H. Peterson, ed., Aspects of Climate Variability in the Pacific and the Western Americas: American Geophysical Union, Geophysical Monograph 55, p. 375-397.

Celsius, A., 1743, Anmärkning om vatnets förminskande, sa i Östersjön som Vesterhafvet: Kongliga Svenska Vetenskaps Akademiens Handlingar, v. 4, p. 33-50.

Cercone, K. R., 1988, Evaporative sea-level drawdown in the Silurian Michigan basin: Geology, v. 16, p. 387-390.

Cess, R. D., G. L. Potter, J. P. Blanchet, G. J. Boer, S. J. Ghan, J. T. Kiehl, H. Le Treut, Z. X. Li, X. Z. Liang, J. F. B. Mitchell, J. J. Morcrette, D. A. Randall, M. R. Riches, E. Roeckner, A. Slingo, K. E. Taylor, W. M. Washington, R. T. Wetherald, and I. Yagai, 1989, Interpretation of cloud climate feedback as produced by 14 atmospheric general circulation models: Science, v. 245, p. 513-516.

Chao, B. F., 1988, Correlation of interannual length-of-day variation with El Niño/Southern Oscillation, 1972-1986: Journal of Geophysical Research, v. 93, p. 7709-7715.

Chapman, D. C., and G. S. Giese, 1990, A model for the generation of coastal seiches on the Caribbean coast of Puerto Rico: Journal of Physical Oceanography, v. 20, p. 1459-1467.

Chappell, J., 1974, Geology of coral terraces, Huon Peninsula, New Guinea—A study of Quaternary tectonic movements and sea-level changes: Geological Society of America, Bulletin, v. 85, p. 553-570.

Chappell, J., 1976, On possible relationships between Upper Quaternary glaciations, geomagnetism, and volcanism: Earth and Planetary Science Letters, v. 26, p. 370-376.

Chappell, J., 1983, Evidence for smoothly falling sea level relative to north Queensland, Australia, during the past 6000 yr.: Nature, v. 302, p. 406-408.

Chappell, J., 1987, Ocean volume change and the history of sea water: in R. J. N. Devoy, ed., Sea Surface Studies—A Global View: London, Croom Helm, p. 33-56.

Chappell, J., E. G. Rhodes, B. G. Thom, and E. Wallensky, 1982, Hydro-isostasy and the sea-level isobase of 5500 B.P. in north Queensland, Australia: Marine Geology, v. 49, p. 81-90.

Chase, C. G., 1979, Subduction, the geoid, and lower mantle convection: Nature, v. 282, p. 464-468.

Chelton, D. B., and R. E. Davis, 1982, Monthly mean sea-level variability along the west coast of North America: Journal of Physical Oceanography, v. 12, p. 757-784.

Chelton, D., and D. Enfield, 1986, Ocean signals in tide gauge records: Journal of Geophysical Research, v. 91, p. 9081-9098.

Chen Jiyu, Yun Caixing, Xu Haigen, and Dong Yougfa, 1979, The development model of the Chang Jiang River estuary during the last 2000 years: Acta Oceanologia Sinica, v. 1, p. 103-111.

Chen, X. Q., in press, Sea level changes since early 1920s from the long records of two tidal gauges in Shanghai, China: Journal of Coastal Research.

Cheney, R. E., and J. G. Marsh, 1981, Oceanic eddy variability measured by GEOS 3 altimeter crossover differences: EOS, American Geophysical Union, Transactions, v. 62, p. 743.

Cheney, R. E., and L. Miller, 1988, Mapping the 1986-1987 El Niño with GEOSAT altimeter data: EOS, American Geophysical Union, Transactions, v. 69, p. 754-755.

Cheney, R. E., B. C. Douglas, and L. Miller, 1989, Evaluation of GEOSAT altimeter data with application to tropical Pacific sea level variability: Journal of Geophysical Research, v. 94, p. 4737-4747.

Cherven, V. B., and A. F. Jacob, 1985, Evolution of Paleogene depositional systems, Williston Basin, in response to global sea level changes: in R. M. Flores and S. S. Kaplan, eds., Cenozoic Paleogeography of the West-Central United States: Rocky Mountain Section Society of Economic Paleontologists and Mineralogists, Rocky Mountain Paleogeography Symposium 3, p. 127-170.

Choubert, G., and A. Faure-Muret (general co-ordinators), 1981, Geological World Atlas: Paris, United Nations Educational, Scientific and Cultural Organization, Africa—sheets 6, 7, 8.

Christiansen, B., 1983, Mangroves—What are they worth? Unasylva, v. 35, p. 2-15.

Christie-Blick, N., J. P. Grotzinger, and C. C. von der Borch, 1988a, Sequence stratigraphy in Proterozoic successions: Geology, v. 16, p. 100-104.

Christie-Blick, N., G. S. Mountain, and K. G. Miller, 1988b, Sea-level history: Science, v. 241, p. 596.

Chugh, R. S., 1961, The Indian mean sea level: International Geophysical Year Symposium, Proceedings, v. 11.

Churchill, Winston S., 1968, The Island Race: London, Gorgi Books, 309 p.

Churkin, M., Jr., and G. H. Packham, 1973, Volcanic rocks and volcanic constituents in sediments, Leg 21, Deep Sea Drilling Project: in R. E. Burns, J. E. Andrews, et al., eds., Initial Reports of the Deep Sea Drilling Project: Washington D. C., U. S. Government Printing Office, v. 21, p. 481-493.

Clague, D. A., and R. D. Jarrard, 1973, Tertiary Pacific plate motion deduced from the Hawaiian-Emperor Chain: Geological Society of America, Bulletin, v. 84, p. 1135-1154.

Clark, J. A., 1980, The reconstruction of the Laurentide ice sheet of North America from sea level data—Method and preliminary results: Journal of Geophysical Research, v. 895, p. 4307-4323.

Clark, J. A., and C. S. Lingle, 1977, Future sea-level changes due to West Antarctic ice sheet fluctuations: Nature, v. 269, p. 206-209.

Clark, J. A., and C. S. Lingle, 1979, Predicted relative sea-level changes (18,000 years B.P. to present) caused by late-glacial retreat of the Antarctic Ice Sheet: Quaternary Research, v. 11, p. 279-298.

Clark, J. A., W. E. Farrell, and W. R. Peltier, 1978, Global changes in post-glacial sea level—A numerical calculation: Quaternary Research, v. 9, p. 265-287.

Clifford, T. N., 1970, The structural framework of Africa: in T. N. Clifford and I. G. Gass, eds., African Magmatism and Tectonics: Darien, Connecticut, Hafner Publishing Co., p. 1-26.

CLIMAT Project Members, 1976, The surface of the ice-age Earth: Science, v. 191, p. 1131-1137.

Cloud, P. E., Jr., 1968, Atmospheric and hydrospheric evolution on the primitive Earth: Science, v. 160, p. 729-736.

Cloud, P., 1976, Beginnings of biospheric evolution and their biochemical consequences: Paleobiology, v. 2, p. 351-387.

Cloud, P. E., and M. F. Glaessner, 1982, The Ediacarian Period and system—Metazoa inherit the Earth: Science, v. 217, p. 783-792.

Coast and Geodetic Survey, 1964, Tide tables, high and low water predictions, 1962—East coast North and South America including Greenland: Washington, D. C., U. S. Department of Commerce, 285 p.

Coast and Geodetic Survey, 1970a, Tide tables, high and low water predictions, east coast North and South America including Greenland: Washington, D. C., U. S. Department of Commerce, 290 p.

Coast and Geodetic Survey, 1970b, Tide tables, high and low water predictions, west coast North and South America including the Hawaiian Islands: Washington, D. C., U. S. Department of Commerce, 226 p.

Coates, D. R., 1983, Large-scale land subsidence: in R. Gardner and H. Scoging, eds., Mega-Geomorphology: Oxford, England, Clarendon Press, p. 212-234.

Cochran, J. R., 1986, Variations in subsidence rates along intermediate and fast spreading mid-ocean ridges: Royal Astronomical Society, v. 87, Geophysical Journal, p. 421-454.

Coghill, N., 1978, Geoffrey Chaucer, the Canterbury Tales: New York, Penguin Books, 526 p.

Colman, S. M., J. P. Halka, C. H. Hobbs III, R. B. Mixon, and D. S. Foster, 1990, Ancient channels of the Susquehanna River beneath Chesapeake Bay and the Delmarva Peninsula: Geological Society of America Bulletin, v. 102, p. 1268-1279.

Commonwealth Secretariat, 1989, Climate Change—Meeting the Challenge: London, British Commonwealth.

Condie, K. C., 1982, Plate Tectonics and Crustal Evolution: New York, Pergamon Press, 310 p.

Coney, P. J., D. L. Jones, and J. W. H. Monger, 1980, Cordilleran suspect terranes: Nature, v. 288, p. 329-333.

Conolly, J. R., and C. C. von der Borch, 1967, Sedimentation and physiography of the sea floor south of Australia: Sedimentary Geology, v. 1, p. 181-220.

Coque, R., 1962, La Tunisie Pre-saharienne: Étude Geomorphologique: Paris, Armand Colin, 476 p.

Corcoran, E., 1988, A different engine? An unbuilt prototype may get another chance: Scientific American, v. 259, p. 111.

Corkan, R. H. (secretary), 1950, Monthly and Annual Mean Heights of Sea Level 1937 to 1946 and Unpublished Data for Earlier Years: Association d'Océanographie Physique, Union Géodésique et Géophysique Internationale, v. 10, 82 p.

Cotton, C. A., 1945, Geomorphology—An Introduction to the Study of Landforms: New York, John Wiley & Sons, 505 p.

Cotton, C. A., 1955, New Zealand Geomorphology: Wellington, New Zealand University Press, 281 p.

Cottrell, A., 1979, The Minoan World: New York, Scribners, 191 p.

Coutellier V., and D. J. Stanley, 1987, Late Quaternary stratigraphy and paleogeography of the eastern Nile Delta, Egypt: Marine Geology, v. 77, p. 257-275.

Cronin, T. M., 1981, Rates and possible causes of neotectonic vertical crustal movements of the emerged southeastern United States coastal plain: Geological Society of America, Bulletin, v. 92, p. 812-833.

Crough, S. T., and D. M. Jurdy, 1980, Subducted lithosphere, hotspots, and the geoid: Earth and Planetary Science Letters, v. 48, p. 15-22.

Culliton, T. J., M. A. Warren, T. R. Goodspeed, D. G. Remer, C. M. Blackwell, and J. J. McDonough III, 1990, 50 years of population change along the nation's coasts, 1960-2010: Rockville, Maryland, Strategic Assessment Branch, National Ocean Service, National Oceanic and Atmospheric Administration, 41 p.

Curray, J. R., F. J. Emmel, D. J. Moore, and R. J. Raitt, 1982a, Structure, tectonics, and geological history of the northeastern Indian Ocean: in A. E. M. Nairn and F. G. Stehli, eds., The Ocean Basins and Margins—v. 6, The Indian Ocean: New York, Plenum Press, v. 6, p. 399-450.

Curray, J. R., D. G. Moore, K. Kelts, and G. Einsele, 1982b, Tectonics and geological history of the passive continental margin at the tip of Baja California: Initial Reports of the Deep Sea Drilling Project: Washington, D. C., U. S. Government Printing Office, v. 64, p. 1089-1116.

Daly, D. J., G. H. Groenewold, and C. R. Schmit, 1985, Paleoenvironments of the Paleocene Sentinel Butte Formation, Knife River area, west central North Dakota: in R. M. Flores and S. S. Kaplan, eds., Cenozoic Paleogeography of the West-Central United States: Society of Economic Paleontologists and Mineralogists, Rocky Mountain Section, Paleogeography Symposium 3, p. 171-185.

Daly, R. A., 1915, The glacial-control theory of coral reefs: American Academy of Arts and Sciences, Proceedings, v. 51, p. 155-251.

Dambara, T., 1971, Synthetic vertical movements in Japan during the recent 70 years: Japanese Society of Geodesy, Bulletin, v. 17, p. 100-108.

Dansgaard, W., S. J. Johnsen, H. B. Clausen, and C. C. Langway, Jr., 1971, Climatic record revealed by the Camp Century ice core: in K. K. Turekian, ed., The Late Cenozoic Glacial Ages: New Haven, Connecticut, Yale University Press, p. 37-56.

Darwin, C., 1889, The Structure and Distribution of Coral Reefs (3rd. ed.): New York, D. Appleton and Co., 344 p.

Davis, G. H., 1987, Land subsidence and sea-level rise on the coastal-plain of the United States: Environmental Geology Water Sciences, v. 10, p. 67-80.

Davis, G. H., and J. R. Rollo, 1969. Land subsidence related to decline of artesian head at Baton Rouge, lower Mississippi Valley, U.S.A: in L. J. Tison, ed., International Symposium on Land Subsidence, Tokyo, Japan: International Association of Hydrological Sciences, Publication 88, v. 1, p. 174-184.

Davis, G. H., J. B. Small, and H. B. Counts, 1963, Land subsidence related to decline of artesian pressure in the Ocala Limestone at Savannah, Georgia: in P. D. Trask and G. A. Kiersch, eds., Geological Society of America, Engineering Geology Case Histories, Number 4, p. 1-8.

Davis, W. M., 1896, The outline of Cape Cod: American Academy of Arts and Sciences, Proceedings, v. 31, p. 303-332.

Davis, W. M., 1899, The geographical cycle: Geographical Journal, v. 14, p. 481-504.

Davis, W. M., 1909, Geographical Essays, edited by D. W. Johnson: Boston, 777 p. (reprinted by Dover Publ., Inc.).

Deacon, R., 1966, Madoc and the Discovery of America: New York, George Braziller, 269 p.

De Almeida, F. F. M., F. Martin, C. Furque, and E. O. Ferreira, 1978, Tectonic map of South America, scale 1:5,000,000: Geological Society of America, Map and Chart Series MC-32, 2 sheets.

Deegan, L. A., H. M. Kennedy, and C. Neill, 1984, Natural factors and human modifications contributing to marsh loss in Louisiana's Mississippi River deltaic plain: Environmental Management, v. 8, p. 519-528.

de Geer, G., 1888-1890, Om Skandinaviens nivafföreändringar under quartärperioden: Geologiska Föreningens i Stockholm Förhandlingar, v. 10 (1888), p. 336-379; v. 12 (1890), p. 61-110.

Dennis, R. E., and E. E. Long, 1971, A user's guide to a computer program for harmonic analysis of data at tidal frequencies: National Oceanographic and Atmospheric Administration, Technical Report NOS 41, 31 p.

Detrick, R. S., J. G. Sclater, and J. Thiede, 1977, The subsidence of aseismic ridges: Earth and Planetary Science Letters, v. 34, p. 185-196.

Devoy, R. J. N. (editor), 1987, Sea Surface Studies—A Global View: London, Croom Helm, 649 p.

de Vries, H. L., and G. W. Barendsen, 1954, Measurements of age by the carbon-14 technique: Nature, v. 174, p. 1138-1141.

de Wit, M. J., R. A. Hart, and R. J. Hart, 1987, The Jamestown Ophiolite Complex, Barberton mountain belt—A section through 3.5 Ga oceanic crust: Journal of African Earth Sciences, v. 6, p. 681-730.

de Wit, M., M. Jeffery, H. Bergh, and L. Nicolaysen, 1989, Geological Map of Sectors of Gondwana, scale 1: 10,000,000: American Association of Petroleum Geologists, 2 sheets.

Diaz, H. F., R. S. Bradley, and J. K. Eischeid, 1989, Precipitation fluctuations over global land areas since the late 1800s: Journal of Geophysical Research, v. 94, p. 1195-1210.

Dietz, R. S., 1961, Continent and ocean basin evolution by spreading of the sea floor: Nature, v. 190, p. 854-857.

Dietz, R. S., and M. J. Sheehy, 1953, Transpacific detection by underwater sound of Myojin volcanic explosions: Oceanographic Society of Japan, Journal, v. 9, p. 53-83.

Dingle, R. V., 1981, Continental margin subsidence—A comparison between the east and west coasts of Africa: American Geophysical Union, Geodynamics Series, Dynamics of Passive Margins, v. 6, p. 59-71.

Dingle, R. V., W. G. Siesser, and A. R. Newton, 1983, Mesozoic and Tertiary Geology of Southern Africa: Rotterdam, A. A. Balkema, 375 p.

Director of Geological Survey of India (coordinator), 1958, Geological Map of Asia and the Far East, scale 1: 5,000,000: Bangkok, Thailand, United Nations Economic Commission for Asia and the Far East, 6 sheets.

Disney, L. P., 1955, Tide heights along the coasts of the United States: American Society of Civil Engineers, Proceedings, v. 81, 9 p.

Dolan, R., and H. G. Goodell, 1986, Sinking cities: American Scientist, v. 74, p. 38-47.

Dolezal, R., and M. Petersen, 1969, Subsidence in the North German coastal region: in L. J. Tison (ed.), International Symposium on Land Subsidence, Tokyo, Japan: International Association of Hydrological Sciences, Publication 88, v. 1, p. 35-42.

Don Zobin, 1956, Chronological Tables of Chinese History: Hong Kong, University Press, v. 2, 417 p.

Donn, W. L., W. R. Farrand, and M. Ewing, 1962, Pleistocene ice volumes and sea-level lowering: Journal of Geology, v. 70, p. 206-214.

Donoghue, J. F., in press, Trends in Chesapeake Bay sedimentation rates during the Late Holocene: Quaternary Research.

Donovan, D. T., and E. J. W. Jones, 1979, Causes of world-wide changes in sea level: Geological Society of London, Journal, v. 136, p. 187-192.

Doodson, A. T. (secretary), 1953, Monthly and Annual Mean Heights of Sea Level 1947 to 1951 and Unpublished Data for Earlier Years: Association d'Océanographie Physique, Union Géodésique et Géophysique Internationale, v. 12, 61 p.

Doodson, A. T., (secretary), 1954, Secular variation of sea-level: Association d'Oceanographie Physique, Union Géodésique et Géophysique Internationale, v. 13, 21 p.

Dumas, B., P. Gueremy, P. J. Hearty, R. Lhenaff, and J. Raffy, 1988, Morphometric analysis and amino acid geochronology of uplifted shorelines in a tectonic region near Reggio Calabria, South Italy: Palaeogeography, Palaeoclimatology, Palaeoecology, v. 68, p. 273–289.

Duncan, R. A., and D. A. Clague, 1985, Pacific plate motion recorded by linear volcanic chains: in A. E. M. Nairn, F. G. Stehli, and S. Uyeda, eds., The Ocean Basins and Margins—v. 7A, The Pacific Ocean: New York, Plenum Press, p. 89–121.

Dupont, J., 1988, The Topnga and Kermadec ridges: in A. E. M. Nairn, F. G. Stehli, and S. Uyeda, eds., The Ocean Basins and Margins—v. 7B, The Pacific Ocean, p. 375–409.

Duque-Caro, H., 1979, Major structural elements and evolution of northwestern Colombia: in J. S. Watkins, L. Montadert, and P. W. Dickerson, eds., Geological and Geophysical Investigations of Continental Margins: American Association of Petroleum Geologists, Memoir 29, p. 329–351.

Dutton, C. E., 1889, The Charleston earthquake of August 31, 1886: U. S. Geological Survey 9th Annual Report, p. 203–528.

Edelman, T., 1954, Tectonic movements as resulting from the comparison of two precision levelings: Geologie en Mijnbouw, v. 16, p. 209–212.

Einarsson, P., 1979, Seismicity and earthquake focal mechanisms along the Mid-Atlantic plate boundary between Iceland and the Azores: Tectonophysics, v. 55, p. 127–153.

Emery, K. O., 1955, Submarine topography south of Hawaii: Pacific Science, v. 11, p. 286–291.

Emery, K. O., 1956, Marine geology of Johnston Island and its surrounding shallows, central Pacific Ocean. Geological Society of America, Bulletin, v. 67, p. 1505–1520.

Emery, K. O., 1960, The Sea off Southern California—A Modern Habitat of Petroleum: New York, John Wiley & Sons, 366 p.

Emery, K. O., 1980, Relative sea levels from tide-gauge records: National Academy of Sciences, Proceedings, v. 77, p. 6968–6972.

Emery, K. O., 1981, Low marine terraces of Cayman Island: Estuarine and Coastal Shelf Science, v. 12, p. 569–578.

Emery, K. O., 1983, Tectonic evolution of East China Sea: in Jin Qingming and J. D. Milliman, eds., Sedimentation on the Continental Shelf, with Special Reference to the East China Sea: International Symposium at Hangzhou: Beijing, China Ocean Press, p. 74–83; see also simplified version in Oceanus, v. 26, no. 4, p. 26–32.

Emery, K. O., 1987, Stick charts of the Marshall Islands: Epigraphic Society, Occasional Papers, v. 16, p. 28–50.

Emery, K. O., and D. G. Aubrey, 1985, Glacial rebound and relative sea levels in Europe from tide-gauge records: Tectonophysics, v. 120, p. 239–255.

Emery, K. O., and D. G. Aubrey, 1986a, Relative sea-level changes from tide-gauge records of eastern Asia mainland: Marine Geology, v. 72, p. 33–45.

Emery, K. O., and D. G. Aubrey, 1986b, Relative sea level change from tide gauge records of western North America: Journal of Geophysical Research, v. 92, p. 13,941–13,953.

Emery, K. O., and D. G. Aubrey, 1989, Tide gauges of India: Journal of Coastal Research, v. 5, p. 489–501.

Emery, K. O., and E. Uchupi, 1972, Western Atlantic Ocean—Topography, Rocks, Structure, Water, Life, and Sediments: American Association of Petroleum Geologists, Memoir 17, 532 p.

Emery, K. O., and E. Uchupi, 1984, The Geology of the Atlantic Ocean: New York, Springer-Verlag, 1050 p.

Emery, K. O., and You Fanghu, 1981, Sea-level changes in the western Pacific with emphasis on China: Oceanologia et Limnologia Sinica, v. 12, p. 297–310.

Emery, K. O., D. G. Aubrey, and V. Goldsmith, 1988a, Coastal neotectonics of the Mediterranean from tide-gauge records: Marine Geology, v. 81, p. 41–52.

Emery, K. O., A. S. Merrill, and E. R. M. Druffel, 1988b, Changed late Quaternary marine environments on Atlantic continental shelf and upper slope: Quaternary Research, v. 30, p. 251–269.

Emery, K. O., H. Niino, and B. Sullivan, 1971, Post-Pleistocene levels of the East China Sea: in K. K. Turekian, ed., The Late Cenozoic Glacial Ages: Silliman Lecture Volume, New Haven, Connecticut, Yale University Press, p. 381–390.

Emery, K. O., J. I. Tracey, Jr., and H. S. Ladd, 1954, Geology of Bikini and Nearby Atolls, Bikini and Nearby Atolls: Part 1, Geology, U. S. Geological Survey, Professional Paper 260-A, 265 p.

Emery, K. O., E. Uchupi, J. Sunderland, H. L. Uktoseja, and E. M. Young, 1972, Geological structure and some water characteristics of the Java Sea and adjacent continental shelf: United Nations Economic Commission for Asia and the Far East, Committee for Coordination of Offshore Prospecting, Technical Bulletin, v. 6, p. 197–223.

Emiliani, C., 1955, Pleistocene temperatures: Journal of Geology, v. 63, p. 538–578.

Emiliani, C., 1978, The cause of the ice ages: Earth and Planetary Science Letters, v. 37, p. 349–352.

Enfield, D. B., and J. S. Allen, 1980, On the structure and dynamics of monthly mean sea level anomalies along the Pacific coast of North and South America: Journal of Physical Oceanography, v. 10, p. 557–578.

Espinosa, A. F., W. Rinehart, and M. Tharp, 1981, Seismicity of the Earth, 1960–1980, scale 1:46,460,600: U. S. Navy through Office of Naval Research, 1 sheet.

Etkins, R., and E. S. Epstein, 1982, The rise of global mean sea level as an indication of climate change: Science, v. 215, p. 287–289.

Fairbanks, R. G., 1989, A 17,000-year glacio-eustatic sea level record—Influence of glacial melting rates on the Younger Dryas event and deep-ocean circulation: Nature, v. 342, p. 637–642.

Fairbridge, R. W., 1961, Eustatic changes in sea level: in L. H. Ahrens, F. Press, R. Kalervo, and S. K. Runcorn, eds., Physics and Chemistry of the Earth: New York, Pergamon Press, v. 4, p. 99–185.

Fairbridge, R. W., and O. A. Krebs, Jr., 1962, Sea level and the southern oscillation: Royal Astronomical Society, Geophysical Journal, v. 6, p. 532–545.

Farmer, G., T. M. L. Wigley, P. D. Jones, and M. Salmon, 1989. Documenting and explaining recent global-mean temperature changes: Norwich, England, Climatic Research Unit, Final Report to National Environmental Research Council, U. K., Contract GR3/6565.

Farrell, W. E., and J. A. Clark, 1976, On postglacial sea level: Royal Astronomical Society, Geophysical Journal, v. 46, p. 647–667.

Faure, H., 1980, Late Cenozoic vertical movements in Africa: in N. A. Mörner, ed., Earth Rheology, Isostasy, and Eustasy: New York, John Wiley & Sons, p. 465–469.

Faure, H., and P. Elouard, 1967, Schéma des variations due niveau de l'Océan Atlantique sur la côte de l'Ouest de l'Afrique depuis 40,000 ans: Académie des Sciences de Paris, Comptes Rendus, v. 265, p. 784–487.

Faure, H., and L. Hebrard, 1977, Variations des lignes de rivages au Sénégal et en Mauritanie au course de L'Holocene: Studia Geologica Polonica, Warszawa, v. 52, p. 144–157.

Fischer, A. G., 1984, The two Phanerozoic supercycles: in W. A. Berggren and J. A. van Couvering, eds., Catastrophes and Earth History; The New Uniformitarianism: Princeton, New Jersey, Princeton University Press, p. 129–150.

Fisher, R. L., G. L. Johnson, and B. C. Heezen, 1967, Mascarene Plateau, western Indian Ocean: Geological Society of America, Bulletin, v. 78, p. 1247–1266.

Fisk, H. N., and E. McFarlan, Jr., 1955, Late Quaternary deltaic deposits of the Mississippi River: in A. Poldervaart, ed., Crust of the Earth: Geological Society of America, Special Paper 62, p. 279–302.

Fisk, H. N., E. McFarlan, C. R. Kolb, and L. J. Wilbert, Jr., 1954, Sedimentary framework of the modern Mississippi delta: Journal of Sedimentary Petrology, v. 24, p. 76–99.

Flemming, N. C., 1968, Archeological evidence of eustatic changes of sea level and earth movements in the western Mediterranean: Geological Society of America, Special Paper 109, 125 p.

Flemming, N. C., and C. O. Webb, 1986, Tectonic and eustatic coastal changes during the last 10,000 years derived from archaeological data: Zeitschrift für Geomorphologie, N. F., supplement 62, p. 1–29.

Flemming, N. C., and P. L. Woodworth, 1988, Monthly mean sea levels in Greece during 1969–1983 compared to relative vertical land movements measured over different timescales: Tectonophysics, v. 148, p. 59–72.

Flemming, N. C., A. Raban, and C. Goetschel, 1978, Tectonic and eustatic changes on the Mediterranean coast of Israel in the last 9000 years: Progress in Underwater Science, v. 3, p. 33–93.

Fletcher, G. L., and R. A. Soeparjadi, 1984, Oil and gas developments in Far East in 1983: American Association of Petroleum Geologists, Bulletin, v. 68, p. 1622–1675.

Flick, P. D. A., and L. Pervinquière, 1904, Sur les plages soulevees de Monastir et de Sfax (Tunisie): Société Geologique de France, Bulletin 4, p. 195–206.

Flint, R. F., 1971, Glacial and Quaternary Geology: New York, John Wiley & Sons, 892 p.

Flint, R. F., R. B. Colton, R. P. Goldthwait, and H. B. Willman (eds.), 1959, Glacial Map of the United States East of the Rocky Mountains, scale 1:1,750,000: Geological Society of America, 2 sheets.

Fontes, J. Ch., and G. Bortolami, 1972, Subsidence of the area of Venice during the past 40,000 years: Consiglio Nazionale Ricerca, Instituto por lo Studio della Dinamica delle Grande Masse, Tech. Rep. 54, Venezia, 11 p.

Forel, F.A., 1895. Le Leman, 2: Lausanne, Switzerland, Librairie de l'Universite, 651 p.

Foreman, M. G. G., 1978, Manual for tidal heights analysis and prediction: Canadian Pacific Marine Science, Report 77-10, 10 p.

Forsythe, R., 1982, The Late Paleozoic to Early Mesozoic evolution of southern South America—A plate tectonic interpretation: Geological Society of London, Journal, v. 139, p. 671–682.

Fox, H. L., 1974, Alexander the Great: New York, Dial Publishing Co., 568 p.

Freund, R., Z. Garfunkel, I. Zak, M. Goldberg, T. Weissbrod, and B. Derin, 1970, The shear along the Dead Sea rift: Royal Society of London, Philosophical Transactions, series A, v. 267, p. 107–130.

Friedrichs, C. T., D. G. Aubrey, and P. E. Speer, 1990, Impacts of relative sea-level rise on evolution of shallow estuaries: in R. T. Cheng, ed., Residual currents and long-term transport: New York, Springer-Verlag, p. 105–122.

Frihy, O. E., 1988, Nile Delta shoreline changes—Aerial photographic study of a 28-year period: Journal of Coastal Research, v. 4, p. 597–606.

Fu, L., and D. B. Chelton, 1985, Observing large-scale temporal variability of ocean currents by satellite altimetry, with application to the Antarctic Circumpolar Current: Journal of Geophysical Research, v. 90, p. 4721–4739.

Fujii, S., C. C. Lin, and H. D. Tija, 1971, Sea level changes in Asia during the past 11,000 years: Quaternaria, v. 14, p. 211–216.

Fulfaro, V. J., and K. Suguio, 1980, Vertical movements in continental southern Brazil during the Cenozoic: in N. A. Mörner, ed., Earth Rheology, Isostasy and Eustasy: New York, John Wiley & Sons, p. 419–425.

Gable, F. J., and D. G. Aubrey, 1989, Changing climate and Pacific archipelagic coastlines: Oceanus, v. 32, p. 71–73.

Gable, F. J., and D. G. Aubrey, 1990a, Potential impacts of contemporary changing climate on the Caribbean coastlines: Ocean and Shoreline Management, v. 13, p. 35–67.

Gable, F. J., and D. G. Aubrey, 1990b, Potential coastal impacts of contemporary changing climate on South Asian seas states: Environmental Management, v. 14, p. 33–46.

Gable, F. J., J. Gentile, and D. G. Aubrey, 1990, Global climatic issues in the coastal wider Caribbean: Environmental Conservation, v. 17, p. 51–60.

Gabrysch, R. K., 1969, Land-surface subsidence in the Houston–Galveston region, Texas: in L. J. Tison, ed., Proceedings of the First International Symposium on Land Subsidence, Tokyo, Japan, publication 88, v. 1, p. 43–54.

Gallagher, B. S., and W. H. Munk, 1971, Tides in shallow water—Spectroscopy: Tellus, v. 23, p. 346–363.

Gardner, J., and J. Maier, 1984, Gilgamesh: New York, Random House, 304 p.

Garfunkel, Z., 1981, Internal structure of the Dead Sea leaky transform (rift) in relation to plate tectonics: Tectonophysics, v. 80, p. 81–108.

Garfunkel, Z., and G. Almagor, 1985, Geology and structure of the continental margin off northern Israel and the adjacent part of the Levantine Basin: Marine Geology, v. 62, p. 105–131.

Gass, I. G., 1970, Tectonic and magmatic evolution of the Afro-Arabian dome: in T. N. Clifford and I. G. Gass, eds., African Magmatism and Tectonics: Darien, Connecticut, Hafner Publishing Co., p. 284–300.

Gatto, P., and L. Carbognin, 1981, The Lagoon of Venice–Natural environmental trend and man-induced modification: Hydrological Sciences Bulletin, v. 26, no. 4, p. 379–391.

Geological Society of Australia, 1971, Tectonic map of Australia and New Guinea, scale 1:5,000,000: Sydney, 1 sheet.

Giese, G. S., and D. G. Aubrey, 1987, Losing coastal upland to relative sea-level rise–3 scenarios for Massachusetts: Oceanus, v. 30, p. 16–27.

Giese, ese, G. S., and R. B. Hollander, 1987, The relationship between coastal seiches at Palawan Island and tide-generated internal waves in the Sulu Sea: Journal of Geophysical Research, v. 92, p. 5151–5156.

Giese, G. S., D. G. Aubrey, and P. Zeeb, 1987, Passive retreat of Massachusetts' coastal upland due to relative sea-level rise: Boston, Commonwealth of Massachusetts, Coastal Zone Management Office, 17 p. + app.

Giese, G. S., D. C. Chapman, P. G. Black, and J. A. Fornshell, 1990, Causation of large-amplitude coastal seiches on the Caribbean coast of Puerto Rico: Journal of Physical Oceanography, v. 20, p. 1448–1458.

Gill, E. D., and D. Hopley, 1972, Holocene sea levels in eastern Australia—A discussion: Marine Geology, v. 12, p. 223–233.

Gilluly, J., 1949, Distribution of mountain building in geologic time: Geological Society of America, Bulletin 60, p. 561–590.

Giovinetto, M. B., and C. R. Bentley, 1985, Surface balance in ice drainage systems of Antarctica: Antarctic Journal of the Unitred States, v. 20, p. 6–13.

Girdler, R. W., 1958, The relationship of the Red Sea to the East African rift system: Geological Society of London, Quarterly Journal, v. 114, p. 79–105.

Gloe, C. S., 1984, Case History no. 9.1. Latrobe Valley, Victoria, Australia: in J. F. Poland, ed., International Hydrological Programme, Working Group 8.4: Paris, United Nations Educational, Scientific and Cultural Organization, p. 145–153.

Goldsmith, V., and M. Gilboa, 1985, Development of an Israeli tidal atlas and comparison with other Mediterranean tidal data: Israel National Oceanographic Institute, Report H8/85, 28 p.

Goldsmith, V., and M. Gilboa, 1987, Mediterranean sea level changes from tide gauges: Proceedings of International Conference on Coastal Engineering, 20th, Taipei, Taiwan, Nov., 1985, American Society of Civil Engineering, p. 223–231.

Gornitz, V., in press, Mean sea-level changes in the recent past: International Workshop on the Effects of Climatic Change on Sea Level, Severe Tropical Storms and their Associated Impacts: Norwich, England, University of East Anglia.

Gornitz, V., and S. Lebedeff, 1987, Global sea-level changes during the past century: in D. Nummedal, O. H. Pilkey, and J. D. Howard, eds., Sea-Level Fluctuation and Coastal Evolution: Society of Economic Paleontologists and Mineralogists, Special Publication 41, p. 3–16.

Gornitz, V., and L. Seeber, 1990, Vertical crustal movements along the East Coast, North America, from historic and late Holocene sea level data: Tectonophysics, v 178, p. 127–150.

Gornitz, V., and A. R. Solow, in press, Observations of long-term tide-gauge records for indications of accelerated sea-level rise.

Gornitz, V., S. Lebedeff, and J. Hansen, 1982, Global sea level trend in the past century: Science, v. 215, p. 1611–1614.

Gortani, M., 1961, Extraction of methane waters and the Po Delta: Natura e Montagna, ser. 2, v. 1, p. 4–13.

Gow, A. J., S. Epstein, and R. P. Sharp, 1973, Climatological implications of stable isotope variations in deep ice cores from Byrd Station, Antarctica: Geological Society of America, Memoir 136, p. 323–326.

Gradstein, F. M., F. P. Agterberg, M.-P. Aubry, W. A. Berggren, J. J. Flynn, R. Hewitt, D. V. Kent, K. D. Klitgord, K. G. Miller, J. Obradovich, J. G. Ogg, D. R. Prothero, and G. E. G. Westerman, 1988, Sea level history: Science, v. 241, p. 599-601.

Graham, N. E., and W. B. White, 1988, The El Niño cycle—A natural oscillator of the Pacific Ocean-Atmosphere system: Science, v. 240, p. 1293-1302.

Grant, D. R., 1980, Quaternary sea-level change in Atlantic Canada as an indication of crustal delevelling: in N.-A. Mörner, ed., Earth Rheology, Isostasy and Eustasy: New York, John Wiley & Sons, p. 201-214.

Griggs, R. F., 1922, The Valley of Ten Thousand Smokes: National Geographic Society, 340 p.

Grotch, S. L., 1988, Regional intercomparisons of general circulation model predictions and historical climate data: Washington D. C., U. S. Department of Energy, Report DOE/NBB-0084 (TR041), 291 p.

Gulliver, F. P., 1899, Shoreline topography: American Academy of Arts and Sciences, Proceedings, v. 34, p. 151-258.

Güst, D. A., K. T. Biddle, D. W. Phelps, and M. A. Uliana, 1985, Associated Middle and Late Jurassic volcanism and extension in southern South America: Tectonophysics, v. 116, p. 223-253.

Gutenberg, B., 1941, Changes in sea level, postglacial uplift, and mobility of the earth's interior: Geological Society of America, Bulletin, v. 52, p. 721-772.

Haile, N. S., 1971, Quaternary shorelines in West Malaysia and adjacent parts of the Sunda Shelf: Quaternaria, v. 15, p. 333-343.

Hakluyt, R., 1600, The Principal Navigations of the English Nation Made by Sea or Overland to the Remote and Farthest Distant Quarters of the Earth at Any Time Within the Compasse of These 1600 Yeares: London, 8 vols. (reprinted in 1907 by J. M. Dent & Co., London).

Hallam, A., 1977, Secular changes in marine inundation of USSR and North America through the Phanerozoic: Nature, v. 269, p. 769-712.

Hallam, A., 1978, Eustatic cycles in the Jurassic: Palaeoceanography, Palaeoclimatology, Palaeoecology, v. 23, p. 1-32.

Hallam, A., 1984, Pre-Quaternary sea-level changes: Earth and Planetary Sciences, Annual Review, v. 12, p. 205-243.

Hamilton, E. L., 1956, Sunken Islands of the Mid-Pacific Mountains: Geological Society of America, Memoir 64, 97 p.

Hamilton, H. C., and W. Falconer, 1854, Geography of Strabo: London, H. G. Bohn, 3 vols., 517, 410, and 422 p.

Han, M. K., 1989, Global warming-induced sea level rise in China—Response and Strategies: Presentation to World Conference on Preparing for Climate Change, Cairo, Egypt, December 19, 1989.

Hansen, J., and S. Lebedeff, 1987, Global trends of measured surface air temperature: Journal of Geophysical Research, v. 92, p. 13345-13372.

Hansen, J., and T. Takahashi, 1984, Climate Processes and Climate Sensitivity: Geophysical Monograph Series 29, American Geophysical Union, Washington D. C., 368 p.

Hansen, J., D. Johnson, A. Lacis, S. Lebedeff, P. Lee, D. Rind, and G. Russell, 1981, Climate impact of increasing atmospheric carbon dioxide: Science, v. 213, p. 957-966.

Haq, B. U., J. Hardenbol, and P. R. Vail, 1987, Chronology of fluctuating sea levels since the Triassic: Science, v. 235, p. 1156-1167.

Harland, W. B., 1983, The Proterozoic glacial record: in L. G. Medaris, Jr., C. W. Byers, D. M. Mickelson, and W. C. Shanks, eds., Proterozoic Geology—Selected Papers from an International Proterozoic Symposium: Geological Society of America, Memoir 161, p. 279-288.

Harmon, R. S., L. S. Land, R. M. Mitterer, P. Garrett, H. P. Schwarcz, and G. J. Larson, 1981, Bermuda sea level during the last interglacial: Nature, v. 289, p. 481-483.

Hayashi, T., 1969, A study on the vertical movements of the earth's crust by means of the precise leveling: Geographical Survey Institute, Bulletin, v. 15, p. 1-67.

Hayes, D. E., and J. Ringis, 1972, The early opening of the central Tasman Sea (abstract): American Geophysical Union, Transactions, v. 53, p. 413.

Hayes, D. E., and B. Taylor, 1978, Tectonics, scale 1:6,442,194 at equator: in D. E. Hayes, ed., A Geophysical Atlas of the East and Southeast Asian Seas: Geological Society of America, MC-25, 8 p. + 6 charts.

Hays, J. D., and W. C. Pitman, III, 1973, Lithospheric motion, sea level changes and climatic and ecological consequences: Nature, v. 246, p. 18-22.

Hays, J. D., J. Imbrie, and N. J. Shackleton, 1976, Variations in the Earth's orbit, pacemaker of the ice ages: Science, v. 194, p. 1121-1132.

Hazel, J. E., L. E. Edwards, and L. M. Bybell, 1984, Significant unconformities and the hiatuses represented by them in the Paleogene of the Atlantic and Gulf Coastal Province: in J. S. Schlee, ed., Interregional Unconformities and Hydrocarbon Accumulation: American Association of Petroleum Geologists, Memoir 36, p. 59-66.

Healy, J., 1962, Structure and volcanism in the Taupo volcanic zone, New Zealand: in G. A. Macdonald and H. Kuno, eds., The Crust of the Pacific Basin: American Geophysical Union, Geophysical Monograph 6, p. 151-157.

Heezen, B. C., and D. J. Fornari, no date, Geological Map of the Pacific Ocean, 1: 35,000,000: United Nations Educational, Scientific and Cultural Organization, sheet 20 of Geological World Atlas.

Heezen, B. C., and M. Tharp, 1965, Descriptive sheet to accompany physiographic diagram of the Indian Ocean, scale 1:1,400,000: Geological Society of America, 1 sheet.

Heezen, B. C., and M. Tharp, 1966, Physiography of the Indian Ocean: Royal Society of London, series A, v. 259, Philosophical Transactions, p. 137-149.

Heirtzler, J. R., P. Cameron, P. J. Cook, T. Powell, H. A. Roeser, S. Sukardi, and J. J. Veevers, 1978, The Argo Abyssal Plain: Earth and Planetary Science Letters, v. 41, p. 21-31.

Herz, N., 1977, Timing of spreading in the South Atlantic—Information from Brazilian alkalic rocks: Geological Society of America, Bulletin, v. 88, p. 101-112.

Hess, H. H., 1946, Drowned ancient islands of the Pacific basin: American Journal of Science, v. 244, p. 772-791.

Hess, H. H., 1962, History of the ocean basins: in A. E. J. Engel, H. L. James, and B. P. Leonard, eds., Petrologic Studies, A Volume in Honor of A. F. Buddington: Geological Society of America, p. 599-620.

Hicks, S. D., 1967, The tide prediction centenary of the United States Coast and Geodetic Survey: International Hydrographic Review, v. 44, p. 121-131.

Hicks, S. D., 1972. Changes in tidal characteristics and tidal datum planes: in The Great Alaska Earthquake of 1964. Oceanography and Coastal Engineering: Washington, D.C., National Academy of Sciences, p. 310-314.

Hicks, S. D., 1978, An average geopotential sea level series for the United States: Journal of Geophysical Research, v. 83, p. 1377-1379.

Hicks, S. D., and J. E. Crosby, 1974, Trends and variability of yearly mean sea level 1893-1972: Washington, D. C., National Oceanic and Atmospheric Administration, National Ocean Survey, NOAA Tech. Memo. 13, 14 p.

Hicks, S. D., and L. E. Hickman, Jr., 1986, United States sea level variations through 1986: Shore and Beach, July, p. 3-7.

Hicks, S. D., and W. Shofnos, 1965, Yearly sea level variations for the United States: Proceedings of the American Society of Civil Engineers, Journal of the Hydraulics Division, HY 5, (4468), p. 23-32.

Higgins, C. G., 1965, Causes of relative sea-level changes: American Scientist, v. 53, p. 464-476.

Hilde, T. W. C., 1983, Sediment subduction versus accretion around the Pacific: Tectonophysics, v. 99, p. 381-397.

Hilde, T. W. C., and S. Uyeda, 1983, Trench depth, variation and significance: in T. W. C. Hilde and S. Uyeda, eds, Geodynamics of the Western Pacific-Indonesian Region: American Geophysical Union, Geodynamics Series, v. 11, p. 75-89.

Hirono, T., 1969, Niigata ground subsidence and ground water change: in L. J. Tison, ed., International Symposium on Land Subsidence: Tokyo, Japan, International Association of Hydrological Sciences, Publication 88, v. 1, p. 144-161.

Hoffert, M. I., and B. P. Flannery, 1985, Model projections of the time-dependent response to increasing carbon dioxide: in M. C. MacCracken and F. M. Luther, eds., Projecting the Climatic Effects of Increasing Carbon Dioxide: Washington, D. C., U. S. Department of Energy, Carbon Dioxide Research Division, p. 149-190.

Hoffman, J. S., D. Keyes, and J. G. Titus, 1983, Projecting future sea level rise—Methodology, estimates to the year 2100, and research needs: Washington, D. C., U. S. Government Printing Office, 055-000-0236-3, 121 p.

Hoffman, J. S., J. B. Wells, and J. G. Titus, 1986, Future global warming and sea level rise: *in* G. Sigbjarnason, ed., Iceland Coastal and River Symposium, Reykjavik: National Energy Authority, p. 245-266.

Holdahl, S. R. and N. L. Morrison, 1974, Regional investigations of vertical crustal movements in the U.S., using precise relevelings and mareograph data: Tectonophysics, v. 23, p. 373-290.

Holland, H. D., 1984, The Chemical Evolution of the Atmosphere and Oceans: Princeton, New Jersey, Princeton University Press, 582 p.

Holtzscherer, J. J., and A. Bauer, 1954, Contribution à la connaissance de l'Inlandis du Groenland: Association Internationale d'Hydrologie, Publication 39, p. 244-296.

Honda, K., T. Terada, Y. Yoshida, and D. Isitani, 1908, Secondary undulations of oceanic tides: Tokyo, Journal of the College of Science, Imperial University, v. 24, p. 1-110.

Horsfield, W., 1975, Quaternary vertical movements in the Greater Antilles: Geological Society of America, Bulletin, v. 86, p. 933-938.

Horsfield, W., 1976, Quaternary crustal movements in the Greater Antilles: Transactions 7th Caribbean Geological Conference, Guadaloupe, 1974, p. 107-113.

Hurley, P. M., and J. R. Rand, 1969, Pre-rift continental nuclei: Science, v. 164, p. 1229-1242.

Huyer, A., and R. L. Smith, 1985, The signature of El Niño off Oregon in 1982-1983: Journal of Geophysical Research, v. 90, p. 7133-7142.

Hwang, J. M., and C. M. Wu, 1969, Land subsidence problems in Taipei Basin: *in* L. J. Tison, ed., International Symposium on Land Subsidence: Tokyo, Japan, International Association of Hydrological Sciences, Publication 88, v. 1, p. 21-34.

Hyndman, R. D., G. K. Muecke, and F. Aumento, 1974, Deep drill 1972—Heat flow and heat production in Bermuda: Canadian Journal of Earth Sciences, v. 11, p. 809-818.

IAPSO, 1985, Advisory Committee on Tides and Mean Sea Level: EOS, American Geophysical Union, Transactions, v. 66, p. 754-756.

Inaba, Y., I. Abe, S. Iwasaki, S. Aoki, T. Endo, and R. Kaido, 1969, Reviews of land subsidence researches in Tokyo: *in* L. J. Tison, ed., International Symposium on Land Subsidence: Tokyo, Japan, International Association of Hydrological Sciences, Publication 88, v. 1, p. 87-98.

Inman, D. L., 1980, Shore processes and marine archaeology: Oceanography in China, CSC PRC Report no. 9 (in China, Ch. 6, p. 47-65); Washington, D. C., National Academy of Sciences, 106 p.

Inman, D. L., D. G. Aubrey, and S. L. Pawka, 1976, Application of nearshore processes to the Nile Delta: Proceedings of Seminar of Nile Delta Sedimentology, Alexandria, 25-29 October 1975 by UNDP/UNESCO Project for Coastal Studies, published by the project under Academy of Scientific Research and Technology, 257 p.

Irving, E., 1979, Pole positions and continental drift since the Devonian: *in* M. W. McElhinny, ed., The Earth—Its Origin, Structure, and Evolution: New York, Academic Press, p. 567-597.

Jackson, E. D., 1976, Linear volcanic chains on the Pacific plate: *in* G. H. Sutton, M. H. Manghnani, and R. Moberly, eds., The Geophysics of the Pacific Ocean Basin and Its Margin; A Volume in Honor of George P. Woollard: American Geophysical Union Geophysical Monograph 19, p. 319-335.

Jamieson, T. F., 1865, On the history of the last geological changes in Scotland: Geological Society of London, Quarterly Journal, v. 21, p. 161-203.

Jones, P. D., S. C. B. Raper, R. S. Bradley, H. F. Diaz, P. M. Kelly, and T. M. L. Wigley, 1986a, Northern Hemisphere surface air temperature variations, 1851-1984: Journal of Climate and Applied Meteorology, v. 25, p. 161-179.

Jones, P. D., S. C. B. Raper, R. S. Bradley, H. F. Diaz, P. M. Kelly, and T. M. L. Wigley, 1986b, Southern Hemisphere surface air temperature variations, 1851-1984: Journal of Climate and Applied Meteorology, v. 25, p. 1213-1230.

Jones, P. D., 1988, Hemispheric surface air temperature variations—Recent trends and an update to 1987: Journal of Climate, v. 1, p. 654-660.

Jones, P. D., 1989, The influence of ENSO on global temperatures: Climate Monitor, v. 17, p. 80-89.

Jones, P. D., P. M. Kelly, G. B. Goodess, and T. R. Karl, 1989, The effect of urban warming on the Northern Hemisphere temperature average: Journal of Climate, v. 2, p. 285-290.

Johnson, D. W., 1919, Shore Processes and Shoreline Development: New York, John Wiley & Sons, 584 p.

Johnson, D. W., 1925, The New England-Acadian Shoreline: New York, John Wiley & Sons, 608 p.

Jouannic, C., C. T. Hoang, W. S. Hantoro, and R. M. Delinom, 1988, Uplift rate of coral reef terraces in the area of Kupang, West Timor—Preliminary results: Palaeogeography, Palaeoclimatology, Palaeoecology, v. 68, p. 259-272.

Kafri, U., 1969, Recent crustal movements in northern Israel: Journal of Geophysical Research, v. 74, p. 4246-4258.

Kafri, U., and I. Karcz, 1975, On the stability of the Mediterranean coast of Israel since Roman times. A further contribution to the discussion: Israel Journal of Earth-Science, v. 24, p. 114-116.

Kailasam, L. N., 1975, Epeirogenic studies in India with reference to recent vertical movements: Tectonophysics, v. 29, p. 505-521.

Kailasam, L. N., 1980, Late Cenozoic vertical movements in the Indian subcontinent and their relation to isostasy: *in* N. A. Mörner, ed., Earth Rheology, Isostasy and Eustasy: New York, John Wiley & Sons, p. 407-418.

Kana, T. W., J. Michel, M. O. Hayes, and J. R. Jensen, 1984, The physical impact of sea level rise in the area of Charleston, South Carolina: *in* M. C. Barth and J. G. Titus, eds., Greenhouse Effect and Sea Level Rise: New York, Van Nostrand Reinhold Company, p. 105-150.

Karl, T. R., H. F. Diaz, and G. Kukla, 1988, Urbanization—Its detection in the United States climate record: Journal of Climate, v. 1, p. 1099-1123.

Karl, T. R., and P. D. Jones, 1990, Reply to comments on "Urban bias in area-averaged surface temperature trends": American Meteorological Society, Bulletin, v. 71.

Kato, T., and K. Tsumura, 1979, Vertical land movements in Japan as deduced from tidal records (1951-1978): Earthquake Research Institute, Bulletin, v. 54, p. 559-628.

Katupotha, J., and K. Fujiwara, 1988, Holocene sea level change on the southwest and south coasts of Sri Lanka: Palaeogeography, Palaeoclimatology, and Palaeoecology, v. 68, p. 189-203.

Katz, H.-R., 1974, Margins of the southwest Pacific: *in* C. A. Burk and C. L. Drake, eds., The Geology of Continental Margins: New York, Springer-Verlag, p. 549-565.

Kawana, T., and P. A. Pirazzoli, 1985, Holocene coastline changes and seismic uplift in Okinawa Island, the Ryukyus, Japan: Zeitschrift für Geomorphologie, Supplement, v. 57, p. 11-51.

Kawana, T., and P. A. Pirazzoli, 1990, Re-examination on the Holocene emerged shorelines in Irabu and Shimoji islands, the south Ryukyus, Japan: The Quaternary Research, v. 28, p. 419-426.

Kayan, I., 1988, Late Holocene sea-level changes on the western Anatolian coast: Palaeogeography, Palaeoclimatology, Palaeoecology, v. 68, p. 205-218.

Kaye, C. A., 1959, Shoreline features and Quaternary shoreline changes, Puerto Rico: U. S. Geological Survey, Professional Paper 317-BB, 140 p.

Keigwin, L. D., and G. A. Jones, in preparation, Glacial-Holocene stratigraphy, chronology, and paleoceanographic observations on some North Atlantic sediment drifts.

Keller, W., 1956, The Bible as History: New York, William Morrow and Co., 452 p.

Kennedy, D. J. L., 1961, A study of failures of liners for oil wells associated with the compaction of oil producing strata: University of Illinois, unpublished doctoral dissertation, 270 p.

Kennett, J. P., 1984, Marine Geology: Englewood Cliffs, New Jersey, Prentice-Hall, 813 p.

Kerr, R. A., 1988a, The weather in the wake of El Niño: Science, v. 240, p. 883.

Kerr, R. A., 1988b, La Niña's big chill replaces El Niño: Science, v. 241, p. 1037-1038.

Kidson, C., 1982, Sea level changes in the Holocene: Quaternary Science Reviews, v. 1, p. 121-151.

Kienle, J., K. G. Dean, and H. Garbeil, 1990, Satellite surveillance of volcanic ash plumes, application to aircraft safety: EOS, American Geophysical Union, Transactions, v. 71, p. 266.

King, P. B., 1969, Tectonic Map of North America, scale 1:5,000,000: U. S. Geological Survey, 2 sheets.

Klige, R. K., O. K. Leontiev, S. A. Lukyanova, L. G. Nikiforov, and V. A. Schleinikov, 1978, Uroven' Berega I Dno Okeana (Levels, Shores, and the Bottom of the Ocean), Moscow, Izdatelstvo Nauka, 191 p.

Knudsen, M., 1901, Hydrographical tables: Copenhagen, G. E. C. Gad, 63 p.

Kolb, C. R., and J. R. van Lopik, 1966, Depositional environments of the Mississippi River deltaic plain–southeastern Louisiana: in M. L. Shirley and J. A. Ragsdale, eds., Deltas in Their Geologic Framework: Houston Geological Society, p. 17–61.

Kolla, V., and E. Coumes, 1987, Morphology, internal structure, seismic stratigraphy, and sedimentation of Indus Fan: American Association of Petroleum Geologists, Bulletin, v. 71, p. 650–677.

Kolla, V., R. Buffler, and J. Ladd, 1984, Seismic stratigraphy and sedimentation of Magdalena Fan, southern Colombian Basin, Caribbean Sea: American Association of Petroleum Geologists, Bulletin, v. 68, p. 316–322.

Komaki, S., 1969, On the variation of artesian head and land-surface subsidence due to groundwater withdrawal: in L. J. Tison, ed., International Symposium on Land Subsidence: Tokyo, Japan, International Association of Hydrological Sciences, Publication 88, v. 1, p. 256–271.

Korsch, R. J., and H. W. Wellman, 1988, The geological evolution of New Zealand and the New Zealand region: in A. E. M. Nairn, F. G. Stehli, and S. Uyeda, eds., The Ocean Basins and Margins–v. 7B, The Pacific Ocean, p. 411–482.

Koyanagi, R. Y., E. T. Endo, and P. L. Ward, 1976, Seismic activity on the island of Hawaii, 1970-1973: in G. H. Sutton, M. H. Manghnani, and R. Moberly, eds., The Geophysics of the Pacific Ocean Basin and Its Margin; A Volume in Honor of George P. Woollard: American Geophysical Union, Geophysical Monograph 19, p. 169–175.

Kuenen, P. H., 1945, De zeespiegeldrijzing der laatste decennia: Tijdschrift van het Koninktlijk Nederlandsch Aardrijkskundig Genootschap, v. 62, p. 159-169.

Kuenen, P. H., 1950, Marine Geology: New York, John Wiley & Sons, 568 p.

Kuenen, P. H., 1954, Eustatic changes of sea-level: Geologie en Mijnbouw, N. S., v. 16, p. 148–155.

Kuenen, P. H., 1955, Sea level and crustal warping: Geological Society of America, Special Paper 62, p. 193–203.

Kugler, H. G., 1933, Contribution to the knowledge of sedimentary volcanism in Trinidad (with discussion): Institute of Petroleum Technologists, Journal, v. 19, p. 743–772.

Kuhn, G. G., and F. P. Shepard, 1984, Sea Cliffs, Beaches, and Coastal Valleys of San Diego County–Some Amazing Histories and Some Horrifying Implications: Berkeley, University of California Press, 193 p.

Kulp, J. L., H. W. Feely, and L. E. Tryon, 1951, Lamont natural radiocarbon measurements, I: Science, v. 114, p. 565–568.

Kumar, R., 1985, Fundamentals of Historical Geology and Stratigraphy of India: New York, Wiley, 254 p.

Ladd, J. W., T. C. Shih, and C. J. Tsai, 1981, Cenozoic tectonics of central Hispaniola and adjacent Caribbean sea: American Association of Petroleum Geologists, Bulletin, v. 65, p. 466–489.

Lallemand, F., 1956, Journal de Bord de Pythéas: Éditions de Paris, 258 p.

Lambert A., J. O. Liard, N. Courtier, A. K. Goodacre, and R. K. McConnell, 1989, Canadian absolute gravity program: EOS, American Geophysical Union, Transactions, v. 70, p. 1447, 1459–1460.

Land, L. S., F. T. Mackenzie, and S. J. Gould, 1967, Pleistocene history of Bermuda: Geological Society of America, Bulletin, v. 78, p. 993–1006.

Landström, B., 1969, Histoire du Voilier, Paris, Albin Michel, 185 p.

Lange, A., 1989, Results of the WMO/CAS NWP data study and intercomparison project for forecasts for the Northern Hemisphere in 1988: World Meteorological Organization, World Weather Watch, Technical Report no. 7.

Larson, R. L., W. C. Pitman III, X. Golovchenko, S. C. Cande, J. F. Dewey, W. H. Haxby, J. L. LaBreque, 1985, The Bedrock Geology of the World: New York, W. H. Freeman & Co., Inc., 1 sheet.

Larue, D. K., 1990, Toa Baja drilling project, Puerto Rico: EOS, American Geophysical Union, Transactions, v. 71, p. 233–234.

Leatherman, S. P., 1985, Effect of accelerated sea-level rise on Ocean City, Maryland: in J. G. Titus, ed., Potential Impacts of Sea Level Rise on the Beach at Ocean City, Maryland: Washington, D. C., U.S. Environmental Protection Agency, 34 p.

Lennon, G. W. (director), 1976-1978, Monthly and annual mean heights of sea level: Bidston, Merseyside, England, Institute of Oceanographic Sciences, Permanent Service for Mean Sea Level, 3 v. (multilithed).

Lensen, G. J., 1974, On the nature of vertical deformation and on the frequency of earth deformation in general: Tectonophysics, v. 23, p. 391–406.

Leonardi, P., 1960, Cause geologiche del graduale profondamento di Venezia e della sua Laguna: Istituto Veneto di Scienza, Lettere Arti, 21 p.

Levitus, S., 1990, Interpentadal variability of steric sea level and geopotential thickness of the North Atlantic Ocean, 1970-1974 versus 1955-1959: Journal of Geophysical Research, v. 95, p. 5233–5238.

Lewis, J., 1988, The implication of sea level rise for island and low lying countries: Report for the Commonwealth Secretariat, Britain.

Li, D., 1984, Geologic evolution of petroliferous basins on continental shelf of China: American Association of Petroleum Geologists, Bulletin, v. 68, p. 993–1003.

Lirer, L., G. Luongo, and R. Scandone, 1987, On the volcanological evolution of Campi Flegrei: EOS, American Geophysical Union, Transactions, v. 68, p. 266–234.

Lisitzin, E., 1955, Les variations annuelles du niveau des oceans: Comité Central d'Océanographie et d'Étude des côtes, Bulletin d'Information, v. 7, p. 235–250.

Lisitzin, E., 1958, Le niveau moyen de la mer: Comité Central d'Océanographie et d'Étude des côtes, Bulletin d'Information, v. 10, p. 254–262.

Lisitzin, E., 1974, Sea-Level Changes: Amsterdam, Elsevier Scientific Publishing Company, 286 p.

Locker, S. D., and D. L. Sahagian, 1984, Tectonic features: in R. T. Buffler, S. D. Locker, W. R. Bryant, S. A. Hall, and R. H. Pilger, Jr., eds., Ocean Margin Drilling Program Regional Atlas Series. Gulf of Mexico, Marine Science International, v. 6, 36 p.

Lofgren, B. E., 1969, Land subsidence due to the application of water: in D. J. Varnes and G. Kiersch, eds., Geological Society of America, Reviews in Engineering Geology, v. II., p. 271–303.

Logo, A. E., and M. M. Brinson, 1979, Calculations of the value of salt water wetlands: in P. E. Greeson, J. R. Clark, and J. E. Clark, eds., Wetland Functions and Values–The State of Our Understanding: Minneapolis, Minnesota, American Water Resources Association, p. 120–130.

Lorius, C., J. Jouzel, C. Ritz, L. Merlivat, N. I. Barkov, Y. S. Korotkevich, and V. M. Kotlyakov, 1985, A 150,000-year climatic record from Antarctic ice: Nature, v. 316, p. 591–596.

Lorius, C., N. I. Barkov, J. Jouzel, Y. S. Korotkevich, V. M. Kotlyakov, and D. Raynaud, 1988, Antarctic ice core–CO_2 and climatic change over the last climatic cycle: EOS, American Geophysical Union, Transactions, v. 69, p. 681, 683–684.

Loutit, T. S., and J. P., Kennett, 1981, Australasian Cenozoic sedimentary cycles, global sea level changes and the deep sea sedimentary record: 26th International Geological Congress, Paris, July 7-17, 1980, Colloquium C_3, Proc.: Oceanologica Acta, p. 45–63.

Lu, R. A., and K. J. McMillen, 1982, Multichannel seismic survey of the Colombia Basin and adjacent margin: in J. S. Watkins and C. L. Drake, eds., Studies in Continental Margins: American Association of Petroleum Geologists, Memoir 34, p. 395–410.

Luyendyk, B. P., 1977, Deep sea drilling on the Ninetyeast Ridge–Synthesis and a tectonic model: in J. E. Heirtzler, H. M. Bolli, T. A. Davies, J. B. Saunders, and J. G. Sclater, eds., Indian Ocean Geology and Biostratigraphy; Studies Following Deep-Sea Drilling Legs 22-29: American Geophysical Union, p. 165–187.

Lyell, C., 1850, Principles of Geology, or the Modern Changes of the Earth and its Inhabitants Considered as Illustrative of Geology (8th ed.): London, Murray, 811 p.

Macintyre, I. G., and W. H. Adey, 1990, Buck Island bar, St. Croix, USVI—A reef that cannot catch up with sea level: Atoll Research Bulletin, v. 336, 7 p.

Madden, R. A., and V. Ramanathan, 1980, Detecting climate change due to increasing carbon dioxide: Science, v. 209, p. 763–768.

Maher, J. C., 1971, Geologic framework and petroleum potential of the Atlantic Coastal Plain and Continental Shelf: U. S. Geological Survey, Professional Paper 659, 98 p.

Malinverno, A., and W. B. F. Ryan, 1986, Extension in the Tyrrhenian Sea and shortening in the Apennines as result of arc migration driven by sinking of the lithosphere: Tectonics, v. 5, p. 227–245.

Mansfield, T., and S. Nishioka, 1989, An assessment of the research on the effects of a rise in sea level by the relevant Japanese ministries: Unpublished draft submitted to the IPCC, Tsukuba, Ibaraki, Japan, National Institute for Environmental Studies.

Manton, W. L., 1987, Tectonic interpretation of the morphology of Honduras: Tectonics, v. 6, p. 633–651.

Marcus, L. F., and W. S. Newman, 1983, Hominid migrations and the eustatic sea level paradigm—A critique: in P. M. Masters and N. C., Flemming, eds., Quaternary Coastlines and Marine Archaeology—Towards the Prehistory of Land Bridges and Continental Shelves: New York, Academic Press, p. 63–85.

Marmer, H. A., 1926, The Tide: New York, D. Appleton and Co., 282 p.

Marmer, H. A., 1927, Tidal Datum Planes: U. S. Coast and Geodetic Survey, Special Publication 135, 142 p.

Marmer, H. A., 1949, Sea level changes along the coasts of the United States in recent years: American Geophysical Union, Transactions, v. 30, p. 201–204.

Marshack, A., 1972a, The Roots of Civilization: New York, McGraw-Hill, 413 p.

Marshack, A., 1972b, Upper Paleolithic notation and symbol: Science, v. 178, p. 817–828.

Mart, Y., 1984, The tectonic regime of the southeastern Mediterranean continental margin: Marine Geology, v. 55, p. 365–386.

Mart, Y., 1987, Superpositional tectonic patterns along the continental margin of the southeastern Mediterranean—A review: Tectonophysics, v. 140, p. 213–232.

Martin, R. G., and J. E. Case, 1975, Geophysical studies in the Gulf of Mexico: in A. E. M. Nairn and G. Stehli, eds., The Ocean Basins and Margins—v. 3, The Gulf of Mexico and the Caribbean: New York, Plenum Press, p. 65–106.

Martinson, D. G., N. G. Pisias, J. D. Hayes, J. Imbrie, T. C. Moore, Jr., and N. J. Shackleton, 1987, Age dating and the orbital theory of the ice ages—Development of a high-resolution 0 to 300,000 year chronostratigraphy: Quaternary Research, v. 27, p. 1–29.

Mascle, J., A. Le Cleac'h, and D. Jongsma, 1986, The eastern Hellenic margin from Crete to Rhodes—Example of progressive collision: Marine Geology, v. 73, p. 145–168.

Masters, P. M., and N. C. Flemming, eds., 1983, Quaternary Coastlines and Marine Archaeology—Towards the Prehistory of Land Bridges and Continental Shelves: New York, Academic Press, 641 p.

Mather, K. F., and S. L. Mason, 1939, A Source Book in Geology: New York, McGraw-Hill, 702 p.

Mavor, J. W., 1969, Voyage to Atlantis: New York, Putnam, 320 p.

Maximov, I. V., 1959, The long period luni-solar tide in the world oceans: Doklady Akademia Nauk SSSR, v. 108, p. 799–801.

Maximov, I. V., 1965a, The solar semi-annual tide in the oceans: Doklady Akademia Nauk SSSR., v. 161, p. 347–350.

Maximov, I. V., 1965b, The long period luni-solar tides in the oceans: Okeanologia, v. 6, p. 26–37.

Maxwell, J. C., 1974, Early western margin of the United States: in C. A. Burk and C. L. Drake, eds., The Geology of Continental Margins: New York, Springer-Verlag, p. 831–852.

Mayuga, M. N. and D. R. Allen, 1969, Subsidence in the Wilmington oil field, Long Beach, California, U.S.A.: in L. J. Tison, ed., International Association of Hydrological Science, Publication 88, v. 1, p. 66–79.

McCoy, F. W., 1980, The upper Thera (Minoan) ash in deep-sea sediments—Distribution and comparison with other ash layers: in C. Doumas, ed., Thera and the Aegean World II: Second International Science Congress, Santorini, Greece, August 1978, Papers and Proceedings, p. 57–78.

McElhinny, M. W., B. J. J. Embleton, L. Daly, and J.-P. Pozzi, 1976, Paleomagnetic evidence for the location of Madagascar in Gondwanaland: Geology, v. 4, p. 455–457.

McFarlan, E., Jr., 1961, Radiocarbon dating of late Quaternary deposits, South Louisiana: Geological Society of America, Bulletin, v. 72, p. 129–158.

McIntyre, A., N. G. Kipp, with A. W. H. Bé, T. Crowley, T. Kellogg, J. V. Gardner, W. Prell, and W. F. Ruddiman, 1976, Glacial North Atlantic 18,000 years ago; A CLIMAT reconstruction: in R. M. Cline and J. D. Hays, eds., Investigation of Late Quaternary Paleoceanography and Paleoclimatology: Geological Society of America, Memoir 145, p. 43–76.

McKenzie, D., and J. G. Sclater, 1971, The evolution of the Indian Ocean since the Late Cretaceous: Royal Astronomical Society, Geophysical Journal, v. 24, p. 437–528.

Meade, R. H., and K. O. Emery, 1971, Sea level as affected by river runoff, eastern United States: Science, v. 173, p. 425–428.

Mehta, A. J., and R. M. Cushman, 1989, Workshop on sea level rise and coastal processes: Washington, D.C., U.S. Department of Energy, Contract number DOE/NBB-0086, 289 p.

Meier, M. F., 1989, Reduced rise in sea level: Nature, v. 343, p. 115–116.

Menard, H. W., 1969, Elevation and subsidence of oceanic crust: Earth and Planetary Science Letters, v. 6, p. 275–284.

Meo, M., 1988, Institutional response to sea-level rise—The case of Louisiana: in M. H. Glantz, ed., Societal Responses to Climate Change—Forecasting by Analogy: Boulder, Colorado, Westview Press, p. 215–242.

Mercer, J. H., 1978, West Antarctic ice sheet and CO_2 greenhouse effect—A threat of disaster: Nature, v. 271, p. 321–325.

Miall, A. D., 1986, Eustatic sea level changes interpreted from seismic stratigraphy—A critique of the methodology with particular reference to the North Sea Jurassic record: American Association of Petroleum Geologists, Bulletin, v. 70, p. 131–137.

Michaelsen, J., 1989, Long-period fluctuations in El Niño amplitude and frequency reconstructed from tree rings: in D. H. Peterson, ed., Aspects of Climate Variability in the Pacific and the Western Americas: American Geophysical Union, Geophysical Monograph, v. 55, p. 69–74.

Milankovitch, M., 1938, Astronomische mittel erforschung der erdgeschichtlichten klimate: Handbuch der Geophysik, v. 9, p. 593–698.

Milliman, J. D., and R. H. Meade, 1983, World-wide delivery of river sediment to the oceans: Journal of Geology, v. 92, p. 1–21.

Milliman, J. D., J. M. Broadus, and F. Gable, 1989, Environmental and economic implications of rising sea level and subsiding deltas—The Nile and Bengal examples: Ambio, v. 18, p. 340–345.

Mills, J. V. G., 1970, Ma Huan—Ying-Yai Sheng-Lan (The Overall Survey of the Ocean's Shores): Cambridge, University Press, 393 p.

Ministry of Economic Affairs of Taiwan, 1978, Tectonic Map of Taiwan, scale 1:500,000: Republic of China, 1 sheet.

Mitchell, J. F. B., C. A. Wilson, and W. M. Cunnington, 1987, On CO2 climate sensitivity and model dependence of results: Royal Meteorological Society, Quarterly Journal, v. 113, p. 293–322.

Mitchum, G. T., and K. Wyrtki, 1988, Overview of Pacific sea level variability: Marine Geodesy, v. 12, p. 235–245.

Mohr, P., 1977, 1974, Ethiopian rift geodimeter survey: Smithsonian Astrophysical Observatory Special Report 376, 111 p.

Molengraaff, G. A. F., 1921, Modern deep-sea research in the East Indian Archipelago: Geographical Journal, v. 57, p. 95–121.

Molnar, P., 1989, The geologic evolution of the Tibetan Plateau: American Scientist, v. 77, p. 350–360.

Molnar, P., and L. R. Sykes, 1969, Tectonics of the Caribbean and Middle America regions from focal mechanisms and seismicity: Geological Society of America, Bulletin, v. 80, p. 1639–1684.

Molnar, P., and P. Tapponier, 1975, Cenozoic tectonics of Asia—Effects of a continental collision: Science, v. 189, p. 419–426.

Molnar, P., and P. Tapponier, 1981, A possible dependence of tectonic strength on the age of the crust in Asia: Earth and Planetary Science Letters, v. 52, p. 107–114.

Monroe, W. H., 1968, High-level Quaternary beach deposits in northwestern Puerto Rico: U. S. Geological Survey, Professional Paper, v. 600-C, p. C140–C143.

Moore, J. G., and R. K. Reed, 1963, Pillow structures of submarine basalts east of Hawaii: U. S. Geological Survey, Professional Paper, v. 475-B, p. B153–B157.

Moore, K. G., and D. V. Kent, 1987, Testing Cenozoic eustatic changes—The critical role of stratigraphic resolution: Cushman Foundation for Foraminiferal Research, Special Publication 24, p. 51–56.

Moore, T. C., Jr., N. G. Pisias, and L. D. Keigwin, Jr., 1982, Cenozoic variability of oxygen isotopes in benthic foraminfera: in Climate in Earth History: Washington, D. C., National Academy Press, p. 172–182.

Mörner, N.-A., 1969, The Late Quaternary History of the Kattegatt Sea and the Swedish West Coast—Deglaciation, Shorelevel Displacement, Chronology, Isostasy and Eustasy: Sveriges Geologiska Undersökning, serial C, no. 640, 487 p.

Mörner, N.-A., 1971, The Holocene eustatic sea level problem: Geologie en Mijnbouw, v. 50, p. 699–702.

Mörner, N.-A., 1976, Eustatic changes during the last 8,000 years in view of radiocarbon calibration and new information from Kattegatt region and other northwestern European coastal areas: Palaeogeography, Palaeoclimatologey, Palaeoecology, v. 19, p. 63–85.

Mörner, N.-A., 1978, Eustasy and geoid changes as a function of core/mantle changes: in N.-A. Mörner, ed., Earth Rheology Isostasy and Eustasy: New York, John Wiley & Sons, p. 535–567.

Mörner, N.-A., 1979, The Fennoscandian uplift and Late Cenozoic geodynamics—Geological evidence: GeoJournal, v. 3.3, p. 287–318.

Mörner, N.-A., 1980, The Fennoscandian uplift—Geological data and their geodynamical implication: in N.-A. Mörner, ed., Earth Rheology, Isostasy and Eustasy: New York, John Wiley & Sons, p. 251–284.

Mowatt, F., 1958, The Grey Seas Under—A History of a Tug: New York, Ballantine Books, 255 p.

Moyer, G., 1982, The Gregorian calendar: Scientific American, v. 246, no. 5, p. 144–152.

Munk, W. H., 1962, Long ocean waves: in M. N. Hill, ed., The Sea: New York, Interscience Publishers, v. 1, p. 647–663.

Munk, W. H., and D. E. Cartwright, 1966, Tidal spectroscopy and prediction: Royal Society of London, Philosophical Transactions, series A, v. 259, p. 533–581.

Munk, W., and R. Revelle, 1952, On the geophysical interpretation of irregularities in the rotation of the Earth: Royal Astronomical Society, Monthly Notices, Geophysical Supplement, v. 6, p. 331–347.

Murayama, S., 1969, Land subsidence in Osaka: in L. J. Tison, ed., International Symposium on Land Subsidence: Tokyo, Japan, International Association of Hydrological Sciences, Publication 88, v. 1, p. 105–130.

Murray, G. E., 1961, Geology of the Atlantic and Gulf Coastal Province of North America: New York, Harper & Bros., 692 p.

Murty, T. S., 1984, Storm surges—Meteorological ocean tides: Canadian Bulletin of Fisheries and Aquatic Science, 212 p.

Nance, R. D., T. R. Worsley, and J. B. Moody, 1986, Post-Archean biogeochemical cycles and long-term episodicity in tectonic processes: Geology, v. 14, p. 514–518.

Nansen, F., 1922, The strandflat and isostasy: Skrifter Utgit av Videnskapsselskapet i Kristiania, 1921, I. Matematisk-Naturvidenskabelig Klasse, v. 2, 313 p.

National Academy of Sciences, 1985, Glaciers, ice sheets, and sea level—Effect of a CO_2-induced climatic change: Washington, D. C., U. S. Department of Energy, DOE/EV/60235-1, 330 p.

National Academy of Sciences, 1987, Responding to changes in sea level, engineering implications: Washington, D. C., Marine Board, National Research Council, National Academy Press, 148 p.

National Academy of Sciences, 1989, Carbon dioxide and climate—A scientific assessment: Washington, D. C., 222 p.

National Ocean Survey, 1980, Tide Tables 1981, High and Low Water Predictions, Central and Western Pacific Ocean and Indian Ocean: Washington, D. C., U. S. Department of Commerce, National Oceanic and Atmospheric Administration, 386 p.

Nautical Almanac Office, 1979, The Nautical Almanac for the Year 1981: Washington, D. C., Superintendent of Documents, U. S. Government Printing Office, 276 p.

Needham, J., 1959, Science and Civilization in China—v. 3, Mathematics and the Sciences of the Heavens and the Earth: Cambridge, University Press, 877 p.

Neev, D., 1977, The Pelusium Line—A major trans-continental shear: Tectonophysics, v. 38, p. T1–T8.

Neev, D., and Z. Ben-Avraham, 1977, The Levantine countries—The Israel coastal region: in A. E. M. Nairn and W. H. Kanes, eds., The Ocean Basins and Margins—v. 4A, The Eastern Mediterranean: New York, Plenum, p. 355–378.

Neev, D., N. Bakler, and K. O. Emery, 1987, Mediterranean Coasts of Israel and Sinai; Holocene Tectonism from Geology, Geophysics, and Archaeology: New York, Taylor and Francis, 130 p.

Newman, W. S., L. J. Cinquemani, R. R. Pardi, and L. F. Marcus, 1980a, Holocene develeling of the United States' east coast: in N.-A. Mörner, ed., Earth Rheology, Isostasy and Eustasy: New York, John Wiley & Sons, p. 449–463.

Newman, W. S., L. F. Marcus, R. R. Pardi, J. A. Paccione, and S. M. Tomecek, 1980b, Eustasy and deformation of the geoid—1000–6000 radiocarbon years BP: in N.-A. Mörner, ed., Earth Rheology, Isostasy and Eustasy: New York, John Wiley & Sons, p. 555–567.

Newton, J. G., 1986, Natural and induced sinkhole development in the eastern United States—Land Subsidence: Proceedings of the 3rd International Symposium on Land Subsidence, Venice, Italy; Wallingford, England, International Association of Hydrological Sciences, Publ. no. 151, p. 549–564.

Nichols, M. M., 1989, Sediment accumulation rates and relative sea-level rise in lagoons: Marine Geology, v. 88, p. 201–219.

Nichols, M., and J. Boon, in press, Sediment transport processes in coastal lagoons: in B. Kuerfue, ed., Coastal Lagoon Processes: Amsterdam, Elsevier Science Publishers.

Niino, H., and K. O. Emery, 1961, Sediments of shallow portions of East China Sea and South China Sea: Geological Society of America, Bulletin, v. 72, p. 731–762.

Nir, Y., and I. Eldar, 1987, Ancient wells and their geoarchaeological significance in detecting tectonics of the Israel Mediterranean coastline region: Geology, v. 15, p. 3–6.

Nomitsu, T., and M. Okamoto, 1927, The causes of the annual variation of the mean sea level along the Japanese coast: Kyoto University, Memoirs of the College of Science, series A, v. 10, p. 125–161.

Norman, S. E., and C. G. Chase, 1986, Uplift of the shores of the western Mediterranean due to Messinian desiccation and flexural isostasy: Nature, v. 322, p. 450–451.

Nur, A., and Z. Ben-Avraham, 1982, Oceanic plateaus, the fragmentation of continents, and mountain building: Journal of Geophysical Research, v. 87, p. 3644–3661.

Oda, M., and S. Kuramoto, 1989, Tide at Showa Base: in H. Kusunoki, chief ed., Science in Antarctica, v. 8, Oceanography, National Institute of Polar Research: Tokyo, Kokon-shyoin, 136 p. (in Japanese).

Oerlemans, J., 1989, A projection of future sea level: Climatic Change, v. 15, p. 151–174.

Officer, C. B., W. S. Newman, J. M. Sullivan, and D. R. Lynch, 1988, Glacial isostatic adjustment and mantle viscosity: Journal of Geophysical Research, v. 93, p. 6397–6409.

Ohmura, A., and N. Reeh, in press, New precipitation and accumulation maps for Greenland: Journal of Glaciology.

Oldale, R. N., 1982, Pleistocene stratigraphy of Nantucket, Martha's Vineyard, the Elizabeth Islands, and Cape Cod: in G. J. Larson and B. D. Stone, eds., Late Wisconsinan Glaciation of New England: Dubuque, Iowa, Kendell/Hunt, p. 1–34.

Oldale, R. N., and C. J. O'Hara, 1980, New radiocarbon dates from the inner continental shelf off southeastern Massachusetts and a local sea-level rise curve for the past 12,000 yr: Geology, v. 8, p. 102–106.

Oleson, J. P. (general editor), 1989, The Harbours of Caesarea Maritima, Results of the Caesarea Ancient Harbour Excavation Project, 1980–1985: Oxford, BAR International Series 491, v. 1, 517 p.

Ota, Y., 1987, Sea-level changes during the Holocene—The Northwest Pacific: in R. J. N. Devoy, ed., Sea Surface Studies—A Global View: London, Croom Helm, p. 348–374.

Ota, Y., Y. Matsushima, and H. Moriwaki, 1982, Notes on the Holocene sea-level study in Japan—On the basis of Atlas of Holocene sea-level records in Japan: Quaternary Research, v. 10, p. 133–143.

Ota, Y., K. R. Berryman, A. G. Hull, T. Miyauchi, and N. Iso, 1988, Age and height distribution of Holocene transgressive deposits in eastern North Island, New Zealand: Palaeogeography, Palaeoclimatology, Palaeoecology, v. 68, p. 135–151.

Pacht, J. A., and B. E. Bowen, 1990, Sequence stratigraphy along an unstable progradional continental margin—Pliocene-Pleistocene offshore Louisiana: American Association of Petroleum Geology, Annual Meeting, San Francisco, Abstracts, p. 149.

Palmason, G., and G. E. Sigvaldason, 1976, Iceland—Oceanic or continental?: in C. L. Drake, ed., Geodynamics, Progress and Prospects: American Geophysical Union, p. 56–58.

Pannikar, N. K., and T. M. Pannikar, 1971, The concept of tides in ancient India: Indian Journal of the History of Science, v. 6, p. 36–50.

Papadopoulos, G. A., and B. J. Chalkis, 1984, Tsunamis observed in Greece and the surrounding area from antiquity up to the present times: Marine Geology, v. 56, p. 309–317.

Pardi, R. R., and W. S. Newman, 1987, Late Quaternary sea levels along the Atlantic coast of North America: Journal of Coastal Research, v. 3, p. 325–330.

Park, R. A., T. V. Armentano, and C. L. Coolan, 1986, Predicting the effects of sea-level rise on coastal wetlands: in J. G. Titus, ed., Effects of Changes in Stratospheric Ozone and Global Climate—vol. 4, Sea Level Rise: Washington D.C., United National Environment Programme and U. S. Environmental Protection Agency, p. 129–152.

Park, R. A., M. S. Trehan, P. W. Mausel, and R. C. Howe, 1989, The effects of sea level rise on U.S. coastal wetlands: in J. B. Smith and D. Tirpak, eds., Potential effects of global climate change on the United States: Washington D. C., Environmental Protection Agency, no. 10, p. 1–45.

Paskoff, R. P., 1980, Late Cenozoic crustal movements and sea level variations in the coastal area of northern Chile: in N. A. Mörner, ed., Earth Rheology, Isostasy and Eustasy: New York, John Wiley & Sons, p. 487–495.

Paskoff, R., and P. Sanlaville, 1983, Les côtes de la Tunisie, variations du niveau marin depuis le Tyrrhenien: Collection de la Maison de l'Orient, Mediterranéen, Lyon, France, no. 14, p. 153–189.

Patterson, C., 1956, Age of meteorites and the Earth: Geochimica et Cosmochimica Acta, v. 10, p. 230–237.

Pattullo, J., W. Munk, R. Revelle, and E. Strong, 1955, The seasonal oscillation in sea level: Journal of Marine Research, v. 14, p. 88–155.

Peltier, W. R., 1976, Glacial isostatic adjustment—II. The inverse problem: Royal Astronomical Society, Geophysical Journal, v. 46, p. 669–705.

Peltier, W. R., 1980, Ice sheets, oceans, and the earth's shape: in N. A. Mörner, ed., Earth Rheology, Isostasy, and Eustasy: New York, John Wiley & Sons, p. 45–63.

Peltier, W. R., 1984, The thickness of the continental lithosphere: Journal of Geophysical Research, v. 89, p. 11,303-11,311, 11,316.

Peltier, W. R., 1985, Climatic implications of isostatic adjustment constraints on current variations of eustatic sea level: in Meier, M. et al., eds., Glaciers, Ice Sheets and Sea Level—Effect of a CO_2-Induced Climatic Change: Washington, D. C., National Research Council, p. 92–103.

Peltier, W. R., 1986, Deglaciation-induced vertical motion of the North American continent and transient lower mantle rheology: Journal of Geophysical Research, v. 91, p. 9099–9123.

Peltier, W. R., 1987, Mechanisms of relative sea-level change and the geophysical responses to ice-water loading: in R. J. N. Devoy, ed., Sea Surface Studies—A Global View: London, Croom Helm, p. 57–94.

Peltier, W. R., 1988, Global sea level and earth rotation: Science, v. 240, p. 895–901.

Peltier, W. R., and J. T. Andrews, 1976, Glacial-isostatic adjustment—I. The forward problem: Royal Astronomical Society, Geophysical Journal, v. 46, p. 605–646.

Peltier, W. R., and A. M. Tushingham, 1989, Global sea level rise and the greenhouse effect—Might they be connected?: Science, v. 244, p. 806–810.

Peltier, W.R., and A. M. Tushingham, in press, The influence of glacial isostatic adjustment on tide gauge measurements of secular sea level: Journal of Geophysical Research.

Peltier, W. R., W. E. Farrell, and J. A. Clark, 1978, Glacial isostasy and relative sea level—A global finite element model: Tectonophysics, v. 50, p. 81–110.

Perthuisot, J.-P., 1979, Quelques propositions contradictoires concernant la synthese du Tyrrhenien de Tunisie presentee par R. Paskoff et P. Sanlaville: in R. Paskoff and P. Sanlaville, eds., Excursion-table ronde sur le Tyrrhenien de Tunisie: International Association for Quaternary Research, Commission des lignes de rivage, Tunis, p. 47–51.

Philander, S. G. H., 1985, El Niño and La Niña: Journal of Atmospheric Science, v. 42, p. 2652–2662.

Philander, S. G. H., N. C. Lau, R. C. Pacanowski, and M. J. Nath, 1989, Two different simulations of Southern Oscillation and El Nino with coupled ocean-atmosphere general circulation models: Royal Society, Philosophical Transactions, series A, v. 329, p. 167–198.

Phipps, C. V. G., 1966, Evidence of Pleistocene warping of the New South Wales continental shelf: Geological Survey of Canada, Paper 66-15, p. 280–293.

Pilkey, O. H., Jr., O. H. Pilkey, Sr., and Robb Turner, 1975, How to Live with An Island—A Handbook to Bogue Banks, North Carolina: Raleigh, North Carolina Department of Natural and Economic Resources, 119 p.

Pinet, P. R., 1975, Structural evolution of the Honduras continental margin and the sea floor south of the western Cayman Trough: Geological Society of America, Bulletin, v. 86, p. 830–838.

Pirazzoli, P., 1974, Dati storici sulmedio mare a Venezia: Accademia delle Scienze dell'Istitutro di Bologna, anno 262, ser. 13, v. 1, p. 125–148.

Pirazzoli, P. A., 1976a, Les variations du niveau marin depuis 2000 ans: École Pratique des Hautes Études, Laboratoire du Géomorphologie, Memoire 30, 421 p.

Pirazzoli, P. A., 1976b, Sea level variations in the northwest Mediterranean during Roman times: Science, v. 194, p. 519–521.

Pirazzoli, P. A., 1977, Sea level relative variations in the world during the last 2000 years: Zeitschrift für Geomorphologie, N. F., v. 21, p. 284–296.

Pirazzoli, P. A., 1979-1980, Les viviers a poissons Romains en Méditerranée: Oceanus, v. 5, p. 191–201.

Pirazzoli, P. A., 1982, Marée estreme à Venezia (periodo 1872–1891): Acqua-Aria, v. 10, p. 1023–1039.

Pirazzoli, P. A., 1986, Secular trends of relative sea-level (RSL) changes indicated by tide-gauge records: Journal of Coastal Research, v. 1, p. 1–26.

Pirazzoli, P. A., 1987, The Mediterranean: in M. J. Tooley and I. Shennan, eds., Sea-Level Changes: New York, Basil Blackwell, p. 152–181.

Pirazzoli, P. A., 1989, Present and near-future global sea-level changes: Palaeogeography, Palaeoclimatology, Palaeoecology (Global and Planetary Change Section), v. 75, p. 241–258.

Pirazzoli, P. A., and G. Delibrias, 1983, Late Holocene and Recent sea level changes and crustal movements in Kume Island, the Ryukyus, Japan: Department of Geography, University of Tokyo, Bulletin, v. 15, p. 63–76.

Pirazzoli, P. and M. Koba, 1989, Late Holocene sea-level changes in Iheya and Noho islands, the Ryukyus, Japan: Association for the Geological Collaboration in Japan, Journal, v. 43, p. 1–6.

Pirazzoli, P. A., and L. F. Montaggioni, 1988, Holocene sea-level changes in French Polynesia: Palaeogeography, Palaeoclimatology, Palaeoecology, v. 68, p. 153–175.

Pirazzoli, P., and J. Thommeret, 1973, Une donnée nouvelle sur le niveau marine à Marseille à l'époque romaine: Académie des Sciences Paris, Comptes Rendus, v. 277, p. 2125-2128.

Pirazzoli, P. A., L. F. Montaggioni, J. F. Saliège, G. Segonzac, Y. Thommeret, and C. Vergnaud-Grazzini, 1989, Crustal block movements from Holocene shorelines: Rhodes Island (Greece): Tectonophysics, v. 170, p. 89-114.

Pirazzoli, P. A., L. F. Montaggioni, J. Thommeret, Y. Thommeret, and J. Laborel, 1982a, Sur les lignes de rivage et la néotectonique à Rhodes (Grece) à l'Holocene: Institiute Océanographique (Paris), Annals, N. Ser., v. 58, p. 89-102.

Pirazzoli, P. A., J. Thommeret, Y. Thommeret, J. Laborel, and L. F. Montaggioni, 1982b, Crustal block movements from Holocene shorelines—Crete and Antikythira (Greece): Tectonophysics, v. 8, p. 27-43.

Pisias, N. G., 1976, Late Quaternary sediment of Panama Basin—Sedimentation rates, periodicities, and controls of carbonate and opal accumulation: Geological Society of America, Memoir 145, p. 375-391.

Pisias, N. G., D. G. Martinsen, T. C. Moore, Jr., N. J. Shackleton, W. Prell, J. Hays, and G. Boolen, 1984, High resolution stratigraphic correlation of benthic oxygen isotope records spanning the last 300,000 years: Marine Geology, v. 56, p. 119-136.

Pitman, W. C., III, 1978, Relationship between eustacy and stratigraphic sequences of passive margins: Geological Society of America, Bulletin, v. 89, p. 1369-1403.

Pitman, W. C., III, 1979, The effect of eustatic sea level changes on stratigraphic sequences at Atlantic margins: in J. S. Watkins, L. Montadert, and P. W. Dickerson, eds., Geological and Geophysical Investigations of Continental Margins: American Association of Petroleumn Geologists, Memoir 29, p. 453-460.

Pitman, W. C., R. L. Larson, and E. M. Herron, 1974, Magnetic lineations of the oceans: Geological Society of America, 1 chart.

Plafker, G., 1976, Tectonic aspects of the Guatemala earthquake of 4 February 1976: Science, v. 193, p. 1201-1208.

Platzman, G. W., 1958, A numerical computation of the surge of 26 June 1954 on Lake Michigan: Geophysics, v. 6, p. 407-438.

Plumstead, E. P., 1973, The enigmatic Glossopteris flora and uniformitarianism: in D. H. Tarling and S. K. Runcorn, eds., Implications of continental drift in the Earth sciences: North Atlantic Treaty Organization, Advanced Study Institute: London, Academic Press, v. 1, p. 413-424.

Poag, C. W., and J. S. Schlee, 1984, Depositional sequences and stratigraphic gaps on submerged United States Atlantic margin: in J. S. Schlee, ed., Marine Unconformities: American Association of Petroleum Geologists, Memoir 36, p. 165-182.

Poe, Edgar Allan, 1966, Complete Stories and Poems of Edgar Allan Poe: Garden City, New York, Doubleday & Company, Inc., 819 p.

Poirier, J. P., and M. A. Taber, 1980, Historical seismicity in the Near and Middle East, north Africa, and Spain from Arabic documents (VII-VIIIth century): Seismological Society of America, Bulletin, v. 70, p. 2185-2194.

Poland, J. F. (ed.), 1984, Guidebook to Studies of Land Subsidence Due to Ground-water Withdrawal: International Hydrological Programme, Working Group 8.4: Paris, United Nations Educational, Scientific and Cultural Organization, 305 p. + app.

Poland, J. F., and G. H. Davis, 1969, Land subsidence due to withdrawal of fluids: in D. J. Varnes and G. Kiersch, eds., Review in Engineering Geology, v. II: Geological Society of America, p. 187-303.

Poland, J. F., and R. L. Ireland, 1988, Land subsidence in the Santa Clara Valley, California as of 1982: U. S. Geological Survey, Professional Paper 497-F, 61 p.

Polli, S., 1942, L'oscillazione annua dell'Oceano Atlantico: Archvio di Oceanografia e Limnologia, v. 2, p. 190-213.

Polli, S., 1947, Gli attuali movimenti verticali delle coste Italiane: Tecnica Italiana, Rivista D' Ingeneria e Scienza, new series 2, v. 4, p. 5-7. Istituto Geofisico, Trieste, publication 228.

Polli, S., 1948, Il progressivo aumento del livello del mare Mediterraneo: Istituto Geofisico, Trieste, Publication 179, p. 21-27.

Polli, S., 1952, Gli attuali movimenti verticali delle coste continentali: Annali di Geofisica, v. 5, p. 597-602.

Polli, S., 1962a, Il progressivo aumento del livello marino lungo le coste del Mediterraneo: Rapports et Procès Rapports et Proces-verbaux des Commission Internationale pour l'Exploration Scientifique de la Mer Méditerranée, v. 16, p. 649-654. (Istituto Sperimentale Talassografico, Trieste, publication 390.)

Polli, S., 1962b, Il problema della sommersione di Venezia: Roma, XII Convegno della Associazione Geofisica Italiana, p. 1-10.

Ponte, F. C., and H. E. Asmus, 1976, The Brazilian marginal basins, current state of knowledge: Academia Brasileira Ciencias, Anais, v. 48, p. 215-239.

Preisendorfer, R. W., F. W. Zwiers, and T. P. Barnett, 1981, Foundations of principal component selection rules: Scripps Institution of Oceanography, SIO Ref. Series no. 81-14, 192 p.

Premchand, K., and C. M. Harish, 1990, Rising sea level—Concern along southwest coast of India: in J. D. Milliman and S. Sabhasri, eds., Sea-level Rise and Coastal Subsidence—Problems and Strategies: New York, John Wiley & Sons.

Prest, V. K., 1969, Retreat of Wisconsin and recent ice in North America, scale 1:5,000,000: Canada Geological Survey, Map, 1 sheet.

Prins, J. E. (ed.), 1986, Impact of sea-level rise on society: Postbox 152, 8300 AD Emmeloord, The Netherlands, Delft Hydraulics Laboratory, 34 p.

Proudman, J. (secretary), 1940, Monthly and Annual Mean Heights of Sea Level, Up To And Including the Year 1936: Association d'Océanographie Physique, Union Géodésique et Géophysique Internationale, v. 5, 255 p.

Pugh, D. T., 1987, Tides, Surges, and Mean Sea-Level: New York, John Wiley & Sons, 472 p.

Pugh, D. T., and H. E. Faull, 1983, Operational Sea-Level Stations: United Nations Educational, Scientific and Cultural Organization, Intergovermental Oceanographic Commission, technical series 23, 40 p.

Pugh, D. T., N. E. Spencer, and P. L. Woodworth, 1987, Data holdings of the Permanent Service for Mean Sea Level: Bidston, Merseyside, England, Permanent Service for Mean Sea Level, IOS.

Pushcharovsky, Uy. M., and G. B. Udintsev, eds., 1970, Tektonicheskaya Karta Tixookeanokogo Segmenta Zemli, scale 1:10,000,000: Moscow, Akademia Nauk, SSSR, 6 sheets.

Qian Shoyou, 1268, Xan Shen reign—period records of Hangzhou district, v. 31, p. 6.

Quennell, A. M., 1958, The structural and geomorphic evolution of the Dead Sea rift: Geological Society of London, Quarterly Journal, v. 114, p. 1-24.

Quinn, W. H., V. Neal, and S. Antunex de Mayolo, 1987, El Niño occurrences over the past four and a half centuries: Journal of Geophysical Research, v. 92, p. 14,449-14,462.

Quinn, W. H., D. O. Zopf, K. S. Short, and R. T. W. Kuo Yang, 1978, Historical trends and statistics of the Southern Oscillation, El Niño, and Indonesian droughts: Fishery Bulletin, no. 76, no. 663-678.

Raban, A., 1983, Submerged prehistoric sites off the Mediterranean coast of Israel: in P. M. Masters and N. C. Flemming, eds., Quaternary Coastlines and Marine Archaeology—Towards the Prehistory of Land Bridges and Continental Shelves: New York, Academic Press, p. 215-232.

Raban, A., 1989, The Harbours of Caesarea Maritima, Results of the Caesarea Ancient Harbour Excavation Project, 1980-1985: J. P. Oleson, gen. ed., v. 1—The Site and the Excavations: Oxford, BAR International Series 491, v. 1, 517 p.

Radok, U., R. G. Barry, D. Jenssen, R. A. Keen, G. N. Kiladis, and B. McInnes, 1982, Climatic and Physical Characteristics of the Greenland Ice Sheet: Boulder, Colorado, Center for Ice Research in Earth Systems, University of Colorado, 88 p.

Ramaraju, V. S., and V. Hariharan, 1967, Sea level variations along the west coast of India: National Geophysical Research Institute, Hyderabad, India, Bulletin 5, p. 11-20.

Rao, S. R., 1987, Marine archaeological explorations off Dwarka, northwest coast of India: Indian Journal of Marine Sciences, v. 16, p. 22-30.

Raper, S. C. B., R. A. Warwick, and T. M. L. Wigley, in press, Global sea level rise—Past and future: in J. D. Milliman, ed., Sea Level Rise and Coastal Subsidence: New York, John Wiley & Sons, SCOPE.

Rappleye, H. S., 1933, Recent areal subsidence found in releveling: Engineering News-Record, v. 110, p. 848.

Redfield, A. C., 1967, Postglacial change in sea level in the western North Atlantic Ocean: Science, v. 157, p. 687–692.

Redfield, A. C., 1980, Introduction to Tides; The Tides of the Waters of New England and of New York: Woods Hole, Massachusetts, Marine Science International, 108 p.

Reid, H. F., 1914, The Lisbon earthquake of November 1, 1755: Seismological Society of America, v. 4, p. 53–80.

Ren Meie, and Zeng Chenkai, 1980, Late Quaternary continental shelf of East China: Acta Oceanologia Sinica, v. 2, p. 93–105.

Revelle, R., 1983, Probable future changes in sea level resulting from increased atmospheric carbon dioxide: in National Academy of Sciences, Changing Climate: Washington, D. C., National Academy Press, p. 433–447.

Revelle, R., and H. E. Suess, 1957, Carbon dioxide exchange between atmosphere and ocean and the generation of an increase of atmospheric CO_2 during the past decades: Tellus, v. 9, p. 18–27.

Reynolds, P. H., and F. Aumento, 1974, Deep drill 1972, potassium-argon dating of Bermuda drill core: Canadian Journal of Earth Sciences, v. 11, p. 1269–1273.

Richards, G. W., 1986, Late Quaternary deformed shorelines in Tunisia: Zeitschrift für Geomorphologie, v. 62, p. 183–195.

Richardson, P. L., and R. A. Goldsmith, 1987, The Columbus landfall—Voyage track corrected for winds and currents: Oceanus, v. 30, p. 2–10.

Rietveld, H., 1986, Land subsidence in the Netherlands—Land Subsidence: Venice, Italy, 3rd International Symposium on Land Subsidence: Wallingford, England, International Association of Hydrological Sciences, Publ. no. 151, p. 455–464.

Rijkwaterstaat, 1990, A global survey of coastal wetlands, their functions and threats in relation to adaptive responses to sea level rise: Dutch IPCC/RSWG/CZM delegation, Tidal Water Division, Note no. GWWS-90, The Netherlands.

Rizzini, A., F. Vezzani, V. Cococcetta, and G. Milad, 1978, Stratigraphy and sedimentation of a Neogene–Quaternary section in the Nile Delta area (A.R.E.): Marine Geology, v. 27, p. 327–348.

Roberts, J. E., 1969, Sand compression as a factor in oil field subsidence: in L. J. Tison, ed., International Symposium on Land Subsidence: Tokyo, Japan, International Association of Hydrological Sciences, Publication 88, v. 2, p. 368–376.

Robin, G. de Q., 1986, Changing sea level: in B. Bolin, B. A. Doos, J. Jager, and R. A. Warrick, eds., Greenhouse Effect, Climate Change, and Ecosystems: New York, John Wiley & Sons, p. 323–359.

Roden, G. I., 1960, On the nonseasonal variations in sea level along the west coast of North America: Journal of Geophysical Research, v. 65, p. 2809–2826.

Roden, G. I., 1966, Low-frequency sea level oscillations along the Pacific coast of North America: Journal of Geophysical Research, v. 71, p. 4755–4776.

Rodhe, H., 1990, A comparison of the contribution of various gases to the greenhouse effect: Science, v. 248, p. 1217–1219.

Roe, F. W. (coordinator), 1962, Oil and Natural Gas Map of Asia and the Far East, scale 1:5,000,000: Bangkok, Thailand, United Nations Economic Commission for Asia and the Far East, 4 sheets.

Roemmich, D., and C. Wunsch, 1984, Apparent changes in the climatic state of the deep North Atlantic Ocean: Nature, v. 307, p. 447–450.

Rognan, P., A. Levy, J. L. Ballais, G. Goude, and J. Riser, 1983, Essai d'interpretation des coupes due Quaternaire recente de l'ouest El Akarit (Sud Tunisien): Geologique de Mediterranée, v. 10, p. 71–91.

Rohde, H., 1975, Wasserstandsbeobachtungen im Bereich der deutschen Nordseeküste von der Mitte des 19. Jahrhunderts: Die Küste, v. 28, p. 1–96.

Rohde, H., 1977, Sturmfluthöhen und säkularer Wasserstandsanstieg an der deutschen Nordseeküste: Die Küste, v. 30, p. 52–143.

Ronov, A. B., V. E. Khain, A. N. Balukhovsky, and K. B. Seslavinsky, 1980, Quantitative analysis of Phanerozoic sedimentation: Sedimentary Geology, v. 25, p. 311–325.

Rosenberg, N. J., P. Crosson, W. Easterling III, K. Frederick, and R. Seejo, 1989, Policy options for adaptations to climate change: in The Beijer Institute, ed., The Full Range of Responses to Climatic Change: United National Environment Programme, p. 69–100.

Ross, D. A., and E. Uchupi, 1977, The structure and sedimentary history of the southeastern Mediterranean Sea: American Association of Petroleum Geologists, Bulletin, v. 61, p. 872–902.

Ross, W. D., 1913, The Works of Aristotle: Oxford, Clarenden Press, 11 vols., 1462 p.

Rossiter, J. R., 1954, Report on the investigation of secular variations of sea level on the coasts of the British Isles, the Canaries and Azore, Egypt, the Gulf Coast, Australia, and on the shores of the Indian Ocean; also for Poland: in Secular Variations of Sea-Level: Association d'Océanographie Physique: Union Géodésique et Géophysique Internationale, Publication Scientifique, v. 13, p. 16–21.

Rossiter, J. R. (secretary), 1958, Monthly and Annual Mean Heights of Sea Level 1952 to 1956 and Unpublished Data for Earlier Years: Association d'Océanographie Physique, Union Géodésique et Géophysique Internationale, v. 19, 78 p.

Rossiter, J. R. (secretary), 1959, Monthly and Annual Mean Heights of Sea Level 1957 to 1958 and Unpublished Data for Earlier Years: Association d'Océanographie Physique, Union Géodésique et Géophysique Internationale, v. 20, 65 p.

Rossiter, J. R. (secretary), 1963, Monthly and Annual Mean Heights of Sea Level 1959 to 1961 and Unpublished Data for Earlier Years: Association d'Océanographie Physique, Union Géodésique et Géophysique Internationale, v. 21, 59 p.

Rossiter, J. R. (secretary), 1968, Monthly and Annual Mean Heights of Sea Level 1962 to 1964: Association d'Océanographie Physique, Union Géodésique et Géophysique Internationale, v. 30, 109 p.

Ruddiman, W. F., and M. E. Raymo, 1988, Northern Hemisphere climate régimes during the past 3 Ma—Possible tectonic connections: Royal Society of London, Philosophical Transactions, series B, v. 318, p. 411–430.

Ryan, W. B. F., 1978, Messinian badlands on the southeastern margin of the Mediterranean Sea: Marine Geology, v. 27, p. 349–363.

Saarloos, J. M., 1951, De geringe nauwkeurigheid van de bodemdalingsgetallen ten opzichte van zeeniveau, afgeleid uit de aflezingen op de Nederlandse kustpeilschalen: Tijdschrift van het Koninklijk Nederlandsch Aardrijkskundig Genootschap, v. 68, p. 101–122.

Sahagian, D., 1987, Epeirogeny and Eustatic sea level changes as inferred from Cretaceous shoreline deposits—Applications to the central and western United States: Journal of Geophysical Research, v. 92, p. 4895–4904.

Said, R., 1962, The Geology of Egypt, Amsterdam, Elsevier, 277 p.

Salisbury, R. D., and W. W. Atwood, 1908, The Interpretation of Topographic Maps: U. S. Geological Survey, Professional Paper 60, 84 p. + 170 plates.

Sarton, G., 1927, 1931, 1947, Introduction to the History of Science: Carnegie Institution of Washington, Publication 376, 3 vols.

Satake, K., and K. Shimazaki, 1988, Detectability of very slow earthquake from tide gauge records: Geophysical Research Letters, v. 13, p. 665–668.

Saucier, R. T., 1963, Recent geomorphic history of the Ponchartrain basin: Louisiana State University, Coastal Studies Series, no. 9, 114 p.

Sauramo, M., 1939, The mode of the land upheaval in Fennoscandia during late-Quaternary time: Commission Géologique de Finlande Bulletin no. 125; Société Géologique de Finlande, Compes Rendus, Bulletin, v. 13, p. 39–63.

Savage, J. C., M. Lisowski, and W. H. Prescott, 1986, Strain accumulation in the Shumagin and Yakataga seismic gaps, Alaska: Science, v. 231, p. 585–587.

Savin, S. M., 1982, Stable isotopes in climatic reconstructions: in Climate in Earth History: Washington, D. C., National Academy Press, p. 164–171.

Sayles, R. W., 1931, Bermuda during the ice age: American Academy of Arts and Sciences Proceedings, v. 66, p. 381–467.

Scaife, W. W., R. E. Turner, R. Costanza, 1983, Coastal Louisiana recent land loss and canal impacts: Environmental Management, v. 7, p. 433–442.

Scammell, G. V., 1981, The World Encompassed: Berkeley, California, University of California Press, 538 p.

Schafer, J. P., and J. H. Hartshorn, 1965, The Quaternary of New England: in H. E. Wright and D. G. Frey, eds., The Quaternary of the United States: Princeton, New Jersey, Princeton University Press, p. 113–128.

Schermerhorn, L. J. G., 1983, Late Proterozoic glaciation in the light of CO_2 depletion in the atmosphere: in L. G. Medaris, Jr., C. W. Byers, D. M. Mickelson, and W. C. Shanks, eds., Proterozoic Geology: Selected Papers from an International Proterozoic Symposium: Geological Society of America, Memoir 161, p. 309–315.

Schlanger, S. O., 1963, Subsurface geology of Eniwetok Atoll: U. S. Geological Survey, Professional Paper 260-BB, p. 991–1066.

Schlanger, S. O., and Y. Ozima, 1977, ^{40}Ar–^{39}Ar geochronological studies on submarine rocks from the western Pacific area: Earth and Planetary Science Letters, v. 33, p. 353–369.

Schlanger, S. O., M. O. Garcia, B. M. Keating, J. J. Naughton, W. W. Sager, J. A. Haggerty, and J. A. Philpotts, 1984, Geology and geochronology of the Line Islands: Journal of Geophysical Research, v. 89, p. 11,261–11,272.

Schlesinger, M. E., 1986, Equilibrium and transient climatic warming induced by increasing CO_2: Climate Dynamics, v. 1, p. 35–51.

Scholl, D. W., E. D. Buffington, and M. S. Marlow, 1975, Plate tectonics and the structural evolution of the Aleutian-Bering Sea region: Geological Society of America, Special Paper 131, p. 1–31.

Schopf, J. W., and D. Z. Oehler, 1976, How old are the eukaryotes?: Science, v. 193, p. 47–49.

Schopf, T. J. M., 1974, Permo-Triassic extinctions, relation to sea-floor spreading: Journal of Geology, v. 82, p. 129–143.

Schureman, P., 1976, Manual of Harmonic Analysis and Prediction of Tides: Washington, D. C., U. S. Government Printing Office, 317 p. (1st edition 1924; reprinted 1940, 1958, 1976).

Schwiderski, E. W., 1979, Global Oceanic Tides, Part II: The Semidiurnal Principal Lunar Tide (M_2), Atlas of Tidal Charts and Maps: Dahlgren, Virginia Naval Surface Weapons Center, TR 79-414, 15 p.

Schwing, F. B., and J. G. Norton, 1990, Earthquake and Bay–Response of Monterey Bay to the Loma Prieta Earthquake: EOS, American Geophysical Union, Transactions, v. 71, p. 250–251, 262.

Sclater, J. G., and R. L. Fisher, 1974, Evolution of the east central Indian Ocean, with emphasis on the tectonic setting of the Ninetyeast Ridge: Geological Society of America, Bulletin, v. 85, p. 683–702.

Sclater, J. G., R. N. Anderson, and M. L. Bell, 1971, Elevation of ridges and evolution of the central eastern Pacific: Journal of Geophysical Research, v. 76, p. 7888–7915.

Sclater, J. G., J. Crowe, and R. N. Anderson, 1976, On the reliability of oceanic heat flow averages: Journal of Geophysical Research, v. 81, p. 2997–3006.

Sclater, J. G., S. Hellinger, and C. Tapscott, 1977, The paleobathymetry of the Atlantic Ocean from the Jurassic to the present: Journal of Geology, v. 85, p. 509–552.

Semeniuk, V., and D. J. Searle, 1986, Variability of Holocene sea level history along the southwestern coast of Australia–Evidence for the effect of significant local tectonism: Marine Geology, v. 72, p. 47–52.

Sengupta, S., 1966, Geological and geophysical studies in western part of Bengal Basin, India: American Association of Petroleum Geologists, Bulletin, v. 50, p. 1001–1017.

Shackleton, N. J., 1987, Oxygen isotopes, ice volume and sea level: Quaternary Science Reviews, v. 6, p. 183–190.

Shackleton, N. J., and N. D. Opdyke, 1973, Oxygen isotope and palaeomagnetic stratigraphy of equatorial Pacific cores V28-238–Oxygen isotope temperatures and ice volumes on a 105 year and 106 year scale: Quaternary Research, v. 3, p. 39–55.

Shackleton, N. J., and N. D. Opdyke, 1976, Oxygen-isotope and paleomagnetic stratigraphy of Pacific core V28-239, late Pliocene to latest Pleistocene: Geological Society of America, Memoir 145, p. 449–464.

Shalem, N., 1956, On seismic sea waves (tsunamis) in the Middle East: Israel Exploration Journal, v. 20, nos. 3-4, p. 159–170 (in Hebrew).

Sharaf El Din, S. H., and Z. A. Moursy, 1977, Tide and storm surges on the Egyptian Mediterranean coast: Rapports et Procès-verbaux des Commission Internationale pour l'Exploration Scientifique de la Mer Méditerranée, v. 24, p. 33–37.

Shaw, H. R., and J. G. Moore, 1988. Magmatic heat and the El Niño cycle: EOS, American Geophysical Union, Transactions, v. 45, p. 1553–1565.

Shepard, F. P., 1937, Revised classification of marine shorelines: Journal of Geology, v. 45, p. 602–624.

Shepard, F. P., 1948, Submarine Geology: New York, Harper & Row, 348 p.

Shepard, F. P., 1963, Submarine Geology, 2nd. ed.: New York, Harper & Row, 557 p.

Shepard, F. P., and H. E., Suess, 1956, Rate of postglacial rise of sea level: Science, v. 123, p. 1082–1083.

Shepard, F. P., G. A. Macdonald, and D. C. Cox, 1950, The tsunami of April 1, 1946: Scripps Institution of Oceanography, University of California, Bulletin, v. 5, p. 391–528.

Shi, E., 1252, Shunyou Reign–Period Records of Hangzhou District, v. 10, p. 16.

Shi, L. X., and M. F. Bao, 1984, in J. F. Poland, ed., International Hydrological Programme, Working Group 8.4: Paris, United Nations Educational, Scientific and Cultural Organization, p. 155–160.

Shor, G. G., H. K. Kirk, and H. W. Menard, 1971, Crustal structure of the Melanesian area: Journal of Geophysical Research, v. 76, p. 2562–2586.

Sillitoe, R. H., 1974, Tectonic segmentation of the Andes–Implication for magmatism and metallogeny: Nature, v. 250, p. 542–545.

Simkin, T., and R. S. Fiske, 1983, Krakatau 1883–The Volcanic Eruption and Its Effects: Washington, D. C., Smithsonian Institution Press, 464 p.

Sjöberg, L. E., 1984, Determination of the land uplift from old water marks and tide gauge data at Ratan and Lövgrundet/Bjorn: Stockholms Universitet Geologiska Institutet, 16th Nordiska Geologiska Vintermötet, Meddelelser, no. 255.

Sloss, L. L., 1963, Sequences in the cratonic interior of North America: Geological Society of America, Bulletin, v. 74, p. 93–113.

Snowden, J. O., 1986. Drainage-induced land subsidence in metropolitan New Orleans, Louisiana, U.S.A.: Land Subsidence, Proceedings of the 3rd International Symposium on Land Subsidence, Venice, Italy: Wallingford, England, International Association of Hydrological Sciences, Publ. no. 151, p. 507–527.

Soegiarto, A., 1985, The mangrove ecosystem in Indonesia–Its problems and management: in K. N. Bardsley, J. D. S. Davie, and C. D. Woodroffe, eds., Coastal and Tidal Wetlands of the Australian Monsoon Region, Mangrove: Australian National University, North Australia Research Unit, Monograph 1, p. 313–326.

Sœmundsson, K., 1967, An outline of the structure of SW-Iceland: in S. Björnsson, ed., Iceland and Mid-Ocean Ridges: Societas Scientiarum Islandica, v. 38, p. 151–161.

Solow, A. R., 1987a, The application of eigenanalysis to tide-gauge records of relative sea level: Continental Shelf Research, v. 7, p. 629–641.

Solow, A. R., 1987b, Testing for climate change: An application of the two-phase regression model: Journal of Climate and Applied Meteorology, v. 26, p. 1401–1405.

Solow, A. R., 1990, Discriminating between models: An application to relative sea level at Brest: Journal of Climate, v. 3, p. 792–796.

Spaeth, M. G., and S. C. Berkman, 1967, The Tsunami of March 28, 1964, as Recorded at Tide Stations: Washington, D. C., U. S. Department of Commerce, Environmental Science Services Administration, ESSA Technical Report CGS 33, 86 p.

Spencer, N. E., P. L. Woodworth, and D. T. Pugh, 1988, Ancillary time series of mean sea level measurements: Bidston, Merseyside, England, Permanent Service for Mean Sea Level, IOS, 89 p.

Stanley, D. J., 1988a, Subsidence in the northeastern Nile Delta–Rapid rates, possible causes, and consequences: Science, v. 240, p. 497–500.

Stanley, D. J., 1988b, Low sediment accumulation rates and erosion on the middle and outer Nile delta shelf off Egypt: Marine Geology, v. 84, p. 111–117.

Stearns, H. T., 1935, Pleistocene shorelines on the islands of Oahu and Maui, Hawaii: Geological Society of America, Bulletin, v. 46, p. 1927-1956.

Stearns, H. T., 1978, Quaternary shorelines in the Hawaiian Islands: Bernice P. Bishop Museum, Bulletin, v. 237, 57 p.

Steckler, M. S., F. Berthelot, N. Lyberis, and X. Le Pichon, 1988, Subsidence in the Gulf of Suez—Implications for rifting and plate kinematics: Tectonophysics, v. 153, p. 249-270.

Steininger, F. F., J. Senes, K. Kleemann, and F. Rügl, 1985, Neogene of the Mediterranean Tethys and Paratethys—Stratigraphic Correlation Tables and Sediment Distribution Maps: Austria, University of Vienna, Institute for Paleontology Press, 2 v., 752 p.

Stewart, R. W., 1989, Causes and estimates of sea-level rise with changing climate: in A. Ayala-Castanares, W. S. Wooster, and A. Yanez-Arancibia, eds., Oceanography 1988, Universidad Nacional Autonoma de Mexico and Consejo Nacional de Ciencia y Technologia, UNAM Press, Mexico, p. 65-68.

Stiros, S. C., 1988, Archaeology—A tool to study active tectonics: EOS, American Geophysical Union, Transactions, v. 69, p. 1633, 1639.

Stommel, H., 1984, Lost Islands—The Story of Islands that Have Vanished from Nautical Charts: Vancouver, University of British Columbia Press, 146 p.

Stommell, H., and E. Stommel, 1983, Volcano Weather—The Story of 1816, The Year Without a Summer: Newport, Rhode Island, Seven Seas Press, 177 p.

Stone, D. B., 1988, Bering Sea—Aleutian Arc, Alaska: in A. E. M. Nairn, F. G. Stehli, and S. Uyeda, eds., The Ocean Basins and Margins—v. 7B, The Pacific Ocean, p. 1-84.

Stotts, V. D., 1985, Values and functions of Chesapeake Bay wetlands for waterfowl: in H. A. Groman, ed., Wetlands of the Chesapeake: Washington, D.C., Environmental Law Institute, p. 129-142.

Striem, H. L., and T. Miloh, 1975, Tsunamis induced by submarine slumpings off the coast of Israel: Licensing Division, Israel Atomic Energy Commission, IA-LD-1-102, 23 p.

Sturges, W., 1987, Large-scale coherence of sea-level at very low frequencies: Journal of Physical Oceanography, v. 17, p. 2084-2094.

Su, H.Y., 1986, Mechanism of land subsidence and deformation of soil layers in Shanghai.: Venice, Italy, Proceedings of the 3rd International Symposium on Land Subsidence: Wallingford, England, International Association of Hydrological Sciences, Publ. no. 151, p. 425-433.

Sugi, N., K. Chinzei, and S. Uyeda, 1983, Vertical crustal movement of northeast Japan since middle Miocene: in T. W. C. Hilde, and S. Uyeda, eds., Geodynamics of the Western Pacific-Indonesian Region: American Geophysical Union and Geological Society of America, Geodynamics Series, v. 11, p. 317-329.

Suguio, K., L. Martin, and J.-M. Flexor, 1980, Sea level fluctuations during the past 6000 years along the coast of the state of São Paulo, Brazil: in N. A. Mörner, ed., Earth Rheology, Isostasy and Eustasy: New York, John Wiley & Sons, p. 471-486.

Summerhayes, C. P., 1986, Sea level curves based on seismic stratigraphy—Their chronostratigraphic significance: Palaeogeography, Palaeoclimatology, and Palaeontology, v. 57, p. 27-42.

Sverdrup, H. U., M. W. Johnson, and R. H. Fleming, 1942, The Oceans—Their Physics, Chemistry, and General Biology: New York, Prentice-Hall, 1087 p.

Sykes, S., M. McCann, and A. Kafka, 1982, Motion of the Caribbean plate during the last 7 m.y. and implications for earlier Cenozoic movements: Journal of Geophysical Research, v. 87, p. 10656-10676.

Talandier, J., 1989, Submarine volcanic activity—Detection, monitoring, and interpretation: EOS, American Geophysical Union, Transactions, v. 70, p. 561, 568-569.

Tarr, R. S., and L. Martin, 1912, The earthquake at Yakutat Bay, Alaska in September, 1899: U. S. Geological Survey, Professional Paper 69, 135 p.

Taylor, E. G. R., 1971, The Haven-Finding Art—A History of Navigation from Odysseus to Captain Cook: London, Hollis and Carter Co., 310 p.

Taylor, E. G. R., and M. W. Richey, 1969, Le Marine Geometrique, les Premiers Instruments Nautiques: Paris, Éditions Maritimes et d'Outre-Mer, 160 p.

Taylor, F. W., 1980, Quaternary vertical tectonism in Hispaniola—Data from multichannel lines and emerged coral reefs: University of Texas Marine Science Institute, Industrial Association Meeting, Oct. 23, 1980, Proceedings, 2 p.

Thatcher, W., 1984, The earthquake deformation cycle at the Nankai Trough, southwest Japan: Journal of Geophysical Research, v. 89, p. 3087-3101.

Thom, B. G., and J. Chappell, 1978, Holocene sea-level change—An interpretation: Royal Society of London, Philosophical Transactions, series A, v. 291, p. 187-194.

Thom, B. G., J. R. Hails, and A. R. H. Martin, 1969, Radiocarbon evidence against higher post-glacial sea levels in eastern Australia: Marine Geology, v. 7, p. 161-168.

Thomas, R. H., 1985, Responses of the polar ice sheets to climatic warming: in Glaciers, Ice Sheets and Sea Level: Washington, D.C., National Academy Press, p. 301-316.

Thompson, K. R., 1986, North Atlantic sea-level and circulation: Royal Astronomical Society, Geophysical Journal, v. 63, p. 57-73.

Thompson, L. G., E. Mosley-Thompson, and B. M. Arnao, 1984, El Niño-Southern Oscillation events recorded in the stratigraphy of the tropical Quelccaya ice cap, Peru: Science, v. 226, p. 50-53.

Thompson, R. E., and S. Tabata, 1987, Steric height trends of ocean station PAPA in the northeast Pacific Ocean: Marine Geodesy, v. 11, p. 103-113.

Thorarinsson, S., 1940, Present glacier shrinkage and eustatic changes in sea level: Geografiska Annaler, v. 12, p. 131-159.

Thorarinsson, S., S. Steinthorsson, T. Einarsson, H. Kristmannsdottir, and N. Oakarsson, 1973, The eruption on Heimaey, Iceland: Nature, v. 241, p. 372-375.

Thorhaug, A., and B. Miller, 1986, Stemming the loss of coastal wetland habitats—Jamaica as a model for the tropical developing countries?: Environmental Conservation, v. 13, p. 72-73.

Thorne, J., and A. B. Watts, 1984, Seismic reflectors and unconformities on passive continental margins: Nature, v. 311, p. 365-368.

Thorne, J. A., and A. B. Watts, 1989, Quantitative analysis of North Sea subsidence: American Association of Petroleum Geologists, Bulletin, v. 73, p. 88-116.

Titus, J. G.(ed.), 1986a, Effects of changes in stratospheric ozone and global climate—v. 4, Sea Level Rise: Washington D.C., United National Environment Programme and United States Environmental Protection Agency, 193 p.

Titus, J. G., 1986b, Greenhouse effect, sea level rise, and coastal zone management: Coastal Management, v. 14, p. 76-82.

Titus, J. G., 1990, Greenhouse effect, sea level rise, and barrier islands: Coastal Management, v. 18, p. 1-20.

Titus, J. G., T. R. Henderson, and J. M. Teal, 1984, Sea level rise and wetlands loss in the United States: National Wetlands Newsletter, v. 6, p. 3-6.

Toynbee, A., 1976, Mankind and Mother Earth—A Narrative History of the World: New York, Oxford University Press, 641 p.

Tracey, J. I., Jr., S. O. Schlanger, J. T. Stark, D. B. Doan, and H. G. May, 1964, General geology of Guam: U. S. Geological Survey, Professional Paper 403-A, 104 p.

Tucholke, B. E., 1981, Geologic significance of seismic reflectors in the deep Western North Atlantic Basin: in J. E. Warme, R. G. Douglas, and E. L. Winterer, eds., The Deep Sea Drilling Project—A Decade of Progress: Society of Economic Paleontologists and Mineralogists, Special Publication 32, p. 23-37.

Tyler, D. A., J. Ladd, and H. W. Borns, 1979, Crustal subsidence in eastern Maine: Maine Geological Survey, NUREG/CR-0887, 12 p.

Uchupi, E., 1988, The Mesozoic-Cenozoic geologic evolution of Iberia, a tectonic link between Africa and Europe: Sociedad Geologica de España, Revista, v. 1, p. 257-294.

Uchupi, E., 1989, The tectonic style of the Atlantic Mesozoic rift system: African Earth Sciences, Journal, v. 8, p. 143-164.

Uchupi, E., and D. G. Aubrey, 1988, Suspect terranes in the North American margins and relative sea-levels: Journal of Geology, v. 96, p. 79–90.

Udintsev, G. B. (chief editor), 1975, Geological-geophysical Atlas of the Indian Ocean: Moscow, Academy of Sciences USSR, 151 p. (folio).

Um, S. H., and H. Y. Chun, 1983, Geology of Korea: Seoul, Korean Institute of Energy and Resources, 74 p.

United Nations Educational, Scientific and Cultural Organization and Bundesanstalt für Bodenforschung, 1962–1980, International Geological Map of Europe and the Mediterranean Regions, scale 1:1,500,000: Hannover, United Nations Educational, Scientific and Cultural Organizationa and Bundesanstalt Bodenforschungen, 49 sheets.

U. S. Geological Survey Staff, 1990, The Loma Prieta, California, earthquake—An anticipated event: Science, v. 247, p. 286–293.

Untung, M., 1967, Results of a sparker survey for tin ore off Bangka and Belitung islands, Indonesia: United Nations Economic Commission for Asia and the Far East, 4th session of the Committee for Coordination of Offshore Prospecting, Report, p. 61–67.

Urien, C. M., J. J. Zambrano, and L. R. Martins, 1981, The basins of southeastern South America (southern Brazil, Uruguay & eastern Argentina) including the Malvinas Plateau and southern South Atlantic paleogeographic evolution: Cuencas Sedimentarias del Jurassico y Cretacico de America del Sur, v. 1, p. 45–125.

Ursell, F., 1953, The long wave paradox in the theory of gravity waves: Cambridge Philosophical Society, Proceedings, v. 49, p. 685–694.

Uziel, J., 1968, Sea level at Ashdod and Elat—Differences between prediction and observations: Israel Journal of Earth-Sciences, v. 17, p. 137–151.

Vail, P. R., R. M. Mitchum, Jr., and S. Thompson, III, 1977a, Seismic stratigraphy and global changes of sea level—Pt. 3, Relative changes of sea level from coastal onlap: in C. E. Payton, ed., Seismic Stratigraphy—Application to Hydrocarbon Exploration: American Association of Petroleum Geologists, Memoir 26, p. 63–81.

Vail, P. R., R. M. Mitchum, Jr., and S. Thompson, III, 1977b, Seismic stratigraphy and global changes of sea level—Pt. 4, Global cycles of relative changes of sea level: in C. E. Payton, ed., Seismic Stratigraphy—Application to Hydrocarbon Exploration: American Association of Petroleum Geologists, Memoir 26, p. 83–97.

Vail, P. R., R. M. Mitchum, Jr., R. G. Todd, J. M. Widmier, S. Thompson, III, J. B. Sangree, J. N. Bubb, and W. G., Hatlied, 1977c, in C. E. Payton, ed., Seismic Stratigraphy—Applications to Hydrocarbon Explorations: American Association of Petroleum Geologists, Memoir 26, p. 49–204.

Valdiya, K. S., 1984, Evolution of the Himalaya: Tectonophysics, v. 105, p. 229–248.

Valentin, H., 1952, Die Küsten der Erde: Beiträge zur Allgemeinen und Regionalen Küstenmorphologie: Petermanns Geographische Mitteilungen, Ergänzungschaft 246, 118 p.

van Andel, Tj. H., and J. J. Veevers, 1965, Submarine morphology of the Sahul Shelf, Northwestern Australia: Geological Society of America, Bulletin, v. 76, p. 695–700.

van Hinte, J. E., S. W. Wise, Jr., B. N. M. Biart, J. M. Covington, D. A. Dunn, J. A. Haggerty, M. W. Johns, P. A., Meyers, M. R. Moullade, J. P. Muza, J. G. Ogg, M. Okamura, M. Sarti, and U. von Rad, 1985, Deep-sea drilling on the upper continental rise off New Jersey, DSDP sites 604 and 605: Geology, v. 13, p. 397–400.

Vanícek, P., 1976, Pattern of recent vertical crustal movements in Maritime Canada: Canadian Journal of Earth Science, v. 13, p. 661–667.

van Pedang, M. N., A. F. Richards, F. Machado, T. Bravo, P. E. Baker, and R. W. le Maitre, 1967, Atlantic Ocean—Catalogue of the Active Volcanoes of the World: International Association of Volcanology: Rome, Istituto Geologia Applicata, pt. 21, 128 p.

van Veen, J., 1945, Bestaat er een geologische bodemdaling te Amsterdam seder 1700?: Tijdschrift van het Koninklijk Nederlandsch Aardrijkskundig Genootschap, v. 62, p. 2–36.

van Veen, J., 1954, Tide-gauges, subsidence-gauges and flood-stones in the Netherlands: Geologie en Mijnbouw, v. 16, p. 214–219.

Veeh, H. H., and J. J. Veevers, 1970, Sea level at -175 m off the Great Barrier Reef 13,600 to 17,000 years ago: Nature, v. 226, p. 536–537.

Veevers, J. J., 1969, Paleogeography of the Timor Sea region: Palaeogeography, Palaeoclimatology, Palaeoecology, v. 67, p. 125–140.

Veevers, J. J., 1974, Regional site surveys, sites 259-262, 263: in J. J. Veevers, J. R. Heirtzler et al., eds., Initial Reports of the Deep Sea Drilling Project: Washington, D. C., U. S. Government Printing Office, v. 27, p. 561–566.

Veevers, J. J., J. R. Heirtzler et al. (eds.), 1974, Initial Reports of the Deep Sea Drilling Project: Washington, D. C., U. S. Government Printing Office, v. 27, p. 1049–1054.

Verbeek, R. D. M., 1885, Krakatau: Batavia, Landsdrukkerij, 546 p.

Villach, 1987, International Conference on the assessment of the role of carbon dioxide and of other greenhouse gases in climate variations and associated impacts: United Nations Environment Programme—International Council of Scientific Unions, World Meteorological Organization.

Viniegra, O. F., 1971, Age and evolution of salt basins of southeastern Mexico: American Association of Petroleum Geologists, Bulletin, v. 55, p. 478–494.

Vinnikov, K. Ya, P. Ya. Groisman, and K. M. Lugina, in press, The empirical data on modern global climate changes (temperature and precipitation): Journal of Climate.

Vitali, C., A. Mauffret, N. Kenyon, and V. Renard, 1985, Panamanian and Columbian deformed belts—An integrated study using Gloria and seabeam transits and seismic profiles: in A. Mascle, B. Biju-Duval, B. Blanchet, and J. F. Stephan, eds., Geodynamique des Caraibes: Paris, Technip, p. 451–460.

von Huene, R., 1972, Structure of the continental margin and tectonism at the eastern Aleutian Trench: Geological Society of America, Bulletin, v. 83, p. 3613–3636.

von Huene, R., and S. Lallemand, 1990, Tectonic erosion along the Japan and Peru convergente margins: Geological Society of America, Bulletin, v. 102, p. 704–720.

Wadge, G., and K. Burke, 1983, Neogene Caribbean plate rotation and associated Central American tectonic evolution: Tectonics, v. 2, p. 633–643.

Wadia, D. N. B., 1963, Geology of India, 4th ed.: London, McMillan, 536 p.

Wageman, J. M., T. W. C. Hilde, and K. O. Emery, 1970, Structural framework of East China Sea and Yellow Sea: American Association of Petroleum Geologists, Bulletin, v. 54, p. 1611–1643.

Wahrhaftig, C., and J. H. Birman, 1965, The Quaternary of the Pacific Mountain system in California: in H. E. Wright, Jr., and D. G. Frey, eds., The Quaternary of the United States: Princeton, New Jersey, Princeton University Press, p. 299–340.

Walcott, R. I., 1972a, Gravity, flexure, and the growth of sedimentary basins at continental edge: Geological Society of America, Bulletin, v. 83, p. 1845–1848.

Walcott, R. I., 1972b, Late Quaternary vertical movements in eastern North America—Quantitative evidence of glacio-isostatic rebound: Reviews of Geophysics and Space Physics, v. 10, p. 849–884.

Walker, D. A., 1988, Seismicity of the East Pacific Rise, Correlations with the Southern Oscillation Index?: EOS, American Geophysical Union, Transactions, v. 69, p. 857–867.

Walker, D. A., 1989, Seismicity of the interiors of plates in the Pacific basin: EOS, American Geophysical Union, Transactions, v. 70, p. 1543–1544.

Walker, D. A., and C. S. McCreery, 1988, Deep-ocean seismology—Seismicity of the northwestern Pacific basin interior: EOS, American Geophysical Union, Transactions, v. 69, p. 737, 742–743.

Wang, C. H., and W. C. Burnett, 1990, Holocene mean uplift rates across an active plate-collision boundary in Taiwan: Science, v. 248, p. 204–206.

Wang, Y., and D. G. Aubrey, 1987, The characteristics of the China coastline: Continental Shelf Research, v. 7, p. 329–349.

Ward, C. M., 1988, New Zealand marine terraces—Uplift rates: Science, v. 240, p. 803.

Ward, S. N., 1988, North America-Pacific plate boundary, an elastic-plastic megashear—Evidence from Very Long Baseline Interferometry: Journal of Geophysical Research, v. 93, p. 7716–7728.

Ward, W. T., and R. W. Jessup, 1965, Changes of sea levels in southern Australia: Nature, v. 205, p. 791–792.

Watts, A. B., and J. Thorne, 1984, Tectonics, global changes in sea level and their relationship to stratigraphical sequences at the U. S. Atlantic continental margin: Marine and Petroleum Geology, v. 1, p. 319–339.

Weaver, J. D., 1968, Terraces in western Puerto Rico: 4th Caribbean Geological Conference, Trinidad, 1965, Transactions, p. 243–245.

Wegener, K., 1929, The Origin of Continents and Oceans: New York, Dover Publications, 246 p.

Weimer, R. J., 1983, Relation of unconformities, tectonics, and sea level changes, Cretaceous of the Denver Basin and adjacent areas: in M. W. Reynolds and E. D. Dolly, eds., Mesozoic Paleogeography of the West-Central United States: Rocky Mountain Section, Society of Economic Paleontologists and Mineralogists, Symposium 2, p. 359–376.

Wells, J. T., 1987, Effects of sea-level rise on deltaic sedimentation in south-central Louisiana: in D. Nummedal, O. H. Pilkey, and J. D. Howard, eds., Sea-Level Fluctuation and Coastal Evolution: Society of Economic Paleontologists and Mineralogists, Special Publication 41, p. 157–166.

Wells, J. T., and J. M. Coleman, 1987, Wetland loss and the subdelta life cycle: Estuarine, Coastal and Shelf Science, v. 25, p. 111–125.

White, W. B., and K. Hasunuma, 1980, Interannual variability in the baroclinic gyre structure of the western North Pacific from 1954–1974: Journal of Marine Research, v. 38, p. 651–672.

Wiessel, J. K., and D. E. Hayes, 1972, Magnetic anomalies in the southeast Indian Ocean: in D. E. Hayes, ed., Antarctic Oceanology II–The Australian–New Zealand Sector: American Geophysical Union, Antarctic Research Series, v. 19, p. 165–196.

Wigley, T. M. L., and S. C. B. Raper, 1987, Thermal expansion of sea water associated with global warming: Nature, v. 330, p. 127–131.

Wigley, T. M. L., and S. C. B. Raper, in press, Future changes in global-mean temperature and thermal-expansion-related sea level rise: in R. A. Warrick and T. M. L. Wigley, eds., Climate and Sea Level Change–Observations, Projections and Implications: Cambridge, England, Cambridge University Press.

Wilkinson, J. H., and C. Reinsch, 1971, Linear Algebra: New York, Springer-Verlag, 439 p.

Williams, D. F., and B. H. Corliss, 1982, The south Australian continental margin and the Australian–Antarctic sector of the Southern Ocean: in A. E. M. Nairn and F. G. Stehli, eds., The Ocean Basins and Margins–v. 6, The Indian Ocean: New York, Plenum Press, v. 6, p. 545–584.

Williams, G. E., 1989, Precambrian tidal sedimentary cycles and Earth's rotation: EOS, American Geophysical Union, Transactions, v. 70, p. 33, 40–41.

Wilson, G., and H. Grace, 1942, The settlement of London due to subsidence of the London Clay: Institute of Civil Engineers, Journal, v. 19, no. 2, paper no. 5294, p. 100–127.

Wilson, J. T., 1966, Did the Atlantic close and then reopen?: Nature, v. 211, p. 676–681.

Windley, B. F., 1983, A tectonic review of the Proterozoic: in L. G. Medaris, Jr., C. W. Byers, D. M. Mickelson, and W. C. Shanks, eds., Proterozoic Geology–Selected Papers from an International Proterozoic Symposium: Geological Society of America, Memoir 161, p. 1–10.

Winograd, I. J., B. J. Szabo, T. B. Coplen, and A. C. Riggs, 1988, A 250,000-year climatic record from Great Basin vein calcite–Implications for Milankovitch theory: Science, v. 242, p. 1275–1280.

Wise, D. U., 1974, Continental margins, freeboard and the volumes of continents and oceans through time: in C. A. Burk and C. L. Drake, eds., The Geology of Continental Margins: New York, Springer-Verlag, p. 45–58.

Witting, R., 1918, Hafsytan, geodytan och landhöjningen utmed Baltiska havet och vid Nordsjön: Fennia, v. 39 (5), 347 p.

Witting, R., 1922, Le soulèvement récent de la Fennoscandie: Geografiska Annaler, v. 4, p. 458–487.

Wood, M., 1985, In Search of the Trojan War: Oxford, Facts On File Publications, 272 p.

Woodwell, G. M., G. J. MacDonald, R. Revelle, and C. D. Kelling, 1979, The carbon dioxide report: Atomic Scientists, Bulletin, v. 35, p. 56–57.

Woodworth, P. L., and D. E. Cartwright, 1986, Extraction of the M_2 ocean tide from SEASAT altimeter data: Royal Astronomical Society, Geophysical Journal, v. 84, p. 227–255.

World Meteorological Organization, 1986, Workshop on comparison of simulations by numerical models of the sensitivity of the atmospheric circulation to sea surface temperature anomalies: Geneva, World Meteorological Organization, WMO/TD-No. 138, WCRP-15, 188 p.

World Meteorological Organization, 1988, Modelling the sensitivity and variations of the ocean-atmosphere system: Geneva, World Meteorological Organization, WMO/TD-No. 254, WCRP-15, 289 p.

Worsley, T. R., R. D. Nance, and J. B. Moody, 1984, Global tectonics and eustasy for the past 2 billion years: Marine Geology, v. 58, p. 373–400.

Worsley, T. R., R. D. Nance, and J. B. Moody, 1986, Tectonic cycles and the history of the Earth's biogeochemical and paleoceanographic record: Paleoceanography, v. 1, p. 233–263.

Wu, P., and W. R. Peltier, 1983, Glacial isostatic adjustment and the free air gravity anomaly as a constraint on deep mantle viscosity: Royal Astronomical Society, Geophysical Journal, v. 74, p. 377–449.

Wyrtki, K., 1975, El Niño–The dynamic response of the equatorial Pacific Ocean to atmospheric forcing: Journal of Physical Oceanography, v. 5, p. 572–584.

Wyrtki, K., 1977, Sea level during the 1972 El Niño: Journal of Physical Oceanography, v. 7, p. 779–787.

Xui Yui, 1978, Dou Sumong, the physical oceanographer in Tang Dynasty and his book *On Tides* (Haitao zhi): Journal of History Research, v. 6, p. 63–67.

Xui Yui, and Li Wen Wei, 1979, Tidal theory and method of predicting tide by Zhang Junfang in Northern Song Dynasty: Shandong College of Oceanology, Journal, v. 2, p. 108–112.

Xui Yui, Li Wen Wei, and Bien Jia-Xi, 1980, Interpretations of works on tides selected from ancient Chinese books: Kexue (Science) Publishing House, 246 p.

Yabe, H., and R. Tayama, 1929, On some remarkable examples of drowned valleys found around the Japanese Islands: Oceanographic Works in Japan, Records, v. 2, p. 11–15.

Yamaguti, S., 1965, On the changes in the heights of mean sea levels before and after the great Niigata earthquake on June 16, 1964: Earthquake Research Institute, Bulletin, v. 45, p. 167–172.

Yamamoto, S., 1984a, Case history no. 9.4. Tokyo: in J. F. Poland, ed., International Hydrological Programme, Working Group 8.4: Paris, United Nations Educational, Scientific and Cultural Organization, p. 175–184.

Yamamoto, S., 1984b, Case history no. 9.5. Osaka, Japan: in J. F. Poland, ed., International Hydrological Programme, Working Group 8.4: Paris, United Nations Educational, Scientific and Cultural Organization, p. 185–194.

Yamamoto, S., 1984c, Case history no. 9.6. Nobi Plain, Japan: in J. F. Poland, ed., International Hydrological Programme, Working Group 8.4: Paris, United Nations Educational, Scientific and Cultural Organization, p. 195–204.

Yamamoto, S., 1984d, Case history no. 9.7. Niigata, Japan: in J. F. Poland, ed., International Hydrological Programme, Working Group 8.4: Paris, United Nations Educational, Scientific and Cultural Organization, p. 205–216.

Yamamoto, S., and A. Kobayashi, 1986, Groundwater resources in Japan with special reference to its use and conservation: Land Subsidence, Proceedings of the 3rd International Symposium on Land Subsidence, Venice, Italy: Wallingford, England, International Association of Hydrological Sciences, Publ. no. 151, p. 381–389.

Yang, H., and X. G. Chen, 1987, Quaternary transgressions, eustatic changes and movements of shorelines in north and east China: in V. Gardiner, ed., International Geomorphology 1986 part II: New York, John Wiley & Sons, p. 807–827.

Yang, H., and Z. Xie, 1984, Sea-level changes along the east coast of China over the last 20,000 years: Oceanologia et Limnologia Sinica, v. 15, p. 1–13 (in Chinese with English abstract).

Yang Zuosheng, K. O. Emery, and Xui Yui, 1989, Historical development and use of thousand-year-old tide prediction tables: Limnology and Oceanography, v. 34, p. 953–957.

Yanshin, A. L. (chief editor), 1966, Tektonicheskaya Karta Evrazii, scale 1:1,500,000: Ministervo Geologii Akademia Nauk, SSSR, 12 sheets.

Yohe, G. W., 1990, The cost of not holding back the sea—A national sample of environmental vulnerability: Middletown, Connecticut, Wesleyan University Publications.

Yonekura, N., R. Ishii, Y. Saito, Y. Maeda, Y. Matsushima, E. Matsumoto, and H. Hayanne, 1988, Holocene fringing reefs and sea-level change in Mangaia Island, southern Cook Islands: Palaeogeography, Palaeoclimatology, Palaeoecology, v. 68, p. 177-188.

Zans, V. A., 1958, The Pedro Cays and Pedro Banks: Jamaica Geological Survey Department, Bulletin, v. 3, 7 p.

Zebiak, S. E., and M. A. Cane, 1987, A model for El Niño-Southern Oscillation: Monthly Weather Review, v. 115, p. 2262-2278.

Zhang, W.-Y. (chief compiler), 1983, The Marine and Continental Tectonic Map of China and Its Environs, scale 1:5,000,000: Beijing, Science Press, 6 sheets.

Zhang, W. Z., and X. J. Niu, 1986, Analysis of the cause of land subsidence in Tianjin, China: in Land Subsidence, Proceedings of the 3rd International Symposium on Land Subsidence, Venice, Italy: Wallingford, England, International Association of Hydrological Sciences, Publ. no. 151, p. 435-444.

Zhang, Y., and W.-C. Wang, 1990, The surface temperature in China during the mid-Holocene: in Zhu Kezhen Centennial Memorial Collections, Beijing, Science Press.

Zhang, Z. M., J. G. Liou, and R. G. Coleman, 1984, An outline of the plate tectonics of China: Geological Society of America, Bulletin, v. 95, p. 295-312.

Ziegler, A. M., 1981, Paleozoic paleogeography: in M. W. McElhinny and D. A. Valencio, eds., Paleoreconstruction of the Continents: American Geophysical Union and Geological Society of America, Geodynamics series, v. 2, p. 31-37.

Zoback, M. L., and M. D. Zoback, 1980, State of stress in the conterminous United States: Journal of Geophysical Research, v. 85, p. 6113-6156.

Zoback, M. L., M. D. Zoback, J. Adams, M. Assumpção, S. Bell, E. A. Bergman, P. Blümling, N. R. Brereton, D. Denham, J. Ding, K. Fuchs, N. Gay, S. Gregersen, H. K. Gupta, A. Gvishiani, K. Jacob, R. Klein, P. Knoll, M. Magee, J. L. Mercier, B. C. Müller, C. Paquin, K. Rajendran, O. Stephansson, G. Suarez, M. Suter, A. Udias, Z. H. Xu, and M. Zhizhin, 1989, Global patterns of tectonic stress: Nature, v. 341, p. 291-298.

Zshehiv, D. E. (ed.), 1966, Geologicheska Karta Mira, scale 1:15,000,000: Moscow, Ministerstva Geologii SSSR, 6 sheets.

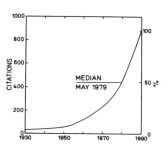

Author Index

A

Abdel-Monem, 137
Adams, 2
Agar, 135
Albani, 102
Aleem, 6
Allen, 31, 32
Alverson, 107
Ambach, W., 26
Ambraseys, 91
Anderson, D. L., 61
Anderson, T. H., 120
Anderson, W. A., 131
Armentrout, J. M., 55
Arur, M. G., 96
Asian Institute of Technology, 2, 35, 100
Atwater, T., 117
Aubouin, J. J., 118, 120, 121
Aubrey, D. G., 26, 43, 44, 46, 47, 50, 102–104, 107, 108, 110–113, 118–123, 127, 128, 147, 165, 228
Aveni, A. F., 5

B

Babbage, C., 56
Badash, L., 53
Balling, N., 88
Bally, A. W., 57, 59
Barazangi, M., 124
Barber, R. T., 42
Barnes, D. F., 116
Barnett, T. P., 44, 46, 77, 93, 150, 165–166, 181
Barth, M. C., 169
Battistini, R., 138
Beavan, J. R., 141
Belknap, D. F., 131
Belperio, A. P., 102
Ben-Avraham, Z., 92, 118
Ben-Menahem, A., 92, 93
Bennema, J., 31
Berger, W. H., 64
Bergsten, F., 12
Berry, L., 107
Berry, R. G., 26
Berryman, K., 118
Biju-Duval, B., 121, 124, 125

Biswas, B., 101
Bjerknes, J., 43
Blaha, J., 26
Blake, M. C., Jr., 105
Bloom, A. L., 55, 65, 74, 76, 102, 130, 158, 181
Bolin, B., 74
Boote, D. R. D., 101
Born, M. A., 12
Bostock, J., 6
Bott, M. H. P., 136
Bottomly, M., 163
Bovis, M., 121, 124
Bowin, C. O., 97, 120, 121, 135, 138
Braatz, B. V., 128, 133, 147–150, 165, 181, 228
Bradley, R. S., 164
Brandini, A. A., 125
Brandt, J. C., 8
Breaker, L., 42
Bretz, J. H., 136
Brink, K. H., 50
Broadus, J. M., 98
Broecker, W. S., 58, 64, 65, 74
Brown, D. A., 95, 101, 102
Bruckner, H., 98, 142
Bryant, E. A., 102, 103, 104
Budyko, M. I., 164
Bull, W. B., 105
Bunbury, E. H., 4
Bunce, E. T., 138
Burke, K., 58, 125
Butcher, S. H., 3

C

Cailleux, A., 73
Cande, S. C., 101
Cane, M. A., 42–44
Carbognin, L., 34
Carrera, G., 133
Carter, H., 4
Cartwright, D. E., 14
Case, J. E., 121
Casey, R. E., 41
Castilla, J. C., 174
Cayan, D. R., 45
Celsius, A., 72

Cercone, K. R., 58
Cess, R. D., 164
Chao, B. F., 44
Chapman, D. C., 51, 54–55, 63, 74, 102, 105, 130, 158
Chappell, J., 25
Chase, C. G., 61
Chelton, D., 42, 50
Chen, J. Y., 7
Chen, X. Q., 36
Cheney, R. E., 14
Cherven, V. B., 54
Choubert, G., 96
Christiansen, B., 174
Christie-Blick, N., 59, 62
Chugh, R. S., 96
Churchill, W. S., 5
Churkin, M., Jr., 101
Clague, D. A., 143
Clark, J. A., 25, 74, 77, 159
Clifford, T. N., 96
Cloud, P. E., 53, 63
Coast and Geodetic Survey, 10, 17
Coates, D. R., 126
Cochran, J. R., 142
Coghill, N., 6
Commonwealth Secretariat, 174
Condie, K. C., 61, 62
Coney, P. J., 118
Conolly, J. R., 102
Coque, R., 95
Corcoran, E., 56
Corkan, R. H., 73, 93
Cotton, C. A., 104
Cottrell, A., 1
Coutellier, V., 91
Cronin, T. M., 56
Crough, S. T., 61
Culliton, T. J., 1, 167
Curray, J. R., 97–98, 118

D

Daly, D. J., 54, 172
Dambara, T., 109
Dansgaard, W., 55
Darwin, C., 172
Davis, G. H., 33, 40, 131

Davis, W. M., 41, 55
Deacon, R., 4
De Almeida, F. F., 121, 124
Deegan, L. A., 126
de Geer, G., 82
Dennis, R. E., 22
Detrick, R. S., 142
Devoy, R. J. N., 169
de Vries, H. L., 56
de Wit, M. J., 94
Diaz, H. F., 164
Dietz, R. S., 54, 141
Dingle, R. V., 93–94, 96, 138
Disney, L. P., 69, 71, 74
Dolan, R., 91–92
Donn, W. L., 67
Donoghue, J. F., 131
Donovan, D. T., 53
Doodson, A. T., 73–74, 93
Dumas, B., 92
Duncan, R. A., 142–143
Dupont, J., 105
Duque-Caro, H., 123
Dutton, C. E., 131

E
Edelman, T., 33
Einarsson, P., 136
Emery, K. O., 1, 3, 5, 10, 16, 19, 20, 24, 34–36, 39, 43, 46, 51, 55–56, 58, 59, 64–65, 66, 69, 70, 71, 72, 75–77, 82, 84–88, 90–93, 96, 98, 99, 101, 105–109, 111, 113, 115, 117, 118, 120, 125, 131, 133, 135, 137, 141, 143–144, 159, 165, 170, 172, 174
Emiliani, C., 54, 58, 64–65
Enfield, D. B., 42, 50, 123
Espinosa, A. F., 92, 97, 105, 107, 114, 135, 139–140
Etkins, R., 76

F
Fairbanks, R. G., 68, 172
Fairbridge, R. W., 42, 67, 74–76, 165
Farmer, G., 163
Farrell, W. E., 74
Faure, H., 94
Fisher, A. G., 59
Fisher, R. L., 58, 138
Fisk, H. N., 31, 126, 171
Flemming, N. C., 13, 56, 90–91
Fletcher, G. L., 106
Flick, P. D. A., 95
Flint, R. F., 26, 114, 129, 136
Fontes, J. Ch., 34
Forel, F. A., 50
Foreman, M. G. G., 22
Forsythe, R., 121, 124
Fox, H. L., 4
Freund, R., 92

Friedrichs, C. T., 172
Frihy, O. E., 92
Fu, L., 14
Fujii, S., 110
Fulfaro, V. J., 125

G
Gable, F. J., 169, 172
Gabrysch, R. K., 39
Gallagher, B. S., 20
Gardner, J., 2, 53
Garfunkel, Z., 92
Gass, I. G., 93, 96
Gatto, P., 34
Giese, G. S., 51, 170
Gill, E. D., 101
Gilluly, J., 118
Giovinetto, M. B., 26
Girdler, R. W., 92
Gloe, C. S., 38
Goldsmith, V., 72, 90
Gornitz, V., 76–78, 148, 150, 165–166
Gortani, M., 34
Gow, A. J., 55
Gradstein, F. M., 59
Graham, N. E., 44
Grant, D. R., 133
Griggs, R. F., 140
Grotch, S. L., 165
Gulliver, F. P., 55
Güst, D. A., 124
Gutenberg, B., 72, 75, 165

H
Haile, N. S., 101
Hakluyt, R., 4
Hallam, A., 54, 57–58, 63
Hamilton, E. L., 172
Hamilton, H. C., 6, 57, 92
Han, M. K., 171
Hansen, J., 76, 163, 165
Haq, B. U., 59, 60
Harland, W. B., 62
Harmon, R. S., 137
Hayashi, T., 109
Hayes, D. E., 64, 101, 102, 107, 137, 139
Hays, J. D., 41, 64
Hazel, J. E., 58–59
Healy, J., 105
Heezen, B. C., 97, 107, 138
Heirtzler, J. R., 102
Hess, H. H., 54, 172
Herz, N., 121
Hicks, S. D., 22, 31, 69, 71
Higgins, C. G., 53
Hilde, T. W. C., 112, 117
Hirono, T., 38
Hoffert, M. I., 165
Hoffman, J. S., 165, 166
Holdahl, S. R., 126, 131

Holland, H. D., 62
Holtzscherer, J. J., 26
Honda, K., 51
Horsfield, W., 125, 136
Hurley, P. M., 97, 101, 157
Huyer, A., 42
Hwang, J. M., 33
Hyndman, R. D., 136

I
IAPSO, 74
Inaba, Y., 37d
Inman, D. L., 1, 35, 92
Irving, E., 61

J
Jackson, E. D., 141, 143
Jamieson, T. F., 82
Jones, P. D., 163
Johnson, D. W., 41, 55
Jouannic, C., 137

K
Kafri, U., 92
Kailasam, L. N., 98
Kana, T. W., 174
Karl, T. R., 163
Kato, T., 109
Katupotha, J., 99
Katz, H.-R., 105
Kawana, T., 111, 140
Kayan, I., 92
Kaye, C. A., 125
Keigwin, L. D., 68
Keller, W., 2
Kennedy, D. J. L., 38
Kennett, J. P., 25
Kerr, R. A., 42
Kidson, C., 67
Kienle, J., 141
King, P. B., 108, 120, 133
Klige, R. K., 74–75, 165
Knudsen, M., 46
Kolb, C. R., 126
Kolla, V., 98, 124
Komaki, S., 38
Korsch, R. J., 105
Koyanagi, R. Y., 141
Kuenen, P. H., 53, 72, 75, 101, 156, 165
Kugler, H. G., 38, 124
Kuhn, G. G., 174
Kulp, J. L., 56
Kumar, R., 99

L
Ladd, J. W., 125
Lallemand, F., 4
Lambert, A., 133

Land, L. S., 136
Landström, B., 6
Lange, A., 165
Larson, R. L., 101, 138, 141
Larue, D. K., 135
Leatherman, S. P., 174
Lennon, G. W., 72-73, 75, 93, 96, 127
Lensen, G. J., 93, 109-111
Leonardi, P., 34
Levitus, S., 47, 164, 166
Lewis, J., 172
Li, D., 108
Lirer, L., 56
Lisitzin, E., 31, 48, 51, 73-75, 165
Locker, S. D., 120
Lofgren, B. E., 32
Logo, A. E., 174
Lorius, C., 55, 64
Loutit, T. S., 57-58
Lu, R. A., 120, 124
Luyendyk, B. P., 138
Lyell, C., 51, 98

M
Macintyre, I. G., 172
Madden, R. A., 74
Maher, J. C., 125, 131, 133
Malinverno, A., 92
Mansfield, T., 172
Manton, W. L., 120
Marcus, L. F., 65
Marmer, H. A., 10, 20, 69, 71, 75
Marshack, A., 5
Mart, Y., 92
Martin, R. G., 120
Martinson, D. G., 55, 64
Mascle, J., 92
Masters, P. M., 56
Mather, K. F., 53-54, 57
Mavor, J. W., 1
Maximov, I. V., 48
Maxwell, J. C., 118
Mayuga, M. N., 32
McCoy, F. W., 1
McElhinny, M. W. B., 138
McFarlan, E., Jr., 126
McIntyre, A., 68
McKenzie, D., 97, 101
Meade, R. H., 45, 50, 98
Mehta, A. J., 169
Meier, M. F., 165-166
Menard, H. W., 141
Meo, M., 173
Mercer, J. H., 74
Miall, A. D., 57, 58
Michaelsen, J., 41
Milankovitch, M., 41
Milliman, J. D., 35, 98, 100, 106, 170, 171, 174
Mills, J. V. G., 6, 8
Ministry of Economic Affairs of Taiwan, 107

Mitchell, J. F. B., 165
Mitchum, G. T., 42, 43, 50
Mohr, P., 94
Molengraaff, G. A. F., 101
Molnar, P., 97, 135
Monroe, W. H., 125
Moore, J. G., 141
Moore, K. G., 64
Moore, T. C., 64
Mörner, N.-A., 12, 25, 26, 55, 57, 67, 75, 82
Mowatt, F., viii
Moyer, G., 9
Munk, W. H., 20, 22, 51, 73, 74, 75, 76
Murayama, S., 38
Murray, G. E., 125
Murty, T. S., 1, 170

N
Nance, R. D., 58, 61, 62, 63
Nansen, F., 82
National Academy of Sciences, 165, 166, 169
National Ocean Survey, 7, 9
Nautical Almanac Office, 7
Needham, J., 4, 5, 6, 7, 8, 9
Neev, D., 1, 13, 16, 56, 92, 93, 110
Newman, W. S., 65, 77, 129, 130, 131
Newton, J. G., 40
Nichols, M. M., 165
Niino, H., 106
Nir, Y., 91
Nomitsu, T., 74
Norman. S. E., 92
Nur, A., 118, 124

O
Oda, M., 95, 105
Oerlemans, J., 165-166
Officer, C. B., 83
Ohmura, A., 26
Oldale, R. N., 25, 129
Oleson, J. P., 1
Ota, Y., 107, 110

P
Pacht, J. A., 55
Palmason, G., 136
Pannikar, N. K., 5
Papadopoulos, G. A., 92
Pardi, R. R., 67, 131
Park, R. A., 172, 174
Paskoff, R. P., 124
Patterson, C., 53
Pattullo, J., 46, 48, 74, 75
Peltier, W. R., 25, 74, 118, 127, 130, 133, 135, 159, 162, 165
Perthuisot, J.-P., 95
Philander, S. G. H., 42, 165
Phipps, C. V. G., 101

Pilkey, O. H., 174
Pinet, P. R., 120
Pirazzoli, P., 65, 72, 90, 91, 92, 93, 96, 111, 140, 141, 165
Pisias, N. G., 55, 64
Pitman, W. C., 23, 41, 61, 101, 102, 118
Plafker, G., 120
Platzman, G. W., 51
Plumstead, E. P., 64
Poag, C. W., 57, 58
Poe, Edgar Allan, 4
Poland, J. F., 31, 33, 34, 38, 39, 124
Polli, S., 72, 73, 74, 90, 92
Ponte, F. C., 121
Preisendorfer, R. W., 215
Premchand, K., 96, 98
Prest, V. K., 129
Prins, J. E., 169
Proudman, J., 73, 74, 93, 105, 137, 138
Pugh, D. T., 5, 14, 21, 26, 51, 73, 93, 114, 169
Puscharovsky, Uy. M., 107

Q
Qian Shoyou, 8
Quennell, A. M., 92
Quinn, W. H., 41, 43-45

R
Raban, A., 13, 91
Radok, U., 26
Ramaraju, V. S., 96
Rao, S. R., 98
Raper, S., 165-166
Rappleye, H. S., 39
Redfield, A. C., 19, 136
Reid, H. F., 51
Ren Meie, 7
Revelle, R., 74, 165-166
Reynolds, P. H., 163
Richards, G. W., 95
Richardson, P. L., 1
Rietveld, H., 33
Rijkwaterstaat, 174
Rizzini, A., 91
Roberts, J. E., 32
Robin, G. de Q., 165-166
Roden, G. I., 42, 46, 50
Rodhe, H., 164
Roe, F. W., 99
Roemmich, D., 47, 164, 166
Rognan, P., 95
Rohde, H., 12
Ronov, A. B., 54, 59
Rosenberg, N. J., 173
Ross, D. A., 92
Ross, W. D., 4
Rossiter, J. R., 73, 93, 96, 105
Ruddiman, W. F., 64
Ryan, W. B. F., 92

S

Saarloos, J. M., 72
Sahagian, D., 54
Said, R., 93
Salisbury, R. D., 55
Sarton, G., 6
Satake, K., 30
Saucier, R. T., 40
Sauramo, M., 72, 82, 88
Savage, J. C., 141
Savin, S. M., 64
Sayles, R. W., 136
Scaife, W. W., 126
Scammell, G. V., 6
Schafer, J. P., 129
Schermerhorn, L. J. G., 62
Schlanger, S. O., 143, 172
Schlesinger, M. E., 166
Scholl, D. W., 141
Schopf, J. W., 63
Schopf, T. J. M., 64
Schureman, P., 22, 47
Schwiderski, E. W., 18, 19
Schwing, F. B., 118
Sclater, J. G., 24, 61, 138, 141
Semeniuk, V., 104
Sengupta, S., 99
Shackleton, N. J., 55, 64, 67
Shalem, N., 51, 92
Sharaf El Din, S. H., 92
Shaw, H. R., 42
Shepard, F. P., 51, 55, 56, 67, 140, 174
Shi, E., 5
Shi, L. X., 36
Shor, G. G., 101
Sillitoe, R. H., 124
Simkin, T., 137
Sjöberg, L. E., 12
Sloss, L. L., 58
Snowden, J. O., 40
Soegiarto, A., 173
Somundsson, K., 136
Solow, A. R., 147, 148, 216, 228
Spaeth, M. G., 51
Spencer, N. E., 73
Stanley, D. J., 35, 91, 92, 171
Stearns, H. T., 142
Steckler, M. S., 92
Steininger, F. F., 92
Stewart, R. W., 165-166
Stiros, S. C., 56, 92
Stommel, H., 137, 141
Stone, D. B., 141
Stotts, V. D., 173
Striem, H. L., 92
Sturges, W., 149, 151
Su, H. Y., 36
Sugi, N., 111
Suguio, K., 125
Summerhayes, C. P., 57, 59

Sverdrup, H. U., 46
Sykes, S., 125

T

Talandier, J., 141
Tarr, R. S., 31
Taylor, E. G. R., 6, 7
Taylor, F. W., 125, 136
Thatcher, W., 93, 110
Thom, B. G., 102
Thomas, R. H., 165
Thompson, K. R., 50
Thompson, L. G., 41
Thompson, R. E., 166
Thorarinsson, S., 165
Thorhaug, A., 174
Thorne, J., 57, 58, 88
Titus, J. G., 169, 172, 174
Toynbee, A., 1
Tracey, J. I., 140
Tucholke, B. E., 58
Tyler, D. A., 131

U

Uchupi, E., 29, 30, 93, 118, 128, 133, 138
Udintsev, G. B., 98, 99, 138
Um, S.H., 107
United Nations Educational, Scientific and Cultural Organization and Bundesanstalt für Bodenforschung, 92
U.S. Geological Survey Staff, 118
Untung, M., 101
Urien, C. M., 121
Ursell, F., 20
Uziel, J., 90

V

Vail, P. R., 25, 54, 57, 58, 59, 101
Valdiya, K. S., 97
Valentin, H., 73, 75
van Andel, Tj. H., 101
van Hinte, J. E., 57, 58
Vaníček, P., 133
van Pedang, M. N., 137
van Veen, J., 12, 72
Veeh, H. H., 101
Veevers, J. J., 101
Verbeek, R. D. M., 137
Villach, 166
Viniegra, O. F., 120
Vinnikov, K., 163, 164
Vitali, C., 124
von Huene, R., 112, 116, 124

W

Wadge, G., 124
Wadia, D. N. B., 98

Wageman, J. M., 106-108
Wahrhaftig, C., 118
Walcott, R. I., 25, 74
Walker, D. A., 42, 141
Wang, C. H., 107
Wang, Y., 1, 107
Ward, C. M., 105
Ward, S. N., 117
Ward, W. T., 101
Watts, A. B., 57
Weaver, J. D., 125
Wegener, K., 54
Weimer, R. J., 54
Wells, J. T., 171
White, W. B., 26
Wiessel, J. K., 102
Wigley, T. M. L., 165-166
Wilkinson, J. H., 215
Williams, D. F., 101
Williams, G. E., 62
Wilson, G., 33
Wilson, J. T., 57
Windley, B. F., 61
Winograd, I. J., 55
Wise, D. U., 54
Witting, R., 12, 72
Wood, M., 57
Woodwell, G. M., 74
Woodworth, P. L., 14
Worsley, T. R., 58, 61, 62, 63
Wu, P., 74
Wyrtki, K., 42, 43

X

Xui Yui, 6, 7

Y

Yabe, H., 106
Yamaguti, S., 30
Yamamoto, S., 36, 37, 38, 173
Yang, H., 107
Yang Zuosheng, 7
Yanshin, A. L., 8, 92, 99, 107, 136
Yohe, G. W., 174
Yonekura, N., 141

Z

Zans, V. A., 125
Zebiak, S. E., 44
Zhang, W.-Y., 107
Zhang, W. Z., 36
Zhang, Y., 163
Zhang, Z. M., 107
Ziegler, A. M., 61
Zoback, M. L., 27, 28
Zshehiv, D. E., 100

Subject Index

A

Aberdeen Island, U.K., 145
Aburatsubo, Japan, 30, 145
Aden, 6, 93–95, 100
Adriatic Sea, 33
Aesthenospheric "bumps," 24
Africa, 8, 60–61, 72, 77, 89–90, 93–97, 99–101, 138, 144, 152–153, 155
Agassiz, L., 54
Ahmedabad, 5
Akhziv Harbor, Israel, 56
Alaska, 154
Albany, Australia, 103–104
Aleutian Islands, U.S.A., 27, 51, 116, 140–141, 153–154
Aleutian Trench, 116, 140
Alexandria, Egypt, 5, 9, 91, 93
Alexander III (The Great), 4
Alger, Algeria, 95
Algeria, 89, 93
Alicante, Spain, 91
Allochthonous block or terrane, *see* Suspect terrane
Almirante Brown, Antarctica, 95
Alpine fault, New Zealand, 105, 153
Alpine orogeny, 93, 152–153
Amazon River, 155, 172
Amsterdam, The Netherlands, 33, 154
Anchorage, U.S.A., 31, 140–141
Andaman Islands, 137–138, 154
Andes mountains, 61, 154
Angola, 61
Annapolis, U.S.A., 131, 146
Antalya, 91–92
Antarctic
  Antarctica, 55, 61, 64, 74, 76, 95, 97, 101, 152
  Circumpolar Current, 14
  Ice sheet, 166
Anti-Atlas, 93
Antofagasta, Chile, 45, 123–124
Appalachian mountain range, 61–62, 153–154
Appenine mountains, 154
Apra, Guam, 139
Arakan-Yoma Mountains, 98
Arakawa River, 36
Arava, 92

Archaeology, 56–57, 90, 92, 98, 175
Arctic Sea, 20, 60, 114, 133
Argentina, 20, 46
Arica, Chile, 45, 123–124
Aristotle, 4
Asamuchi, Japan, 37
Ashdod, Israel, 91
Asia, 27, 75, 77, 79, 100, 105–114, 145, 155
Asia Minor, 90
Association d'Océanographie Physique, 73, 93
Astoria, OR, U.S.A., 146
Aswan High Dam, 35, 91, 171
Aswan Low Dam, 35, 171
Athens, Greece, 4
Atlantic City, NJ, U.S.A., 128, 130–131, 146
Atlantic Ocean, 47, 50, 60–61, 67, 74, 84, 136–137, 152
Atlas Mountains, 96, 153–154
Atoll, 172
Auckland, New Zealand, 104–105
Australia, 26–27, 38, 44, 61, 75, 79, 95, 97, 101–104, 130, 137, 144, 153–155, 158
Ayukawa, Japan, 37

B

Bab-al-Mandeb, 93
Baffin Island, 25
Bahama Outer Ridge, 68
Bakar, Yugoslavia, 91, 145
Balboa, Panama, 45, 120
Balearic Islands, 51
Baltic Sea, 51, 114
Baltic Shield, 83
Baltimore, U.S.A., 77, 146, 154
Bangkok, Thailand, 2, 16, 35–36, 100, 154
Bangladesh, 1, 3, 99, 170–171, 174
Barahona, Dominican Republic, 125
Barbados, 55, 136–137, 172
Bar Harbor, U.S.A., 133
Barentsburg, Norway, 136–137
Baton Rouge, U.S.A., 40
Battle of Hastings, 6
Bay of Bengal, 46, 48, 74

Bay of Fundy, 20, 131, 133
Bayou Rigaud, U.S.A., 40, 126
Beaches, raised, 104
Beachrock, 65, 140
Beijing, P.R.C., 36
Belem, Brazil, 45, 123–124
Belgium, 86
Bengal Basin, 99
Benioff Zone, 105, 112, 124, 135, 137, 140, 153
Benkoelen, Indonesia, 137
Bergen, Norway, 145
Bering Sea, 174
Bering Strait, 1
Bermuda
  Island, 46, 136, 144, 154–155
  Rise, 68
  Swell, 24
Bhavnagar, India, 96, 98
Bhutan, 97
Biodiversity, 174
Bjorn, Sweden, 145
Black Sea, 4
Block Island, U.S.A., 129
Bogoslof, U.S.A., 140
Bohai Gulf, P.R.C., 36, 48, 107
Bombay, India, 41, 96, 98, 145, 156
Bône, Algeria, 93, 95
Bonin-Japan-Kuril Trench, 112
Boston, U.S.A., 128, 146, 154
Boutlier Point, Canada, 133
Brahmaputra River, 1, 25, 96–98, 154, 170
Brazil, 125
Breccan's Caldron, 4
Brest, France, 89, 144–145
Bridgeport, U.S.A., 128
Brittany, France, 7
Buck Island, 172
Buenaventura, Colombia, 45, 123–124
Buenos Aires, Argentina, 123, 125, 156
Bunbury, Australia, 103–104
Burma, 4, 98, 137
Buzzards Bay, U.S.A., 128

C

Cadiz, 51
Caesarea, Israel, 1

231

232 Subject Index

Cagliari, Italy, 91
Cairns, Australia, 103–104
Calcutta, India, 96, 98, 145, 154
Caldera, 45, 123
Calendars, 9
California, 16, 118
Cambridge, MA, U.S.A., 131
Camp Cove, Australia, 104
Canada, 72–73, 75, 127, 131–134, 144, 167
Cananeia, Brazil, 123, 156
Canary Islands, 24, 144, 154
Canary Island Ridge, 137
Canavieras, 123
Canton Island, 143
Cape Cod, MA, U.S.A., 127, 131, 170
Cape Cod Canal, MA, U.S.A., 128
Cape Fear Arch, MA, U.S.A., 131
Cape foldbelt, South Africa, 93
Cape Hatteras, U.S.A., 25
Caribbean Plate, 135
Caribbean Sea, 60, 77, 79, 118–122, 125, 134
Carlsbad Ridge, 138
Carmel uplift, Israel, 92
Carmen, Mexico, 118, 120
Caroline Islands, 143, 154
Cartagena, Columbia, 45, 123–124
Carthage, 5
Cascais, Portugal, 89, 145
Catania, Italy, 91
Cayman Islands, 61
Cayman Trough, 120, 135
Cebu, Philippines, 139, 146
Cedar Keys, FL, U.S.A., 126, 156
Central America, 5, 29, 45, 75, 77, 118–123
Ceuta, 91, 95
Chad, 94
Chandeleur Islands, LA, U.S.A., 171
ChangJiang, P.R.C., 6, 106, 155, 171
  Approach, 7, 9
Chao Phrya River, 35, 100, 155
Charleston, SC, U.S.A., 131, 146, 156
Charlottetown, Canada, 131, 133, 146
Charybdis, 3
Chatham Harbor, MA, U.S.A., 50
Chaucer, G., 6
Chesapeake Bay, U.S.A., 127, 130–132, 158, 173
Chi Ma Wan, Hong Kong, 107
Chile, 124
Chilung, Taiwan, 107
Chimbote, 124
China
  People's Republic of (P.R.C.), 1, 5, 7–10, 36, 46, 75, 97, 100, 171
  Republic of, see Taiwan
Chinhai, South Korea, 107
Chios, Greece, 3
Chortis block, 120–121

Christmas Island, 142–143
Churchill, Canada, 131, 133
Civitavecchia, Italy, 91
Climate, 40–47, 151, 163–165
Clouds, 163
Cnut, 5
Coal mining, 38
Cochin, India, 96, 98, 145, 156
Coconut Island, Oahu, U.S.A., 141
Colombia, 124
Colonia, Uruguay, 45–46, 123, 156
Comodoro Rivadavia, Argentina, 123–125
Constantinople, 4
Continental drift, 54
Cook Inlet, AK, U.S.A., 20
Coquimbo Bay, 124
Coral, 99, 107, 125–126, 138, 140, 172
Cordova, AK, U.S.A., 31, 115–116
Costa Rica, 120
Crescent City, CA, U.S.A., 117, 146
Cresques, A., 7
Cristobal, Panama, 45, 120
Cro-Magnon man, 5
Cuba, 120, 135

D
Dakar, Senegal, 93–94
Dardenelles, 4
Darius, 4
Darwin, Australia, 41, 43–44
Davao, Philippines, 139
da Vinci, L., 5, 53
Davis Strait, 20
Daytona Beach, FL, U.S.A., 156
Dead Sea, 12, 16, 93, 153
Deccan Trap, India, 98
Degerby, Finland, 145
Delaware Bay, U.S.A., 130–131, 158, 172
Deltas, 154–155
Denmark, 5, 86
Deuterium, 64
Diamond Harbor, India, 96, 98
Djakarta, Indonesia, 41
Dominican Republic, 135–136
Dou, S. M., 6–7
Dover, DE, U.S.A., 131
Draghallan, Sweden, 145–146
Drake Passage, 61
Dubrovnik, Yugoslavia, 91
Dutch Harbor, AK, U.S.A., 140

E
Earthquakes, 29–31, 36, 51, 91–93, 97–98, 103–104, 107, 110, 113–114, 117–118, 131, 136, 138, 140–141, 154, 174
East African Rift, 93–94, 96, 153
East China Sea, 107

East Indies, 153–154
East Pacific Rise, 42
East Scotia, 61
Easter Island, 41, 74
Eastern Ghats orogeny, 62
Eastport, ME, U.S.A., 131, 133, 146
Ecuador, 43, 124
Egypt, 5, 12, 90, 171, 174
Eigenanalysis, 83, 108, 147, 215–216
El Niño, 14, 41–45, 50, 74, 78, 103, 151, 174
Emperor-Hawaiian Seamount Chain, 113
England, 5, 12, 20, 86, 88, 144
English Channel, 83
Eniwetok Island, 143
Ensenada, Mexico, 120
Epeirogenesis, 54
Erosion, see Shoreline retreat
Eucaryotes, 63
Eudoxas, 47
Eugene Island, LA, U.S.A., 40, 126
Euphrates River, 2, 3, 5
Eurasia, 114
Europe, 27, 60–61, 72, 74–75, 77–79, 82–89, 129, 145, 151–152
Eustasy,, 81, 92, 110, 133, 158–160, 165–166, 175
Eutrophication, 173
Everglades, FL, U.S.A., 31
Exotic terranes, see Suspect terranes

F
Falkland Islands, 60
Farakka, India, 99
Faulting and folding, 27–31, 79, 95, 103, 105, 108, 115–116, 118, 123, 125–126, 133–134, 151, 154
Fennoscandia, 12, 25, 55, 67, 72–76, 86, 88, 144, 147–149, 154, 157, 160–162, 167
Fernandina Beach, FL, U.S.A., 156
Finland, 83, 88, 144
Florida, U.S.A., 16, 40, 125, 127, 130–131, 158
Florida Peninsular Arch, see Ocala Arch
Fong, Y., 6
Fort Phrachula, Thailand, 36, 100–101
Fort Pulaski, GA, U.S.A., 40, 125, 133, 146, 156
Fortaleza, Brazil, 123
Fossa Magna, 111
Fracture zones, 143
France, 12, 20, 25, 72–73, 83, 86, 88–89, 144
Franklin, VA, U.S.A., 131
Freemantle, Austalia, 103–104, 156
Freeport, TX, U.S.A., 39, 126
French Polynesia, 141
Friday Harbor, WA, U.S.A., 146
Furuogrund, Sweden, 145

G

Galapagos Islands, Ecuador, 42
Galveston, TX, U.S.A., 39, 72, 76, 126, 145–146, 149, 154–155
Ganges Cone, 98
Ganges River, 1, 25, 46, 96, 98–99, 154, 170
General Circulation Model (GCM), 164–165
Genoa, Italy, 91, 145
Geodetic leveling, 109, 118, 126, 131, 133
Geographic Information System (GIS), 170
Geoid, 26–27, 61, 94, 125, 160
Geostationary Earth Orbiting Earth Satellite (GOES), 13
Gerald the Welshman, 6
Germany, 25, 73, 83, 86
Gibraltar, Strait of, 6, 89, 91, 93, 95, 153
Gilgamesh, 2, 53
Glacier Bay, AK, U.S.A., 26, 116
Glaciers, 53
   cycles, 24, 54, 62, 75
   rebound, 75, 79, 82, 88, 114, 116, 128, 130–131, 133–134, 151–152
   stadials/interstadials, 24
   stages, 15
   surge/melting, 25–26, 127
Global Positioning Satellite (GPS), 83, 104
Gondwana, 155
Goose Creek, TX, U.S.A., 39
Granger Bay, South Africa, 95
Gravity
   measurements, 97–98, 104, 116, 133, 135
   meters, 83
Great Basin, U.S.A., 55
Great Lakes, 174
Greater Antilles, 134–136, 141
Greece, 13
Greenhouse effect, 146, 151, 164–165, 167, 175
Greenland, 55, 60–61, 74, 76, 151–152, 157
Greenland Ice Sheet, 166
Grenville orogeny, 62
Groningen gas field, The Netherlands, 33
Groundwater
   intrusion, 34–40, 167, 172
   withdrawal, 2, 31–40, 72, 78, 92–93, 100, 124, 126, 131, 145, 151, 155, 172
Guam, 140
Guanglin Chau (tidal bore), P.R.C., 7
Guangzhou, P.R.C., 1, 7
Guantanamo Bay, Cuba, 45, 125
Guatamala, 120
Guaymas, Mexico, 120
Gulf of Bothnia, 72, 83
Gulf of California, 117–118, 120, 152–153
Gulf of Elat, 93, 153
Gulf of Finland, 83
Gulf of Honduras, 120
Gulf of Kutch, 98
Gulf of Maine, 131
Gulf of Mexico, 51, 55, 61, 72, 125–127
Gulf of Suez, 92
Gulf of Venice, 33
Gulf Stream, 26
Guyots, 172
Gwennan's Bane, Northern Ireland Currents of, 4

H

Hai Nan Island, P.R.C., 107
Haifa, Israel, 91–92
Haimen, P.R.C., 7–8
Halifax, Canada, 133, 146
Halley, E., 13
Hamina, Finland, 145
Hampton Roads, VA, U.S.A., 146
Hangö, Finland, 145
Hangzhou, P.R.C., 4, 7–8, 10
Harmonic analysis, 47
Harrington Harbor, Canada, 133
Hawaiian Islands, 24, 51, 141–143, 154
Heat flow, 97
Helios, 3
Hellespont, 4
Helsinki, Finland, 77, 145–146
Hengchun Peninsula, Taiwan, 107
Herodotus, 4
Hicks, S.D., 12
Hilo, HI, U.S.A., 141–142
Himalaya Mountains, 97–99, 152–153
Hinge zone, 24, 132, 158
Hispaniola, 125, 135
Hokkaido Island, Japan, 46, 112, 114
Homer, 3, 4
Honduras, 120
Hong Kong, 105–107
Honolulu, HI, U.S.A., 141–142, 146
Honshu Island, Japan, 30, 110
Honsu-Kokkaido-Kuril Islands Ridge, 112
Hooghly River, India, 99
Hooke, R., 13
Hosojima, Japan, 77, 145
Hot spots, 24, 139, 141–143, 154
Houston, TX, U.S.A., 39, 126
HuangHe, P.R.C., 25, 106, 154, 171
Hudson Bay, Canada, 130, 133
Hudson Strait, Canada, 20
Huludao, P.R.C., 105–107
Hyderabad, India, 154
Hydrocarbon extraction, 31–32, 38–40, 72, 115, 124, 126, 145, 151, 155
Hydrocompaction, 31–32
Hydrodynamics, 48
Hypsometric curve, 23, 170

I

Iceland, 60, 154
Imbituba, Brazil, 45, 123, 156
Inchon, South Korea, 107
India, 5, 13, 26, 95, 96–101, 138, 144, 153, 170
Indian Ocean, 41, 60, 74, 93, 137
Indo-Gangetic Plain, 97–98
Indonesia, 8, 137–138, 141, 153
Indonesian plate, 101
Indus Delta, 4, 98, 154, 171
Inland Sea, Japan, see Seto
Irrawaddy River, Burma, 100, 154, 171
Ise Bay, Japan, 37
Isla Martin Garcia, Argentina, 45–46
Island arcs, 139–141, 153
Isostasy
   glacial, 24–25, 78, 99, 118, 132
   hydro-, 25, 74, 94
   sediment, 25
Isotopes, 63, 95, 143
Israel, 2, 13, 16, 56, 90, 110
Italy, 1, 12, 33–35, 72, 90, 92, 144, 154
Izmir, Turkey, 91–92, 154
Izuhara, Japan, 105, 107

J

Jakobstad, Finland, 145
James River, VA, U.S.A., 130
Japan, 8, 16, 26, 29–31, 36–38, 72, 74–76, 79, 107–114, 140–141, 144, 153–154, 172–173
Japan Sea, 30, 46, 108, 113
Java, 74
Java Trench, 137–138
Jeju, South Korea, 107
Jiangsu Province, P.R.C., 6
Johnston Island, 143, 155–156
Jolo, Philippines, 139
Jordan, 16
Jordan River and valley, 16, 93
Juneau, AK, U.S.A., 116, 144, 146
Jurong, Indonesia, 101, 137

K

Kabret, Egypt, 91–92, 95
Kahului Harbor, HI, U.S.A., 142
Kaliningrad, U.S.S.R., 145
Kandla, India, 96, 98
Kaneohe Bay, HI, U.S.A., 141
Kaosiung, Taiwan, 107
Karachi, Pakistan, 100
Karelian orogeny, 62
Kashiwazaki, Japan, 37–38
Kasko, Finland, 145
Katmai, AK, U.S.A., 140
Katsuura, Japan, 37
Ke, 7
Kelvin wave, 42–43, 50

Kemi, Finland, 145
Kenya, 138
Kerala coast, India, 98
Kermadec-Tonga Trench, 105
Ketchikan, AK, U.S.A., 146
Key West, Florida, 125–126, 146, 156
Kibaren orogeny, 93
Kidderpore, India, 96, 99
Kilauea, HI, U.S.A., 141
Kings Bay, GA, U.S.A., 48
Kinneret Fault, Israel, 93
Kiptopeke Beach, VA, U.S.A., 131
Kiribati, 172
Klagshamn, Sweden, 145
Knossos, Greece, 57
Knysna, South Africa, 95
Ko Sichang, Thailand, 101
Ko Taphao Noi, Thailand, 156
Kobe, Japan, 37–38
Krakatoa, 51, 137
Kublai Khan, 5
Kula Plate, 116–117
Kungsholmsfort, Sweden, 145
Kuril Islands, 112
Kuril Trench, 114
Kuroshio Current, 26
Kwajalein Island, 143, 154, 156
Kyparissi, Greece, 56
Kyushu Island, Japan, 108
Kyushu-Nankai Trough, 111
Kyushu-Ryukyu Trench, 107

L
La Coruña, Spain, 89
Laccodive Ridge, 97, 99
Lagoons, 167–169, 172
Lagos, Portugal, 89, 145
La Goulette, 95
La Guaira, Venezuela, 123
La Jolla, CA, U.S.A., 43, 146
Lake Erie, U.S.A., 129
Lake Maracaibo, Venezuela, 38, 123–124, 155
Lake Superior, U.S.A., 16
La Libertad, Ecuador, 45, 123
La Madelena, Italy, 91
La Manche, *see* English Channel
Lamia, Greece, 56
La Niña, 42
Lanyi Island, Taiwan, 107
La Pallice, France, 89
La Paloma, Uruguay, 123, 156
La Paz, Mexico, 120
Latrobe Valley, Australia, 38
La Union, El Salvador, 45, 120
Laurentian ice sheet, 74
Legaspi, Philippines, 139
Length-of-Day, 44, 74
Lesser Antilles, 134–135
Li, J. P., 6

Line Islands, 24, 143, 154
Lisbon, Portugal, 51
Little Ice Age, 41, 163
Lofoten Peninsula, Norway, 3
Logan's Line Fault, 133
Loihi, HI, U.S.A., 141–142
Loma-Prieta earthquake, CA, U. S. A., 118
London, England, 33, 154
Long Beach, CA, U.S.A., 32, 39, 48, 53, 115, 155
Long Island, NY, U.S.A., 127–129
Long Island Sound, 127–130, 132
Lord Howe Rise, 101
Los Angeles, CA, U.S.A., 39, 146
Louisiana, U.S.A., 55, 172
Lu, S., 7
Lüderitz, Namibia, 93, 95, 156
Lui, C. M., 7–8
Lunar month, 7, 9
Lutau Island, Taiwan, 107

M
Ma, H., 8
Macau, 105, 107, 144–145
Mackay, Australia, 103–104
Madagascar, 138
Madras, India, 96, 98
Maelström, 3, 4
Magadan, U.S.S.R., 114
Magdalena Fault, 124
Magdalena River, Colombia, 172
Magellan Strait, 20
Magnus, Olaus, 3
Magueyes Island, Puerto Rico, 45, 125
Maine, 131
Majuro, 26
Malaysia, 101
Maldive Islands, 172
Malvinas-Agulhas Fracture Zone, 138
Mangalore, India, 98
Mangroves, 173–174
Manila, Philippines, 139, 145–146
Mantyluota, Finland, 145
Manzanillo, Mexico, 45, 118, 120
Marathon, 4
Mar del Plata, Argentina, 123–125
Marghera, Italy, 34
Mariana Islands, 139–141, 154
Mariana Trench, 140
Marine terrace, 82, 98, 105, 110, 118, 124–125, 133, 135–137
Marseille, France, 91–92
Marshall Islands, 5, 51, 143, 154, 172
Martha's Vineyard, MA, U.S.A., 129
Massachusetts, U.S.A., 130, 170
Massacre Bay, AK, U.S.A., 140
Massif de la Hotte, Haiti, 135
Matarani, Peru, 45, 123–124
Mauritanides foldbelt, 93
Mauritius, 154–155

Mayport, FL, U.S.A., 146, 156
Median Line, Japan, 111, 113
Mediterranean Sea, 1, 3–4, 51, 56, 60, 72, 79, 89–92, 171
Megna River, Bangladesh, 1
Mei, C., 6–7
Mekong River, Thailand, 171
Meltwater, 25, 75–76, 78, 103, 165–166
Mendocino Fracture Zone, CA, U.S.A., 117
Mera, Japan, 37, 145
Messina Strait, Italy, 3, 4, 31, 91–92
Methane, 34
Mexico, 118, 121, 125
Miami Beach, FL, U.S.A., 131–132, 146, 156
Midway Islands, 141–142
Milankovitch cycle, 41, 55, 58, 65
Mindanao Trench, *see* Philippine Trench
Ming, C.Z., Emperor, 8
Mississippi River and Delta, 25, 31, 40, 72, 76, 126, 155, 171
Misumi, Japan, 37
Moan Island, Caroline Islands, 143
Mohammed, 6
Mokpo, Korea, 107
Mokuoloe Island, HI, U.S.A., 141–142
Monaco, 91
Montagua fault, 120
Montauk, NY, U.S.A., 128
Monterey, CA, U.S.A., 42
Monterey Bay, CA, U.S.A., 118
Montevideo, Uruguay, 45–46, 123, 156
Mosken Island, Norway, 3
Mossel Bay, South Africa, 95
Moulmein, Burma, 48
Mount Etna, Sicily, 92
Mount Redoubt, AK, U.S.A., 141
Muertos Trough, 135
Mugho, South Korea, 107
Murmansk, U.S.S.R., 114
Murray Fracture Zone, 118

N
Nagaeva Bay, U.S.S.R., 114
Nagasaki, Japan, 51
Nagoya, Japan, 37
Nantucket Island, MA, U.S.A., 128–129
Naos Island, Panama, 120
Napier, New Zealand, 104
Naples, FL, U.S.A., 126
Naples, Italy, 56, 92
National Ocean Survey, NOAA, 7, 9
Nawiliwili Bay, HI, U.S.A., 142
Nazca Ridge, 124
Neah Bay, WA, U.S.A., 146
Nedre Soder, Sweden, 145
Neo-tectonics, 78, 95, 103, 107
Nepal, 97

Netherlands, The, 12, 31, 33, 48, 72, 75, 83, 86
Netherlands East Indies, 137
Newcastle, Australia, 102–104, 145
Newfoundland, Canada, 131, 133
New Guinea, 55, 105, 136–137
New London, CT, U.S.A., 128
Newlyn, U.K., 145, 147
New Orleans, LA, U.S.A., 40, 154
Newport, RI, U.S.A., 128, 146
New Rochelle, NY, U.S.A., 128
Newton, I., Sir, 5
New York, NY, U.S.A., 128–129, 131, 146, 154
New Zealand, 101, 104–105, 110, 153–154
Next Generation Water Level Measurement System (NGWLMS), 14
Nezugaseki, Japan, 30–31
Nicaragua, 120
Nicaragua Rise, 120
Niger River, Nigeria, 171
Niigata, Japan, 30, 38
Nile River, Egypt, 25, 35, 57, 92, 154, 171
Ninety-east Ridge, 97, 99, 137
Nobi Plain, Japan, 37
Nordauslandet ice sheet, 26
Normans, 6
North America, 29, 60–61, 77–78, 114–118, 151–152, 155, 157
North Point, Hong Kong, 105, 107
North Pole, 26
North Sea, 83, 88
North Shields, U.K., 145
Norway, 5, 83, 86, 144
Norwegian Sea, 83
Nova Scotia, Canada, 16, 131

O
Oahu, HI, U.S.A., 141–142
Ocala Arch, FL, U.S.A., 125
Oceania, 75
Odysseus, 3
Ogi, Japan, 37–38
Okinawa, 140
Okinawa Trough, 112
Ölands norra udde landsort, Sweden, 145
Old Bahama Channel, 135
Ominato, Japan, 37
Onahama, Japan, 37
Onisaki, Japan, 37
Oran, Algeria, 93, 95
Orinoco River, Venezuela, 172
Orogenies, 61–62
Osaka, Japan, 37–38
Oslo, Norway, 145
Oulu, Finland, 145
Oura, Japan, 37
Oxygen isotopes, 55, 64, 67, 137

P
Pacific Countercurrent, 42
Pacific Ocean, 42, 60–61, 74, 85, 118, 135, 139, 142, 153
Padre Island, TX, U.S.A., 126
Pago Pago, 143
Pakistan, 97
Palawan Island, Philippines, 51
Palermo, Argentina, 123, 156
Palermo, Italy, 91
Palmer Peninsula, Antarctica, 95
Pan-African orogeny, 62, 93
Panama, 120, 123
Pangaea, 24, 61, 97
Paraná River, Brazil, 45–46, 172
Paris, M., 7
Pasajes, Spain, 89
Pearl River, see ZhuJiang
Peat, 40
Pelusium fault, 93
Pensacola, FL, U.S.A., 125–126, 133, 146, 156
Peripheral bulge, 25, 83, 89, 129, 133, 151
Permanent Service for Mean Sea Level (PSMSL), 31, 69, 73
Pernambuca Fault, 124
Persia, 4
Persian Gulf, 6, 8, 48
Peru, 43, 50, 124
Petropavlovsk, U.S.S.R., 114
Pharos lighthouse, Egypt, 93
Philadelphia, PA, U.S.A., 128, 146, 154
Philippine
  Islands, 107, 135, 139–141, 144, 153–154
  Plate, 108, 113
  Sea, 112
  Trench, 139
Phlegrian fields, Italy, 56
Phoenicians, 5
Phoenix Island, 143
Phrachuap Kiri Khan, Thailand, 100–101, 155–156
Phuket, Thailand, 100–101, 155
Pictou, Canada, 131, 133
Pius, A., 57
Plaine du Cul-de-Sac, Haiti, 135
Plate tectonics, 23–24, 54, 59–61, 152
  convergence, 57, 81, 99, 101, 108, 118, 125, 153–154
  divergence, 57, 81
Pliny, 6
Plum Island, MA, U.S.A., 130
Po River, Italy, 33–34, 92, 154
Point Atkinson, Canada, 146
Point-au-Père, Canada, 133, 146
Point Lonsdale, Australia, 104
Pointe St. Gilda, France, 89
Poland, 25, 86
Pollen, 67
Polynesians, 5

Port Adelaide, Australia, 102–104
Port-au-Prince, Haiti, 125, 135–136
Port-aux-Basques, Canada, 131–133
Port Blair, Andaman Islands, 137–138
Port Isabel, TX, U.S.A., 126
Port Jefferson, NY, U.S.A., 128
Portland, ME, U.S.A. 133,, 145–146
Port Lincoln, Australia, 102–104
Port Louis, Mauritius, 137–138, 146
Port Nolloth, South Africa, 93, 95, 156
Porto Corsini, Italy, 34, 91
Porto Maurizio, 91
Port Royal, Jamaica, 125
Port Said, Egypt, 35, 91, 95
Portsmouth, NH, U.S.A., 127, 133, 146
Port Thewfik, Egypt, 91–92, 95
Port Tuapse, U.S.S.R., 114, 145
Portugal, 89, 144
Poseidonius, 6
Pozzuoli, Italy, 53, 56, 92
Precipitation, 163–164
Prince Edward Island, Canada, 131
Prince Rupert, Canada, 146
Prince William Sound, AK, U.S.A., 31
Progreso, Mexico, 45, 118, 120, 126
Providence, RI, U.S.A., 128
Ptolemy, 9
Puerto Armuelles, Panama, 120
Puerto Castilla, Honduras, 120
Puerto Cortes, Honduras, 45, 120
Puerto Limon, Guatamala, 45, 120
Puerto Madryn, Argentina, 45, 123, 125, 156
Puerto Plata, Dominican Republic, 125
Puerto Rico, 51, 135
Puerto Rico Trench, 135–136
Puntarenas, Costa Rica, 45, 120
Pusan, South Korea, 107
Pyramides, Argentina, 123
Pyrénées Mountains, 89, 154
Pytheas, 4

Q
Quebec, Canada, 133
Quequen, Argentina, 123–124
Qingdao, P.R.C., 105, 107
Qinhuangdao, P.R.C., 105, 107
Qiantang River, P.R.C., 4, 6–8, 10
Qin, S. H., 5
Quan, M., 4
Quelccaya ice sheet, Peru, 41
Quintana Roo basin, 120

R
Raahe, Finland, 145
Radiocarbon, 56, 58, 65, 67, 78, 82, 88, 92, 94, 99, 107, 110, 128–131, 136, 138, 159, 172, 175
Rangoon, Burma, 100–101, 154

Ratan, Sweden, 145
Rathlin Island, Ireland, 4
Ravenna, Italy, 34
Recife, Brazil, 45, 123–124
Red Sea, 96, 152
  rift, 93
Resolute, Canada, 133
Reykjavik, Iceland, 136–137
Rhine River, The Netherlands, 33, 72, 83, 154
Rhône River and Delta, France, 91, 154
Richmond, VA, U.S.A., 130
Rio de Janeiro, Brazil, 123
Riohacha, Colombia, 45, 123
River
  dams, 78, 99
  run-off, 45–46, 50, 57, 99, 126
Rivière-du-Loup, Canada, 133
Rockall Bank, 60–61
Rockport, ME, U.S.A., 126
Rome, Italy, 12
Rovinj, 91
Ruskaya Gavan, U.S.S.R., 114
Russian platform, 83
Ryukyu Islands, 107, 111–112, 140
Ryukyu-Kyushu-Nankai Trough, 112–113, 140

S

Sabine Pass, TX, U.S.A., 39, 126
Saint Bede the Venerable, 6
St. Francis, Canada, 133
St. Georges, Bermuda, 136–137, 146, 156
St. Jean-de-Luz, France, 89
St. John, Canada, 133, 146
St. John's, Canada, 133
St. Lawrence River Valley, Canada, 133
St. Nazaire, France, 89
St. Petersburg, FL, U.S.A., 126, 156
Ste. Anne des Monts, Canada, 133
Saipan, 140
Salamis, 4
Salina Cruz, Mexico, 45, 120
Salinity, 46
Salvador, Brazil, 45, 123, 156
Samoa, 143
San Andreas Fault, CA, U.S.A., 117–118, 152
San Diego, CA, U.S.A., 43, 145–146
Sandy Hook, NY, U.S.A., 128, 146
San Francisco Bay, CA, U.S.A., 38, 50, 118, 144–146, 154
San Jacinto belt, 123
San Jose, Guatamala, 120
San Juan, Puerto Rico, 125
Santa Clara Valley, CA, U.S.A., 38–39
Santa Cruz de Tenerife, 136–137, 146
Santa Monica, CA, U.S.A., 146
Santander, Spain, 89
Santiago, Chile, 41
Santo Tomas, Guatamala, 120
Santorini, Greece, 1, 57
Sao Paulo, Brazil, 61
Satellite altimetry, 14, 83
Saugor, India, 96, 98–99
Savannah, GA, U.S.A., 40, 50, 131
Scandinavia, see Fennoscandia
Scheldt Estuary, The Netherlands, 1
Scotian Sea, 134
Scotland, 5, 72, 82–83, 86, 88
Sea-floor spreading, see Plate tectonics
Sea level
  curves, 59–60, 64, 66
  rates-of-rise, relative, 69, 71–79, 84–88, 143–150, 168
  scenarios, 165–166
Seamounts, 24
Sea of Okhotsk, 114
Sea-surface temperature (SST), 42, 44, 76, 164
Seattle, WA, U.S.A. 43., 145–146
Sediment, 53
  compaction, 31–40, 83, 92, 96, 98–100, 124, 126, 133, 154–155, 171
  deep-ocean, 24
  loading, 78
  rates, 24, 40, 131
  subsidence, 31–40, 92, 98, 100, 110, 124, 126, 171
Seiches, see Waves
  shelf
Seismic stratigraphy, 57–59
Seleucas, 6
Sembawang, Indonesia, 101, 137
Serapis, Temple of, Italy, 1, 56
Seto, Japan, 8, 108, 113, 136
Seward, AK, U.S.A., 140
Seychelles Bank, 138
Sfax, Tunisia, 95
Shandong Peninsula, P.R.C., 106
Shang Dynasty, 9
Shanghai, P.R.C., 36, 107, 154, 171
Sheerness, England, 33
Shen, G., 6
Shi, E., 7–8
Shikoku Island, Japan, 8, 110
Shillong Plateau, India, 98
Shoreline, 82, 133
  classification, 55–56
  retreat, 34, 167
Showa Base, Antarctica, 95
Sibaura, Japan, 37
Siberia, U.S.S.R., 114
Sicily, Italy, 3, 72
Simons Bay, South Africa, 95
Sinai, Egypt, 56, 110
Singapore, 100, 137
Sinkholes, 40
Sinu Belt, 123
Sitka, AK, U.S.A., 115, 117
Skagway, AK, U.S.A., 115–116
Smogen, Sweden, 145
Soma, Japan, 37
Somalia, 138
Song Dynasty, 7
Sousse, Tunisia, 95
South Africa, 93
South America, 29, 42, 45, 60–61, 75, 77, 79, 93, 119, 121, 123–125, 144, 152–155, 160
South Equatorial Current, 42
South Korea, 105–107, 113
South Pole, 26
South Sandwich Island arc, 134, 153
Southend, England, 33, 145
Southern Oscillation, see El Niño
Spain, 12, 89
Spitsbergen, 26, 136
Split, Yugoslavia, 91
Sri Lanka, 98–99
Steric response, 46–50, 163–166, 175
Stockholm, Sweden, 145
Storms
  surge, 1, 167, 170
  typhoons, 1, 170
Stress distributions, 27–28
Stromatolites, 63
Subduction, 79, 107, 124, 135, 141, 151–154
Submarine canyons, 102, 106
Sulu Sea, 51
Sumida River, Japan, 36
Sunda Shelf, Indonesia, 101
Supercontinents, 62
Suspect terranes, 23, 29, 118, 120–121
Susquehanna River, MD, U.S.A., 131
Sweden, 83, 144
Sweeper Cove, AK, U.S.A., 140
Sweyn, 5
Sydney, Australia, 43–44, 102–104, 145

T

Tacloban, Philippines, 139
Tadoussac, Canada, 133
Tahiti, 141
Tai Po, Hong Kong, 107
Taiwan, 46, 105, 107, 140
Takoradi, Ghana, 77, 93–95
Talara, Peru, 45, 123–124
Tamatave, Madagascar, 137–138
Tamayo transform fault, 120
Tamboro, Indonesia, 51, 137
Tanakura Fault, Japan, 108, 111
Tang Dynasty, 6–7
Tanggu, P.R.C., 36, 106–107
Tangshan, P.R.C., 107
Tan-nowa, Japan, 37–38
Tanzania, 138
Tarshish, 5

Tectonism, 29–30, 53, 81, 93, 100, 105, 107–108, 114, 118, 120, 123–124, 128, 131, 134, 162
Tema, Ghana, 93, 95, 156
Temperature, 159, 163
  sea, 46
Tethys Ocean, 98
Texas, 16, 126
Texas City, TX, U.S.A., 39, 155
Thailand, 100, 137, 174
Thames River, 33
Thera, *see* Santorini
Thermal expansion, *see* Steric response
Thermopylae, 4
Thessaloniki, Greece, 91
Thevanard, Australia, 102–104, 156
Thompson, W., Sir, 47
Thrinacia, 3
Tianjin, P.R.C., 31, 35, 171
Tibetian Plateau, 97–98
Tides
  amphidromes, 19
  benchmarks, 150
  bores, 7–10
  charts, 18–20
  constituents (frequencies), 16–22, 47–48
  gauges, 12–14, 57, 144
  lunar influence, 5–6
  machines for prediction, 21–22
  meridian rule, 6
  quarter rule, 6
  semi-diurnal, 16–18
  set-up, 49
  stations, 70–71, 82
  tables, 7–11
Tigris River, Iraq, 3
Timor Trench, 101, 137
Toba, Japan, 37
Tokyo, Japan, 36–37
Tonoura, Japan, 77, 145
TOPEX/Poseiden, 14
Topography
  ocean surface (dynamic), 26, 46
Transverse ranges, 118
Trieste, Italy, 72, 91, 145
Troy, Turkey, 3, 56
Truk Island, 143
Tsunamis, 1, 51, 57, 93, 118, 151
Tsushima Strait, 107
Tumaco, Colombia, 123
Tunisia, 89, 93, 95
Turku, Finland, 145
Tuticorin, India, 98

Tuvalu, 172
Typhoons, *see* storms

U
Ubatuba, Brazil, 123, 156
Ulysses, *see* Odysseus
Ulsan, South Korea, 107
Unalaska, AK, U.S.A., 140
United States of America, 15, 25, 38–40, 65, 69–72, 74–76, 125–134, 144, 153, 161–162, 167, 174
Ural Mountains, U.S.S.R., 114
Uruguay, 46
U.S.S.R., 86, 114, 144

V
Vaasa, Finland, 145
Valdez, AK, U.S.A., 31, 51
Valparaiso, Chile, 123–124
Vancouver, Canada, 146
Varberg, Sweden, 145
Varves, 67
Venezuela, 16, 38
Venice, Italy, 16, 34, 53, 72–73, 91, 154–155
Veracruz, Mexico, 118, 120, 126
Veracruz basin, 120
Very Long Baseline Interferometry (VLBI), 83, 104, 117
Victoria, Canada, 146
Vigo, France, 89
Vikings, 6
Vishakhapatnam, India, 96, 98, 156
Volcanism, 79, 105, 107, 112, 118, 125, 135–141, 143, 151, 153–156, 164, 172

W
Wajima, Japan, 37, 146
Wakayama, Japan, 37–38
Wake Island, 143
Walvis Bay, Namibia, 93, 95, 156
Wang, C., 6
Washington D.C., U.S.A., 146
Waves
  set-up, 52
  shelf, 50–51
  wind, 16, 51–52
Weizhou, P.R.C., 105–107
Wellington Harbor, New Zealand, 104–105

West Antarctic Ice Sheet, 26
West Indies, 51, 153
West Point, VA, U.S.A., 131
West Timor, 137
Wetlands, 171–172
Whirlpools, 3–4
Willets Pt., NY, U.S.A., 128, 146
Williamstown, Australia, 102–104
Wilmington, DE, U.S.A., 131, 146, 156
Wilson cycle, 57, 61
Woods Hole, MA, U.S.A., 128, 146
Wu, D., King, 9
Wushau, P.R.C., 7

X
Xerxes, 4
Xiamen, P.R.C., 105–107
Xou, O. Y., 7

Y
Yafo, (Jaffa) Israel, 91
Yakutat, AK, U.S.A., 31, 43, 115–116
Yan, S., 7
Yang, Q., 6
Yangtze River, *see* Changjiang, P.R.C.
Yanguan, P.R.C., 7–10
Yangzhou, P.R.C., 7
Yantai, P.R.C., 106–107
Yarmouth, Canada, 133
Yellow River, *see* HuangHe, P.R.C.
Yellow Sea, 46
Yokkaichi, Japan, 38
Yokosuka, Japan, 37
Yosu, South Korea, 107
Ystad, Sweden, 145, 148
Yucatan Peninsula, Mexico, 120–121
Yugoslavia, 90, 144
Yui, Q., 7
Yuzhno Kurilsk, U.S.S.R., 114

Z
Zambesi graben, Mozambique, 93
Zhan, H., Admiral, 8
Zhang, J. F., 7
Zhapo, P.R.C., 105, 107
Zhejiang
  Province, P.R.C., 6–7
  Ting (Pavilion), 7
Zhenjiang, P.R.C., 7
ZhuJiang, 7, 171
Zuider Zee, The Netherlands, 33